Industrial Environmental Health

THE WORKER AND THE COMMUNITY

ENVIRONMENTAL SCIENCES

An Interdisciplinary Monograph Series

ARTHUR C. STERN, editor, AIR POLLUTION, Second Edition. In three volumes, 1968

L. FISHBEIN, W. G. FLAMM, and H. L. FALK, CHEMICAL MUTAGENS: Environmental Effects on Biological Systems, 1970

DOUGLAS H. K. LEE and DAVID MINARD, editors, PHYSIOLOGY, ENVIRONMENT, AND MAN, 1970

KARL D. KRYTER, THE EFFECTS OF NOISE ON MAN, 1970

R. E. MUNN, BIOMETEOROLOGICAL METHODS, 1970

M. M. KEY, L. E. KERR, and M. BUNDY, PULMONARY REACTIONS TO COAL DUST: "A Review of U. S. Experience," 1971

DOUGLAS H. K. LEE, editor, METALLIC CONTAMINANTS AND HUMAN HEALTH, 1972

DOUGLAS H. K. LEE, editor, ENVIRONMENTAL FACTORS IN RESPIRATORY DISEASE, 1972

H. ELDON SUTTON and MAUREEN I. HARRIS, editors, MUTAGENIC EFFECTS OF ENVIRONMENTAL CONTAMINANTS, 1972

RAY T. OGLESBY, CLARENCE A. CARLSON, and JAMES A. MCCANN, editors, RIVER ECOLOGY AND MAN, 1972

LESTER V. CRALLEY, LEWIS T. CRALLEY, GEORGE D. CLAYTON, and JOHN A. JURGIEL, editors, INDUSTRIAL ENVIRONMENTAL HEALTH: The Worker and the Community, 1972

MOHAMED K. YOUSEF, STEVEN M. HORVATH, and ROBERT W. BULLARD, PHYSIOLOGICAL ADAPTATIONS: Desert and Mountain, 1972

In preparation

DOUGLAS H. K. LEE and PAUL KOTIN, editors, MULTIPLE FACTORS IN THE CAUSATION OF ENVIRONMENTALLY INDUCED DISEASE

MERRIL EISENBUD, ENVIRONMENTAL RADIOACTIVITY, Second Edition

Industrial Environmental Health

THE WORKER AND THE COMMUNITY

Sponsored by Industrial Health Foundation, Inc.
Pittsburgh, Pennsylvania

EDITOR

LESTER V. CRALLEY
Environmental Health Services
Aluminum Company of America
Pittsburgh, Pennsylvania

ASSOCIATE EDITORS

Lewis J. Cralley
Department of Health, Education and Welfare
Public Health Service
Health Services and Mental Health Administration
National Institute of Occupational Safety and Health
Cincinnati, Ohio

George D. Clayton
George D. Clayton and Associates
Southfield, Michigan

John A. Jurgiel
Industrial Health Foundation, Inc.
Pittsburgh, Pennsylvania

ACADEMIC PRESS *New York and London 1972*

ACADEMIC PRESS, INC.
111 Fifth Avenue, New York, New York 10003

United Kingdom Edition published by
ACADEMIC PRESS, INC. (LONDON) LTD.
24/28 Oval Road, London NW1

LIBRARY OF CONGRESS CATALOG CARD NUMBER: 70-187243

PRINTED IN THE UNITED STATES OF AMERICA

Contents

Epidemiologic Studies of Occupational Diseases

Lewis J. Cralley

Toxicology

Emil A. Pfitzer

Noise

Paul L. Michael

Nonionizing Radiation

David H. Sliney

Ionizing Radiation

Harry F. Schulte

Engineering Approach to Analysis and Control of Heat Exposures

Harwood S. Belding

Evaluation of Chemical Hazards in the Environment

Robert G. Keenan

Hazard Evaluation and Control

Bernard D. Bloomfield and James C. Barrett

Personal Protective Devices

William A. Burgess

New and Recurring Health Hazards in Industrial Processes

J. A. Houghton

Contributions of Ergonomics in the Practice of Industrial Hygiene

Bruce A. Hertig

Air Pollution

George D. Clayton

An Empirical Approach to the Selection of Chimney Height or Estimation of the Performance of a Chimney

R. C. Wanta

Water Pollution

S. Charles Caruso, Francis Clay McMichael, and William R. Samples

Agriculture Now

Keith R. Long and David L. Mick

List of Contributors

Numbers in parentheses indicate the pages on which the authors' contributions begin.

JAMES C. BARRETT, Division of Occupational Health, Michigan Department of Public Health, Lansing, Michigan (303)

HARWOOD S. BELDING, Department of Occupational Health, Graduate School of Public Health, University of Pittsburgh, Pittsburgh, Pennsylvania (271)

BERNARD D. BLOOMFIELD,* Department of Public Health, State of Michigan, Lansing, Michigan (303)

WILLIAM A. BURGESS, Harvard School of Public Health, Department of Environmental Health Sciences, Boston, Massachusetts (355)

S. CHARLES CARUSO, Carnegie-Mellon University, Pittsburgh, Pennsylvania (467)

GEORGE D. CLAYTON, George D. Clayton and Associates, Southfield, Michigan (411)

LEWIS J. CRALLEY, Department of Health, Education and Welfare, Public Health Service, Health Services and Mental Health Administration, National Institute of Occupational Safety and Health, Cincinnati, Ohio (1)

BRUCE A. HERTIG, Laboratory for Ergonomics Research, Department of Mechanical and Industrial Engineering, University of Illinois, Urbana, Illinois (401)

J. A. HOUGHTON, Industrial Chemical Service, Liberty Mutual Insurance Co., Boston, Massachusetts (381)

ROBERT G. KEENAN, Laboratory Services, George D. Clayton and Associates, Southfield, Michigan (281)

KEITH R. LONG, Department of Preventive Medicine and Environmental Health, College of Medicine, The University of Iowa, Iowa City, Iowa (495)

*Deceased.

FRANCIS CLAY McMICHAEL, Fellow of Mellon Institute, Carnegie-Mellon University, Pittsburgh, Pennsylvania (467)

PAUL L. MICHAEL, Environmental Acoustics Laboratory, The Pennsylvania State University, University Park, Pennsylvania (145)

DAVID L. MICK, Institute of Agricultural Medicine, The University of Iowa, Iowa City, Iowa (495)

EMIL A. PFITZER, Department of Environmental Health, College of Medicine, University of Cincinnati, Cincinnati, Ohio (71)

WILLIAM R. SAMPLES,* Carnegie-Mellon University, Pittsburgh, Pennsylvania (467)

HARRY F. SCHULTE, Industrial Hygiene Group, Los Alamos Scientific Laboratory, University of California, Los Alamos, New Mexico (243)

DAVID H. SLINEY, Laser-Microwave Division, U. S. Army Environmental Hygiene Agency, Department of the Army, Edgewood Arsenal, Maryland (171)

R. C. WANTA, 28 Hayden Lane, Bedford, Massachusetts (455)

*Present address: Environmental Control, Wheeling-Pittsburgh Steel Corporation, Wheeling, West Virginia.

Foreword

The passage of Federal legislation specifically covering safety and health aspects of industrial operations, which has occurred in the interval since publication of "Industrial Hygiene Highlights," has created a situation in which the safety and health standards area has been likened to being "present at creation," so that at the national level it becomes essential to use good judgment in identifying and using wisely our existing resources. These resources, which include technical information, facilities, and trained professional, technical, and scientific personnel, must be conserved as we make plans to extend them as rapidly and effectively as practicable to reach a level more nearly consistent (in the case of qualified specialists in the field of occupational health alone conservatively estimated to involve a tenfold increase) with the concurrent upgrading of safety and health practices in facilities of some 4,000,000 employers engaged in industrial and business pursuits. Under such conditions, it is important to consider the nature of the resource represented in the Industrial Health Foundation (IHF), sponsor of this volume.

A nonprofit, research and education association of industries established in 1935 for the advancement of safe and healthful working conditions in industry, IHF changed its name in 1970 from Industrial Hygiene Foundation to reflect its concern for all health and safety-related problems of industry. Problems such as rehabilitation from nonoccupational diseases or injuries and employee mental health conservation, neither of which is related ordinarily to industrial environmental factors, have been among the pioneering programs of the Foundation. Other membership services have involved medical, nursing, biostatistical, and epidemiological studies, and industrial hygiene and safety engineering surveys and controls.

IHF sincerely believes in cooperation with all agencies, in and out of government, whose missions are similar or related to its own. The unique relationship of IHF to its members provides the means of gaining cooperation – so essential to employee health research – and also a means for obtaining joint funding.

Recently, Mr. William D. Ruckelshaus, head of the Environmental Protection Agency, stated that there is need to communicate more widely and effectively to our communities the vast and growing body of scientific information on environmental quality control and its importance to human progress in our increasingly industrialized society. Mr. Ruckelshaus called for an "ecology reference shelf" in every public and school library. The publications of our Foundation, which have been serving for many years the needs of the community as well as of its industrial members, will become increasingly important in the future in meeting the needs of a public whose accelerating interest and expectancy in environmental quality areas can be termed "explosive."

IHF is most fortunate in having the valuable assistance of a Publications Committee. The Committee's assistance in preparation of this volume and review of all Foundation publications helps ensure maximum utilization within the total community. Chaired by Dr. Lester V. Cralley, Chief Editor of this volume, the Publications Committee is composed of the following outstanding members in the field of information sciences: Mrs. Madeline C. Carey, Industrial Health Foundation; Mr. Kenneth H. Fagerhaugh, Carnegie-Mellon University; Mr. Daniel R. Pfoutz, Carnegie Library of Pittsburgh; Mrs. Glenora E. Rossell, University of Pittsburgh Libraries. The earnest thanks of the Foundation's members and all others who use our publications are extended to the Publications Committee. We look forward to receiving their continued assistance and guidance in carrying out our Foundation's increasingly important mission in facing a future full of challenge.

Daniel C. Braun, M.D.
President, Industrial Health Foundation
Pittsburgh, Pennsylvania

Foreword

Increasing public concern over environmental forces and hazards has led, predictably, to awareness of the need for improved conditions for all who work. As a result, we are seeing the beginnings of a great new era for safety and health in America's workplaces.

The Occupational Safety and Health Act of 1970 is the Congressional response to a nationwide demand for tighter measures to assure employee safety and health on the job. This landmark legislation broadens the Federal government's jurisdiction in this area to over 57 million workers in 4.1 million places of employment. But government, of itself, can serve only as the pathfinder and the watchdog. The essential functions of any good safety and health operation must be performed by employers and employees accepting their mutual responsibilities.

We in the U. S. Department of Labor's Occupational Safety and Health Administration are charged with carrying out the new law through setting and enforcing safety and health standards for places of employment, training professional safety experts as compliance officers to inspect conditions in the workplace, compiling complete injury and illness statistics, and informing the public of progress made. All of these – in combination with the research function delegated to the Department of Health, Education and Welfare's National Institute of Occupational Safety and Health – will be of inestimable value to industry in discovering the causes and prevention of occupational injuries and illnesses. The task before us is enormous. But the potential saving in human suffering as well as in money cost is a goal well worth the effort.

Industry and labor have already shown a recognition of their responsibilities regarding safety and health in the workplace. Ongoing programs of research, education, and self-regulation are daily supplementing government- and academic-sponsored efforts. Not surprisingly, some industries and employers are in the vanguard of programs protecting workers' safety and health, while others have not yet accepted the challenge.

This volume represents an industry commitment and a response to the demonstrated need. Thoughtful research by topflight experts from academia, government, and industry provides the springboard that, properly implemented, can lead to a safer, healthier in-plant environment. It is increasingly recognized by industry that good working conditions are good business.

In its statement of purpose, the Occupational Safety and Health Act of 1970 says it seeks to encourage "joint labor-management efforts to reduce injuries and disease arising out of employment," and "to stimulate employers and employees to institute new and to prefect existing programs." This book consists of a broad-based series of expert analyses which will assist everyone concerned – management, labor, and government – with increasing safety and health in the nation's workplaces.

George C. Guenther
Assistant Secretary of Labor
Occupational Safety and
Health Administration
Washington, D.C.

Preface

Industrial hygiene has come a very long way since its beginnings as an applied science about the time of World War I when an aroused social consciousness sought to improve working conditions in the sweat shops and the dusty trades. The initial empirical controls, primarily engineering in nature, showed the resourcefulness of the industrial hygienist. Later the chemists, physicists, toxicologists, and biological scientists made their contributions to the practice of industrial hygiene. Air pollution control and radiologic health were once within the province of industrial hygiene but were split off during the 1950's because of rising public concern. For reasons explained in Dr. Cralley's Preface, environmental conditions have continued to deteriorate so that now the public is being "struck" by the impact of the total environment. Dr. Cralley's philosophy of considering man's total exposure – at work, in the community, and at home – is commendable, as is his resolve to update information in this rapidly expanding field of environmental health.

Under the Occupational Safety and Health Act of 1970, the National Institute for Occupational Safety and Health is charged with responsibilities for research, technical services, and training relative to on-the-job health and safety. The information which is produced must be widely disseminated, and will be useful not only to those in occupational safety and health but also to those in the other environmental health sciences. Conversely, much of the research and related information developed in the other environmental health areas will be directly applicable to occupational health. Because of this general applicability within a broad area and because of the overlap of professional disciplines, the overall usefulness of "Industrial Environmental Health: The Worker and the Community" will be readily appreciated.

Marcus M. Key, M.D.
Assistant Surgeon General
Director, National Institute for
Occupational Safety and Health
Washington, D.C.

Preface

Today's environmental health problems are a cominglement of complex stresses arising primarily from (1) a population density within a specific shed greater than that which can be properly handled by the existing controls, (2) tardiness in recognizing and controlling significant sources of pollution, and (3) the reluctance of individuals to practice self-restraint voluntarily. Consequently, stress situations have built up to a point that a major organized program is required to resolve the problem. This is an inevitable by-product of social evolution.

One of the first indications of impending environmental problems occurred in the early 1940's with the advent of tremendously increased production associated with World War II. Specific air basins with limited self-cleansing characteristics, such as Los Angeles, gave clear sensory evidence that a progressively worsening condition was developing. Since that time, air pollution, water pollution, and solid waste disposal problems have grown so great they no longer can be ignored. A series of state and federal acts have been passed and regulations announced to give guidance to the tremendous effort now being expended. The following are among the important federal events.

1. Clear Air Amendments of 1970
2. The Army Corps of Engineers Federal Permit Program, under the Refuse Act of 1899, covering waste discharge into navigable waters
3. The Water Quality Improvement Act of 1970
4. The Occupational Safety and Health Act of 1970
5. The Federal Coal Mine Health Safety Act of 1969
6. The Construction Safety Act

Additional state and federal action can be anticipated in these areas as well as in solid waste disposal.

Another area that demands a place high on the list of environmental problems requiring attention is in-plant environment, that is, the environmental health hazards of the man on the job. We must now consider the whole environment of man – on the job, in the community, and at home. This is a most important development because the stresses of man, regardless of origin, must be evaluated and controlled in terms of their totality.

With continued improvement of the working and community environment, the time is approaching when man's greatest environmental stress may well be that of his own individual creation – his home. Many of his hobbies are associated with hazards that would not be tolerated on the job. His exposure to do-it-yourself solvents, sensitizing materials, high noise levels such as power tools or high music levels, and failure to wear safety protection equipment, to cite a few, constitute serious hazards. To make matters worse, man's insult to himself from smoking, self-medication, improvident working habits beyond that to which he is acclimated, overeating, etc., may actually dwarf environmental stresses on the job and in the community.

However, man's reaction is in terms of his total stress, not merely its component parts. For this reason, those concerned with environmental health cannot ignore the temptation to treat lightly man's most vulnerable spot – his home and his voluntary acts of indiscretion. Added to this is the increasing amount of time that will be available off the job. The time is coming when man's stress will be largely dependent upon his own acts of discretion or indiscretion.

Were man to actively pursue those things within his own reach that would reduce pollution – greater use of public transportation, segregation of solid waste into components that lend themselves to better disposal, avoidance of unnecessary use of water that in turn requires waste treatment, avoidance of the needless scatter of litter – an immediate impact upon the environment would result. To phrase it differently, education will become a dominant factor in man's determining his own degree of stress. If, added to this, an organized approach were used in the community to establish land usage regulations to prevent a population density beyond the load that can be properly managed, a real basis would be laid for a long-range solution.

Technological progress is being made in securing the tools needed to measure the effect of single or interacting stresses on the environment. But the fact remains that information is inadequate to state with any degree of confidence that air pollution has an ill effect upon health except under acute conditions, that thermal pollution within the parameters usually discussed results in an inferior environment, that the increment of community noise plays a really significant role in loss of hearing, that minimal measurable physiological change associated with a chemical or physical stress is anything other than an indication that a successful adjustment has been made.

One thing is certain: man is demanding improvement in his environment. It is essential, therefore, that we have a clear picture of the reasons behind that demand. Sooner or later, society will insist upon a cost-benefit basis for determining the degree of improvement sought. A means must be employed that will define a level of environmental air quality in which the cost of control and the benefits of control are related. However difficult, this approach must provide the basis for predetermining the quality of the environment to be achieved.

Presently, there is uncertainty regarding the degree of environmental improvement needed that will be commensurate with society's total need. It is important then that our knowledge be current, that we be aware of new developments, and that our evaluations be kept under continual scrutiny.

"Industrial Environmental Health: The Worker and The Community," provides the profession with new information associated with the period 1968-1970. It is a continuation of the period 1965-1967 that was covered in "Industrial Hygiene Highlights." The scope of the latter volume has been enlarged and care has been exercised to avoid duplicate coverage. The title also was changed to more accurately describe the contents. Air and water pollution have been treated only in a general way to bring these areas into overall focus since detailed coverage is now available in excellent textbooks and professional journals. With the exception of meteorology, in which a new approach has been presented which we believe will more properly predict tall stack dilution, the major emphasis has been placed on the in-plant environment. The time has come when man's environmental health stress must be considered in terms of both its totality and its increments. A future volume will give major attention to man's off-the-job stress as a component part of his overall environmental health stress.

Lester V. Cralley

Epidemiologic Studies of Occupational Diseases

LEWIS J. CRALLEY

Epidemiologic studies of occupational diseases are undertaken in order to determine the nature and prevalence of disease in a working population under risk and to associate the parameters of the etiologic agent(s) with the observed biologic response. Though these studies are often long and tedious, no better way has been devised for studying the effects of the environment on the worker over a lifetime of employment. These data are necessary as a means of evaluating the ultimate criteria of a safe working environment; i.e., the positive health of the worker in that environment.

The epidemiologic study of occupational diseases has expanded tremendously over the past decade; keeping abreast of findings from epidemiologic research in the study of occupational diseases is becoming an ever-increasing problem. This review covers the period 1968–1970.

The single greatest impact on the epidemiologic study of occupational disease in the United States was the enactment of Public Law 91-596 "Occupational Safety and Health Act of 1970."[257] The Act establishes the Institute of Occupational Safety and Health in the Department of Health, Education, and Welfare with an assigned responsibility to "conduct and publish industry-wide studies of the effect of chronic or low-level exposure to industrial materials, processes, and stresses on the potential illness, disease, or loss of functional capacity in aging adults." The Act further stipulates "The Secretary of Health, Education, and Welfare, in order to comply with his responsibilities under paragraph (2), and in order to develop needed information regarding potentially toxic substances or harmful physical agents, may prescribe regulations requiring employers to measure, record, and make reports on the exposure of employees to substances or physical agents which the Secretary of Health, Education, and Welfare reasonably believes may endanger the health or safety of employees. The

Secretary of Health, Education, and Welfare also is authorized to establish such programs of medical examinations and tests as may be necessary for determining the incidence of occupational illnesses and the susceptibility of employees to such illnesses."

It is predicted that the Occupational Safety and Health Act of 1970 will effect a vast increase in the epidemiologic study of occupational diseases by both governmental and nongovernmental sponsorship.

The different approaches to the epidemiologic study of occupational diseases were covered in "Industrial Hygiene Highlights"[1] which stressed that in studying the health status of the worker at risk, the parameters used must measure the total interrelated stresses over the full 24-hour day, 7-day week of the overall environment. For this reason adequate control studies are needed to derive comparative data on workers with similarities but without involvement in the stress under study. Because of the additive and synergistic effects of many chemical and physical stresses, it has become necessary in the cross-sectional and long-term study of occupational diseases to obtain information on the social habits of the worker under study, including smoking history, consumption of alcoholic beverages, and use of self-medicaments and drugs. Increased attention must also be given to multiple on-the-job stresses including the simultaneous stresses of vibration, temperature, noise, energy requirements of the job, and chemical agents, since these may have additive, synergistic, or antagonistic effects on the chemical stress under study.

Over the past several years increased recognition has been given that occupational diseases are international and that an effective use of international talent and experience through good communications and frequent interchange of research findings is essential in establishing knowledge for their understanding and control. It is predicted that this trend will be firmly established and promoted.

Good progress has been made in preventing worker exposure to toxic materials through the engineering control of contaminants at points of dissemination and the appropriate use of personal protective equipment. In most instances where a high prevalence of occupational disease exists, there has been a failure to use already available knowledge and procedures in controlling, monitoring, and maintaining exposure levels within safe and rational limits.

This review, following the pattern of the previous one,[1] covers only epidemiologic studies of the chemical stresses to which workers are subjected. Neither biologic nor physical agents are covered. The review is aimed at updating information from epidemiologic studies of occupational diseases from chemical stresses reported in the technical journals since the publication of "Industrial Hygiene Highlights."[1] In some instances, background information is presented. Also, only reports in the English language are reviewed. It is obvious that in such an enormous undertaking many reports of studies were not included either because they were not at hand or for other reasons could not be utilized. Throughout, when data are available, emphasis is on the environmental aspects and etiologic agents rather than on the biologic response, though in epidemiology neither should be considered exclusive from the other.

Many of the governmental agencies responsible for occupational health in various countries publish annual reports which are an excellent source of information for on-going epidemiologic studies.[2-7] Other studies present general epidemiologic information on working populations, especially methodologies that have broad applications.[8-12]

ALDRIN

Auar and Czegledi-Janko[13] studied the relation of neurologic symptoms and electroencephalographic findings on workers occupationally exposed to aldrin in a manufacturing plant. Fifteen workers out of a total of forty in the plant (three chosen because of symptoms and twelve chosen at random) were given a neurologic examination, including history and symptoms, and an electroencephalogram. A blood sample was taken for determining the HEOD (active ingredient in aldrin) level. The findings showed signs of slight poisoning when the HEOD concentrations in the whole blood exceeded roughly 0.1 ppm and no poisoning was noted when the HEOD concentration was under 0.05 ppm. There was variation in susceptibility and response as evidenced in one worker having no symptoms with an HEOD blood level of 0.25 ppm and another worker showing symptoms and electroencephalographic changes with a HEOD blood level of 0.05 ppm. In three workers severely affected, symptoms ceased within 7 months after cessation of exposure. The HEOD concentration in the blood fell slowly but was still higher than the HEOD blood level in the general population after 2 years following cessation of exposure.

The threshold limit value adopted by the American Conference of Governmental Industrial Hygienists for aldrin is 0.25 mg/m^3.[50]

ARSENIC

Lee and Fraumeni[251] studied the mortality experience of 8047 white, male smelter workers exposed during 1938–1963 to arsenic trioxide. These data were compared with similar data of the white male population of the same states.

Information on the date and place of birth, Social Security number, jobs by type and duration in the smelter previous to 1964, and vital data for each person was obtained through company records. Additional follow-up information was obtained from death registers of state health departments, social security claims, and other governmental agencies. Death certificates were obtained on the workers who had died during 1938–1963 and the underlying causes of death were classified according to the latest revision of the International Lists of Diseases and Causes of Death. Detailed information is presented on the source of material and methods used.

No environmental exposure data were presented. For analyses, the work group was classified according to heavy, medium, and light exposures to arsenic and sulfur dioxide. No smoking history was available on any of the workers.

Of the entire study group of 8047 there were 1877 deaths as compared to 1634 expected. The excess deaths were essentially due to respiratory cancer and diseases of the heart; from respiratory cancer 44.7 deaths were expected as compared to 147 observed (SMR 329) and from diseases of the heart 614.1 deaths were expected as compared to 725 observed (SMR 118). The higher SMR (Standard Mortality Ratio) correlated with the degree of exposure to arsenic. Although the same relation also existed between respiratory cancer (SMR) and degree of exposure to sulfur dioxide, the authors believed that arsenic was the major cause but that the sulfur dioxide may have had enhanced the effect. Among 317 smelter workers heavily exposed to ferromanganese dust, five died from lung cancer as compared to 1.25 expected deaths.

It is significant that neither qualitative nor quantitative exposure data were given for sulfur dioxide, arsenic, or other trace metals known to be present. It is quite feasible that the same relation of deaths from lung cancer (SMR) and extent of exposure to other trace metals individually or combined may be true. Although the causative agent(s) were not specifically identified, it appears quite definite that the smelter workers studied do have excess deaths from lung cancer and other causes which should be further defined.

ASBESTOS

Research on the health effects of asbestos has continued at a high level of activity over the past several years. A number of reasons account for this interest. Asbestos is an important mineral, both economically and in modern industrial technology. As an inert and durable mineral fiber, it has great potential in future industrial fiber technology, expecially when it can be used in contained or bound form. Excessive exposure, through inhalation, however, is associated with fibrosis and cancer though the specific etiologic agent(s) relating to the latter are still unknown. A long latent period of 20 years or more usually exists from initial exposure to recognition of cancer; the latent period for asbestosis is often much shorter, depending on the criteria used for its recognition.

As evidence of the continuing interest in asbestos, six international conferences were held during 1968–1970 in which the health effects of asbestos were major subjects: (*a*) The Second International Conference on the Biologic Effects of Asbestos, 1968, in Dresden[14]; proceedings of the conference have not as yet been published. (*b*) The International Conference on Pneumoconiosis, Johannesburg, South Africa, 1969[15–28]; the Proceedings of this conference have been published. (*c*) XVI International Congress on Occupational Health, 1969, Tokyo, Japan[29–33]; the proceedings of this conference have been published. (*d*) The Symposium on the Tissue Response to Asbestos, University of South Wales and Monmouthshire, 1970, Cardiff, Wales[34]; papers presented at this symposium will not be published as proceedings. This was primarily a discussional not aimed at the formal presentation of research data. (*e*) The Third International Symposium on Inhaled Particles, 1970, London, England[35–38]; the proceedings of this conference will be published. (*f*) Colloquium on Asbestos, 1970, Sardinia[39]; proceedings of this colloquium will be published.

From the above and other scientific meetings, well over 250 papers have been presented during 1968–1970 on various aspects of asbestos and health. Many of these, though, are repetitive and cover essentially the same research data which were presented at earlier meetings. However, notable advancements in knowledge of the epidemiology of asbestos-related diseases have been reported in a number of areas during this period.

Procedures have been developed for the digestion of body tissue and for the concentration of bare fibers and ferruginous bodies from these tissues[40, 41] which now makes it possible to determine the fiber burden of different tissues such as lungs, lymph nodes, and tumors, to identify and quantify the fibers present, and to correlate these data with the presence or absence of pathology.

These new procedures will be especially rewarding in comparing similar data of this type on industrially exposed workers and nonworkers. Gross *et al.*[38] studied the lungs of twenty-three residents of Pittsburgh coming to autopsy for the number of inorganic fibers and ferruginous bodies present. The optical microscopy used did not detect fibers much under 1.0 μm in diameter and thus the submicron fibers, including many of the asbestiform ones, were not counted. Report on the identification of the fibers was left to a subsequent publication. The number of inorganic fibers per gram of dry lung in the twenty-three residents ranged from 15,000–380,000. The total dust content ranged from 0.6–6.1% of the dry weight of the lungs. There was no correlation of number of fibers to total dust. Also work histories were not available on the twenty-three residents. The number of ferruginous bodies did not relate to the number of fibers nor to the total dust content of the lungs. The residue from the burning of wood and paper products gave appearances and fiber measurements that resembled many of the fibers observed in the lung and may suggest an apparent source of the fibers noted. Research of this nature should be increased.

Dunn *et al.*[48] reported on a mortality study of several occupational groups, using data including information from union membership lists, to determine the risk to lung cancer. Data on occupational length of time on job, exposures to asbestos, other known hazards associated with job, cigarette smoking history, etc., were obtained by mailed questionnaires from the union members. Data on intensity of exposure or the type of asbestos used were not presented. Of the fourteen occupational groupings, only asbestos workers were found to have an increased risk of lung cancer, with the risk increasing with length of time in the occupation.

Knox *et al.*[42] reported on a study of mortality from lung cancer and other causes among workers in an asbestos textile factory. This continuation of a previous study[43] extended the time of observation from 1961–1966. Chrysotile was used predominantly in the factories studied, but some crocidolite was used during all relevant periods. Though dust control was constantly being improved since 1933, it was not likely until about 1961 that concentration levels of asbestos dust were maintained at a low level plant-wide. Dust levels in 1933 and earlier were massive. Since 1961, the asbestos fiber yearly mean dust levels in the plant air (5–500 μm in length with length at least three times the diameter) remained under 8 fibers/cm^3 and more nearly 5–6 fibers/cm^3. Medical data revealed that the trend in lung cancer mortality decreased with reduction of length of employment before 1933. Among those who have worked only since 1933, the observed mortality from all causes was close to that expected. The authors were optimistic that with the improved dust controls and resultant lower asbestos fiber levels in the air, the occupational hazard of bronchial carcinoma had been largely eliminated.

Newhouse[44] reported on a mortality study of workers in an asbestos factory which included some 4500 workers employed between 1933 and 1964. Three comparative groups were established: mortality from beginning of second to end of fourth year of the study, mortality from beginning of fifth to end of sixteenth year, and mortality from seventeenth to end of thirty-first year. Each group was subdivided into those who had worked under 2 years and over 2 years in the factory. Environmental data on asbestos dust levels were not given. Through observation and data on the amount of time spent in dusty areas, extent of dust, percent of asbestos, etc., exposure groupings were established on a graded basis from light to severe. Chrysotile, amosite, and crocidolite were being processed. No significant difference was noted in the mortality observed and expected until after an elapse time of 16 years or longer from the time of first exposure. Of the group that had worked 16 years and longer, there was no mortality increase in those with low to moderate exposure. However, in those with heavy exposure there was a significant excess in deaths from cancer of the lung, of the pleura, and of other sites for those who had been employed less than 2 years as well as for those who had worked longer.

Dixon *et al.*[49] showed that the metals extracted from chrysotile, i.e., nickel, cobalt, chromium, and manganese strongly inhibited the activity of benzpyrene hydroxylase and they postulated that this permitted a longer time of residency of carcinogenic polycyclic aromatic compounds in the tissues with thus a greater risk of cancer.

Gibbs *et al.*[36] investigated the qualitative aspects of dust exposure in the Quebec asbestos mining and milling industry with particular reference to the presence of organic constituents, trace metals, dust mineralogy, and radioactivity. Detailed data are given on the geology and mineralogy of Quebec chrysotile, mining and milling procedures, and qualitative and quantitative aspects of total and respirable dust. Airborne dust samples did not contain appreciable aromatic hydrocarbons. The trace metals (manganese, nickel, cobalt, and chromium) content of the respirable dust was usually greater than in that of the total dust and varied considerably depending on the process involved and the source of the asbestos. The radioactivity (radon) in the mines ranged from 0.0–0.08 of a work level limit and was not considered an important factor in the lung cancer mortality.

Much of our knowledge of the effects of exposure to asbestos comes from the health experience of shipyard asbestos workers. There has been, however, a dearth of information on the nature and extent of past asbestos exposures to workers, either directly or surreptitiously in the refitting and repairing of ships. Harries[252] gives considerable insight to this problem. Though the study is continuing and no time weighted exposure data are given, measurements during operations in the removal of old asbestos insulating materials and the application of new materials showed ranges from 0.1 fiber/cm^3 in the store rooms to 3815 fibers/cm^3 in cleanup operations (e.g., application amosite, 9–40 fibers/cm^3; removal of amosite sections, 29–1040 fibers/cm^3; removal of blue sprayed asbestos, 112–1906 fibers/cm^3; bagging asbestos debris, 106–3815 fibers/cm^3). Thus the potential for massive exposures was great even though employment may have been of relatively short duration.

McDonald *et al.*[253] reported on a comprehensive study of mortality from lung cancer and other causes in the workers in the chrysotile asbestos mines and mills of Quebec. The study cohort consisted of 11,788 mine and mill workers born between the years 1891 and 1920 who had been employed in the asbestos mine and mills one calendar month or longer. This age group was selected to assure that the cohort spanned a lifetime of employment and thus would include the age of high mortality. Age specific death rates for the Province of Quebec were used for controls. Detailed information is given on the materials and methods used.

Levels of asbestos dust exposure were established for designated job classifications. Dust measurements, using the standard and midget impingers, expressed as millions of particles per cubic foot of air, were available since 1949 throughout the drying, crushing, and milling operations of the industry. Similar data were available since 1946 in the small factory from a number of independent investigations. Where needed, extrapolation of present-day dust levels to conditions and practices of the past were used to estimate previous exposures. Data on the asbestos dust exposure and other environmental factors for the group of workers comprising the mortality study will be reported at a future date.

Mortality data on the cohort is presented in tables using a number of parameters for measuring effects and relationships. In general, when the workers were classified into six increments of increasing levels of dust exposure (< 10, 10, 100, 200, 400, and over 800), only the highest dust index category showed an increase in standardized mortality ratio – 20% above the others. Of this excess, two-thirds was probably due to pulmonary fibrosis and related diseases, and one-third to cancer, mainly of the respiratory tract. The authors believed that the true difference in rates for respiratory cancer between those maximally and minimally exposed to asbestos dust to be around the threefold increase and to be both dust and time related. The present study bore out the finding of a previous study[23] that primary malignant mesothelial tumors were rarely associated with chrysotile asbestos production in Quebec. The authors state "These findings strongly suggest either that chrysotile is less likely to cause malignant disease of the lung and pleura than other forms of asbestos such as crocidolite or that workers engaged on insulation and processing are exposed to additional factors which explain the difference."

Vigliani's study of asbestos exposure and its results in Italy[22] appears to have many parallels to McDonald's study. In this respect, it should be stressed that the different types of asbestos may have a different mineralogic derivation and that different processes are used in freeing and opening the fibers so that the metals associated with different processes in the mining and use of asbestos will vary considerably. If the metals associated with asbestos are an important etiologic factor in the resultant cancer, then different types of asbestos under different circumstances of processing would also show a different dose–response relationship. Thus the observations of McDonald and Vigliani seem plausible.

Additional research findings reported during 1968–1970 on areas of investigations already under way include Selikoff's work on asbestos exposure, smoking, and neoplasia,[45] Noro's appraisal of occupational and nonoccupational asbestosis in Finland,[46] and Balzar's and Cooper's study of the work environment of insulating workers.[47]

Realizing that much of the confusion arising out of differences in research findings resulted from the lack of uniformity in characterizing the type of asbestos under study, the Sub-Committee on Asbestosis, Permanent Commission and International Association on Occupational Health, established a work group to develop uniform procedures that could be used internationally to characterize and measure asbestos dust in air. This work group met as a part of the Colloquium on Asbestos in Sardinia.[39] It is anticipated that the report of the recommended uniform procedures will be available during 1971.

The American Conference of Governmental Industrial Hygienists in 1970 revised the threshold limit value of asbestos downward[50] and in the Notice of Intended Changes listed the threshold limit value for asbestos at "five fibers/ml $> 5\ \mu m$ in length as determined by the membrane filter method at $430 \times$ magnification phase contrast illumination. Concentrations > 5 fibers/ml, but not to exceed 10, may be permitted for 15 minute periods each hour up to 5 times daily."

A major obstacle to further progress in defining safe levels of exposure to asbestos fibers is the lack of knowledge on the etiologic agents involved and their interrelations. Until this information is obtained and uniform procedures are used for characterizing the asbestos under research, major differences will continue to exist in research findings, both in the nature of the biologic response and the prevalence of disease observed. There is good evidence, however, that asbestos can be handled safely and that maintaining exposure levels within recommended limits will control or appreciably reduce to a low level associated fibrosis and lung cancer.

BERYLLIUM

In 1969 Mancuso and El-Attar[51] reported on an epidemiologic study of the beryllium industry using mortality data. In data from the Social Security Administration, a group of 3685 white males were identified who had worked in two companies during 1937–1948 manufacturing beryllium products. Of the 3685 employees, 2863 (77.7%) had worked 2 years or less in the companies and 822 (22.3%) had worked for more than 2 years. The group was divided into four cohorts of 3-year spreads and each worker was placed into the cohort corresponding to the first date of his employment in these companies. Each cohort was followed for the same number of years to obtain comparative data. Causes of death were obtained from death certificates and were classified. The authors acknowledged many of the limiting factors in their approach: limited information on the worker and on exposure periods and intensities, and the nature of death certificate data. Although limited interim data were presented, the study was designed for a much longer follow-up period with a subsequent report.

Mancuso[52] in 1970 gave a further follow-up report of the study described above.[51] A major objective was to determine the influence, if any, of duration of employment in the industry and of the occurrence of previous occupational respiratory illness on causes of death. A comprehensive analysis of the data is presented. It showed a relationship between the duration of employment and specific causes of death. Workers shown through Social Security Administration records to have been employed from 1 to 5 reporting quarters had a higher lung cancer rate than those employed for 6 quarters and longer; no reason was given for this. There was also a much higher cancer rate among the workers with prior chemical respiratory illness. It may have been that the group employed for 1–5 quarters worked during the period of greater exposure to beryllium or that in this group were those that were the most susceptible and who could have experienced the greatest discomfort and earliest illness and thus might have left the plants for other employment.

Lieben and Williams[53] reported on a follow-up study of cases of respiratory disease associated with beryllium refining and alloy fabrication. A case was defined as berylliosis on the basis of a positive exposure history, significant x-ray changes which had persisted over several years, compatible histologic changes in the lungs, the clinical course of illness, and a consensus of physicians familiar with the disease. The seventy-five cases were divided into twenty-three in-plant cases, twenty-three contact exposure cases, and twenty-nine community cases. A perplexing question was why there were fewer community cases of berylliosis in the immediate vicinity of the plant where concentration levels were the heaviest than in outlying districts with lower levels of beryllium. A restudy of the community cases revealed that several of those so classified had exposures to beryllium not previously known, e.g., visiting or working in the immediate vicinity of the plant, or contact with beryllium workers who wore excessively contaminated work clothing outside the plant. Of the ninety-five individuals with chronic lung disease believed at one time to have berylliosis, it was shown later that many should have been classified otherwise. These are important observations that show the need for careful investigations when in-plant and community disease may have common etiologic factors and where the disease may mimic a similar response from other causes.

The current threshold limit value of 0.002 mg/m^3 for beryllium recommended by the American Conference of Governmental Industrial Hygienists[50] was established several years ago and was extrapolated from minimal data. The value was very useful in reducing the heavy exposures to beryllium of the past. However, inadequate epidemiologic data are available since the heavy exposures of the past have been reduced to permit a reevaluation of the current threshold limit value.

The mortality study and longitudinal epidemiologic study[54] now underway in the beryllium producing industry by the Bureau of Occupational Safety and

Health Service, Department of Health, Education, and Welfare, should provide additional definitive data on the prevalence and trend of occupational disease in the industry and on safe levels of exposure to beryllium.

CADMIUM

Lehnert *et al.*[55] reported on a study of eighteen workers in a zinc smelter to evaluate the usefulness of cadmium determination in urine by atomic absorption spectrometry as a screening test in industrial exposures. Medical examinations of the workers since 1967 had not revealed any clinical symptoms attributable to cadmium, though there was known to be exposure. Estimated air concentrations of cadmium were not presented.

The dust in the roasting plant contained 0.05–3.0% cadmium. Zinc, lead, and mercury were also present. The cadmium was usually present as the sulfide but existed in some areas as the oxide.

Cadmium levels were determined in both the serum and urine of the eighteen workers. Cadmium levels in the urine were corrected to creatinine content. All of the cadmium serum levels were within 3 standard deviations of the average and were not regarded as abnormal. The urine from fourteen of the eighteen workers showed cadmium levels exceeding the normal average urine concentration by more than 3 standard deviations.

The increased urinary cadmium level did not indicate cadmium poisoning since no clinical evidence was present in the workers. Nevertheless, cadmium urine analysis on workers exposed to cadmium was recommended as a means of monitoring working conditions. When elevated levels are present, further evaluation of protective measures should be made.

The American Conference of Governmental Industrial Hygienists has adopted a threshold limit value of 0.2 mg/m^3 for metal cadmium and soluble cadmium salts, and a "C" value of 0.1 mg/m^3 for cadmium oxide fume (as cadmium).[50]

CARBON DISULFIDE

Much of the current interest on the health effects of exposure to carbon disulfide has stemmed from the presentation and recommendations of the International Symposium on the Toxicology of Carbon Disulphide, held in Prague, Czechoslovakia, September 1966, under the sponsorship of the Subcommittee for Occupational Health in the Production of Artificial Fibers, Permanent Commission and International Association on Occupational Health. A major recommendation of the symposium was that comprehensive epidemiologic studies on the morbidity and mortality of workers exposed to toxic vapors in the viscose rayon industry for several years be either continued or started.[56, 57]

Tiller *et al.*[58] in 1968 reported on the relation between carbon disulfide exposure and mortality from coronary heart disease. Of workers exposed to carbon disulfide in three viscose rayon factories in England and Wales, between 1933–1962, 42% of the 223 deaths in male workers were certified to be coronary heart disease – in contrast to 24% deaths in other workers of the same age, and 17% of the deaths in other men unexposed to carbon disulfide in the same industry. In one factory, death rates from coronary heart disease between 1950-1964 were 2½ times greater in the employees exposed to carbon disulfide who had worked more than 10 years. In the viscose-making department, 17% of the measurements for atmospheric exposure levels to carbon disulfide in the churn room were over 20 ppm, the threshold limit value established by the Ministry of Labour, 1965. There was no exposure to hydrogen sulfide. There was, however, general rotation of workers to jobs in other areas with no exposure to carbon disulfide so that the time-weighted exposure was much less than this level. In the spinning department about 50% of the measurements for atmospheric exposure levels to carbon disulfide were over 20 ppm. Exposure to hydrogen sulfide was generally within the threshold limit of 10 ppm established by the Ministry of Labour. The excessive deaths from coronary heart disease were in the workers employed in the spinning department. The authors recommended further biochemical and morbidity studies which they believed would give additional information on the relation of arteriosclerosis and coronary heart disease.

Martinez and Farina,[59] in clinical examinations of seventy-five workers exposed to high levels of carbon disulfide for many years in viscose factories, reported 83% workers having vasculopathy in the encephalic region, 22% having disturbance of renal activity, 5% having renal hypertension, and 7% having diabetic symptoms.

Herneberg *et al.*[70] studied coronary heart disease and mortality among workers exposed to carbon disulfide. The study group consisted of 343 out of a total of 365 male workers with at least 5 years exposure to carbon disulfide in a viscose rayon plant between 1942–1967 and forty-five deceased workers. Suitably matched controls were selected from workers in a paper mill in the same city. Air levels of carbon disulfide and hydrogen sulfide had been measured since 1950. The concentrations were very high in the 1940's, had lowered to around 20–40 ppm in the 1950's and to 10–30 ppm in the 1960's. Questionnaire data on each person still living provided pertinent information on social and medical histories, smoking habits, clinical symptoms, present medication, etc. A medical examination included blood pressure measurements, electrocardiograph readings during and after exercise, and maximal oxygen uptake.

Statistically significant higher mean systolic and diastolic blood pressures were observed in the exposed workers in contrast to the control group. The exposed workers had a slightly higher frequency of angina history. The electrocardiographic finding of the exposed group as a whole showed a slightly higher prevalence of several pathologic conditions in comparison with the controls.

Of the forty-eight workers who had been exposed to carbon disulfide for at least 5 years and had died under 65 years of age, twenty-five had died of coronary disease in comparison to an expected 15.2 deaths.

Additional clinical studies of workers exposed to high levels of carbon disulfide have revealed disturbance of the pupilary reaction to light,[60] retinal vascular changes,[61-62] reduction in urinary excretion of total 17-ketosteroids, 17-hydroxycorticosteroids, and 3α-11-deoxy-17-ketosteroids,[63-65] increased incidence of diabetes,[66] hematochemical changes,[67,68] and an increase in the excretion of the ethereal fraction of urinary sulfate.[69]

Four different epidemiologic studies[58,70,71,72] have thus far related a higher coronary heart disease among workers exposed to carbon disulfide. Although the air level of carbon disulfide was around 20 ppm at the times of the studies, all of the workers had much higher previous levels of exposure to carbon disulfide. There is still insufficient epidemiologic data on workers exposed over prolonged periods to carbon disulfide at a time-weighted level of 20 ppm to judge further this level as a threshold limit value.

CHLORINE

The prevalence of chronic obstructive pulmonary disease in chlorine gas workers was investigated by Chester *et al.*[73] Of the 139 workers examined, 55 had additional accidental exposure to chlorine sufficient to require oxygen therapy. Continuous air monitoring data showed that approximately 99% of the chlorine air levels in the working environment were under 1.0 ppm. It was not possible to obtain information on the chlorine air levels during the periods of accidental exposures. The examination of the 139 workers included questionnaire information on social, work, and medical histories; physical examination; chest films; and ventilatory tests [vital capacity (VC); forced vital capacity (FVC); forced expiratory volume by time in seconds ($FEV_{0.5}$, $FEV_{0.75}$, FEV_1, FEV_2, and FEV_3); maximal expiratory flow (MEF); maximal midexpiratory flow (MMF); and calculated maximal breathing capacity (MBC)]. A detailed discussion is presented interrelating the many parameters of the medical examinations.

Of the 139 workers examined, the chest films of 82 were normal and 56 showed changes considered to be consistent with old granulomatous disease. Only one of the 56 had an abnormal ventilatory function. The FVC and FEV_1

values of smokers were significantly reduced in comparison to the values of the nonsmokers. The MMF was not significantly different between the smokers and nonsmokers when chlorine exposures were not considered. The group of smokers with the additive chlorine exposures when compared with the group of nonsmokers and only background chlorine exposures, showed a significant reduction in the MMF of the doubly exposed group. The authors concluded, "Only 3 workers had significant impairment of ventilatory function. Radiographic appearance of the lungs and the questionnaires correlated poorly with abnormal ventilatory function. The immediate effect of acute chlorine exposure was an obstructive ventilatory defect that cleared rapidly. Smokers who were exposed to chlorine demonstrated decreased maximal midexpiratory flow."

Patil *et al.*[74] reported on an epidemiologic study of the health of 600 diaphragm cell male workers exposed to chlorine while employed in over twenty-five plants producing chlorine. The study included a control group of 382 workers. Detailed occupational and medical history data were obtained on each worker as well as clinical data from hematologic examinations, chest x rays, electrocardiograms, and pulmonary function tests, with special attention directed to clinical signs, symptoms, and respiratory complaints.

The mean time-weighted exposure to chlorine was 0.15 ppm with a range from 0.006–1.42 ppm. Slightly over 97% of the workers studied had time-weighted exposures to chlorine under 1.0 ppm.

Clinical data on the workers obtained through chest x rays, electrocardiograms, pulmonary function tests, and hematologic tests were normal in comparison to the control group and did not show any dose–response relation. The group of workers exposed to chlorine did have more complaints of nervousness, frequent colds, chest pains, shyness, tooth decay, and anxiety than the workers in the control group. These complaints, however, could not be attributed to a dose–response relation with chlorine and were probably related to multiple factors. The authors concluded that the present threshold limit value of 1.0 ppm adopted by the American Conference of Governmental Industrial Hygienists[50] is adequate for its intended purpose.

CHROMITE

Sluis-Cremer and DuToit[75] investigated pneumoconiosis in chromite miners in South Africa. The chromite ore consisted of the oxides of chromium and iron. The free silica in the ore was less than 5%, and mostly under 1%. Vanadium (0.05%–0.81%), as well as trace levels of nickel, titanium, and manganese oxides were also present.

Dust concentrations were measured by the konimeter. Counting was done under dark-field illumination at a magnification of 150 X after ignition-immersion to remove combustibles and soluble salts. The particle sizes were approximately 0.5–5 μm. Prior to 1949 the mines were very dusty and mean dust levels of 1000 particles/cm^3 often existed. By 1954 when wet mining methods and other dust controls were introduced, the average count had fallen to 180 particles/cm^3 and has remained at that level. Of a labor force of some 1500, ten chromite miners showed radiologic evidence of fine nodulation. Few of the miners had remained underground longer than 5 years and most of these had also been exposed to other mining and industrial dust. None of the ten workers complained of dyspnea on exertion. One had scattered crepitations. Otherwise, the clinical examinations revealed nothing abnormal.

The authors concluded that the pneumoconiosis in the chromite miners was due to the deposition of a radioopaque chromite dust in the tissue and that the dust caused no fibrosis. Their conclusion was supported by experiments conducted on rats by the Pneumoconiosis Research Unit which showed that chromite injected intratracheally was nonfibrogenic.

COAL

The dramatic increase in attention to coal workers' pneumoconiosis over the past 3 years is attested to on many fronts, including public attention through news media, scientific symposia, research, and legislation. The problem is an international one and, as with other occupational diseases, this recognition along with an increased effectiveness in the coordination of international research talent and experience will stimulate programs and afford more rapid progress in understanding and controlling the disease.

Five symposia had entire or major attention devoted to coal workers' pneumoconiosis during which epidemiologic aspects of the disease were covered: (*a*) The International Conference on Pneumoconiosis, Johannesburg, South Africa, 1969[76–89]; (*b*) International Congress on Occupational Health, Tokyo, Japan, 1969[90–95]; (*c*) International Conference on Coal Workers' Pneumoconiosis, Lexington, Kentucky, 1969[96]; (*d*) the Third International Symposium on Inhaled Particles, London, England, 1970[97–106]; (*e*) Symposium on Respirable Coal Mine Dust, Washington, D.C., 1970.[107] There were likely many other symposia or smaller scientific sessions on coal workers' pneumoconiosis included in the programs of other major scientific meetings.

In the United States a prevalence survey[108] involving x rays and pulmonary function tests of miners was performed in the Appalachian bituminous coal fields from 1962–1965. The survey revealed that 9.5% of the miners currently working and 18.6% of the formerly employed coal miners showed x-ray evidence of pneumoconiosis. For miners currently working, the percent ranged from 2.0 for those who had worked 0–9 years underground to 24.2 for miners who had worked over 40 years underground. The report gives detailed information on the methodology used and on the findings.

Lainhart[255] presented data on roentgenographic evidence of coal workers' pneumoconiosis in three geographic areas in the United States: Appalachia, Illinois–Indiana, and Utah. In the study, a chest roentgenogram (14 X 17 in.) was taken on each of 3579 miners out of 3602 (a 4.0% sample) working coal miners from the three geographic areas. Participation ranged from 91.7% in the Illinois–Indiana area to 97.5% in the Utah area with an overall participation of 93.2%. The USPHS modification of the International Radiological Classification of Chest Films[256] was used.

Definite roentgenographic evidence of pneumoconiosis was reported in 10% of the miners examined in the Appalachian coal fields, 6% in the Illinois–Indiana coal fields, and 4% in the Utah coal fields. There was little roentgenographic change with less than 10 years underground. A marked change occurred after 20 years of underground mining. The change was greatest among miners at face activities.

In 1968, in the United States, the Bureau of Mines and the Public Health Service began an extensive cooperative national study of coal workers' pneumoconiosis in which the Bureau of Mines was responsible for developing environmental exposure data, and the Public Health Service for clinical and epidemiologic data on the miners. The Bureau of Mines, in 1968 and 1969, obtained detailed exposure data on sixteen underground coal mining occupations in twenty-nine mines. Gravimetric measurements were made of the respirable fraction of the airborne dust in the breathing zone of the miners studied. The MRE instrument[109] was used basically for the study. However, because it was not considered amenable to full-shift measurements in the breathing zone of active workers, actual measurements were made with a 10 mm nylon size selector, membrane filter, and a belt-mounted, battery powered pump. At the instrument sampling flow rates used, the two devices yielded different results. A conversion factor was experimentally determined and applied.[110]

This study indicated that at the time of measurement, most face workers were exposed to dust concentrations from about 4.5 to over 8.4 mg of respirable dust per cubic meter of air (MRE).

The clinical and epidemiologic study of the miners by the Public Health Service, now under way with the cooperation of labor and management, involves the workers at thirty-one coal mines across the United States. Many of these mines are the ones at which the Bureau of Mines obtained the detailed dustiness data mentioned above. The National Study includes chest x rays, pulmonary function tests, and detailed medical and occupational histories of workers at the selected mines. Field work was expected to be completed by June 30, 1971.

Preliminary data from the first half of the mines show that 94% of the miners participated voluntarily. Prevalence of the disease appears to be similar to that previously reported by Lainhart,[108] especially if age and years in mining are taken into account.

A new Federal Coal Mine Health and Safety Act of 1969 went into effect in the United States on December 30, 1969. Among the pertinent provisions of this Act are:

1. A gravimetric dust standard based on the MRE instrument is adopted.
2. A dustiness limit of 3 mg/m^3 of respirable dust is established at the outset, except that for an interim period, by special permit, mines may not exceed 4.5 mg.
3. The dustiness limit is to be further reduced from time to time by regulation to reach 2.0 mg/m^3 of respirable dust by December 30, 1972.
4. Every coal miner in or at an underground coal mine is to be afforded the opportunity to have a chest x ray before June 30, 1971, and periodically thereafter.
5. Every miner who enters the industry shall be x rayed within 6 months of employment, and periodically thereafter.
6. Persons who show definite evidence of pneumoconiosis may elect to be assigned immediately to work in atmospheres containing not more than 2.0 mg/m^3 of respirable dust without loss of wages. (By regulation, this reassignment option applies to all persons with category 2/1 or higher, with simple[111] or complicated pneumoconiosis, and to persons who have developed category 1/0 or higher, simple pneumoconiosis in less than 10 years in mining).

As a result of the Federal Coal Mine Health and Safety Act of 1969, dustiness levels in the mines have rapidly decreased, and many employees who had previously been working at significantly higher dust levels are now being exposed at, or below, maxima provided in the Act.

It is immediately apparent that for present and future progression study purposes, the x ray and other information being obtained at the time of transition by the National Study of Coal Workers' Pneumoconiosis and from the x-ray program called for in the Act, will contribute especially valuable data since it occurs at the critical point of a curve reflecting the relationship of dust measurements and dust exposure vs elapsed time.

There are presently reported to be 135,000 coal miners in or at underground coal mines in the United States. From the two studies, combined, it is expected to have more than 100,000 men examined between August 1969 to June 30, 1971, with over 97% of them examined subsequent to enactment of the Federal Coal Mine Health and Safety Act of 1969.

The new British standard, adopted in April 1970, makes a difference between general coal mining and driving openings in hard rock. The MRE instrument is used in both cases, and the limits are 8 mg/m^3 and 3 mg/m^3, respectively. In the case of longwall coal mining, which predominates in England, the sampling point is in the return roadway, 70 m from the face. In room and pillar mining, it is adjacent to the mine worker. According to the reports from England, in practice, because the measurement made in the return roadway is a summation of the dust produced over the entire longwall coal face in that type of mining, the men will not be exposed to more than 4.3 mg/m^3 of respirable dust. Jacobsen *et al.*[112,113] describe the new standards and methods of conversion for comparison from counting techniques to gravimetric measurements.

Events during the past several years will probably have a more profound effect upon the study and control of coal workers' pneumoconiosis than the occurrences in any similar period. These include the development of new dust measuring and control techniques, the study of parameters of coal dust exposure and their biologic importance, the application of new statutary dust limits which will upset the continuum in studying the development and progression of the disease; they also include greatly expanded chest x-ray programs, and clinical programs which should provide more reliable data concerning the prevalence and etiology of the disease.

COTTON – FLAX, HEMP, JUTE

Though respiratory disease from working in cotton textile mills has been known for some time, byssinosis did not gain worldwide attention until the last decade. To attest further to the growing interest in this subject, one has only to scan the technical papers on byssinosis over the past 5 years. In epidemiology alone, there have been over a score published in technical journals.[114-140] The Second International Conference on Respiratory Disease in Textile Workers was held September 29–October 2, 1968, in Alicanti, Spain. Over twenty papers were presented at the conference on the epidemiologic aspects of byssinosis. The XVI International Congress on Occupational Health, 1969, Tokyo had a number of papers on the epidemiology of respiratory diseases associated with exposure to vegetable dusts.[137-139] A national conference on cotton dust and health held in Charlotte, North Carolina, May 1970, was attended by over a hundred persons representing government, industry, labor, and research bodies. Proceedings of the conference have been published.[136]

Through epidemiologic studies, increased attention is being directed to developing methods for measuring exposures in terms that relate to biologic response, to defining effects of different exposure levels on biologic response, to determining safe levels of exposure, and to establishing recommended procedures for controlling exposure and maintaining medical surveillance over workers at risk. The following discussion is indicative of the epidemiologic research in the past 3 years.

Valic *et al.*[118] studied byssinosis, chronic bronchitis, and ventilatory capacities in workers exposed to soft hemp dust in the spinning department of a textile factory in Yugoslavia. The study population group of one hundred and six workers (ninety-three women and thirteen men) made up over 95% of the employment in the department. A control group consisted of forty-nine workers (thirty-eight women and eleven men) employed in other departments. A modified British Medical Research Council questionnaire on respiratory symptoms was used to obtain information for each person on work history, medical history, respiratory symptoms, and smoking habits. Ventilatory tests included forced vital capacity (FVC) and forced expiratory volume 0.75 second ($FEV_{0.75}$) before the beginning and at the end of each work shift. Air samples for total dust were taken with a modified MSA electrostatic precipitator and a filtration method which compared to the Hexhlet method. Total airborne hemp dust in the spinning department ranged from 2.9 mg/m^3 at the wet spinners, to 19.5 mg/m^3 at the carders. The average total dust concentrations in the other departments where the control group worked was under 1.0 mg/m^3. In the spinning department 40.6% of the workers had byssinosis and 15.1% had chronic bronchitis. The range of disease by jobs in the spinning department was 57% byssinosis and 35.7% chronic bronchitis in carders (average total dust 19.5 mg/m^3) to 20% byssinosis and 0.0% chronic bronchitis in wet spinners (average total dust 2.9 mg/m^3). There was no disease in the control group. The ventilatory data are discussed in detail as related to the many parameters measured.

Bouhuys *et al.*[123] conducted a mill byssinosis study in cotton textile workers comprised of 217 men working in carding and spinning rooms and a control of 241 men working in other mill areas. Data were obtained on a standard questionnaire form for each person through interviews. Lung function tests on each person included FVC, FEV_1, and maximum expiratory flow volume (MEFV) curves. Air samples for respirable dust taken at the time of the survey showed the following average concentrations: picker, 0.72 mg/m^3; carding, 0.48 mg/m^3; slubbing, 0.43 mg/m^3; spinning, 1.07 mg/m^3; and winding, 0.47 mg/m^3. Of the 214 men exposed to dust in the carding and spinning rooms, 29% (61) had byssinosis defined as having chest tightness or cough, or both, on Mondays during work. In 31% (66) of the carders and spinners, the FEV_1 decreased significantly on Mondays. Symptom prevalence of byssinosis was higher among the smokers (32%) as compared with the nonsmokers (9%). The FEV_1 values for smokers and nonsmokers, however, were similar. Dust control and periodic medical examinations, including FEV_1 were recommended.

Simpson and Barnes[121] investigated airborne cotton dust concentrations during lint removal and seed crushing, and their effect on the ventilatory capacity of exposed workers. Airborne dust samples were taken with the Greenburg Smith Impinger using 10% alcohol. Dust levels ranged from 15–18 mg/m^3 in work areas near the lint removals to 30–37 mg/m^3 at the bailer. The group studied comprised sixteen workers exposed to cotton lint and a control of eleven unexposed workers. The ventilatory measurements included VC and FEV_1. There was a significant decrease in the ventilatory capacity of the workers exposed to high concentrations of cotton dust whereas the nonexposed control group had no decrease. Workers who smoked had a lower FEV_1.

Exposure to dust in the cottonseed oil extraction industry was studied by Noweir *et al.*[140] In the two plants studied, 147 workers representing approximately one-half of the total work force (110 handling cotton seed and 37 grinding seed and extracting oil) were selected at random. Social, occupational, and medical history (including chest symptoms) questionnaire data were obtained on each worker along with a complete physical examination.

Dust concentrations at cottonseed handling operations ranged from 73 mg/m^3 in the general air to 590 mg/m^3 during evacuation of cottonseed bags. At cottonseed grinding and oil extraction operations, dust concentrations ranged from 3 mg/m^3 in the general air to 29 mg/m^3 at cottonseed grinding machine. Chemical analysis data are presented on settled dust at different operations. Of the 110 workers handling cottonseed, 35 (30.2%) complained of dyspnea, tightness of the chest, and coughing and wheezing during work. None of the thirty-seven workers grinding cottonseed and extracting oil had these complaints.

Zuskin *et al.*[124] studied the prevalence of byssinosis in carding and spinning workers in the cotton textile industry. The worker study group included 120 men and 38 women working at and around carders and spinners. Work force employment stability was good. Residence, work history, medical history, chronic respiratory symptoms, smoking habits, etc., were recorded on a standard questionnaire form. Ventilatory function tests included FVC, FEV_1, and MEFV. Tests were made on all subjects at the time of interview and repeated on a selected number of sixty-seven workers on Monday and Wednesday before and after work. Air sampling was done with a high-volume filter sampier. A cyclone upstream of the filter removed nonrespirable particulates.

In Mill A, the mean dust concentrations for total and respirable dust at the carders was 1.63 mg/m^3 and 0.87 mg/m^3, respectively, and at the spinners 1.91 and 0.92 mg/m^3, respectively. In Mill B the total and respirable dust concentrations at the carders was 1.55 and 0.50 mg/m^3, respectively, and at the spinners 1.54 and 0.55 mg/m^3, respectively. In the carders and spinners at both mills, the FEV_1 decreased significantly during the work shift on Monday. At both mills combined, 17% of the workers tested had symptoms of byssinosis;

25% among carders and 12% among spinners. FEV_1 determinations on workers before and after work were recommended as a medical surveillance procedure along with dust determinations for assessing and controlling exposures.

Popa *et al.*[125] investigated the sensitization of textile workers to antigens of cotton, hemp, flax, and jute in two groups. The first consisted of eighty-two workers from a cotton factory and eighty-three from a hemp, flax, and jute factory, all of whom had been included in a recent byssinosis study. The second group included forty-one workers with byssinosis selected for a more detailed study. Tests included intradermal reaction to prepared allergens and inhalation reaction to prepared allergens and to textile macerate. In intradermal skin tests, immediate reaction to the allergens were seldom seen but delayed reaction was nearly always present.

For cotton workers with chronic bronchitis, seven out of twelve were positive (delayed reaction) to blowing-room allergin; and for cotton workers with byssinosis, seven out of eleven were positive (delayed reaction) to the same allergin. In flax, hemp, and jute workers with chronic bronchitis, two out of eight were positive (delayed reaction) for flax fiber, six out of eight positive for hemp fiber, and seven out of eight to jute fiber. For the flax, hemp, and jute workers with byssinosis, six out of eleven were positive (delayed action) for flax fiber, eight out of eleven to hemp fiber, and ten out of eleven to jute fiber. Of thirteen with byssinosis only one had positive response to inhaling textile macerate.

Khogali[126] studied the prevalence of byssinosis in cotton ginnery workers in the Sudan. The study group comprised 323 full time workers, 35 seasonal farfara workers, and a control of 24 from the fire brigade. Occupational history, medical history, chest symptoms, and smoking history data on each were recorded on a standardized questionnaire form. Each person was given a physical examination and ventilatory tests including FVC and FEV_1. A Hexhlet sampler was used to measure airborne dust. The mean dust concentration (< 7 μm) at the ginnery workshop was 0.11 mg/m^3, at the ginnery hall 0.63 mg/m^3, and at farfara 2.6 mg/m^3. The prevalence of byssinosis had a direct correlation with dusty exposure levels: 19% for workers at ginnery work shop, 20% for workers at the ginnery hall, and 49% for farfara. The mean FEV_1 value of workers exposed to the dust was significantly lower than that of the control group.

Smith *et al.*[127] studied the respiratory symptoms and ventilatory capacities of fifty-four men and twenty-two women workers exposed to hemp and flax in making rope. Information on each person including medical history, respiratory symptoms, and smoking habit was recorded using the Medical Research Council questionnaire for bronchitis. FEV_1 was measured on each worker. Airborne dust samples were collected with a modified Hexhlet sampler. At operations such as winding, twisting, cabling, polishing, etc., the dust concentration was low; mean values were 0.4 mg/m^3 for total and 0.1 mg/m^3 for respirable dust. Higher concentrations were found at opening, hacking, carding, spinning, etc.; mean values were 1.7 mg/m^3 for total and 0.5 mg/m^3 for respirable dust. Thirty-seven percent (six out of sixteen) of the men with high dust exposure had symptoms of byssinosis compared to 0.0% for those with low dust exposure. There was no significant difference in the FEV_1 of workers in either the high or low dust concentrations nor in the ventilatory capacities of workers with or without symptoms of byssinosis. It was noted that the dust concentration in this factory was much less than the other factories where a higher prevalence of byssinosis was reported.

Molyneux and Tombleson[132] reported an epidemiologic study of respiratory symptoms in the Lancashire Mills, 1963–1966. The study included card and blowroom workers in fourteen cotton mills and two synthetic fiber mills over a period of three years, and 1359 cotton workers and 227 synthetic fiber workers. For the synthetic fiber workers, synthetic fiber was incorporated into the cotton in the three mills to the extent of 20%, 25%, and 16%, respectively. Questionnaire data as well as a FEV_1 measurement were obtained on each person. The Hexhlet and total samplers were used to measure mass concentrations of respirable, medium, and total components of the dust. Dust concentrations averaged 3.1 mg/m^3 for total and 0.64 mg/m^3 for respirable dust at coarse mills and 1.2 mg/m^3 for total and 0.25 mg/m^3 for respirable dust at medium mills. The total prevalence for byssinosis in the cotton workers was 26.9% compared to 4.4% for the synthetic fiber workers. The prevalence of byssinosis by occupations varied from 0.0–60.0%. The incidence of bronchitis was higher in workers with symptoms of byssinosis and was also influenced by smoking and age.

Merchant *et al.*[128] reported on the screening for byssinosis of 441 cotton textile employees, using a questionnaire, chest films (PA and lateral), spirometry, and single-breath nitrogen and carbon dioxide volume gradients. The authors found that the diagnosis of byssinosis by history alone appeared to underestimate the true incidence. Pulmonary function testing was found to be the most sensitive test for identifying the byssinotic reactor. There was a correlation between breathlessness and symptoms of chronic bronchitis and byssinosis, but there was no evidence of increased emphysema among byssinotics.

Schrag and Gullet[133] studied byssinosis in textile mills in North Carolina. Classifying as having byssinosis all workers who reported characteristic chest tightness, breathlessness, or both at least occasionally on Mondays, they reported 12% of the total workers, 29% of cardroom workers, 10% of weavers, and 9% of spinners as having byssinosis. Those classified as having byssinosis had significantly lower FEV_1 values than those without byssinosis.

deTreville reported in the Industrial Health Foundation Medical Series Bulletin No. 18-71 (1971) "Proteolytic Enzymes in Cotton Mill Dust – A Possible Cause of Byssinosis"[258] that a limited investigation showed a correlation between exposure to proteolytic enzymes in the cottonseeds, hulls, and bolls and the symptom – complex, byssinosis. This is apparently the first report associating enzyme activity with physiologic findings in cotton workers and is being followed up with more definitive investigations.

It is widely accepted that the effects of cotton dust on the respiratory system are due not to the cotton fiber per se, but to foreign material – primarily the bract, a thin leaf under which the cotton boll forms. The American Conference of Governmental Industrial Hygienists Threshold Limit Value for cotton dust (raw) of 1.0 mg/m^3 is based on total dust.[50] A definition of an inhalable fraction (dust having an aerodynamic diameter < 15 μm) for air sampling instruments has been proposed[136] and a design for a size-separator to remove the > 15 μm diameter suggested. The threshold limit value for the respirable size cotton dust presumably would be somewhat less than the present value for total dust, but a recommended value has not as yet been published.

FIBROUS GLASS

The production and consumer use of fibrous glass and associated products have shown a continued expansion over the past several years. In the form of wool fibers, with diameters ranging from 5–14 μm, it is used in the manufacture of insulation, thermal, and acoustical products. As textile fibers, with diameters ranging to as low as 4 μm, it is used in the manufacture of fabrics, insulation products, and reinforcements for other materials such as plastics. For experimental purposes and use in some specialty products, a small amount of fibrous glass is manufactured with diameters under 1 μm. At the present time, however, the production of the latter is extremely limited and there is no forecast that it will significantly increase in the foreseeable future.

Because of the expanded production and use of fibrous glass products, a continuing interest in any health problems which may arise from associated exposures has been maintained.

Johnson *et al.*[141] in 1969 reported data on worker exposure to fibers in the manufacture of fibrous glass. All exposure levels to both respirable and total fibers in the manufacture of wool and textile fibers were well below the threshold limit value of 10 mg/m^3 adopted by the American Conference of Governmental Industrial Hygienists.[50]

A chest roentgenographic study of workers with prolonged exposure to airborne fibrous glass particles was reported by Wright.[142] The total particulate dust concentration levels ranged from 0.93 mg/m^3 to 13.3 mg/m^3 with an overall plant average of 2.24 mg/m^3. Fibers made up less than 1% of total particulates. The fibers had a median diameter of 6 μm. Formaldehyde and phenol concentrations were under 0.5 ppm. Past fiber concentrations in the plant very likely had greatly exceeded these levels which represent levels achieved after extensive exhaust ventilations and dust control practices had been installed. Comparisons were made of roentgenogram readings of 1389 workers classified by duration and intensity of exposure, and included workers with 10–25 years of exposure. No unusual pattern of radiologic densities was reported.

The roentgenographic study by Wright was subsequently expanded to include extensive information on respiratory illness history and lung function.[143] Neither the respiratory illness history nor test results which included spirometric tracings for mean forced expiratory volume (FEV), FEV_1, FEV_3, and maximal midexpiratory flow (MMF) showed any significant trend which related to degree of fibrous glass exposure.

A subsample of thirty workers reported above[143] was more extensively studied [144,145] through a follow-up study which included a comprehensive general medical history, a special respiratory questionnaire, a comprehensive general physical examination, a battery of ventilatory functioning testing, and measurement of pulmonary diffusing capacity for carbon monoxide. Data from these examinations failed to show any evidence that workers with the highest exposures were less healthy than those with minimal exposures. These data are further supported by a study of the lungs of twenty autopsied persons who had long histories of work in fibrous glass manufacturing which confirmed that fibrous glass is not pathogenic in the lungs.

There are no mortality data on workers with prolonged exposure to fibrous glass. Since such data are important, the Bureau of Occupational Safety and Health, U. S. Public Health Service, Department of Health, Education, and Welfare, is conducting a mortality study of workers in the fibrous glass manufacturing industry.[146]

Milby and Wolf[147] reported a number of cases of respiratory tract irritation from fibrous glass inhalation. Respiratory tract manifestations were tabulated on sixty-six case reports.

Possick *et al.*[148] reported on an investigation in fibrous glass plants to assess the prevalence and significance of work-related dermatitis. From 5–10% of new workers had sufficient irritation and discomfort to seek other employment. The dermatitis was caused by the irritant action of the fibrous glass. Fibers of large diameter were found more likely to cause irritation. Allergy did not play a role in the dermatitis.

Epidemiologic data indicate that fibrous glass in the lung is essentially inert and nonfibrogenic. The major problem in the manufacture and use of fibrous glass products is one of skin and upper respiratory tract irritation.

The American Conference of Governmental Industrial Hygienists classified fibrous glass as an inert material and has adopted an intended change which will establish a threshold limit value for fibrous glass (< 5–7 μm diameter) of 30 million particles per cubic foot of air or 10.0 mg/m^3 (whichever is the smallest) of total dust.[50]

FLUORIDE

A study of the urinary fluoride excretion in twenty-four electric arc welders exposed to low hydrogen electrode fumes was reported by Smith.[149] The mean age of the twenty-four welders was 41.5 years; that of sixteen control workers was 47 years.

The time-weighted fluoride breathing zone concentration for all welders was 0.4 mg/m^3 (range 0.1–10.0 mg/m^3) and varied from 0.3 mg/m^3 (range 0.2–8.2 mg/m^3) at open welding to 1.0 mg/m^3 (range 0.4–10.0 mg/m^3) at welding in confined areas. The mean urinary fluoride concentration for all welders was 2.4 mg/liter (range 0.5–4.8 mg/liter) and varied from 1.8 mg/liter (range 0.5–2.8 mg/liter) at open welding to 2.9 mg/liter (range 1.5–4.8 mg/liter) at welding in confined areas. The mean urinary fluoride level for the control was 1.0 mg/liter (range 0.4–1.5 mg/liter). The author concluded "from the evidence presented in the present studies it appears unlikely the concentration of fluoride in the fume from low hydrogen electrode arc welding is sufficiently high to be retained in the body longer than the period between successive exposures." The American Conference of Governmental Industrial Hygienists Threshold Limit Value for fluoride (as F) is 2.5 mg/m^3.[50]

GRAIN HANDLERS

Kleinfeld *et al.*[150] reported on a clinical and physiologic study of fifty-five grain scoopers and grain elevator operators who removed wheat, oats, barley, corn, and soybeans from the holds of ships, transferred the grains to storage elevators, and performed related clean-up activities. The age of the workers ranged from 22–72 years with a mean of 50. The exposure period ranged from 2–36 years with a mean of 19. Of the fifty-five grain handlers, 60% gave a positive history of smoking.

Air concentration measurements of grain dust were not made. The nature of the operations, however, would predict appreciable exposures. Each worker in the study group was given a detailed medical, social, and occupational history questionnaire, a clinical examination, and pulmonary ventilatory tests. The predominant symptoms were cough, dyspnea, chronic wheezing, and grain fever.

General malaise, chills, and fever several hours after leaving work were the predominant traits of the grain fever. Cough was present in fifteen (27.3%) scoopers, sixteen (29.1%) complained of shortness of breath upon exertion, nine (16.4%) had chronic wheezing, and eighteen (32.7%) gave a history of grain fever. A history of dermatitis was reported in seven (12.7%) and peptic ulcers in thirteen (23.6%) of the workers studied. A significant difference was noted in the clinical and pulmonary function findings for smokers as compared with those for nonsmokers. The authors concluded that of all the predisposing factors, i.e., age, length of exposure, smoking, and allergy, smoking was the most important factor relating to the increase of respiratory symptoms and altered pulmonary ventilation. As with smoking and other exposures, the question arises concerning the possible additive or synergistic effects in combined grain dust and smoking exposures.

IRON ORE (HEMATITE, MAGNETITE)

Lung cancer in workers mining hematite (iron) ore, in West Cumberland, England, was studied by Boyd *et al.*[151] in a follow-up of a previous study in the same area. In the first study[152] covering the period 1930–1935, 10% of 238 necropsies on hematite miners showed carcinoma of the lungs compared to 2.7% in 78 necropsies in nonminers used as controls. Extensive dust control was instituted around 1935. Massive dust exposures were present previous to this time. Indicative of its effectiveness is the reduction in percent of chest abnormalities between 1935 and 1955. In 1935, 30% of the miners with 10–20 years underground had abnormal chest radiographs and 1.5% had progressive massive fibrosis. Twenty years later only 1% had abnormal radiographs.

The follow-up study was undertaken to complement and extend the earlier necropsy study and to see if mortality data confirmed the suspected occupational risk of lung cancer. The death certificates of 5811 men who lived in two of the communities comprising the West Cumberland mining areas were examined. Mortality data on the iron ore miners who died between the years 1948–1967 were compared with the data for nonminers who lived in the mine communities and with the relevant national data.

Prior to 1913 hematite ore, predominantly ferric oxide containing 10–12% silica, was mined by manual methods. In that year the pneumatic drill was introduced and greatly increased the dust levels at the working face until wet drilling methods were introduced in 1925. Dust concentration levels at the face during dry penumatic drilling were often in the many millions of particles per milliliter. Through the application of wet methods and other procedures for dust control, the dust levels at the face were reduced to 2500 particles per milliliter for dust under 5 μm in diameter with a silica content usually less than 10%. No data were presented on the amount of other metals which may have been in the respirable dust, either from the hematite ore or from the abrasive action of the pneumatic drills in mining.

More recently, a survey conducted by the Radiological Protection Services (England) showed three of four hematite mines in the Egremont area of West Cumberland had radon measurements ranging from 30 to over 300 picocuries/liter (pCi/liter) with an average of 100. None of the levels were below the maximum permissible level of 30 pCi/liter recommended by the International Commission on Radiological Health Protection. Since representative samples of exposed rock did not show any radioactivity, the authors speculated that the radon may be carried by the underground water as has been demonstrated elsewhere.

The follow-up study revealed that during the period 1948–1967, of 686 deaths among the miners, 42 (6%) were attributed to cancer of the lung, 74 (11%) to other cancer, 174 (25%) to respiratory causes, and 396 (58%) to other causes. In underground miners there were 36 deaths (21 expected) from lung cancer, 65 deaths (56 expected) from other cancers, 52 deaths (16 expected) from respiratory causes. The miners suffered a lung cancer mortality about 70% higher than normal. No data were available on the smoking habits of the miners studied. The authors concluded that the increased risk to cancer may be due to either the radon or to the iron oxide. It would be interesting to obtain information on the miners' exposure to trace metals in the respirable dust since other research has shown that their presence and smoking may have an additive or synergistic effect.[153–155]

A clinical, roentgenologic, and physiologic study of magnetite workers was reported by Kleinfeld *et al.*[156] The study group consisted of fifty-seven workers of whom forty-one were miners and sixteen were sinterers. The average length of exposure of workers in the mine was 27.5 years and in the sintering plant 27.8 years. A control group of eighteen workers was made up of men of similar age distribution who lived in the same area but who had no prior occupational dust exposure in the industry. Environmental studies showed the magnetite ore to consist of 35–40% iron oxide, 48–52% silicates, and 7–9% free silica. In contrast, the concentrate in the sintering plant averaged 81% iron oxide, 8% silicates, and 2% free silica. In the mine the average dust concentration varied from 5.0–7.5 millions of particles per cubic foot of air (mppcf), and in the sintering plant 5–10 mppcf. No data were presented on other metal constituents of the dust in the mine or sintering plant. Each participant in the study had a detailed medical, social, and occupational history, clinical examination, electrocardiogram, chest roentgenogram, and a battery of pulmonary function tests including pulmonary diffusion. The prevalence of smoking among the miners, sinterers, and the control group was similar (58, 56, and 56%). There was a greater prevalence of abnormal respiratory, radiologic, and pulmonary function findings in the miners compared to the controls. The miners had a significantly greater incidence of dyspnea and positive lung findings compared to the sinterers, who were similar to the controls. It is believed that this was due to the greater percentage of silicates and free silica in the mine dust in comparison to the sintering plant dust. No consistant correlation was apparent between clinical, radiologic and pulmonary findings, and duration of exposure. There was no evidence that long-term low-level exposure to iron oxide dust predisposed to an increased prevalence of respiratory symptoms or signs. In view of the findings of Boyd *et al.*[151] that hematite miners in West Cumberland, England, had a greater risk from lung cancer, similar data on the magnetite miners studied by Kleinfeld *et al.*[156] would be very useful and might give additional insight on whether the radon in the Cumberland mines was an important factor.

Jorgensen and Svensson[157] studied the pulmonary function and respiratory tract symptoms of workers in an iron ore mine where diesel trucks were used underground. A total of 240 workers under study were subdivided into equal numbers of underground and surface workers and each of these in turn into younger and older smokers and nonsmokers. Diesel equipment, both transportation and loading, has been increasingly used since 1959. Underground concentration levels of nitrogen dioxide, nitrogen dioxide plus nitrogen oxide, carbon monoxide, and aldehydes were measured at regular intervals and 3,4-benzpyrene on occasion. In the past, sporadic measurements indicated that the concentration of nitrogen dioxide was around 0.5–1.5 ppm with higher values over short periods. The levels of sulfur dioxide were not measured. Aldehyde levels were low, but at times peaked at 10–20 ppm. Current levels of the above gases were not given, but it was assumed that they were likely of the same order of magnitude. Insufficient data on levels of these gases were given to make an evaluation. Past dust measurements in the mine, using the konimeter, ranged from 300–600 particles/cm^3. In 1968, dust measurements using millipore filters and personal air samplers showed levels ranging from 3–9 mg/m^3. Some 20–30% of the particles were less than 5 μm in diameter. The percent of quartz in the less than 5 μm fraction was 6–7%. The above levels of dust appeared to exceed somewhat the American Conference of Governmental Industrial Hygienists Threshold Limit Value for quartz. Medical data on the workers consisted of a modified British Medical Research Council's bronchitis questionnaire along with spirometer measurements, forced vital capacity, and $FEV_{1.0}$. In general, there was little difference in the spirometric data of any of the groups whether they worked above or below ground or were smokers or nonsmokers. Repeated attacks of productive bronchitis and of simple chronic bronchitis were greater in the underground workers and more so among the smokers than nonsmokers. The authors believed that there was a synergistic action between smoking and some factor of the underground work.

IRON OXIDE

Sentz and Rakow[158] studied the exposure of seventy-three workers to iron oxide fumes at arcair (also known as jet-arc) and powder burning operations encountered in foundry and steel fabrication establishments.

The powder-burning operator was exposed to average iron oxide levels of 31 mg/m^3 with no exhaust ventilation and 5 mg/m^3 when side draft exhaust ventilation was provided. The size and content of metal casting and the duration of burning as well as the operation enclosure and the nature of the exhaust ventilation related to levels of iron oxide exposures. At arcair operations, the worker was exposed to an average of 21 mg/m^3 when only dilution ventilation was provided. This was reduced to around 3 mg/m^3 when side draft ventilation was used. The iron oxide in the air just outside the operator's helmet, with general ventilation only, averaged 40% higher (35 mg/m^3) than in the air inside the helmet.

Although the operators were potentially exposed to varied concentrations of manganese, nickel, chromium, and copper fumes and to carbon monoxide, carbon dioxide, nitrogen oxides, and ozone gases, the study was confined solely to iron oxide fumes. Chest roentgenograms were used to appraise the health of the seventy-three workers in the study group. All of the workers had other previous work experience, and all also welded or did other related operations. The length of time at the arcair or powder-burning operations ranged from 2 months to 12 years. Of the seventy-three workers studied, three had positive findings on the chest roentgenograms. Only one of the three had subjective symptoms of a productive cough and shortness of breath on moderate exercise. None of the positive roentgenograms could be related solely to powder-burning or arcair operations.

A clinical, roentgenographic, and physiologic study of twenty-five welders exposed mainly to iron oxide was reported by Kleinfeld *et al.*[159] The age ranged from 25–70 years with a mean of 48. The duration of exposure at welding ranged from 3–32 years with a mean of 19. A control of twenty men was selected who had the same age distribution and lived in the same area but had no exposure to occupational dust. The exposure to iron oxide resulted from the welding and flame cutting of block iron and stainless steel in the manufacture of sheet metal products. In addition to iron oxide, tests were made for ozone and fluoride as a fluoride flux was used in one of the welding processes. Iron oxide levels inside the welders shields ranged from 0.65–1.7 mg/m^3 and outside the shields from 1.6–12 mg/m^3. The iron oxide concentration in the general air was below a detectable level. Ozone concentrations were negligible; a fluoride level of 2.7 mg/m^3 was measured outside the face shield. The flame cutter had an iron oxide exposure of 3.0 mg/m^3. These levels of exposure are within the threshold limit value of 10.0 mg/m^3 adopted by the American Conference of Governmental Industrial Hygienists.[50] Previous air samples taken at the same plant in 1960 before the ventilation was improved showed iron oxide levels ranging from 30–47 mg/m^3. Since the mean exposure period of the twenty-five welders was 19 years, most of these were exposed for a number of years to the higher iron oxide levels before the controls were improved. Each welder in the study group had a detailed medical, social, and occupational history, clinical examination, chest roentgenogram, and a battery of pulmonary function tests including carbon monoxide diffusion. Correlation of clinical findings were made on welders who had worked less or more than 20 years. The prevalence of smoking in the welders and control groups was similar (56–55%). The clinical and physiologic findings in the welders, average exposure 19 years, did not significantly differ from the control group. There was an increased prevalence in the welders of positive roentgenographic findings consistent with siderosis; the prevalence was 2.8 times higher in the welders with over 20 years exposure in contrast to those with under 20 years exposure. Prolonged exposure to iron oxide did not produce significant clinical or physiologic alterations when compared with the control group.

DIISOCYANATES

The diisocyanates are important industrially because of their use in the making of polyurethane-type polymers and plastics. These products have a very wide use in industry and their production has increased tremendously during the past 20 years. Pulmonary problems were associated at the very beginning with exposure to the diisocyanates. A number of epidemiologic studies have been reported in the interim defining the nature and extent of clinical effects from exposure to these chemicals.[160–167] A symposium on urethanes and isocyanates was sponsored by the Southern California Section of the American Industrial Hygiene Association in Los Angeles, November 1970.

Bruckner *et al.*[168] made a clinical and immunologic appraisal of forty-four workers exposed to diisocyanates during the production of the chemicals and simultaneously determined the exposure levels. Twenty-six of the forty-four had multiple exposures to diisocyanates while eighteen had never worked with nor near isocyanates. Exposures were encountered in three areas: research building with median concentration levels varying from 0.033 (range (0.0–0.24) ppm in 1957 to 0.0 in 1967; application laboratory with median concentration levels varying from 0.035 (range 0.0–0.19) in 1962 to 0.0 in 1967; and pilot plant with median concentration levels varying from 0.077 in 1957 to 0.0 in 1966. Symptoms included headaches; eye, nasal, mouth, and throat irritation; cough; shortness of breath; and chest tightness. All of these clinical parameters were significantly higher in the exposed group in contrast to the nonexposed in which the symptoms were essentially nonexistent. The isocyanates are both potent primary irritants and sensitizers. The latent period has been as short as 2–3 weeks but in most instances has involved repeated exposures extending from months to years. The (NCO) group of the molecule is considered responsible for the biologic effects of isocyanate compounds.

Peters *et al.*[169] studied the acute respiratory effects in workers exposed to low levels of toluene diisocyanate. Thirty-eight workers were included in the study. The plant manufactured flexible, molded polyurethane foam. Toluene diisocyanate concentration levels, in the pouring area, ranged from 0.02–0.03 ppm in 1965 to 0.001–0.003 ppm in 1966; at the unloading area, concentrations in 1966 were 0.0001–0.0004 ppm; packing and shipping, 0.0004 ppm; machine shop, 0.0001 ppm; and rotational casting area, 0.0001 ppm. The workers at the time of the study were thus exposed to concentrations of toluene diisocyanate ranging from 0.0001–0.003 ppm. These levels were generally below the American Conference of Governmental Industrial Hygienist threshold limit "C" value of 0.02 ppm.[50] Each of the thirty-eight workers was given a respiratory questionnaire including an occupational, medical, and smoking history, and a battery of pulmonary function tests. The latter given on Monday at the beginning and end of the shift showed significant decreases in forced vital capacity, forced expiratory volume/one second, peak flow rate, and expiratory flow rates at 50% and 25% of vital capacity. By Friday the FVC had returned to baseline, but the FEV_1 remained depressed and the expiratory flow rates were more depressed.

Peters *et al.*[170] further studied the above workers to determine if exposure to toluene diisocyanate had caused any cumulative or chronic effects on ventilatory capacity, and if a prediction could be made on which workers will be sensitive to this chemical. Twenty-eight of the workers studied 6 months earlier were given similar pulmonary function tests on Monday morning and afternoon and Tuesday morning and afternoon. Six new employees were also studied for comparative data. The air concentration level of toluene diisocyanate was 0.0005 (range 0.0000–0.0120) ppm. A significant change in ventilatory function of the workers over the 6-month interval was indicative of a cumulative effect of exposure to toluene diisocyanate. The 1-day change in FEV_1 and the assessment of respiratory symptoms both appear favorable as procedures in screening workers to detect those likely to be affected by exposure to toluene diisocyanate.

In further follow-up of the workers previously studied,[169,170] Peters *et al.* completed an 18-month observation to obtain information relating to respiratory impairment in workers exposed to levels of toluene diisocyanate within the threshold "C" limit value of 0.02 ppm.[171] The FEV_1 was determined at 6-month intervals for 18 months on the same workers Monday morning and afternoon. The authors concluded: "(1) Acute changes in ventilatory capacity occur in workers exposed to toluene diisocyanate at levels less than the American Conference of Governmental Industrial Hygienists threshold limit 'C' value 0.02 ppm. (2) These acute changes were not reversed overnight. (3) Cumulative changes in ventilatory capacity were observed in workers during the 18 months of observation. (4) Workers with respiratory symptoms demonstrated a greater response to toluene diisocyanate than asymptomatic ones, and (5) a highly significant correlation between one-day and 6-, 12-, and 18-month change was observed."

Paisley[172] reported a respiratory symptoms outbreak from toluene diisocyanate among workers manufacturing electrical tools. In the factory of 700 workers (600 women, 100 men) 40 workers complained of breathlessness and cough, with breathlessness, occurring in spasms, as the main feature. An investigation revealed that a new wire, polyurethane-covered, had recently been introduced. When it was heated to destruction, as when applying a heated soldering iron to the coated wire, diisocyanate vapors were released. The authors cautioned on the danger from fires in enclosed spaces where polyurethane may be present thus releasing high concentrations of isocyanate vapors.

LEAD

The prevalence of anemia in a group of 228 lead burners employed in shipbreaking was studied by McCallum *et al.*[173] The major source of exposure was from the oxyacetylene cutting of metal coated with a lead paint in the salvaging of metal through shipbreaking. A full occupational history, medical history with special attention on symptoms relating to lead poisoning, and a blood and a urine sample were obtained on each of the 228 workers. Of the 228 workers studied, 133 (58.3%) had been employed as burners for less than 5 years and 95 (41.7%) for over 5 years. Eighty-seven percent were under 55 years of age.

Twenty-seven percent (59) of the 225 workers completing the study had symptoms of lead poisoning. Anemia was found in 36% (80) of this group. Of 209 burners supplying urine samples, the lead concentration was over 120 μg/liter in 152 (72%) and over 300 μg/liter in 50 (23%). The data was discussed in detail along with methods of prevention. The American Conference of Governmental Industrial Hygienists threshold limit value for lead is 0.2 mg/m^3.[50]

Rieke[174] describes problems in Portland, Oregon, 1941–1968, from lead intoxication as a result of shipbuilding and shipscrapping operations. During the period of heavy ship construction, 1941–1945, lead poisoning at the shipyard was relatively rare. Through programmed staging of operations, use of exhaust ventilation, and the wearing of respirators, reasonable surveillance and control could be maintained over the working environment. At the end of World War II, the operations changed from shipbuilding to shipscrapping with many of the workers transferring to the new operations. In the postwar transition as many as 150–300 burners were employed at two riverbank locations at the scrapyard of a local steel mill. The shipscrapping was thought to be temporary, but it still existed 20 years later. Few environmental controls were observed and periodic medical examination of the workers was never done. Eight air samples taken during burning operations showed lead levels ranging from 0.02–2.73 mg/m^3 and averaging 1.16 mg/m^3. Paint was the chief source of lead. Many other metals, depending on the alloy components, were also released into the air during metal burning and cutting operations.

Tests on fifteen workers showed 100% of their urine lead levels over 0.18 mg/liter in mid 1963. Through close supervision of the use of respirators, urinary lead levels were reduced by 1966 to only 10% exceeding 0.18 mg/liter. Six detailed case reports are given. The author discussed his experience and gives recommendations for the control of lead in such industrial practices.

Williams *et al.*[175] in investigating lead absorption in workers in an electric accumulator (battery) factory used personal samplers worn on the lapel of each of thirty-nine workers every shift for 2 weeks to assess airborne levels of lead exposure. During the second week blood lead, urinary lead, urinary coproporphyrin, urinary δ-aminolevulinic acid, punctate basophil count, and hemoglobin were estimated daily. The data on exposure and clinical parameters were correlated. The findings showed wide variations of lead exposure for workers doing similar jobs, some differing as much as 4:1. Work habits were thought to account for much of this. Both blood lead and urinary coproporphyrin correlated as highly with levels of lead in air as did the other biochemical tests. Better urinary correlation appeared when the urine was corrected for specific gravity.

Soliman *et al.*[176] conducted biochemical studies on Egyptian workers exposed to lead in an electric accumulator factory. The biochemical tests included delta and gamma globulin and urinary coproporphyrin. The study was based on 112 workers not having clinical signs of lead intoxication and 20 control workers. Previously established parameters for the tests were used and seventy-six of the workers were diagnosed as having chronic plumbism or preclinical lead toxicity.

Jovicic[177] examined a group of thirty-five workers with saturnism and presaturnism and a control group of ninety-four workers for gastroduodenal ulcers, gastritis, and gastric acidity. The lead exposed group had a 12.5%

frequency of ulcer compared to 3.15% for the general population. Firm conclusions could not be drawn from the limited data. Hypoacidity was present in 76.5% of the group exposed to lead compared to 54.4% for the control group. The increased percentage of gastritis and hypoacidity in the workers with saturnism was considered significant.

Rakow and Lieben[178] followed up 10 years later on twenty-four cases of lead poisoning from a battery plant to determine their general state of health, urine lead levels, and any evidence of renal damage. From each person, through an interview, information was elicited on symptoms and illnesses that may have persisted. An occupational history was also taken. Urine samples were taken on all twenty-four and blood samples on four. In eight of the twenty-four still working in the plant, the mean urinary lead level was 0.249 mg/liter with a range of 0.035–0.81 mg/liter. For the group from whom blood samples were taken, the comparative levels were: urine 0.81 mg/liter – blood 0.172 mg/100 gm; urine 0.226 mg/liter – blood 0.081 mg/100 mg; urine 0.162 mg/liter – blood 0.263 mg/100 gm; urine 0.467 mg/liter – blood 0.12 mg/100 gm. Of the group with no further exposure to lead, the urine values ranged from 0.006–0.112 mg lead/liter with a mean of 0.047 mg/liter. Albuminuria was shown in 21.7% (13%, excluding two persons described as drinkers). This would indicate some injury to the kidney. Fifty-four percent described varying symptoms as muscular complaints, abdominal cramps, fatigue, constipation, insomnia, headache, and nose bleed. Five out of the twenty-four had peptic ulcers some 3–8 years after diagnosis of lead poisoning, all of which had healed. Eleven of the twenty-four denied any signs or symptoms.

Linch *et al.*[254] in evaluating tetraalkyl lead exposures found that air concentration levels at fixed station sampling sites did not correlate well with similar data from personal monitoring samplers, with urinary excretion rates, or with clinical data. Breathing zone concentrations of lead using personal monitoring samplers approximately correlated with the urinary lead excretion rate. These data stress the importance of proper air sampling in determining worker exposure to airborne toxic materials.

LINDANE

Milby *et al.*[179] studied the lindane level in the blood of seventy-nine persons, of whom seventy-one worked in plants processing lindane and eight resided in residences where lindane vaporizers were used for pest control. The main objective was to determine if blood lindane levels could serve as a measure of intensity and duration of exposure. The data indicated that blood lindane level is a reflection of recent lindane absorption, but does not appear to increase with duration of exposure only.

Czegledi-Janko and Auar[180] reported on clinical and laboratory findings of thirty-seven workers exposed up to 2 years to lindane in a fertilizer plant. Each was given a neurologic examination and an electroencephalogram. Blood lindane determinations were made. Twenty persons nonoccupationally exposed to lindane were used as controls. The control mean blood level for lindane was 0.008 ppm and ranged from 0.003–0.017 ppm. In the exposed group, fifteen out of seventeen workers with lindane blood levels of 0.02 ppm and over had clinical symptoms and electroencephalographic changes which related to exposure. In instances of the latter changes in workers with lindane blood levels under 0.02 ppm, the workers, except for one case, all had exposures 2 years previously to aldrin. The American Conference of Governmental Industrial Hygienists has adopted a threshold limit value of 0.5 mg/m^3 for lindane.[50]

MANGANESE

Tanaka and Lieben[181] investigated manganese exposures in seventy-five industrial plants in Pennsylvania. Principal types of industries included steel castings, nonferrous metal castings, metal manufacture, ceramics and brick, chemical manufacture, and processing of manganese ore or ferromanganese. Fifty-three percent of the atmospheric levels for manganese at crushing, pulverizing, mixing, weighing, and bagging operations exceeded the threshold limit ceiling value (TLV) for manganese of 5.0 mg/m^3.[50] It was also exceeded by 16% at melting and pouring of steel alloys, and by 11% at welding and burning.

In general, the group of workers having exposure levels to manganese over 5.0 mg/m^3 also had higher urinary manganese concentrations in contrast to the workers exposed to less than 5.0 mg/m^3. Of thirty-eight workers examined out of forty-eight exposed to manganese levels under 5.0 mg/m^3, screening for manganese poisoning was negative. In contrast, of 117 workers examined out of 144 exposed to manganese levels over 5.0 mg/m^3, seven had definite signs and symptoms of manganese poisoning. The authors stated that the present "C" ceiling limit of 5.0 mg/m^3 for manganese may require revision in the future as evidenced by their findings.

MERCURY AND ORGANOMERCURY COMPOUNDS

Mercury poisoning among workers in California mercury mills and mines was studied by West and Lim.[182] Of 281 workers employed, 96 were subjects of a study that considered airborne mercury exposure levels, clinical and subjective symptoms, and urinary mercury levels. The severity of poisoning was determined by the presence of one or more of the classical signs and symptoms. Workers without symptoms of mercury poisoning, who had worked 3–312 months, had urinary mercury levels ranging from 200–1100 μg/liter. On the other hand, workers with mild to moderately severe cases of mercury poisoning, who had worked 1–6 months, had urinary mercury levels ranging from 320–1400 μg/liter. In workers with moderate to severe cases of mercury poisoning who had worked 3–24 months, with airborne exposure levels at times exceeding 1.2 mg/m^3, the urinary mercury levels ranged from 950–7100 μg/liter. At urinary levels of mercury over 800 μg/liter, the presence and severity of symptoms and manifestations of intoxication correlated well with urinary levels. Workers with urinary mercury levels under 800 μg/liter tended to have mild complaints. The authors stated that urinary mercury levels, in spite of variations, were valuable for assessing and documenting the intensity and duration of mercury exposure among the workers studied.

Rentos and Seligman[183] studied the relationship between levels of exposure to airborne mercury and clinical observations on eighty-three workers employed in the mining and milling of cinnibar. Mercury vapor levels in the mines ranged from 0.009–0.03 mg/m^3 and in the mills from 0.01–1.09 mg/m^3. In addition to examining the workers for signs and symptoms, laboratory determinations were made on each worker for glutathione, glucose-6-phosphate dehydrogenase, serum creatinine, serum alkaline phosphatase, blood urea nitrogen, and serum protein electrophoresis; urinary mercury studies were also performed. Nine workers exposed to work area concentrations of mercury averaging 0.02 mg/m^3 and three workers exposed to concentrations averaging 0.16 mg/m^3 showed no clinical symptoms of mercury poisoning; three out of seventeen workers exposed to mercury concentrations averaging 0.21 mg/m^3 and twenty-nine out of fifty-four workers exposed to mercury concentrations averaging 0.31 mg/m^3 had clinical signs of mercury poisoning. Other than urinary mercury levels, the remaining laboratory tests had little correlation with environmental data and symptoms of mercury poisoning. All workers with symptoms of mercury poisoning had urinary mercury levels in excess of 300 μg/liter, the upper normal zone. However, some workers with mercury urinary levels above 300 μg/liter had no signs of mercury poisoning. The authors concluded that urinary mercury levels were valuable in assessing excess body mercury burden and that their data supported the threshold limit value for mercury vapor of 0.1 mg/m^3, though the safety factor of this value was likely not more than 2.

Milne *et al.*[184] studied acute mercurial pneumonitis in workers exposed to high concentrations of mercury vapor. Case histories are presented on four employees working from 2.5–4 hours inside a tank contaminated with metallic mercury. Urinary mercury levels in the four workers, 10 to 14 days after exposure, ranged from 0.10–0.13 mg/liter with the 24-hour excretion rate ranging from 0.10–0.17 mg. The air concentration of mercury vapor at breathing height was estimated to range from 1.1–2.9 mg/m^3. Symptoms included cough, gasping respirations, tightness in the chest, dryness in throat, and fever. There was a latent period of several hours from initial time of exposure to onset of symptoms.

An investigation of a population exposed to organomercurial seed dressing was carried out by Taylor *et al.*[185] The chemicals used consisted of alkyl-aryl mercury compounds in powder and liquid preparations. The seed treatment season extends from October to March of each year. Evaluation of the seed dressing worker's exposure to mercury was underway but data were not available at the time of this report. In the study, clinical data, including a general health history, work history, electrocardiogram, urinary protein excretion, serum glutathione reductase, serum phosphoglucose isomerase, and blood and urine mercury were obtained on thirty-three cereal seed dressers and compared with thirty-three age-matched controls. The exposure of the first group to mercury varied from 2 months to 33 years; the controls, mainly hospital staff and students, had no known exposure to mercury. The clinical data did not show significant abnormalities in the electrocardiograms. The seed dressers did show elevated mercury blood and urinary excretion levels, low grade proteinuria, and partial inhibition of serum phophoglucose isomerase when compared to the controls. The significance of the low level proteinuria and plasma enzyme changes is not known in terms of eventual morbidity and mortality. This was indication, however, that excessive exposure to the mercury compounds was occurring.

Wada *et al.*[186] studied the human porphyrin metabolism response to a low concentration of mercury vapor in forty-seven workers employed in a tungsten rod manufacturing plant, who had no clinical symptoms of mercury poisoning. The concentration of mercury in the plant had been slightly lower than 0.1 mg/m^3 for the past several years. No information was given on the potential exposure of the forty-seven workers to other toxic materials. A significant correlation was found between level of mercury in the urine and decreased activity of δ-aminolevulinic acid dehydratase, cholinesterase, and urinary level of coproporphyrin. A maximum allowable concentration of urinary mercury of 200 μm/gm creatinine was recommended.

Smith *et al.*[187] investigated the effects of exposure to mercury on 642 workers employed in mercury cell rooms in the manufacture of chlorine. A suitable control group was included. A complete medical history along with job description and nature of exposure to mercury was obtained on each worker. In addition, each worker was given a comprehensive physical examination including a neurologic and hematologic examination, a chest x ray, and electrocardiogram, and mercury levels in the blood and urine were determined. An analysis of mercury vapor in the work environment showed that 84.5% of the study group was exposed to less than 0.10 mg mercury/m^3 while 18.5% of the study population exceeded mercury levels of 0.10 mg/m^3 (4.8% in range of 0.24–0.27 mg/m^3). There was nearly always a low background exposure to chlorine. The workers were required to wear respirators during certain periods of peak mercury exposure. Levels during these periods were not used in calculating time-weighted exposure values. Urine, blood, and air mercury levels for the workers are interrelated in the report. These data suggested that an air threshold level of mercury of 0.1 mg/m^3 in general corresponds to a blood mercury level of 6.0 μg/100 ml and a urine level of 0.22 mg/liter uncorrected for specific gravity. Clinical data on the exposed workers, including chest x rays, electrocardiograms, and laboratory hematologic data, were found to be normal. Abnormalities of teeth and gums, tremors and abnormal reflexes, insomnia, shyness and nervousness, along with loss of appetite and weight, did, however, exhibit a dose–response relation. Though many of the clinical parameters studied were normal for workers exposed to a time-weighted airborne level of 0.1 mg/m^3, loss of appetite and weight, and other clinical signs and complaints did not indicate good confidence in this level, especially relative to any margin of safety.

Based on data from recent studies, the threshold limit value committee of the American Conference of Governmental Industrial Hygienists has announced a Notice of Intended Changes for all forms of mercury, excepting alkyl mercury, reducing the level to 0.05 mg/m^3.[50] The threshold limit for alkyl mercury remains at 0.01 mg/m^3.

OIL MIST

An extensive study of oil mist exposure in a printing plant was reported by Lippman and Goldstein.[188] Environmental studies showed exposure levels of ink mist (a suspension of carbon black in mineral oil) at the Goss presses to average 12.2 mg/m^3, with peaks up to 28.5 mg/m^3. The time-weighted exposure at these presses for ink mist of all sizes was 8.6 mg/m^3. The diameter of the ink mists ranged from 9 μm to 30 μm with a median range of 14 μm. An average of 15% of the mist was within the respirable range. The time-weighted average exposure to respirable mist at the Goss presses was 1.4 mg/m^3, and at the Wood presses 1.0 mg/m^3.

Data are presented on factors which influence ink mist exposure levels such as production rate, press running time, ventilation, and use of ink mist suppression equipment.

The time-weighted respirable fraction of the ink mist exposure to pressmen was well within the threshold limit value of 5.0 mg/m^3 adopted by the American Conference of Governmental Industrial Hygienists, though this value was exceeded for the time-weighted average for total ink mist. The authors stress the need for interpreting standards on the basis of exposure to respirable fractions.

A 15-year mortality study by Goldstein *et al.*[189] was conducted on a group of pressmen (2797 man-years* at risk) occupationally exposed to mineral oil mist at the above levels as a companion study to that of Lippman and Goldstein.[188] A control group of compositers (5127 man-years at risk) not exposed to the mist was similarly studied. No significant difference in mortality or morbidity for respiratory diseases was noted between the two groups.

Ely *et al.*[190] studied the mortality, symptomatology and respiratory function in humans occupationally exposed to oil mist at machine shop operations. The predominant oils used were of mineral origin, though some were of vegetable and animal origin. The oil mist concentrations for the mortality and prevalence studies were similar, with exposure levels ranging from 0.07–110 mg/m^3. The mean and median exposure concentrations for the mortality study were 1.5 and 3.7 mg/m^3, respectively, and for the prevalence study 1.0 and 5.2 mg/m^3, respectively. In the mortality study, proportional mortality was analyzed in 343 deaths occurring between 1942 and 1961 among men with 5 or more years of tenure in oil mist departments and, for the same period, in 3122 deaths in men with tenure only in nonoil mist departments. There was no apparent difference in the proportionate mortality of the two groups. A prevalence study was conducted on 1700 individuals of whom 242 were exposed to oil mist. Medical data on each person included a questionnaire based on the British "Short Questionnaire on Respiratory Symptoms (1960)," medical and work history, smoking history, chest roentgenogram, and spirometry.[190] Height, age, and cigarette years were the major factors relating to forced vital capacity and forced expiratory volume 1.0-sec findings. Symptoms also related to smoking variables. The effect of oil mist exposure was not significant.

The following papers were presented in the session on occupational cancer of the XVI International Congress on Occupational Health, Tokyo, 1969[29]; (*a*) Epidemiologic Investigation on Cutting Oil Cancer by C. and J. Thony, Cluses, France; (*b*) Carcinogenicity of Ingot Case Oils by M. Kuratsune, Department of Public Health, Faculty of Medicine, Hyushi University, Fukuoka, Japan; and (*c*) Carcinogenic Substances in Estanion Oil-Shale Industry by P. Boganski, International Agency for Research on Cancer, Lyon, France. These will be published in the Proceedings of the Congress.

ORGANOPHOSPHORUS PESTICIDES

Jager *et al.*[191] studied the neuromuscular function of thirty-six workers exposed to organophosphorus (dimethylvinyl phosphates) and organochlorine

*Man-years = the number of employees times years worked during the 15-year study.

compounds in formulating pesticides, twenty-four workers exposed to organochlorine compounds only, six workers who had each received an acute exposure to organophosphorus compounds, and twenty-eight control male workers. The examination included electromyography and venous blood sampling for whole-blood cholinesterase determinations. Nearly one-half of the thirty-six workers exposed in the formulating of organophosphorus and organochlorine pesticides and six of the workers with single acute exposures to organophosphorus compounds had absorbed amounts adequate to produce abnormal electromyographic responses. Since only one of the twenty-four workers exposed to organochlorine compounds alone showed an abnormal electromyographic response, the organophosphorus compounds were considered the offending agents. The whole-blood cholinesterase levels were very low and did not relate to the electromyographic findings. Several reasons were given for this apparent lack of relationship. The analytical methods employed were not sensitive enough to permit the determination of organophosphorus levels in blood and tissue for correlation with electromyographic data. Since all the workers examined, including those with electromyographic abnormalities, appeared to be healthy and none complained of symptoms or muscular weakness, electromyography appeared to provide a very sensitive early warning of exposure to organophosphorus compounds.

PLATINUM

Parrot *et al.*[192] investigated the cutaneous and respiratory allergenic responses of workers to platinum salts in platinum refining. Of fifty-one workers examined, thirty-five were observed to have platinosis (respiratory and cutaneous manifestations of allergic nature) produced by skin contact or inhalation of chloroplatinic acid or its salts and never by metallic platinum. Cutaneous and respiratory manifestations and precautionary measures to prevent exposures are discussed along with industries where platinum salts may be encountered. The American Conference of Governmental Industrial Hygienists Threshold Limit Value for platinum salts (as platinum) is 0.002 mg/m^3.[50]

PROTEOLYTIC ENZYMES

Proteolytic enzymes have been used or associated with food processing for many years; e.g., for meat tenderizing, baking, and cheese ripening. They are also used in industries such as leather and dry cleaning.[193]

The recent addition of proteolytic enzymes, derived from *Bacillus subtilis*, to washing powders has given rise to epidemiologic studies to define the nature and extent of health problems associated with exposure to the enzyme. Proteolytic enzyme-containing products for presoak were first introduced into Holland around 1963 and became available in Britain and the United States in 1967.

Flindt[194] investigated the occurrence of pulmonary disease due to the inhalation of proteolytic enzyme derivatives of *Bacillus subtilis* in workers manufacturing detergent products. During and subsequent to 1966, when the enzyme was first used in the factory, a number of workers complained of chest symptoms. Other workers had been certified as having bronchitis, bronchial spasm, asthma, influenza, etc. In the study, data on exposed workers were obtained from medical and work histories, chest radiograms, lung function tests, skin testing, and blood and sputum examinations. The most significant symptom observed was breathlessness, which at times was acute and severe. In some instances, response was within a few minutes after initial exposure. More often a lapse period of around 8 hr existed between initial exposure and onset of symptoms. Chest pains, general weakness, and febrile conditions were also reported. Many of the workers had become sensitized to the enzyme as indicated through positive prick skin tests. Strict industrial hygiene procedures were recommended for controlling exposure to the enzyme. Medical surveillance of workers was considered essential.

Pepsy *et al.*[195] reported five cases of allergic reactions of the lungs to enzymes of *Bacillus subtilis* in workers with bronchial disease who had inhaled high concentrations of the enzyme preparation. The workers showed positive immediate response to the skin prick tests, asthmatic reaction to inhalation tests, and the presence of precipitins in the sera.

Greenburg *et al.*[196] studied 121 workers, 46 women and 65 men, in a detergent factory exposed to dust containing derivatives of *Bacillus subtilis*. The employees worked in the department where the blended detergent was packed by automatic equipment equipped with exhaust ventilation. Retrospective data were not available on the intensity or duration of exposure to the enzyme. The clinical data were derived from the Medical Research Council Questionnaire on Respiratory Symptoms, supplemental questions on symptoms specifically related to enzyme exposure, spirometry (FVC and FEV_1), and skin tests by the prick technique. The clinical data confirmed previous observations that dust containing enzymes derived from *Bacillus subtilis* can produce sensitization. The skin test showed 40% of the 121 to be sensitized. Forty-four percent (21 out of 48) of the sensitized workers had a FEV_1/FVC ratio below 70% compared to 14% (10 out of 73) of the nonsensitized workers. The association of sensitization with impairment of ventilatory function was not accounted for by the smoking habits. The questionnaire was most useful in deriving information on complaints and symptoms. In the 6-month follow-up during which extensive improvements were made to reduce exposure to the enzyme, skin tests were given to the negative group of the first examination as well as spirometry on those previously shown to be sensitized by the skin test. There was one additional positive skin test. Twenty-three of the twenty-five workers who had not previously shown ventilatory impairment remained unchanged and two had deteriorated. Eight of the nineteen with previously reduced ventilatory capacity had improved to normal, four had considerable improvement, four remained unchanged, and three were worse. The authors noted that one case of sensitization as shown by skin test appeared in a period of only 6 days exposure and that impaired lung function can be quite severe.

Newhouse *et al.*[197] reported on an epidemiologic study of 271 workers producing enzyme washing powders. The population studied included the workers exposed to enzymes in the packing, warehousing, and shipping of enzyme detergents. Each person was given a questionnaire to obtain data on respiratory symptoms of the past year during exposure to enzymes as well as any symptoms previous to enzyme exposure. Clinical data were obtained through ventilatory tests (FEV_1 and FVC), chest x rays, and a prick skin test. Twenty-one percent of the skin tests (57 out of 271) were positive. No delayed reactions were reported from the skin tests. There was a highly significant correlation between those with symptoms and those having positive skin reaction. Chronic bronchitis did not appear to correlate with the positive skin tests. Of the eleven workers with FEV_1/FVC ratio under 80%, six out of fifty-seven were from workers showing sensitization to skin tests, compared to 5 out of 214 for nonsensitized workers. No radiologic change in the lung fields was evident.

After the initial study, changes were made through increased ventilation, isolation of operations, and vacuum cleaning to reduce exposure to enzyme-containing dust. Filter and supplied air respirators were worn in dusty operations. In measuring airborne levels of enzyme, air was drawn through a filter paper at a rate of around 50 m^3/hour. The enzyme activity was based on its ability to denature protein and was measured in Anson Units. In a 6-month period, May through October, monthly average levels of enzyme in the air ranged from 0.3 to 24×10^{-6} Anson Units/m^3 air. At the levels of enzyme encountered, there was no correlation between the levels of dust in the air and the fall in FEV_1 of sensitized workers. In a subsequent repeat survey of the sixty-one workers then defined as sensitized, thirteen (21%) gave a history of family allergic disease and forty (65%) gave positive response to skin tests with one or more of the common allergens. The FEV_1 of the 103 examined during both test periods showed no significant change.

The American Conference of Governmental Industrial Hygienists[50] has proposed a threshold limit "C" value for Subtilisins (proteolytic enzymes) of 0.0003 mg/m^3 (as 100% pure crystalline enzyme).

SILICA

Only moderate attention appears to have been directed toward the epidemiologic aspect of silicosis when compared to the great economic importance of this disease. The International Conference on Pneumoconiosis, Johannesburg, 1969[15], included a special symposium on silicosis.

Wagner[198] in a study in West Germany reported that the number of initial compensation cases for silicosis and silicotuberculosis had dropped from 7123 in 1954 to 2122 in 1967. Percentage-wise, silicosis and silicotuberculosis in relation to initial compensation cases for all occupational diseases dropped from 68% in 1949 to 42% in 1967.

Gilson *et al.*[199] reported on a study of respiratory responses to duration of foundry work. The study included 1860 workers, age 35–64, in sixty-seven foundries, and a control group of 1777 workers. The examination included an occupational history, medical research council questionnaire on respiratory symptoms, vital capacity (VC) and forced expiratory volume/one second, and sputum collection. Environmental exposure data were not represented in this brief coverage of a large survey.

The effect of smoking on simple pneumoconiosis was evident, the smokers having a 30% higher prevalence of category 1 and above than the nonsmokers and ex-smokers. There was significantly more simple pneumoconiosis among fettling workers than those on the foundry floor. The proportion of workers with bronchitis related in general to the years of work in the foundry. The FEV_1 was not related to the years of work on the foundry floor.

Beadle and Harris[200] studied the relationship between inhaled dust and the incidence of silicosis in South African gold miners. Measurements of dust exposure levels were made on approximately 650 underground miners from twenty mines using the konimeter, standard thermal precipitator, and the modified thermal precipitator. Occupational histories were obtained on a group of 1200 underground workers who had started work between 1934 and 1938, had completed at least 3000 shifts, and had no other known dust exposure. The radiologic study of the workers was made in 1964 and again in 1968. No correlation could be established between mean dust level, number of shifts worked, and the probability of certification for silicosis. Radiologic indications were not an essential criterion in the diagnosis of silicosis. A good correlation, however, was established between mean dust level, number of shifts worked, and the development of radiologic silicosis. Data were presented on the probability of having silicosis for various dust exposure levels and duration using the dust evaluation procedures stipulated.

Prowse[201] interrelated clinical, electrocardiographic, and pulmonary function data, and length of service of 240 workers (representing 6056 years of underground service) of silicosis in South Africa gold mines. Data are presented on the radiologic classification and clinical symptoms of the 240 workers. Twenty-four percent (58) of the 240 workers had evidence of a reduced vital capacity (less than 80% of the predicted normal). A degree of correlation existed between functional lung disease, age, duration of exposure to dust, and the degree of smoking.

The Miners Medical Bureau of the Republic of South Africa has a comprehensive study of obstructive airway disease underway involving some 2000 white gold miners and a control of several hundred matched nonminers to determine the influence of smoking, silica dust, and other factors on the disease. A preliminary report by Wiles *et al.*[202] covering data on 800 mine workers and 120 control nonminers shows a progressive decrease in lung function (maximum midexpiratory flow) with increase in the level of smoking. In heavy smoking the effect is so dominant that it overwhelms any influence of dust exposure alone on the disease. At both high and low dust risk there were more cases with productive cough among smokers than among nonsmokers, with the higher dust risk having the greatest percent. Also, the percentage with productive cough was much higher in miners than nonminers in each of the smoking groups.

Ruttner[203] investigated the relation of lung cancer to silicosis in Switzerland. Out of 688 deaths from silicosis between 1960–1967, sixteen workers also had lung cancer. The 2.33% (16 out of 688) is lower than the 3.99% in all male deaths over 20 years of age. The author stated that the current data confirmed the data of 20 years ago that there is no evidence that silicosis increases the risk of lung cancer in Switzerland.

Papers presenting other epidemiologic data at the silicosis symposium[15] included "Chronic Bronchitis in Banta Mine Workers" by G. K. Sluis-Cremer; "Emphysema in South Africa Gold Mines," by I. Prinsloa and N. F. Laubscher; and "Pulmonary Disability and Pulmonary Function in Gold Miners on the Witwatersrand," by L. D. Erasmus *et al.*

Gregory[204] studied the prevalence of pneumoconiosis in a steel foundry. The diagnosis of pneumoconiosis was based on radiologic evidence and the occupational history. No information was given on the levels of exposure to silica. Over the 6-year period of the study, 1955–1960, the average prevalence rate for silicosis was 6.4%. The risk in the fettling and grinding shop was over seven times that in the main foundry. The latency time from beginning of employment to the stage of nodulation was around 31 years for workers in the fettling and grinding shop, and 36 years in the main foundry. Progressive massive fibrosis appeared around 10 years after nodulation was first noticed by x ray. It was present in 50% of the cases 10 years or more after the x ray first showed nodulation. Though new cases of silicosis appeared at about the same rate throughout the survey period from 1955 to 1960, the duration of exposure was increasing before nodulation became evident. This was attributed to a gradual improvement in dust control and the use of respirators.

Ahlmark[205] reported on a survey of silicosis, dust conditions, and dust control in Sweden. Approximately 25,000 to 30,000 workers were at risk. Around 80–90 new cases per year were appearing, mostly from mining, quarrying, and steel and iron foundries. Silicosis developed on the average after 25 years of exposure with ranges from 13 years for quartz workers to 34 years for granite quarry workers. Detailed information is presented on the techniques of environmental sampling and assessing of dust levels. Based on the threshold limit value for quartz-containing dusts, the following approximate percentages of work sites by industry had excessive dust exposures: quartz 50%, granite 45%, construction 45%, iron foundries 20%, steel works 30%, steel foundries 10%, and ceramics 10%. The investigation presented a generally favorable trend in the reduction of silicosis, but pointed up the need for increased attention to dust control and monitoring wherever silica dust is encountered.

Grundorfer and Raber[206] investigated silicosis in granite workers exposed to high concentrations of dust in lower Austria. Eighteen cases of silicosis was found in a work force of 170 over a period of 10 years. Of the 18 cases, 15 were from the crushing plant where only twenty workers were employed. Silicosis was evident within 11 years from first exposure. The crushing plant had no dust control. Air samples showed dust concentrations of 2100–5000 particles/cm^3 at the big crusher, 1500–3500 particles/cm^3 along the conveyor belt, 4670 particles/cm^3 in the motor room of the second crusher, and 3200–>10,000 particles/cm^3 at work places around the second crusher.

Silicosis remains a serious occupational disease throughout the world. It seems inconceivable that after the research, studies, symposia, etc., of the past 50 years, anyone knowledgeable about occupational disease, including members of management, labor, and government, is unaware of the hazards of exposure to silica. The problem appears mainly in the application of information already available in the prevention of exposure through dust control. Many selective and progressive industries are doing a good job in preventing silicosis. There are obviously many industries where dust control is inadequate or nonexistent. Little further progress will be evident in the prevention of silicosis, however, until increased attention and implementation is given to dust control programs.

STEEL PRODUCTION

To help identify the nature and extent of occupational health problems in the steel making industry, the U. S. Public Health Service and the University of

Pittsburgh, Graduate School of Public Health, in cooperation with three large steel firms initiated in 1962 a comprehensive long-term mortality study of steel workers. The first report of the study by Lloyd and Ciocco[207] covered the methods used.

The study cohort consisted of 50,072 steelworkers employed at seven plants in one county in 1953. The cohort represented approximately 62% of all men working in basic iron and steel production at that time in the county. For men in the cohort who left the industry before 1962, a follow-up was made to determine their vital status. For the workers who had died, information on the underlying cause of death was obtained from death certificates and coded. Deaths were classified according to the Manual of the International Classification System.[208] Deaths between 1953–1961 among the study cohort showed an overall pattern among both whites and nonwhite of lower mortality than would have been predicted in the general population. This was attributed to the selectivity of the employed population. For the age group below 54 in 1953, the mortality for nonwhites was greater than that for whites; the reverse was true for the age group over 55.

Robinson[209] studied the cohort for mortality by level of income in whites and nonwhites. The comparative index used was the job score which is a relative rating based on skill, responsibility, effort, and working conditions. Each job score was given a corresponding hourly wage. A negative association existed between income (job score) and mortality for whites. Nonwhite workers did not exhibit this association. Reasons were presented to explain this difference.

Redmond *et al.*[210] studied the problems associated with the follow-up of a large cohort. The methods, procedures, and experience then reported would be of immense value to others contemplating a study of this magnitude.

Lloyd *et al.*[211] studied the cohort for mortality by work area. In the method used, the mortality for the total cohort was compared to mortality of employees in specified work areas within the industry. Fifty-three work areas were established for comparative purposes. Five areas had significantly higher deaths than were predicted from the total steel worker experience: cold reducing mills, janitorial machine shop, mechanical maintenance assigned, sheet finishing, and shipping. There were less deaths observed than expected in the carpenter shop. These six areas were also studied for observed and expected deaths from malignant neoplasms, vascular lesions affecting central nervous systems, heart disease, accidents, and all other causes. The authors state "Two exceptions which did not appear to be due to selection for health were the excess mortality from vascular lesions of the central nervous system for men employed in sheet finishing and shipping and excess mortality from malignant neoplasms of the respiratory system for non-white coke plant workers." It is suggested that these differences may be due to factors in the working environment.

The Department of Social and Occupational Medicine, Welsh National School of Medicine, in 1964 initiated an investigation of the prevalence and determinants of respiratory disability among men employed in two large integrated steel works in South Wales. A major objective was to investigate the relative importance of occupation, residence, smoking habits, and other environmental influences in the etiology of bronchitis among workers in the group under study. Lowe *et al.*[212] described the procedures used in collecting data on respiratory symptoms, ventilatory capacity, physique, smoking habits, and occupational histories of male employees in two works, one employing about 9500 and the other about 16,500 workers. Production, maintenance, managerial, and office workers were included in the study. Operations included coke ovens, blast furances, steelmaking, heat mills, cold mills, tinplate and galvanizing, limestone quarries, maintenance, laboratories, etc. Data from the battery of tests are presented on the nonbronchitics. These data were used as a baseline for measuring the effects of other parameters and stresses, which were reported in subsequent papers.

Warner *et al.*[213] presented sampling methodology and data on levels of sulfur dioxide and particulates inside the two South Wales works and the surrounding community. The Hexhlet sampler was used for collecting particulates and a Dreschel bubbler bottle train containing hydrogen peroxide, for collecting sulfur dioxide. A portable total sampler, constructed on the same principles, was also used where mobility was required. The concentration of sulfur dioxide at sites within the two works exceeded 5000 $\mu g/m^3$ only 3.2% and 0.9% of the time (5 ppm TLV = 13,000 $\mu g/m^3$ at 25°C and 760 mm Hg pressure). Sulfur dioxide levels in the communities averaged under 80 $\mu g/m^3$. The respirable fraction of particulates seldom exceeded 2–3 mg/m^3 in the main production departments and was often under 1 mg/m^3. The particulates were essentially iron oxide. The community particulate levels were under 77 $\mu g/m^3$ and often under 30 $\mu g/m^3$.

Lowe *et al.*[214] presented data correlating symptoms and ventilatory capacity on 10,449 employees working in one or another of 115 defined working areas. The correlation was conducted in three ways. In the first, each worker was assigned an exposure level of sulfur dioxide and respirable dust in the area he worked. These were related to ventilatory capacity (FEV_1), age, smoking habits, and length of exposure. In the second, the 114 working areas were subdivided into four subgroups based on sulfur dioxide and respirable particulate levels and these were correlated with the prevalence of chronic bronchitis and the mean FEV_1. In the third approach, the mean atmospheric levels of the 114 working areas were correlated with the prevalence of chronic bronchitis, the mean FEV_1, age, and smoking habits. If any relation between respiratory disability and the levels of air pollution in the two plants existed, it was too minimal to be detected by the approaches used. The data demonstrated the overriding importance of cigarette smoking in the etiology of chronic bronchitis.

URANIUM

The earlier reports of Archer *et al.*[215] and Parsons *et al.*[216] have given support that there is a causal relationship between exposure to the daughters of radon and lung cancer. In 1957 the U.S. Public Health Service[217] recommended an exposure limit to radon of one work level (equivalent to the quantity of daughter products in 1 liter of air in complete equilibrium with 100 picocuries of radon – or, 1.3×10^5 MEV of potential alpha energy per liter of air). The American Standards Association in 1960[218] established the first exposure standard for radon at one work level (WL) with the provision that all operations other than ventilation work should cease at 10 WL. Between 1960–1967 the "cease work" level in many of the states mining uranium was reduced to 5 WL and then to 2 WL.

Lundin *et al.*[219] in 1969 reported on a study of mortality in uranium miners in relation to radiation exposure, hard-rock mining, and cigarette smoking for the years 1950–1967. During the period 1950–1967, 398 deaths occurred in white underground miners in contrast to 251 expected out of a total population of 3414. The excessive mortality (147 deaths) was due primarily to violent causes (120 vs. 50.5 expected) and malignant neoplasias of the respiratory system (62 vs. 10 expected). For the first time, an excess in lung cancer was seen in the estimated accumulated exposure in categories less than 120 WLM (working level month) and 120–359 WLM. Since mortality patterns also differed in the hard-rock mining experience and since there was an excess of hard-rock mining experience in the above categories ($<$ 120 WLM and 120–359 WLM) some factor in the hard-rock mining may have been contributory to the excess respiratory cancer cases observed in these groups. Also the excess respiratory cancer deaths per 10,000 person-years of observations were ten times higher in the cigarette smoking than in the nonsmoking uranium miners. The authors, however, believed the evidence pointed toward radon exposure as the causative factor.

Evans[220] had a different interpretation of the data of Lundin *et al.* and expressed the opinion that the excess incidence of lung cancer in the less than 360 WLM categories is compatible with the increased incidence of lung cancer exhibited generally by underground miners. He further stated that the data strongly suggested that exposures which are clearly below 1000 WLM, e.g., in the 300–400 WLM domain, should be well below the practical threshold. He proposed that the annual exposure be restricted to 12 WLM.

Saccomanno[221] stated that he had observed a downward trend in lung cancer incidence and expressed the belief that there is little evidence to indicate that a 1–3 WL is injurious to health.

Cooper[222] in reviewing the evidence implicating radon and radon daughters as causative factors in the excessive lung cancer observed in uranium miners along with the exposure–response data relationship available believed that a 1.0 WL was a justifiable standard for a time-weighted exposure to radon.

On the other hand a report by Boyd *et al.*[151] indicated a lung cancer excess of 70% among Cumberland, England, iron miners while a second report by Duggan *et al.*[223] shows the median WL's in these mines ranging from 0.03–2. The report did not identify the mines where the miners had been employed nor how long they had worked in each mine. The authors also implicated the iron ore as a causative agent. No information was available on the smoking habits of either the miners who died from lung cancer or the underground miners as a group. It is therefore apparent that there is a lack of agreement among authorities with regard to the significance of the epidemiologic findings to date.

In December 1968, the Secretary of the Department of Labor had the following standard written into the *Federal Register.*[224] "Occupation exposure to radon daughters in mines shall be controlled so that no individual will receive an exposure of more than 2 Working Level Months in any consecutive 3 month period and no more than 4 Working Level Months in any consecutive 12 month period. Actual exposures shall be kept as far below these values as practicable." The Department of Labor Standard allowed variations up to 12 WLM per year until January 1, 1971.

On January 12, 1969[225] the Federal Radiation Council made the following recommendations:

"1. Occupational exposure to radon daughters in underground uranium mines should be controlled so that no individual miner will receive an exposure of more than 6 WLM in any consecutive 3-month period and no more than 12 WLM in any consecutive 12-month period. Actual exposures should be kept as far below these values as practicable.

"2. As a policy measure of prudence, the agencies having responsibility for regulating the uranium mining industry should be advised that the Federal Radiation Council recommends an annual exposure level of 4 WLM as of January 1, 1971.

"3. The uranium mining industry is urged to continue efforts to progressively lower exposure levels in the mines so that the anticipated 4 WLM standard can be attained by January 1, 1971."

On January 16, 1969, the Department of the Interior published its standards in the *Federal Register.*[226] These were based on the Federal Radiation Council (FRC) guidance and did not differ greatly from those published earlier by the Department of Labor. They provided a 12 WLM yearly limit until January 1, 1971, when the limit would be reduced to 4 WLM per year. This standard also prohibited smoking where uranium is mined.

The Department of Interior rules were modified to provide that the action to be taken on January 1, 1971, would be based on the recommendations of the Federal Radiation Council. In turn the Federal Radiation Council will make their recommendations on the finding or re-analysis of the mortality data used in the 1967 analysis. On December 1, 1970, the FRC recommended that the action date be changed to July 1, 1971. The President approved this change on December 15, 1970.[233]

Nelson *et al.*[227] point out the need for separate identification of radon daughters and their distribution over the size spectrum from free ions, through condensation nuclei, to sized classes of particulates and how they vary over time in mine conditions. Without such additional understanding of the physics of these elements only a "much rounded" estimate of dose to the lung epithelium is possible.

The need for additional characterization of the mine atmosphere is emphasized by Morken.[228] He states: "We must conclude then that the unit WL is neither appealing nor appropriate as a dose rate estimator. The dose unit WLM is certainly not appropriate for the biologist or epidemiologist who studies the risk involved during exposures to radon products."

Another question concerns the significance of external exposure to γ and β radiation. Several attempts have been made to measure external gamma exposures using film badge and thermoluminescent (TL) dosimeters.[229] The TL dosimeters were tested in a mine near Denver by investigators from Colorado State University. In this mine, with exceptionally high-grade ore, γ exposure rates of 0.5 and 1 mR/hr were found to be not uncommon. Film badges have been used by the Public Health Service and the Bureau of Mines as well as by some of the mining companies. The results of these attempts were not completely satisfactory, but indicated that, except in mines with unusually rich ore, external radiation exposure was probably well below 25% of the MPC.

As the permissible exposure to radon daughters is lowered the need for accurate measurement of the level is increased. The Juno which served well when 10 WL concentrations were common became obsolete when the maximum permissible level was reduced to 1 WL. The scintillation counter became the accepted device. But as the permissible level goes still lower other more sensitive counters will be required, and techniques for collection of larger volumes of air will be desirable. Other possibilities include the development of a bioassay technique, personal dosimeters, and area monitors. Two Instant Working Level Meters have been developed which made it possible to collect and count a greater number of samples per unit of time.

Efforts are being made to supply answers to the problem areas discussed in the foregoing paragraphs. In order to define realistically the deviations from equilibrium that would be expected in an operating mine, a suitable mathematical technique is necessary. The simultaneous equations of Thomas[230] and the least squares of Raabe and Patterson[231] are being considered. Measurements should be made in operating mines with varying degrees of ventilation, including those using filtration and electrostatic precipitation. Techniques for determination of the unattached ions should be examined.

A variety of thermoluminescent chemicals in different packages are being evaluated for measurement of γ and β radiation. Through these investigations it is expected that a definite statement relative to the significance of external γ and β exposure can be made.

A number of dosimeters, designed for measuring cumulative α exposure to radon daughters, have been evaluated by the Atomic Energy Commission New York Health and Safety Laboratory.[232] Although they performed reasonably well during laboratory tests, they failed to give acceptable results when worn by working miners. Additional evaluation of the most attractive of these units is planned using the facilities of the Public Health Service research mine.

In vivo counting of the weak γ ray from ^{210}Pb as a method of estimating total past exposure is being investigated by New York University and University of California Los Alamos Safety Laboratory.

Area monitoring based on measurement of beta emission of ^{210}Pb in large air samples is being tested by the New Mexico Health and Social Services Department.

The Instant Work Level Meters will be evaluated using the facilities of the research mine. Results of sampling will be checked against the Thomas equations and the least squares technique of Raabe and Patterson as well as against the Kusnetz and Rolle methods.

It is expected that the completion of the preceding investigations plus work being performed by the Atomic Energy Commission, the Bureau of Mines, and numerous universities will make possible a far better understanding of the working environment experienced in underground uranium mines and the mechanism by which this environment produces pathologic changes in lung epithelium.

Very little work has been done toward finding new techniques for controlling radon daughter levels in uranium mines. Filters to remove the daughters have been tried with some success. Electrostatic precipitators have recently been used by one mining company. In their experience the precipitators are far superior to filters since they are as effective as filters and require far less power. Ventilation however can still be said to be the prime method of control.

VINYL ACETATE

Deese and Joyner[234] investigated the effects of chronic exposures to vinyl acetate in a chemical plant which had been producing vinyl acetate for 22 years. The study group of twenty-one out of a total of twenty-six employees consisted of those who had long periods of employment and were assigned to the vinyl acetate complex. The twenty-one workers had a mean age of 45.3 years and a mean length of service of 15.2 years. A matching control group was selected that had no exposure to the vinyl acetate.

Time-weighted exposures to the vinyl acetate ranged from 5–10 ppm. Intermittent short exposures were as high as 50 ppm with an occasional level as high as 300 ppm. The workers in the study group were given a detailed physical examination including multiphasic testing, chest x rays, electrocardiograms, audiometry, and spirometry. In tabulating data from the physical examination, mean values were presented for both the exposed and control groups. The medical records and multiphasic examinations did not indicate any chronic effects from long-term exposures to vinyl acetate at time-weighted levels of 5–10 ppm. Some workers were able to detect odor to vinyl acetate as low as 0.4 ppm. Significant eye and upper respiratory irritation were not noted below 10 ppm but were definite at 21.6 ppm. The American Conference of Governmental Industrial Hygienists[50] has proposed a threshold limit value for vinyl acetate of 10.0 ppm.

VINYL – POLYVINYL CHLORIDE

Acroosteolysis occurring in men engaged in the polymerization of vinyl chloride was investigated by Harris and Adams.[235] The workers in the two case reports were autoclave cleaners at a process where vinyl chloride was polymerized and where scraping the walls of the autoclave to remove the polyvinyl chloride afforded maximal hand skin contact.

The hands of 588 polyvinyl chloride workers including 150 autoclave cleaners were x rayed. Two cases were identified and described in which the terminal phalanges of the hands, the patella, and phalanges of the feet were involved. Two other cases, both autoclave cleaners, involved early changes accompanied by Raynaud's phenomenon.

A study of thirty-one cases of occupational acroosteolysis in workers in the vinyl chloride polymerization process was reported by Wilson *et al.*[236] Twenty-seven of the thirty-one cases had worked at some time as polycleaners, i.e., handscraping cleaning of the walls and agitators of the polymerizers. Of over 1000 workers who handled the finished resin or who processed it into plastic products, there were no cases of acroosteolysis. The osteolysis was specific to the distal phalanges of hands and was often associated with Raynaud's phenomenon. The prevalence was reported as less than 3% of the employees who had worked as polycleaners for at least 12 months. Hand soreness had at times caused some partial disability. Improvement in symptoms and radiologic appearance occurred in many without apparent explanation.

The Institute of Environmental and Industrial Health, The University of Michigan, Ann Arbor, Michigan, conducted a very comprehensive study of occupational acroosteolysis in workers engaged in the production and compounding of vinyl chloride and polyvinyl chloride.[237–239] The study was reported in three sections as Occupational Acroosteolysis: I, An Epidemiological Study; II, An Industrial Hygiene Study, and III, A Clinical Study.

The population for the epidemiologic study[237] consisted of 5011 workers employed in thirty-two plants who were at the time of the study, or in the past, working in any capacity in the plants manufacturing vinyl chloride, polyvinyl chloride, or in compounding the resin. Each worker completed a medical questionnaire with special reference to eliciting information on Raynaud's phenomenon or symptoms of peripheral vascular insufficiency, use of power tools, skin or dermal appendage abnormalities, history of hand injury, and work history. Each worker also had x rays of both hands. The epidemiologic data was tabulated in detail. A total of forty-one cases (twenty-five definite and sixteen suspicious) of acroosteolysis was uncovered in the 5011 employees studied. Acroosteolysis appeared to be clearly associated with the hand cleaning of the polymerizers. The extent of degassing the reactor (polymerizer) before entry for cleaning appeared relevant. Work practices rather than a specific agent or combination of agents appeared to relate to the occurrence of the disease. Workers engaged at jobs other than cleaning of the polymerizers did not appear to be at risk in developing acroosteolysis. The disease appeared to be systemic rather than of localized nature. The study was unable to define the etiologic agent or its portal of entry. The data suggested that some idiosyncratic "sensitization" or "susceptibility" factor was important. Medical surveillance of the workers was recommended.

The industrial hygiene study[238] was conducted in thirty-two plants producing and compounding polyvinyl chloride. One plant produced the vinyl chloride monomer, twenty-six were polymer production plants, and five plants conducted compounding and fabrication operations. A detailed description is given of operations in the production of vinyl chloride and polyvinyl chloride, vinyl chloride recovery, and the reactor cleaning. Since the problem of acroosteolysis seemed to relate to exposure in the cleaning of the reactors, special attention was given to the study of this phase. Vinyl chloride concentration in the reactor prior to ventilating was estimated around 3000 ppm. After aeration, the vinyl chloride concentration inside the reactor during scraping was under 100 ppm and usually around 50 ppm. The scraping released some unpolymerized vinyl chloride so that levels of 600–1000 ppm existed close to the hand during these operations. Other constituents of the scrapings from the reactor were believed to be partially polymerized resins, unreacted catalysts and additives, and unpolymerized vinyl chloride. Tables are presented showing differences in reactor cleaning procedures of the plants according to the presence or absence of acroosteolysis cases. Tentative preventive measures based on judgment and good practices are presented.

The clinical study[239] presents the results of intensive clinical investigation of four workers with acroosteolysis. The authors concluded "All (subjects studied) had osteolytic lesions, especially in the distal phalanges of the hands and Raynaud's phenomenon. All had worked as polyvinyl chloride reactor vessel cleaners with hand scraping being the common mode of operation. Raynaud's phenomenon anteceded osteolytic lesions. One of the subjects was in negative calcium and phosphorus balance. Plethysmographic abnormalities were present in three. Esophageal motility was normal. Scintiscans of the hands revealed variable uptakes in the fingers which correlated with the radiographic lesions. A wide variety of clinical laboratory parameters were normal."

WELDING

A study of respiratory symptoms and pulmonary function in 156 welders with exposure to welding fumes of more than 5 years was reported by Fogh *et al.*[240] Of the 156 welders, 121 did full-time welding, 17 about half-time, and 18 something more than half-time. One hundred and fifty-four were electric arc welders, two were oxyacetylene cutters, and some twenty did occasional oxyacetylene welding. All the welding was done essentially on mild steel; stainless steel or other metal welding was rare. No regard was given to types of procedures or to the types of electrodes used. A control group of 152 workers was selected from the same plants but from areas where welding was not normally done. Through questionnaire, information on work and medical history, smoking history, and respiratory symptoms was obtained on each welder and control worker. In addition each worker in the study was given a physical examination including conventional clinical laboratory studies, electrocardiogram, roentgenogram of lungs and heart, and an ophthalmologic examination. Pulmonary function tests included vital capacity, total lung capacity, residual volume, and forced expiratory volume 1 sec (FEV_1).

No significant difference was found in the incidence of chronic bronchitis between the welding and control group. The mean FEV_1 was identical for the two groups; the smoking habits for the two groups were also the same. Welders with increasing use of tobacco appeared to have increasing impairment in pulmonary function. The difference is significant when nonsmokers were compared to smokers. The increased use of tobacco was associated with an increased in the percentage of workers in both groups with cough and sputum, the smoking welders having appreciably the highest percentage.

Additional studies on welding have been covered in the sections "Iron Oxide"[158,159] and "Fluoride."[149]

WOOD

A number of epidemiologic studies have been reported in the last decade on respiratory and skin problems associated with exposure to wood.[241-244] Morgan *et al.*[245] reported on dermatitis due to exposure to wood dusts at three furniture factories. Through patch tests, the dermatitis at two factories was traced to the wood *Khaya anthatheca,* a species of African mahogany, and at the third factory to *Machaerium schleroxeylon,* a rosewood. The authors listed a number of woods commonly reported as having irritating dusts. Guarea, Marsonia, and Western red cedar were classified as having both respiratory and skin effects, while Dahoma and Makore showed only respiratory effects, and Iraka and White peroba showed only skin effects.

Sossman *et al.*[246] investigated four cases of woodworkers hypersensitive to wood dust, who had been exposed to wood dust from 17–25 years with duration of symptoms 4–8 years. Age ranged from 33–54 years. The workers were examined before and after exposure to various wood dusts and wood-dust extracts. The examination included pulmonary function tests and skin reactivity tests. The results of the tests are discussed and information presented on clinical patterns of wood dust hypersensitivity. The causative agent in one was oak dust, in another, mahogany dust, and in two cases, cedar dust.

Gandevia[247] conducted a study on the ventilatory capacity of forty-nine workers during exclusive exposure to western red cedar in a furniture factory. Although the wood working equipment, e.g., nailer, moulder, and sander machines, was equipped with exhaust ventilation, there was evidence of much dust in the air. Dust concentrations, using the thermal precipitator, were around 250–270 particles per cubic centimeter. Most of the particles were under 2 μm in diameter. Smoke and mineral particles appeared to dominate. A modified form of the British Medical Research Council's questionnaire was completed on each of the persons in the study and ventilatory data for forced expiratory volume, 1 sec, and forced vital capacity recorded. Of forty-seven workers completing the study, four had rhinitis and three asthma attributable to the western red cedar dust. The group with maximal exposure, studied before and after work on Tuesday and Friday, showed significant mean decreases in FEV_1 in contrast to the group with minimal exposure which showed no significant change. Smokers observed as having a loose or productive cough had the greater or more consistent decreases in ventilatory capacity.

Gandevia and Milne[248] reported on six case studies of occupational asthma and rhinitis due to western red cedar with special reference to bronchial reactivity. Eye and nasal irritation and nasal obstruction were among the first complaints. This was followed in some cases after some weeks by an irritating cough, episodes of nocturnal cough and wheezing, and a diminished exercise tolerance.

Acheson *et al.*[249] presented data showing that workers in the furniture industry in certain sections of England had a substantial incidence of adenocarcinoma of the nasal cavity and nose. The agent was thought to be a constituent of commonly used woods of the area such as oak and beech. The latent period was estimated around 39 years, though tumors were known to develop after 5 years of exposure.

Hadfield[250] reported on a study of ninety-two cases of carcinoma of nasal sinuses of woodworkers in the Buckinghamshire and Oxfordshire areas. The 1961 census listed 9520 woodworkers in these two areas. Of the ninety-two carcinomas, thirty-four were squamous carcinoma, thirty-five adenocarcinoma, and twenty-three anaplastic, transitional, unclassified. Of the thirty-five cases with adenocarcinoma of the ethmoid sinuses, twenty-nine were known to have occurred in woodworkers in the furniture industry. Smoking did not appear to relate to this type of cancer, but it appeared that the use of snuff could be a contributory factor. The agent, though unknown, appears to relate to the dust from the wood.

REFERENCES

1. Cralley, L. V., Cralley, L. J., and Clayton, G. D., eds., "Industrial Hygiene Highlights," Vol. 1. Industrial Hygiene Foundation of America, Pittsburgh, Pennsylvania, 1968.
2. HM Chief Inspector of Factories, Department of Employment and Productivity. Annual Report. Her Majesty's Stationery Office, London.
3. Ministry of Labour. National Institute of Industrial Health. Annual Reports. Kawasaki.
4. International Labour Office, Occupational Safety and Health Series. Occupational Safety and Health Branch, Geneva.
5. Pneumoconiosis Research Unit. Medical Research Council. Reports and Reviews, Johannesburg.
6. Industrial Health Foundation of America. Medical and Engineering Series Bulletins, Pittsburgh, Pennsylvania.
7. The Occupational Medical Foundation and Institute of Occupational Health. Annual Reports. Helsinki.
8. Froggatt, P., Short term absence from industry. I. Literature, definitions, data and the effect of age and length of service. *Brit. J. Ind. Med. 27*, 199 (1970).
9. Froggatt, P., Short term absence from industry. II. Temporal variation and inter-association with other recorded factors. *Brit. J. Ind. Med. 27*, 211 (1970).
10. Froggatt, P., Short term absence from industry. III. The inference of proneness and a search for causes. *Brit. J. Ind. Med. 27*, 297 (1970).
11. Howard, P., A long-term follow-up of respiratory symptoms and ventilatory function in a group of working men. *Brit. J. Ind. Med. 27*, 326 (1970).
12. Ayer, H. E., and Lynch, J. R., Association of disability and selected occupational hazards. *Arch. Environ. Health 17*, 225 (1968).
13. Auar, P., and Czegledi-Janko, G., Occupational exposure to Aldrin: Clinical and laboratory findings. *Brit. J. Ind. Med. 27*, 279 (1970).
14. Second International Conference on the Biologic Effects of Asbestos. April, 1968, Dresden. Proceedings to be published.

15. Shapiro, H. A., ed., "Pneumoconiosis." Proc. Int. Conf., Johannesburg, 1969. Oxford Univ. Press, 1970.
16. Smithers, W. J., Some observations on asbestosis in a factory population. *In* "Pneumoconiosis" (H. A. Shapiro, ed.), p. 155. Proc. Int. Conf. Johannesburg, 1969. Oxford Univ. Press, New York and London, 1970.
17. Newhouse, M. L., The mortality of asbestos factory workers. *In* "Pneumoconiosis" (H. A. Shapiro, ed.) p. 158. Proc. Int. Conf. Johannesburg, 1969. Oxford Univ. Press, New York and London, 1970.
18. Ashcraft, T., and Heppleston, A. G., Mesothelioma and asbestos on Tyneside. *In* "Pneumoconiosis" (H. A. Shapiro, ed.), p. 177. Proc. Int. Conf. Johannesburg, 1969. Oxford Univ. Press, New York and London, 1970.
19. Selikoff, I. J., Hammond, E. C. and Churg, J., Mortality experience of asbestos insulation workers 1943–1968. *In* "Pneumoconiosis" (H. A. Shapiro, ed.), p. 180. Proc. Int. Conf. Johannesburg, 1969. Oxford Univ. Press, New York and London, 1970.
20. Avril, J., and Champeix, J. Results of asbestos exposure in France. *In* "Pneumoconiosis" (H. A. Shapiro, ed.), p. 187. Proc. Int. Conf. Johannesburg, 1969. Oxford Univ. Press, New York and London, 1970.
21. Kiviluoto, R., and Meurman, L., Results of asbestos exposure in Finland. *In* "Pneumoconiosis" (H. A. Shapiro, ed.), p. 190. Proc. Int. Conf. Johannesburg, 1969. Oxford Univ. Press, New York and London, 1970.
22. Vigliani, E. C., Asbestos exposure and its results in Italy. *In* "Pneumoconiosis" (H. A. Shapiro, ed.), p. 192. Proc. Int. Conf. Johannesburg, 1969. Oxford Univ. Press, New York and London, 1970.
23. McDonald, A. D., Harper, A., El-Attar, O. A., and McDonald, J. C., Epidemiology of primary malignant mesothelial tumors in Canada. *In* "Pneumoconiosis" (H. A. Shapiro, ed.), p. 197. Proc. Int. Conf. Johannesburg, 1969. Oxford Univ. Press, New York and London, 1970.
24. McNulty, J. C., Asbestos exposure in Australia. *In* "Pneumoconiosis" (H. A. Shapiro, ed.), p. 201. Proc. Int. Conf. Johannesburg, 1969. Oxford Univ. Press, New York and London, 1970.
25. Gelfand, M., and Morton, S. A., Asbestosis in Rhodesia. *In* "Pneumoconiosis" (H. A. Shapiro, ed.), p. 204. Proc. Int. Conf. Johannes burg, 1969. Oxford Univ. Press, New York and London, 1970.
26. Webster, I., Asbestos exposure in South Africa. *In* "Pneumoconiosis" (H. A. Shapiro, ed.), p. 209. Proc. Int. Conf. Johannesburg, 1969. Oxford Univ. Press, New York and London, 1970.
27. Gilson, J. C., Asbestos health hazards. Recent observations in the United Kingdom. *In* "Pneumoconiosis" (H. A. Shapiro, ed.), p. 173. Proc. Int. Conf. Johannesburg, 1969. Oxford Univ. Press, New York and London, 1970.
28. Sleggs, C. A., Mesothelioma, including peripheral lung malignancy and tuberculosis in the North West Cape. *In* "Pneumoconiosis" (H. A. Shapiro, ed.), p. 225. Proc. Int. Conf. Johannesburg, 1969. Oxford Univ. Press, New York and London, 1970.
29. "XVI International Congress on Occupational Health," Tokyo, 1969. *In* "XVI International Congress on Occupational Health, Tokyo, 1969." Japan Industrial Safety Association, Tokyo.
30. Rossiter, C. E., Bristol, L. J., Cartier, P. H., Gibbs, G. W., Gilson, J. C., Sluis-Cremer, G. K., and McDonald, J. C., Dust exposure and radiologic appearances in Quebec asbestos workers. *In* "XVI International Congress on Occupational Health, Tokyo, 1969," p. 205. Japan Industrial Safety Association, Tokyo.

31. Noro, L., Epidemiology of asbestosis. *In* "XVI International Congress on Occupational Health, Tokyo, 1969," p. 207. Japan Industrial Safety Association, Tokyo.
32. Ferris, B. G., Jr., Ranadine, M., Peters, J. M., Murphy, R. L. H., and Burgess, W. A., Prevalence of chronic respiratory disease. Asbestosis in ship repair workers. *In* "XVI International Congress on Occupational Health, Tokyo, 1969," p. 209. Japan Industrial Safety Association, Tokyo.
33. Cralley, L. J., Identification and control of asbestos exposures. *In* "XVI International Congress on Occupational Health, Tokyo, 1969," p. 217. Japan Industrial Safety Association, Tokyo.
34. "Symposium on the Tissue Response to Asbestos." University of South Wales and Monmouthshire, Cardiff, 1970.
35. "Third International Symposium on Inhaled Particles. London, 1970." Sponsored by the British Occupational Hygiene Society. Proceedings to be published.
36. Gibbs, G. W., McDonald, J. C., Rossiter, C. E., Cartier, P., and Grainger, T. R., Qualitative aspects of asbestos dust exposure in the Quebec asbestos mining and milling industry. *In* "Third International Symposium on Inhaled Particles, London, 1970." Proceedings to be published.
37. Navratil, M., Pleural plaques due to asbestos exposure compared with the relevant findings in the non-exposed population. *In* "Third International Symposium on Inhaled Particles, London, 1970." Proceedings to be published.
38. Gross, P., Cralley, L. J., Davis, J. M. G., deTreville, R. T. P., and Tuma, J., A quantitative study of fibrous dust in the lungs of city dwellers. "Third International Symposium on Inhaled Particles, London, 1970." Proceedings to be published.
39. Colloquium on Asbestos, Sardinia, 1970. Sub-committee on Asbestosis, Permanent Commission and International Association on Occupational Health.
40. Gross, P., Tuma, J., and deTreville, R. T. P., Fibrous dust particles and ferruginous bodies. *Arch. Environ. Health 21,* 38 (1970).
41. Gold, C., Quantitative methods in the study of asbestos in tissue. *In* "Symposium on the Tissue Response to Asbestos, Cardiff, 1970."
42. Knox, J. F., Holmes, S., Doll, R., and Hill, I. D., Mortality from lung cancer and other causes among workers in an asbestos textile factory. *Brit. J. Ind. Med. 25,* 293 (1968).
43. Hill, I. D.,. Doll, R., and Knox, J. F. Mortality among asbestos workers. *Proc. Roy. Soc. Med. 59,* 59 (1968).
44. Newhouse, M. L., A study of the mortality of workers in an asbestos factory, *Brit. J. Ind. Med. 26,* 294 (1969).
45. Selikoff, I. J., Hammond, E. C., and Churg, J., Asbestos exposure, smoking and neoplasia. *J. Amer. Med. Ass. 204,* 106 (1968).
46. Noro, L., Occupational and non-occupational asbestosis in Finland. *Amer. Ind. Hyg. Ass. J. 29,* 195 (1968).
47. Balzar, J. L., and Cooper, W. C., The work environment of insulating workers. *Amer. Ind. Hyg. Ass. J. 29,* 222 (1968).
48. Dunn, J. E., Jr., and Weir, J. M., A prospective study of mortality of several occupational groups. *Arch. Environ. Health 17,* 71 (1968).
49. Dixon, J. R., Lowe, D. B., Richards, D. E., Cralley, L. J., and Stokinger, H. E., The role of trace metals in chemical carcinogenesis: Asbestos cancer. *Cancer Res. 30*, 1068 (1970).
50. Executive Secretary, American Conference of Governmental Industrial Hygienists. The Threshold Limit Values of Airborne Contaminants and Intended Changes. Cincinnati, Ohio, 1970.

51. Mancuso, T. F., and El-Attar, O. A., Epidemiologic study of the beryllium industry. Cohort methology and mortality studies. *J. Occup. Med. 11,* 422 (1969).
52. Mancuso, T. F., Relation of duration of employment and prior respiratory illness to respiratory cancer among beryllium workers. *Environ. Res. 3,* 251 (1970).
53. Lieben, J., and Williams, R. W., Respiratory disease associated with beryllium refining and alloy fabrication, 1968 follow-up. *J. Occup. Med. 11,* 480 (1969).
54. "Epidemiologic Study in the Beryllium Producing Industry." Bureau of Occupational Safety and Health, Public Health Service, Department of Health, Education, and Welfare, Washington, D.C.
55. Lehnert, G., Klavis, G., Schaller, K. H., and Haas, T., Cadmium determination in urine by atomic absorption spectrometry as a screening test in industrial medicine. *Brit. J. Ind. Med. 26,* 156 (1969).
56. Brieger, H. International symposium on the toxicity of carbon disulphide. *J. Occup. Med. 9,* 185 (1967).
57. Vigliani, E. C., International symposium on the toxicology of carbon disulphide. International recommendations made and accepted. *J. Occup. Med. 9,* 366 (1967).
58. Tiller, J. R., Schilling, R. S. F., and Morris, J. N., Occupational toxic factor in mortality from coronary heart disease. *Brit. Med. J. 4,* 407 (1968).
59. Martinez, N., and Farina, G., A clinical-statistical contribution to the study of carbon disulfide vasculopathy: Analysis of 75 cases. *Med. Lav. 60,* 11 (1969).
60. Savic, S. M., Influence of carbon disulfide on the eye. *Arch. Environ. Health 14,* 325 (1967).
61. Maugeri, U., Visconti, E., and Cavalleri, A., Retinal changes in young workers exposed to carbon disulfide. *Med. Lav. 57,* 741 (1966).
62. Maugeri, U., Cavalleri, A., and Visconti, E., Ophthalmalynamometric findings in occupational carbon disulfide poisoning. *Med. Lav. 57,* 730 (1966).
63. Cavalleri, A., Djuric, D., Maugeri, U., Branksvic, D., Visconti, E., and Rezman, I., Endocrinological findings in young workers exposed to carbon disulfide. I. Urinary excretion of total 17-ketosteroids. *Med. Lav. 57,* 566 (1966).
64. Cavalleri, A., Djuric, D., Maugeri, U., Branksvic, D., Visconti, E., and Rezman, I., Endocrinological findings in young workers exposed to carbon disulfide. II. Urinary excretion of 17-hydroxycorticosteroids. *Med. Lav. 57,* 573 (1966).
65. Cavalleri, A., Djuric, D., Maugeri, U., Branksvic, D., Visconti, E., and Rezman, I., Endocrinological findings in young workers exposed to carbon disulfide. III. Urinary excretion of 3α,11-desoxy-17-ketosteroids. *Med. Lav. 57,* 755 (1966).
66. Ferrero, G. F., The incidence of sugar diabetes among workers exposed to the risk of carbon disulfide poisoning. *Med. Lav. 60,* 38 (1969).
67. Visconti, E., Vidakovic, A., Maugeri, U., Visnjic, V., Calveri, A., and Dodic, S., Haematochemical findings in young workers exposed to carbon disulfide. IV. Thromboelastographic study of the coagulation process. *Med. Lav. 57,* 751 (1966).
68. Visconti, E., Visnjic, V., Cavalleri, A., Vidakovic, A., Maugeri, U., and Rezman, I., Haematochemical findings in young workers exposed to carbon disulfide. III. Behavior of antiheparin activity. *Med. Lav. 57,* 745 (1966).
69. Djerassi, L. S., and Lumbroso, R., Carbon disulfide poisoning with increased ethereal sulphate excretion. *Brit. J. Ind. Med. 25,* 220 (1968).
70. Herneberg, S., Partanen, T., Nordman, C. H., and Sumari, P., Coronary heart disease among workers exposed to carbon disulphide. *Brit. J. Ind. Med. 27,* 313 (1970).
71. Goto, S., and Hotta, R., The medical and hygienic prevention of carbon disulphide poisoning in Japan. "Toxicology of Carbon Disulfide," H. Brieger, and J. Teisinger, eds.), 1st Ed., p. 219. Excerpta Medica Foundation, Amsterdam, 1967.

72. Lilis, R., Gaurilescu, N., Moscovici, B., Teculescu, D., Roventa, A., Nestorescu, B., Senchea, A., and Pilat, L., Effects cardiovasculaires de l'exposition prolongée au sulfure de carbone. *Med. Lav. 59,* 41 (1968).
73. Chester, E. H., Gillespie, D. G., and Krause, F. D., The prevalence of chronic obstructive pulmonary disease in chlorine gas workers. *Amer. Rev. Resp. Dis. 99,* 365 (1969).
74. Patil, L. S., Smith, R. G., Vorwald, A. J., and Mooney, T. F., The health of diaphragm cell workers exposed to chlorine. *Amer. Ind. Hyg. Ass. J. 31*, 678 (1970).
75. Sluis-Cremer, G. K., and duToit, R. S., Pneumoconiosis in chromite miners in South Africa. *Brit. J. Ind. Med. 25,* 63 (1968).
76. Kitson, G. H. J., Dust sampling and dust control on South African colleries. *In* "Pnemoconiosis" (H. A. Shapiro, ed.), p. 283. Proc. Int. Conf. Johannesburg, 1969. Oxford Univ. Press, New York and London, 1970.
77. Reisner, M. T. R., Cumulative dust exposures and pneumoconiosis responses in German coal mines. *In* "Pneumoconiosis" (H. A. Shapiro, ed.) p. 286. Proc. Int. Conf. Johannesburg, 1969. Oxford Univ. Press, New York and London, 1970.
78. Walton, W. H., Progress of the 25-Pit scheme. *In* "Pneumoconiosis" (H. A. Shapiro, ed.), p. 292. Proc. Int. Conf. Johannesburg, 1969. Oxford Univ. Press, New York and London, 1970.
79. Einbrodt, H. J., The influence of dust elimination and the effects on the development of pneumoconiosis. *In* "Pneumoconiosis" (H. A. Shapiro, ed.) p. 299. Proc. Int. Conf. Johannesburg, 1969. Oxford Univ. Press, New York and London, 1970.
80. Wagner, J. C., Complicated coal workers' pneumoconiosis. *In* "Pneumoconiosis" (H. A. Shapiro, ed.), p. 303, Proc. Int. Conf. Johannesburg, 1969. Oxford Univ. Press, New York and London, 1970.
81. Pendergrass, E. P., Coal workers' pneumoconiosis in the United States: Radiologic considerations with special reference to large opacities. *In* "Pneumoconiosis" (H. A. Shapiro, ed.), p. 309. Proc. Int. Conf. Johannesburg, 1969. Oxford Univ. Press, New York and London, 1970.
82. Worth, G., Epidemiological evidence on the relationship between lung function and coal dust exposure. *In* "Pneumoconiosis" (H. A. Shapiro, ed.), p. 322. Proc. Int. Conf. Johannesburg, 1969. Oxford Univ. Press, New York and London, 1970.
83. Frans, A., Portier, N., Veriter, C., Brasseur, L., and Lavenne, F., Lung diffusing capacity for carbon monoxide and alveolar–arterial differences for oxygen and carbon dioxide in coalminers still at work. *In* "Pneumoconiosis" (H. A. Shapiro, ed.), p. 514. Proc. Int. Conf. Johannesburg, 1969. Oxford Univ. Press, New York and London, 1970.
84. Walton, W. H., Conversion factors between particle numbers, surface area and mass concentrations for coal dust. *In* "Pneumoconiosis" (H. A. Shapiro, ed.), p. 584. Proc. Int. Conf. Johannesburg, 1969. Oxford Univ. Press, New York and London, 1970.
85. Kitson, G. H. J., Calculation of the relationship between photoelectric readings and particle number, area and weight concentrations in South African coal mine dust clouds. *In* "Pneumoconiosis" (H. A. Shapiro, ed.), p. 587. Proc. Int. Conf. Johannesburg, 1969. Oxford Univ. Press, New York and London, 1970.
86. Brewer, H. and Reisner, M. T. R., Dust sampling instruments to collect the respirable fraction of coal dust for compositional analysis and animal research. *In* "Pneumoconiosis" (H. A. Shapiro, ed.), p. 595. Proc. Int. Conf. Johannesburg, 1969. Oxford Univ. Press, New York and London, 1970.
87. Schulte, K., Hazard indices in West German miners. *In* "Pneumoconiosis" (H. A. Shapiro, ed.), p. 600. Proc. Int. Conf. Johannesburg, 1969. Oxford Univ. Press, New York and London, 1970.

88. Walton, W. H., Dust sampling instruments and hazard indices in coal dust in the United Kingdom. *In* "Pneumoconiosis" (H. A. Shapiro, ed.), p. 603. Proc. Int. Conf. Johannesburg, 1969. Oxford Univ. Press, New York and London, 1970.
89. Jacobson, M., Environmental dust survey of bituminous coal mines in the United States. *In* "Pneumoconiosis" (H. A. Shapiro, ed.), p. 623. Proc. Int. Conf. Johannesburg, 1969. Oxford Univ. Press, New York and London, 1970.
90. Jarry, J. J., and Amoudru, C., Struggle against pneumoconiosis in coal mines in France. *In* "XVI International Congress on Occupational Health, 1969, Tokyo," p. 667. Japan Industrial Safety Association, Tokyo.
91. Frans, A., Portier, N., Veriter, C. Brasseur, L., and Lavenne, F., Alveolar-arterial tension differences for oxygen and carbon dioxide and lung diffusing capacity for CO in coalminers still at work. *In* "XVI International Congress on Occupational Health, 1969, Tokyo," p. 686. Japan Industrial Safety Association, Tokyo.
92. Worth, G., Visibility of coal dust in x-rays of the lung. *In* "XVI International Congress on Occupational Health, 1969, Tokyo," p. 671. Japan Industrial Safety Association, Tokyo.
93. Worth, G., Lung scanning in pneumoconiosis. *In* "XVI International Congress on Occupational Health, 1969, Tokyo," p. 673. Japan Industrial Safety Association, Tokyo.
94. Milijic, B., Trickovic, K., and Savic, N., Anthracosilicosis in the brown coal mines of Aleksinac and Jelasnica. *In* "XVI International Congress on Occupational Health, 1969, Tokyo," p. 682. Japan Industrial Safety Association, Tokyo.
95. Firtze, E., Some epidemiological aspects in coalminers dust exposure. *In* "XVI International Congress on Occupational Health, 1969, Tokyo," p. 685. Japan Industrial Safety Association, Tokyo.
96. "International Conference on CWP." Synopsis of the Work Session Proceedings, Sept. 10-12, 1969. Spindletop Research, Inc., Lexington, Kentucky, 1969.
97. Bergman, I., Casswell, C., and Rossiter, C. E., The relationship between dust content and radiological appearance in simple pneumoconiosis of coal workers. *In* "Third International Symposium on Inhaled Particles. London, 1970." Proceedings to be published.
98. Leiteritz, H., Bower, H. D., and Bruckmann, E., Mineralogical characteristics of airborne dust in coal mines of Western Europe and their relations to pulmonary changes of coal-hewers. *In* "International Conference on CWP." Spindletop Research, Inc., Lexington, Kentucky, 1969.
99. Jacobson, M., Respirable coal mine dust in the United States. *In* "Third International Symposium on Inhaled Particles, London, 1970." Proceedings to be published.
100. Walton, W. H., Characteristics of respirable dust in British coal mines. *In* "Third International Symposium on Inhaled Particles, London, 1970." Proceedings to be published.
101. Saric, M., The prevalence of respiratory symptoms in groups of coalminers and the relationship between these symptoms and some functional parameters. *In* "Third International Symposium on Inhaled Particles, London, 1970." Proceedings to be published.
102. Minette, A., The role of industrial dust exposure in the production of chronic bronchitis in coalminers. *In* "Third International Symposium on Inhaled Particles, London, 1970." Proceedings to be published.
103. Rae, S., Chronic bronchitis and dust exposure in British coalminers. *In* "Third International Symposium on Inhaled Particles, London, 1970." Proceedings to be published.

104. Reichel, G., and Ulmer, W. T., The inter-relationship of coalminers pneumoconiosis and bronchitis: An epidemiological study on 952 steel workers and 1306 coalminers. *In* "Third International Symposium on Inhaled Particulates, London, 1970." Proceedings to be published.
105. Reisner, M. T. R., Results of epidemiological studies on pneumoconiosis in West German coal mines. *In* "Third International Symposium on Inhaled Particles, London, 1970." Proceedings to be published.
106. McLintock, J. S., Rae, S., and Jacobsen, M., The attack rate of PMF in British coal miners. *In* "Third International Symposium on Inhaled Particles, London, 1970." Proceedings to be published.
107. Proceedings of the Symposium on Respirable Coal Mine Dust, Washington, D.C., Nov. 3-4, 1969. *U. S. Bur. of Mines Inform. Circ.* IC8458 (1970).
108. Lainhart, W. S., Doyle, H. N., Enterliine, P. E., Henschel, A., and Kendrick, M. A., Pneumoconiosis in Appalachian bituminous coal miners. Pub. Health Service Publ. 2000 (1969).
109. Dunmore, J. H., Hamilton, R. J., and Smith, D. S. A., An instrument for the sampling of respirable dust for subsequent gravimetric assessment. *J. Sci. Instrum. 41,* 669 (1964).
110. Doyle, H. N., Dust concentration in the mines. Proceedings of the Symposium on Respirable Coal Mine Dust, Washington, D.C., Nov. 3-4, 1969. *U. S. Bur. Mines Inform. Circ.* IC8458, pp. 27–32 (1970).
111. Gilson, J. C., Chairman UICC Working Group, UICC/Cincinnati certification of the radiographic appearances of pneumoconiosis, a cooperative study by the UICC committee. *Chest 58,* 57 (1970).
112. Jacobsen, M., Rae, S., Walton, W. H., and Ragan, J. M., The relationship between pneumoconiosis and dust exposure in British coal mines. *In* "Third International Symposium on Inhaled Particulates, London, 1970." Proceedings to be published.
113. Jacobsen, M., Rae, S., Walton, W. H., and Ragan, J. M., New dust standards for British coal mines. *Nature (London) 227,* 445 (1970).
114. Elwood, P. C., Pemberton, J., Merret, J. D., Carey, G. C. R., and McAulay, I. R., Byssinosis and other respiratory symptoms in flax workers in Northern Ireland, *Brit. J. Ind. Med. 22,* 27 (1965).
115. McKerrow, C. B., Gilson, J. C., Schilling, R. S. F., and Skidmore, J. W., Respiratory function and symptoms in rope makers. *Brit. J. Ind. Med. 22,* 204 (1965).
116. Gandevia, B., and Milne, J., Ventilatory capacity changes on exposure to cotton dust and their relevance to byssinosis in Australia. *Brit. J. Ind. Med. 22,* 295 (1965).
117. Kondakis, X. G., and Pournaras, N., Byssinosis in cotton ginneries in Greece. *Brit. J. Ind. Med. 22,* 291 (1965).
118. Valic, F., Zuskin, E., Walford, J., Kersic, W., and Paukovic, R., Byssinosis, chronic bronchitis and ventilatory capacities in workers exposed to soft hemp dust. *Brit. J. Ind. Med. 25,* 176 (1968).
119. Mekky, S., Roach, S. A., and Schilling, R. S. F., Byssinosis among winders in the cotton industry. *Brit. J. Ind. Med. 24,* 123 (1967).
120. Bouhuys, A., Barbero, A., Lindel, S. E., Roach, S. A., and Schilling, R. S. F., Byssinosis in hemp workers. *Arch. Environ. Health 14,* 553 (1967).
121. Simpson, G. R., and Barnes, R., Cotton dust exposure during lint removal. *Arch. Environ. Health 17,* 807 (1968).
122. Barnes, R., and Simpson, G. R., Cotton dust exposure. *Med. J. Aust. 1,* 897 (1968).
123. Bouhuys, A., Wolfson, R. L., Horner, D. W., Brain, J. D., and Zuskin, E., Byssinosis in cotton textile workers. *Ann. Intern. Med. 71,* 257 (1969).
124. Zuskin, E., Wolfson, R. L., Harpel, G., Welborn, J. W., and Bouhuys, A., Byssinosis in

carding and spinning workers. *Arch. Environ. Health 19*, 666 (1969).

125. Popa, V., Gavrilescu, N., Preda, V., Teculescu, D., Plecias, M., and Cirstea, M., An investigation of allergy in byssinosis. Sensitization to cotton, hemp flax, and jute antigens. *Brit. J. Ind. Med. 26;* 101 (1969).
126. Khogali, M., A population study in cotton ginnery workers in Sudan. *Brit. J. Ind. Med. 26*, 308 (1969).
127. Smith, G. F., Coles, G. V., Schilling, R. S. F., and Walford, J., A study of rope workers exposed to hemp and flax. *Brit. J. Ind. Med. 26*, 109 (1969).
128. Merchant, J. A., Kilburn, K. H., Rousch, D., Drake, B. M., Hamilton, J. D., and Lumsden, J., Screening for pulmonary dysfunction among cotton textile workers. Paper delivered to American Thoracic Society Meeting, 1969.
129. Bouhuys, A., Barbero, A., Schilling, R. S. F., and van de Woestijne, K. P., Chronic respiratory disease in hemp workers. *Amer. J. Med. 46,* 526 (1969).
130. Bouhuys, A., and Nicholls, P. J., *In* "The Effect of Cotton Dust on Respiratory Mechanics in Man and in Guinea Pigs, Inhaled Particles and Vapours" (C. N. Davies, ed.), Vol. II, p. 75. Pergamon, London, 1967.
131. Bouhuys, A., Byssinosis in textile workers pneumoconiosis. *In* "Pneumoconiosis" (H. A. Shapiro, ed.), p. 412. Proc. Int. Conf. Johannesburg, 1969. Oxford Univ. Press, New York and London, 1970.
132. Molyneux, M. K. B., and Tombleson, J. B. L., An epidemiologic study of respiratory symptoms in Lancashire mills, 1963-66. *Brit. J. Ind. Med. 27*, 225 (1970).
133. Schrag, P. E., and Gullet, A. D., Byssinosis in cotton textile mills. *Amer. Rev. Resp. Dis. 101,* 497 (1970).
134. Bouhuys, A., Gilson, J. C., and Schilling, R. S. Byssinosis in the textile industry. *Arch. Environ. Health 21,* 475 (1970).
135. deTreville, R. T. P., Cotton dust – IHF/ATMI respiratory study. Transactions of 25th Annual Meeting of Industrial Hygiene Foundation, October, 1970.
136. Lynch, J. R., Air sampling for cotton dust. *In* "Proceedings of the National Conference on Cotton Dust and Health, 1970," p. 33. University of North Carolina, School of Public Health, Chapel Hill, North Carolina, 1970.
137. El-Batawai, M. F., Respiratory disease resulting from inhalation of cotton and other vegetable dusts, definition of byssinosis. *In* "XVI International Congress on Occupational Health, 1969, Tokyo." Proceedings to be published.
138. Zuskin, E., Bouhuys, A., and Wolfson, R. L., Byssinosis and ventilatory function in USA cotton workers. *In* "XVI International Congress on Occupational Health, 1969, Tokyo." Proceedings to be published.
139. Baselga-Monte, M., Byssinosis in the textile industry of Catalonia – epidemiological considerations derived from the study of beneficiaries of the professional illnesses insurance in the province of Barcelonia. *In* "XVI International Congress on Occupational Health, 1969, Tokyo." Proceedings to be published.
140. Noweir, M. H., El-Sadek, Y., and El-Dakhakhny, A. A., Exposure to dust in the cottonseed oil extraction industry. *Arch. Environ. Med. 19,* 99 (1969).
141. Johnson, D. L., Healey, J. J., Ayer, H. E., and Lynch, J. R., Exposure to fibers in the manufacture of fibrous glass. *Amer. Ind. Hyg. Ass. J. 30,* 545 (1969).
142. Wright, G. W., Airborne fibrous glass particles. Chest roentgenograms of persons with prolonged exposures. *Arch. Environ. Health 16,* 175 (1968).
143. Utidjion, H. M., Industrial hygiene foundation statistical studies of health of fibrous glass workers. Proceedings of fibrous dust seminar, Nov. 1968. *Ind. Health Found. Med. Ser. Bull. No. 16-70* (1970).
144. Utidjoin, H. M., and deTreville, R. T. P., Fibrous glass manufacturing and health.

Report of epidemiologic study: Part I. Thirty-fifth Annual Meeting, Industrial Health Foundation, Pittsburgh, Pennsylvania, Oct. 1970.

145. deTreville, R. T. P., Hook, H. N., and Morrice, G., Fibrous glass manufacturing and health. Report of a comprehensive physiological study: Part II. Thirty-fifth Annual Meeting, Industrial Health Foundation, Pittsburgh, Pennsylvania, Oct. 1970.
146. Cralley, L. J., Records studies of health of fibrous glass workers. Proceedings of fibrous dust seminar. Nov. 1968. *Ind. Hyg. Found. Med. Ser. Bull. No. 16-70* (1970).
147. Milby, T. H., and Wolf, C. R., Respiratory tract irritation from fibrous glass inhalation. *J. Occup. Med. 11*, 409 (1969).
148. Possick, P. A., Gellin, G. A., and Key, M. M., Fibrous glass dermatitis. *Amer. Ind. Hyg. Ass. J. 31*, 12 (1970).
149. Smith, L. K., Urinary fluoride excretion in electric arc welders exposed to low hydrogen electrode fumes. *Ann. Occup. Hyg. 11*, 203 (1968).
150. Kleinfeld, M., Messite, J., Swencicki, R. E., and Shapiro, J., A clinical and physiologic study of grain handlers. *Arch. Environ. Health 16,* 380 (1968).
151. Boyd, J. T., Doll, R., Faulds, J. S., and Leiper, J., Cancer of the lung in iron ore (haematite) miners. *Brit. J. Ind. Med. 27*, 97 (1970).
152. Faulds, J. S., Haematite pneumoconiosis in Cumberland miners. *J. Clin. Pathol. 10*, 187 (1957).
153. Sunderman, F. W., Jr., Inhibition of induction of benzpyrene hydroxylase by nickel carbonyl. *Cancer Res. 27,* 950 (1967).
154. Selikoff, I. J., Hammond, E. C., and Churg, J., Asbestos exposure, smoking and neoplasia. *J. Amer. Med. Ass. 204,* 106 (1968).
155. Dixon, J. R., Lowe, D. B., Richards, D. E. Cralley, L. J., and Stokinger, H. E., The role of trace metals in chemical carcinogenesis: Asbestos cancers. *Cancer Res. 30*, 1068 (1970).
156. Kleinfeld, M., Messite, J., Shapiro, J., Kooyman, O., and Levin, E., A clinical, roentgenological and physiological study of magnatite workers. *Arch. Environ. Health 16,* 392 (1968).
157. Jorgensen, H., and Svensson, A., Studies on pulmonary function and respiratory tract symptoms of workers in an iron ore mine where diesel trucks are used underground. *J. Occup. Med. 12,* 348 (1970).
158. Sentz, F. C., Jr., and Rakow, A. B., Exposure to iron oxide fumes at arcair and powder-burning operations. *Amer. Ind. Hyg. Ass. J. 30,* 143 (1969).
159. Kleinfeld, M., Messite, J., Kooyman, O., and Shapiro, J., Welders siderosis. *Arch. Environ. Health 19,* 70 (1969).
160. Swensson, A., Holmquist, C., and Lundgren, K., Injury to the respiratory tract by isocyanates used in making lacquers. *Brit. J. Ind. Med. 22*, 50 (1965).
161. Zapp, J. A., Hazards of isocyanates in polyurethane foam plastic production. *Arch. Environ. Health 15,* 324 (1957).
162. Walworth, H. T., and Virchow, W. E., Industrial hygiene experiences with toluene diisocyanate. *Amer. Ind. Hyg. Ass. J. 20*, 205 (1959).
163. Duncan, B., Toluene diisocyanate inhalation toxicity: Pathology and mortality. *Amer. Ind. Hyg. J. 23*, 447 (1962).
164. Brugsch, H. C., and Elkins, H. B., Toluene diisocyanate (TDI) toxicity. *N. Engl. J. Med. 268,* 353 (1963).
165. Gandevia, B., Studies of ventilatory capacity and histamine response during exposure to isocyanate vapour in polyurethane foam manufacture. *Brit. J. Ind. Med. 20*, 204 (1963).
166. Silver, H. M., Toluene diisocyanate asthma. *Arch. Intern. Med. 112,* 155 (1963).

167. Dodson, V. N., Asthma and toluene diisocyanate exposure. *J. Occup. Med. 8,* 81 (1966).
168. Bruckner, H. C., Avery, S. B., Stetson, D. M., Dodson, V. N., and Ronayne, J. J., Clinical and immunologic appraisal of workers exposed to diisocyanates. *Arch. Environ. Health 16,* 619 (1968).
169. Peters, J. M., Murphy, R. L. H., Pagnotto, L. D., and VanGanse, W. F., Acute respiratory effects in workers exposed to low levels of toluene diisocyanate (TDI). *Arch. Environ. Health 16,* 642 (1968).
170. Peters, J. M., Murphy, R. L., and Ferris, B. G., Jr., Ventilatory function in workers exposed to low levels of toluene diisocyanate: A six-months follow-up. *Brit. J. Ind. Med. 26,* 115 (1969).
171. Peters, J. M., Murphy, R. L. H., Pagnatto, L. D., and Whittenberger, J. L., Respiratory impairment in workers exposed to "safe" levels of toluene diisocyanate (TDI). *Arch. Environ. Health 20,* 364 (1970).
172. Paisley, D. P. G., Isocyanate hazard from wire insulation: An old hazard in a new guise. *Brit. J. Ind. Med. 26,* 79 (1969).
173. McCallum, R. I., Sanderson, J. T., and Richards, A. E., The lead hazard in shipbreaking: The prevalence of anemia in burners. *Ann. Occup. Hyg. 11,* 101 (1968).
174. Rieke, F. E., Lead intoxication in shipbuilding and shipscrapping. *Arch. Environ. Health 19,* 521 (1969).
175. Williams, M. K., King, E., and Walford, J., An investigation of lead absorption in an electric accumulator factory with the use of personal samplers. *Brit. J. Ind. Med. 26,* 202 (1969).
176. Soliman, M. H. M., El-Sadik, Y. M., El-Koshlan, K. M., and El-Waseef, A., Biochemical studies on Egyptian workers exposed to lead. *Arch. Environ. Health 21,* 529 (1970).
177. Jovicic, B., Ulcer and gastritis in the professions exposed to lead. *Arch. Environ. Health 21,* 526 (1970).
178. Rakow, A. B., and Lieben, J., Twenty-four cases of lead poisoning, ten years later. *Arch. Environ. Health 16,* 785 (1968).
179. Milby, T. H., Samuels, A. J. and Ottoboni, F., Human exposure to lindane: Blood lindane levels as a function of exposure. *J. Occup. Med. 10,* 584 (1968).
180. Czegledi-Janko, G., and Auar, P., Occupational exposure to lindane: Clinical and laboratory findings. *Brit. J. Ind. Med. 27,* 283 (1970).
181. Tanaka, S., and Lieben, J., Manganese poisoning and exposure in Pennsylvania. *Arch. Environ. Health 19,* 674, (1969).
182. West, I., and Lim, J., Mercury poisoning among workers in California's mercury mills. A preliminary report. *J. Occup. Med. 10,* 697 (1968).
183. Rentos, P. G., and Seligman, E. J., Relationship between environmental exposure to mercury and clinical observation. *Arch. Environ. Health 16,* 794 (1968).
184. Milne, J., Christopher, A., and DeSilva, P., Acute mercurial pneumonitis. *Brit. J. Ind. Med. 27,* 334 (1970).
185. Taylor, W., Guirgis, H. A., and Stewart, W. K., Investigation of a population exposed to organomercurial seed dressing. *Arch. Environ. Health 19,* 505 (1969).
186. Wada, O., Toyokawa, K., Suzuki, T., Suzuki, S., Yano, Y., and Nakao, K., Response to a low concentration of mercury vapor. Relation to human porphyrin metabolism. *Arch. Environ. Health 19,* 485 (1969).
187. Smith, R. G., Vorwald, A. J., Patil, L. S., and Mooney, T. F., A study of the effects of exposure to mercury in the manufacture of chlorine. *Amer. Ind. Hyg. Ass. J. 31,* 687 (1970).
188. Lippman, M., and Goldstein, D. H., Oil mist supplies, environmental evaluation and control. *Arch. Environ. Health 21,* 591 (1970).

189. Goldstein, D. H., Benoit, J. N., Tyroler, H. A., An epidemiologic study of an oil mist exposure. *Arch. Environ. Health 21,* 600 (1970).
190. Ely, T. S., Pedley, S. F., Hearne, F. T., and Stille, W. T., A study of mortality, symptoms and respiratory function in humans occupationally exposed to oil mist. *J. Occup. Med. 12,* 253 (1970).
191. Jager, K. W., Roberts, D. V., and Wilson, A., Neuromuscular function in pesticide workers. *Brit. J. Ind. Med. 27,* 273 (1970).
192. Parrot, J. U., Hebert, R., Saindelle, A., and Ruff, F., Platinum and platinosis. Allergy and histamine release due to some platinum salts. *Arch. Environ. Health 19,* 685 (1969).
193. Rose, A. H., "Industrial Microbiology." Plenum Press, New York, 1961.
194. Flindt, M. L. H., Pulmonary disease due to inhalation of derivitives of *Bacillus subtilis* containing proteolytic enzymes. *Lancet I,* 1177 (1969).
195. Pepys, J., Hargreave, F. E., Langbottom, I. L., and Faux, J., Allergic reaction of the lungs to enzymes of *Bacillus subtilis. Lancet I,* 1181 (1969).
196. Greenburg, M., Milne, J. F., and Watt, A., Survey of workers exposed to dusts containing derivitives of *Bacillus subtilis. Brit. Med. J. 2,* 629 (1970).
197. Newhouse, M. L., Tagg, B., Pocock, S. J., and McEwan, A. C., An epidemiologic study of workers producing enzyme washing powders. *Lancet I,* 689 (1970).
198. Wagner, R., Silicosis and silica-tuberculosis in industry in West Germany. *In* "Pneumoconiosis" (H. A. Shapiro, ed.) p. 450. Proc. Int. Conf. Johannesburg, 1969. Oxford Univ. Press, New York and London, 1970.
199. Gilson, J. C., Lloyd-Davies, T. A., and Oldham, P. D., A study of respiratory responses to duration of foundry work. *In* "Pneumoconiosis" (H. A. Shapiro, ed.), p. 445. Proc. Int. Conf. Johannesburg, 1969. Oxford Univ. Press, New York and London, 1970.
200. Beadle, D. G., and Harris, E., The relationship between the amount of dust breathed and the incidence of silicosis. An epidemiologic study of South African European gold miners. *In* "Pneumoconiosis" (H. A. Shapiro, ed.), p. 473. Proc. Int. Conf. Johannesburg, 1969. Oxford Univ. Press, New York and London, 1970.
201. Prowse, C. M., Aspects of pulmonary function in silicosis in South African gold mines. *In* "Pneumoconiosis" (H. A. Shapiro, ed.), p. 508. Proc. Int. Conf. Johannesburg, 1969. Oxford Univ. Press, New York and London, 1970.
202. Wiles, F. J., Faure, M. H., Sluis-Cremer, G. K., VanDoorn, H. T., and Kruger, W. D. K., A cohort of white gold miners. *In* "Pneumoconiosis" (H. A. Shapiro, ed.), p. 350. Proc. Int. Conf. Johannesburg, 1969. Oxford Univ. Press, New York and London, 1970.
203. Ruttner, J. R., Silicosis and lung cancer in Switzerland. *In* "Pneumoconiosis" (H. A. Shapiro, ed.), p. 512. Proc. Int. Conf. Johannesburg, 1969. Oxford Univ. Press, New York and London, 1970.
204. Gregory, J., A survey of penumoconiosis at a Sheffield steel foundry. *Arch. Environ. Health 20,* 385 (1970).
205. Ahlmark, A., Silicosis, dust concentrations and dust control in Sweden. *Staub 29,* 1 (1969).
206. Grundorfer, W., and Raber, A., Progressive silicosis in granite workers. *Brit. J. Ind. Med. 27,* 110 (1970).
207. Lloyd, J. W., and Ciocco, A., Long-term mortality study of steelworkers. I. Methodology. *J. Occup. Med. 11,* 299 (1969).
208. "Manual of the International Statistical Classification of Diseases, Injuries, and Causes of Death." World Health Organization, Geneva, Sixth Rev. Ed., 1948, Seventh Rev. Ed., 1957.
209. Robinson, H., Long-term mortality study of steelworkers. II. Mortality by level of

income in whites and non-whites. *J. Occup. Med. 11,* 411 (1969).
210. Redmond, C. K., Smith, E. M., Lloyd, J. W., and Rush, H. W., Long-term mortality study of steelworkers. III. follow-up. *J. Occup. Med. 11,* 513 (1969).
211. Lloyd, J. W., Lundin, F. E., Redmond, C. K., and Geiser, P. B., Long-term mortality study of steelworkers. IV. Mortality by work areas. *J. Occup. Med. 12,* 151 (1970).
212. Lowe, C. K., Pelmear, P. L., Campbell, H., Hitchens, R. A. V., Khosla, T., and King, T. C., Bronchitis in two integrated steel works. I. Ventilatory capacity, age, and physique of non-bronchitic men. *Brit. J. Prev. Soc. Med. 22,* 1 (1968).
213. Warner, C. G., Davies, G. M., Jones, J. C., and Lowe, C. R., Bronchitis in two integrated steel works. II. Sulphur dioxide and particulate atmospheric pollution in and around the two works. *Ann. Occup. Hyg. 12,* 151 (1969).
214. Lowe, C. R., Campbell, H., and Khosla, T., Bronchitis in two integrated steel works. Respiratory symptoms and ventilatory capacity related to atmospheric pollution. *Brit. J. Ind. Med. 27,* 121 (1970).
215. Archer, V. E., Magnuson, H. J., Holaday, D. A., and Lawrence, P. A., Hazards to health in uranium mines. *J. Occup. Med. 4,* 55 (1962).
216. Parsons, W. D., deVillers, A. J., Bartlett, L. S., and Becklake, M. R., Lung cancer in a fluorspar mining community, prevalence of respiratory symptoms and disability. *Brit. J. Ind. Med. 21,* 110 (1964).
217. Holaday, D. A., Rushing, D. E., Coleman, R. D., Woolrich, P. F., Kusnetz, H. V., and Bales, W. F., Control of radon and daughters in uranium mines and calculations on biologic effects. *Pub. Health Service Publ.* No. 494 (1967).
218. American Standards Association (now American National Standards Institute), Radiation Protection in Uranium Mines and Mills (concentrators), American Standard N7, p. 31, 1960.
219. Lundin, F. E., Lloyd, J. W., Smith, E. M. Archer, V. E., and Holaday, D. A., Mortality of uranium miners in relation to radiation exposure, hard rock mining and cigarette smoking. *Health Phys. 16,* 571 (1969).
220. Evans, D., Radiation protection of uranium miners. Hearing before the Subcommittee on Research, Development and Radiation, Joint Committee of Atomic Energy, pp. 318-337, March 17-18, 1969.
221. Saccomanno, G., Hearing on Safety and Health Standards. Bureau of Labor Standards, U. S. Department of Labor, November 20, 1968.
222. Cooper, W. C., Uranium mining and lung cancer. *J. Occup. Med. 10,* 82 (1968).
223. Duggan, M. J., Soilleux, P. J., and Strong, J. C., The exposure of United Kingdom miners to radon. *Brit. J. Ind. Med. 27,* 106 (1970).
224. Federal Register, *33,* No. 252, December 28, 1968.
225. Federal Register, *34,* No. 11, January 16, 1969.
226. Federal Register, *34,* No. 10, January 15, 1969.
227. Nelson, I. C., Parker, H. M., Cross, F. T., Craig, D. K., and Stuart, B. O., "A Research Report for the Public Health Service, A Further Appraisal of Dosimetry Related to Uranium Mining," p. 106. Battelle Memorial Institute, Richland, Washington, December 1969.
228. Morken, D. A., The Relation of lung dose rate to working level. *Health Phys. 16,* 796, (1969).
229. McCurdy, D. E., Thermoluminescent dosimetry for personal monitoring of uranium miners. Special Report, Colorado State University, p. 168, June 1969.
230. Thomas, J. W., Determination of radon progeny in air from alpha activity of air samples. USAEC Health and Safety Laboratory Report HASL - 202, pp. 9-15 (1968).
231. Raabe, O. G., and Patterson, F. S., A method of analysis of air sampling data for

particulate alpha emitters in a radon-thoron daughter atmosphere. University of Rochester Report 658, p. 92, February 2, 1965.
232. White, O., An evaluation of six radon dosimeters. USAEC Health and Safety Laboratory, Technical Memorandum 69-23A, p. 21, October 29, 1969.
233. Federal Register *35,* No. 245, December 18, 1970.
234. Deese, D. E., and Joyner, R. E., Vinyl acetate: A study of chronic human exposure. *Amer. Ind. Hyg. Ass. J. 30,* 449 (1969).
235. Harris, D.,K., and Adam, W. G. F., Acroosteolysis occurring among men engaged in the polymerization of vinyl chloride. *Brit. Med. J. 3,* 712 (1967).
236. Wilson, R. H., McCormick, W. E., Tatum, C. F., and Creech, J. L., Occupational acroosteolysis. Report of 31 cases. *J. Amer. Med. Ass. 201,* 577 (1967).
237. Dinman, B. D., Cook, W. A., Whitehouse, W. M., Magnuson, H. J., and Ditcheck, T., Occupational acroosteolysis. I. An epidemiological study. *Arch. Environ. Health 22,* 61 (1971).
238. Cook, W. A., Giever, P. M., Dinman, B. D., and Magnuson, H. J., Occupational acroosteolysis, II. An industrial hygiene study. *Arch. Environ. Health 22,* 74 (1971).
239. Dodson, V. N., Dinman, B. D., Whitehouse, W. M., Nasr, A. N. M., and Magnuson, H. J., Occupational acroosteolysis, III. A clinical study. *Arch. Environ. Health 22,* 83 (1971).
240. Fogh, A., Frost, J., and Georg, J., Respiratory symptoms and pulmonary function in welders. *Ann. Occup. Hyg. 12,* 213 (1969).
241. deMiranda Bastas, A., and deMatos Filho, A., A "jacaranda" timber causing dermatitis. "Proceedings of the Fifth World Forestry Congress," Univ. of Washington, College of Forestry, Seattle, Washington, p. 1414 (1960).
242. Krogh, H. K., Contact eczema caused by true teak (*Tectona grandis*). An epidemiologic investigation in a furniture factory. *Brit. J. Ind. Med. 19,* 42 (1962).
243. Komatsu, F., Respiratory allergy in woodworkers. *Int. Cong. Occup. Health Proc. 14th* 1748 (1964).
244. Krogh, H. K., Contact eczema caused by true teak (*Tectona grandis*). A follow-up study of a previous epidemiologic investigation, and a study into the sensitizing effect of various teak extracts. *Brit. J. Ind. Med. 21,* 65 (1964).
245. Morgan, J. W. W., Orsler, R. J., and Wilkinson, D. S., Dermatitis due to the wood dusts of *Khaya anthatheca* and *Machaerium scleroxylon. Brit. J. Ind. Med. 25,* 119 (1968).
246. Sossman, A. J., Schleuter, D. P., Fink, J. N., and Barboriak, J. J., Hypersensitivity to wood dust. *New Engl. J. Med. 281,* 977 (1969).
247. Gandevia, B., Ventilatory capacity during exposure to Western Red Cedar. *Arch. Environ. Health 20,* 59 (1970).
248. Gandevia, B., and Milne, J., Occupational asthma and rhinitis due to Western Red Cedar (*Thuja plicata*), with special reference to bronchial reactivity. *Brit. J. Ind. Med. 27,* 235 (1970).
249. Acheson, E. D., Caudell, R. H., Hadfield, E. H., and Macbeth, R. G., Nasal cancer in woodworkers in the furniture industry. *Brit. Med. J. 2,* 587 (1968).
250. Hadfield, E. H., Damage to the human mucosa by wood dust. *In* "Third International Symposium on Inhaled Particles, London, 1970." Proceedings to be published.
251. Lee, A. M., and Fraumeni, J. F., Arsenic and respiratory cancer in man: An occupational health study. *J. Nat. Cancer Inst. 42,* 1045 (1969).
252. Harries, P. G., Asbestos hazards in naval dockyards. *Ann. Occup. Hyg. 11,* 135 (1968).
253. McDonald, J. C., McDonald, A. D., Gibbs, G. W., Siemiatycki, J., and Rossiter, C. E., Mortality in the chrysotile asbestos mines and mills of Quebec. *Arch. Environ. Health 22,* 677 (1971).
254. Linch, A. L., Wiest, E. G., and Carter, M. D., Evaluation of tetraalkyl lead exposure by

personal monitor surveys. *Amer. Ind. Hyg. J. 31*, 170 (1970).

255. Lainhart, W. S., Roentgenographic evidence of coal workers' pneumoconiosis. *Arch. Environ. Health 12*, 314 (1966).
256. A Cooperative Study, Radiologic Classification of the Pneumoconioses. *Arch. Environ. Health 12*, 314, (1966).
257. Occupational Safety and Health Standards (Williams-Steiger Occupational Safety and Health Act of 1970). U. S. Dept. of Labor. *Fed. Regist. 36*, 10518 (1971).
258. de Treville, R. T. P., Proteolytic enzymes in cotton mill dust – a possible cause of byssinosis. *Ind. Health Found. Med. Ser. Bull.* 18-71 (1971).

Toxicology

EMIL A. PFITZER

THE AGE OF ENVIRONMENTALISM

For several years the activities of the industrial toxicologist had been increasingly concerned with health risks to the consumer and the community at large. Today this trend has come near to full swing of the pendulum. The toxicity problem of the worker on the job relates mainly to the application of presently available information, as it well should be, and the observation of a new occupational toxicity problem is a relative rarity. The concern for health risks has moved from the industrial worker to the consumer, to the community, and now, to the environment itself. The most dramatic news of the day reports the finding of industrial chemicals in unexpected places, their persistence in the environment, their biologic magnification, and their impact on the nonhuman features of the environment. Ecologic concern is an everyday expression which has broadened the scope of the toxicologist to the effects of trace ingredients which have migrated far afield from their normal initial usage.

The environmentalist has the ear of the public, the concerned citizen, the politician, the statesman, the scientist, and industrial management. As never before the subjects of air pollution, water pollution, chemicals in food, and toxic side effects of drugs are a part of everyone's conversation. And of particular importance is the fact that the tone of the conversation is one that asks for action to safeguard the health of man and to maintain or improve the quality of the environment. Actions are being taken, actions that have enormous economic impact, and actions which would have been considered as impossible only a few years ago.

The changed scene has made the life of the toxicologist at once both frustrating and stimulating. Frustrating because he doesn't have the answers to many of the perplexing questions being asked; frustrating to some toxicologists because their time-honored methods for making judgments about safety evaluation are being challenged; stimulating because suddenly there are a multitude of people interested in his professional activities. Everyone wants to talk with him about the potential for unwanted effects from the chemicals introduced by technologic advances. The realization that we need much more toxicity data about environmental chemicals has an element of strong urgency. There may be concern that the full swing of the pendulum brings with it the hazards of "overkill" or a "backlash" of reactions which may complicate future efforts. The toxicologist can afford little time for such concerns; his immediate challenges are legion. He can expect his data to be scrutinized by nonscientists, to be interpreted in the newspapers and in legal hearings, to face requirements for exactness and statistic validity with increasing rigor. The life of the toxicologist will never be the same. Already his methodologies are undergoing revision and definition; more sensitive, specific, and subtle measurements of effects are in critical need due to long-term, low-level exposures to mixtures of complex chemicals.

NEW TOXICITY PROBLEMS

Occupational Acroosteolysis

A new occupational disease called occupational acroosteolysis has developed among workers associated with the cleaning of polymerizing tanks for vinyl chloride. This disease has the distinctive feature of loss of bone structure in the fingertips. It is manifest by x-ray observation and a marked discomfort on exposure to cold. The bone changes are rarely seen in other parts of the body and several of the patients have had external skin lesions. Somewhat less than 50 cases of the disease have been reported worldwide which yields a low prevalence rate. It has been suggested that the condition may result from three factors: a chemical insult, a physical insult, and a personal idiosyncrasy.[1-3]

Detergent Enzymes

A new dimension was added to the setting of threshold limit value (TLV) levels when it became necessary to have exposure standards for workers involved in the manufacturing of detergent enzyme products. The active ingredients are proteolytic enzymes derived from organisms such as *Bacillus subtilis* and are known as subtilisins. The enzymes are primary skin and respiratory tract irritants, which have caused bronchoconstriction and respiratory allergies.

Extensive animal experiments have been published.[4] Exposed humans have shown airways obstruction (as shown by reduced vital capacity, forced vital capacity, and forced expiratory volume in one second), asthmatic reactions, sore throat, headache, congested nares, persistent cough, and positive intradermal skin reactions.[5-8] The TLV has been set to prevent skin irritation and allergic respiratory sensitization based upon current information; the value of 0.0003 mg/m^3 is expressed in terms of the crystalline active pure enzyme.[9] A finished detergent product containing 0.012% pure enzyme would have a TLV of 2.5 mg/m^3. The TLV is a ceiling limit; it has also been expressed as six additional numbers which are based on the units for assay of the enzyme activity. It is believed that the health problems of primary irritant dermatitis and pulmonary hypersensitivity have been limited to plant exposures and that consumer safety has been established and confirmed.[8]

Bis(chloromethyl) Ether as a Carcinogen

The α-haloethers are very reactive compounds and their hydrolysis products are highly irritant to the skin, eyes, and mucous membranes. Bis(chloromethyl) ether, as a 2% solution in benzene applied to the backs of mice three times a week for more than 46 weeks, was found to be a very potent carcinogen. Thirteen of twenty mice bore papillomas, and twelve of these papillomas progressed to squamous cell carcinomas by 325 days. A similar study with chloromethyl methyl ether showed a lack of activity as a carcinogen, although it was considered that continued testing may indicate weak carcinogenic activity.[10]

METHODOLOGY

Reproduction, Mutagenesis, Teratogenesis

The toxicologic testing procedures for the evaluation of safety of chemicals in the environment have been undergoing intense scrutiny for several years. Reproduction studies have been central to much of the debate with particular concern for tests for teratogenesis and mutagenesis. The Food and Drug Administration has set up several advisory committees to evaluate protocols for safety evaluation. A report from one of their panels has been published. It presents recommendations for a multigeneration reproduction study to include an "effect" dose level, data on teratogenic potential, and data on mutagenic potential.[11]

Carcinogenesis

The methods used to test a chemical agent for carcinogenic potential have been reviewed.[12,13] It was noted that an important factor would be the chemical similarity to known carcinogens. All of our methods for detecting carcinogenic activity, at the present, require the production of neoplastic lesions in animals. It was noted that the term "spontaneous" as applied to tumors in animals is a misnomer; it merely says that we do not know the causative factors in the production of the particular neoplasm.[12,13]

A method for the experimental induction of bronchogenic carcinoma has been developed. Benzo[*a*]pyrene with hematite dust was injected intratracheally fifteen times into laboratory rodents. Respiratory tract tumors, mostly bronchogenic carcinomas, developed in 100% of the cases. It was considered that the morphology of the tumors was close to human lung cancer.[14]

Lung-Clearance Models

One of the mechanisms for alveolar clearance involves the phagocytosis of particles by alveolar macrophages followed by transport up the tracheobronchial tree and into the esophagus. A method for determining the number of macrophages and particles brought from the lungs to the oropharynx combines the collection of sputum from a surgically implanted tube in the esophagus with the separation of cells from the viscous pulmonary secretions by the mucolytic action of *N*-acetyl-L-cysteine. It was reported that this technique makes it possible to assess the role of cellular and noncellular mechanisms in transferring particulates from the lungs to the oropharynx.[15]

Male rats were exposed to radiolabeled lead chromate particles by either inhalation or intratracheal injection. At hourly intervals from 7–20 hours after exposure, the particles cleared from the lung into the esophagus were collected and analyzed. It was assumed that the pulmonary system consisted of a series of compartments which were cleared sequentially in time via the trachea. Based on this assumption it was found that the intratracheal injection led to a greater proportion of the lung burden being deposited deeper in the respiratory tract as compared with inhalation. However, the clearance rates observed were the same for the two methods of exposure.[16]

Rats were exposed to an aerosol of plutonium-239 dioxide and at varying time periods after exposure the alveolar macrophages were washed from the lungs for observation. A large percentage of the washed out particles could be removed by this process. It was recommended that pulmonary lavage may offer a practical method for the therapeutic removal of insoluble radioactive particles accidentally deposited in the lungs of man.[17]

The scanning electron microscope has provided three-dimensional evidence of structures which heretofore had been accepted on faith in the morphologists' drawings. This is also true for pictorial evidence of the deposition of particles on the surface of the lungs. A particularly striking picture shows an alveolar macrophage with its pseudopod in contact with polystyrene particles.[18]

The field of industrial hygiene has made increasing use of the measurement of dust of respirable size as a measurement of improved hygienic significance. Research on improved estimates of the particle sizes which are deposited in various regions of the pulmonary tract has continued. It has been possible to study the deposition of particles in humans by the inhalation of monodisperse insoluble iron oxide spheres, tagged with ^{198}Au ^{51}Cr, or ^{99m}Tc, having unit density diameters of 2.1–12.5 μm. The particles were inhaled through a mouthpiece. It was possible to estimate deposition in the following regions: the mouth itself, the pharynx–larynx, the ciliated tracheobronchial tree, and the nonciliated alveolar spaces. With mouth breathing it was found that deposition in the tracheobronchial tree was more efficient than in the head, and that for particles larger than 4 μm this deposition took place primarily by impaction. There were very large differences in the efficiency of the tracheobronchial tree as a particle collector among the subjects tested.[19] Excellent reviews of the physiologic as well as industrial hygiene aspects of "respirable" particles are available.[20–23]

Inhalation Chambers

Exposure of animals to environmental contaminants by inhalation is a major stumbling block to many investigators. A detailed description of inhalation chamber design has been provided of those used by the Health Effects Research Program of the National Center for Air Pollution Control. A particularly popular one has been the "portable" animal-exposure chamber which is on wheel casters for convenient movement to various locations. It is a 27-in. cube with a pyramidal top; it will conveniently roll through a doorway and is completely self-contained. Good distribution of the air contaminant can be maintained with the air stream entering at the top and exhausting near the bottom.[24] Less elaborate exposure chambers have also been described.[25,26] A comprehensive review of the methods available for generating aerosols to be dispersed in inhalation chambers has been published.[27]

When conducting inhalation experiments it is well known that large quantities of the air contaminant can be lost from the breathing zone by adsorption on the chamber walls or skin and fur of the experimental subjects. A chamber which was being monitored for ferric oxide aerosol with a Sinclair-Phoenix Forward Scattering Aerosol Analyzer showed the presence of substantial quantities of aerosol when dogs and nitrogen dioxide, but no ferric oxide, were introduced to the chamber. It was found that in the presence of either dogs or ferric oxide aerosol, the supply of nitrogen dioxide to chambers resulted in high concentrations of both nitrate and ammonium ions. The extraneous aerosol, identified as ammonium nitrate, markedly influenced the degree of toxicity observed in the experiments.[28]

Eye Irritation

Rabbits have been the most commonly used test animal for the evaluation of a chemical as an eye irritant. Indeed the various regulatory agencies have required such tests on rabbits. However, it was found that eye irritation tests conducted in rhesus monkeys more accurately predicted the nature of the response found in humans.[29]

Prediction of Dose Levels

When testing a new compound, the toxicologist will typically first observe the effects of single exposures and then proceed to increasingly lengthy repeated exposure studies, often up to 2 years. It is important that the 2-year study provide information on a minimum effect level and a maximum no ill-effect level, and it is one of the jobs of the toxicologist to predict the dose levels which will provide such data. A comparison has been made of results from single exposure, 7-day, and 90-day studies to test their ability to predict minimum effect levels for 2-year feeding studies. It was found that single exposure values were not efficient predictors, but formulas for predicting were developed for 90-day feeding studies from 7-day results and 2-year feeding studies from either 7-day or 90-day results.[30]

Statistical Significance

The experimental toxicologist sometimes feels he is a victim of statistical significance. At times results show apparent significant differences when the differences are due to some artifact or very unusual occurrence which the toxicologist recognizes but does not have numerical observations in his experiment to prove. Several cases of a situation in which the statistical difference was due to an abnormal value in the concurrently dosed controls have been cited. It was recommended that in order for changes to be considered as deleterious, they should be dosage related and illustrate a trend away from the norm for that stock of animals.[31]

METALS AND THEIR COMPOUNDS

Lead

LEAD INTOXICATIONS

Although lead can be handled safely, lead intoxication is occurring regularly and it is frequently not diagnosed. The problems associated with the diagnosis of plumbism are discussed and seven illustrative case histories have been presented.[32]

An outdoor painter who did not wear a face mask developed lead poisoning. The case was considered as unusual in that, for the first time, hematemesis was considered as a manifestation of the lead intoxication.[33]

Thirty-five workers in Yugoslavia with lead intoxication or presaturnism were evaluated for gastric acidity and prevalence of ulcer. There was a marked correlation between hypoacidity and exposure to lead, but the frequency of gastroduodenal ulcer was not significantly different from the general population.[34]

In a group of 112 Egyptian workers with chronic lead intoxication or presaturnism, a significant number showed a reduction in the α_1- and α_2-globulin fractions of serum proteins, along with an increase in proteinuria and coproporphyrinuria. Because of variations in the degree and duration of exposure, and possibly individual susceptibility, it was not possible to correlate the biochemical changes with clinical diagnoses.[35]

SCREENING TESTS

A survey in a high-incidence area of lead poisoning in Chicago found 250 children with an initial blood lead value greater than 40 μg/100 ml. Each child was referred to the Chicago Board of Health Lead Clinic for clinical examination of possible increased lead exposure. Chelation therapy was given to 56 of the children after the clinical diagnosis of increased lead exposure, based upon the minimum criteria of two consecutive blood lead values greater than 60 μg/100 ml coupled with either a definite history of pica or x-ray film evidence of lead ingestion. Urinary δ-aminolevulinic acid levels were also measured, but the clinician was not aware of the values when making the diagnosis. It was thus possible to correlate the laboratory findings with the clinical impression of lead exposure. Of the fifty-six children given chelation therapy, forty-eight had urinary δ-aminolevulinic acid levels of 1 mg/100 ml or greater. Of the 194 children not treated, 179 had urinary δ-aminolevulinic acid levels below 1 mg/100 ml. None of the other laboratory tests had a comparable degree of correlation. It was suggested that urinary δ-aminolevulinic acid, when measured as described in this study, may serve as an effective screening test for early lead exposure in asymptomatic individuals.[36]

Coproporphyrinuria was used as a screening test for potentially hazardous lead exposures in a steel plant. Coproporphyrin in urine was detected by visual observation of fluorescence under an ultraviolet light. When a trace or more of coproporphyrin was detected, a second urine sample was analyzed for lead. If the urinary lead concentration was greater than 0.20 mg/liter of urine, the worker was examined by a physician, sampled for blood lead and subject to a 6-week restriction from the exposure area. This screening method exposes a high percentage of false positives in that many workers with detectable coproporphyrinuria had urinary lead values less than 0.20 mg/liter. However very few false negatives occur as evidenced by only 4 of 925 samples that showed the combination of urinary lead in excess of 0.20 mg/liter and no detectable coproporphyrinuria.[37] Numerous other publications are available which have examined the relationship between lead exposure and various biologic tests.[38,39] One such study has included an analysis of the cost of each test and concluded that the measurement of urinary coproporphyrins was the best screening test per unit cost.[40]

LEAD AND ALA-DEHYDRASE

One of the most surprising findings in research with lead has been the marked correlation between increased lead in blood and decreased enzymatic activity of δ-aminolevulinic dehydrase in red blood cells. This enzyme is critical in the synthesis of the porphyrin structure in hemoglobin. The correlation has been so marked that the enzyme measurement has been suggested as a preferred assay for screening for lead absorption.[41,42]

While there can be no doubt that the enzyme activity shows a biochemical change associated with lead exposure, the question remains as to whether this has any clinical significance for the health of the individual. The unraveling of the unanswered questions on this subject may well shed light on the mechanism and dynamics of the effect of lead on heme synthesis. It has been shown that lead acts at a large number of biochemical sites and might contribute to metabolic alterations through several different mechanisms.[43]

ENDOCRINE FUNCTION

A group of twenty-four patients, three with industrial lead exposure and twenty-one with a history of ingesting illicit whiskey, were found to have evidence of excessive exposure to lead by the presence of more than 1 mg lead/24-hr urine sample after the administration of 2 gm calcium disodium edetate given intravenously. Measurements of the 24-hour ^{131}I uptake before and after injections of thyroid-stimulating hormone indicated that there was injury to the iodine concentrating mechanism of the thyroid. It was indicated that this injury was due to the lead exposure.[44]

Eight men with long-term ingestion of illicit distilled whiskey and one man exposed to lead fumes while employed in a battery factory were determined to have an increased body burden of lead as indicated by urinary excretion of lead after EDTA infusion. Their endocrine function was tested by the renin-aldosterone response to salt deprivation. Seven of the nine men failed to show an increase to the normal range of plasma renin activity and aldosterone secretion rate. It was suggested that lead was responsible for the abnormalities observed.[45]

HUMAN TISSUES

Human tissues were obtained from thirty-three cities of the United States and foreign countries for lead analysis. Lead accumulated to some extent in bone, aorta, kidney, liver, lung, pancreas, and spleen of American subjects up to the fifth or sixth decade but not thereafter. Foreign subjects appeared to be in a more or less "steady state" in respect to accumulation of lead with age. It was estimated that approximately 91% of the lead in the body was stored in the bone. There was no evidence of bizarre innate toxicity in these human subjects and the lead in the soft tissues had not replaced the essential trace elements of low concentration: chromium, manganese, cobalt, copper, or molybdenum.[46] A study on human tissues from Great Britain found similar results except that it was reported that the lead concentration in bone increased in both sexes with advancing age and did not "level off" after the seventh decade. The wide scatter in results and the few samples at the higher age levels indicates that this question needs further research.[47]

LUBRICANTS

Lubricants containing lead naphthenate, applied to the clipped back of rabbits, were found to be nonirritating although detectable quantities of lead were absorbed. Similar findings were noted with ten human volunteers. The use of lead naphthenate-containing lubricants should be accompanied by adequate supervision to prevent significant skin contact.[48]

EFFECTS ON RATS

Lead acetate was given to rats in their drinking water up to 10 weeks. Apart from the actual content of lead in urine, the most sensitive change observed was the formation of intranuclear inclusion bodies in the tubular-lining cells of the kidneys. This effect was observed with 40 mg lead/100 ml drinking water. At higher doses the next most sensitive abnormality was decreased body weight, followed by increased δ-aminolevulinic acid excretion, reticulocytosis, renal edema, and aminoaciduria. Anemia only occurred at the highest dose of 1000 mg lead/100 ml of drinking water.[49]

Reticulocytes from phenylhydrazine-injected rats were incubated with lead acetate at low concentrations. Since it was found that lead depressed cellular and mitochondrial respiration, it was suggested that lead has a direct effect on cellular oxidative capacity.[50]

Rats were continuously exposed to airborne particles of lead sesquioxide at concentrations of 10 and 150 $\mu g/m^3$. It was found that the number of alveolar macrophages which could be washed from the lungs of the rats was significantly reduced to a minimum value which did not decrease further after 7 days of exposure. When exposure to lead was stopped, the number of macrophages returned to normal in about 3–4 days. The significance of this finding with respect to the health of the rats is not clear, but an inhibition of the alveolar clearance mechanism provides the opportunity for the invasion of microorganisms and other foreign particles.[51,52]

When male rats were fed 25 ppm of lead nitrate in their drinking water for a lifetime, the only toxic effects observed were poor coats and loss in weight from 24–30 months of age. When chromium was deficient from the diet, there was a shortened life span which indicated that chromium may be antagonistic to lead toxicity.[53]

Lead acetate is a potent calcergen as evidenced by its ability to cause a deposition of hydroxyapatite following subcutaneous injection into rats. When rats were given intravenous injections of neutral lead acetate followed immediately by a 30–90 min exposure to vapors of formaldehyde and then examined after 5 days, it was found that there was selective calcification in the pulmonary hilus region. It was suggested that this technique may furnish a model for revealing otherwise latent degrees of pulmonary irritation by air contaminants.[54]

AIR QUALITY STANDARDS

The health effects of lead with a particular emphasis on the setting of community air quality standards have been extensively reviewed. It seems clear that lead in ambient air is not the major source of our current health problems relating to lead.[55–58]

Mercury

INORGANIC INTOXICATIONS

Workers in a sodium hydroxide producing plant were exposed to an average concentration of mercury in air of 0.3 mg/m^3 (ranging from 0.072–0.88). This study did not provide a dose–response relationship between mercury exposure and health, but a majority of the exposed workers had complaints and signs and symptoms consisting of stomatitis, tremors, and behavioral changes. The clinical findings were alleviated when exposure was discontinued.[59]

A study of mercury exposure among workers employed in mercury mines and mills in California revealed that twenty-three of eighty-three workers had symptoms of mercury poisoning. All with mercury poisoning were from the mills rather than the mines. It was believed that despite wide fluctuations, urinary mercury measurements were valuable. At levels above 800 μg/liter there was a rather good correlation with the severity of overt manifestations of mercury intoxication.[60]

A group of eighty-three workers in seventeen mercury mine and mill locations were evaluated for their exposure to mercury and their clinical status. Eighteen had evidence of mercury poisoning. It was suggested from this study that the current TLV for mercury of 0.1 mg/m^3 of air contains no more than a safety factor of 2. None of the numerous biochemical analyses on the blood were satisfactory for purposes of determining impending or early mercury poisoning.[61]

Forty-seven asymptomatic workers who had been exposed to mercury for varying periods of time were examined for changes in δ-aminolevulinic acid dehydratase activity in red blood cells, cholinesterase activity in serum, and the urinary output of mercury, protein, coproporphyrin, and δ-aminolevulinic acid. There was a statistically significant correlation between the concentration of mercury in the urine and the coproporphyrin content of urine or the δ-aminolevulinic acid dehydratase activity in red blood cells. However the correlations were too weak to be of value in the practical assessment of the response of individuals to mercury. On the other hand there was a marked decrease in the activity of cholinesterase in serum which correlated well with the urinary level of mercury. From these results it was judged that the maximum permissible concentration of urinary mercury would be 200 μg/gm creatinine in chronic exposure to inorganic mercury.[62]

Mercury cells in which the cathode is a flowing sheet of elemental mercury are used for brine electrolysis to produce chlorine. A group of 642 workers from twenty-one chloralkali plants were studied to establish dose–response relationships between their mercury exposure and state of health. Symptoms were only observed when the exposure was in excess of 0.1 mg/m^3 and consisted of loss of appetite, weight loss, tremors, insomnia, and other indications of early effects on the nervous system. There was no indication of impairment of the cardiorespiratory, gastrointestinal, or hepatorenal systems. Although there were no significant signs or symptoms in persons exposed at or below 0.1 mg/m^3, it was felt that there may not be an adequate safety factor for exposure of workers. Accordingly, the current TLV for inorganic mercury vapor and salts has been reduced to 0.05 mg/m.[39,63]

Contact dermatitis due to metallic mercury is relatively rare. Two cases are reported and the relationships between skin sensitivity and mucosal sensitivity to mercury are reviewed.[64]

IN SALIVA

It has been hypothesized that the gingivitis and oral disorders which often accompany mercurialism may be attributed to the mercury in salivary secretions. Analytic developments for measuring trace quantities of mercury and reproducible procedures for collecting saliva have provided the basis for studies of the relationship of salivary levels of mercury to concomitant blood or urine levels. In thirteen normal adults with no known mercury exposure, it was not possible to detect any mercury in their saliva (less than 0.5 μg/100 ml). In a group of forty workers exposed to mixtures of inorganic and organic mercury compounds, the average content of mercury in saliva was 5 μg/100 ml, with a range from 1–15.5 μg/100 ml. The corresponding average values for blood and urine were 13 and 515 μg/100 ml, respectively. A high correlation was observed between blood and saliva mercury contents, suggesting that salivary secretions may be of diagnostic value as an indirect monitor of blood mercury. This suggestion was tested in a second plant on twenty-nine workers and the mercury in blood values predicted from the concentration of mercury in saliva were within plus or minus 20% of the actual values.[65,66]

ORGANOMERCURY EXPOSURE

A group of thirty-three male workers exposed to a variety of organomercurial seed disinfectants were examined for changes in various biochemical parameters. Blood mercury concentrations, urinary mercury concentrations, protein in urine, and activity of serum phosphoglucose isomerase were all affected significantly. Further work is necessary to establish whether the biochemical changes would be a useful indicator of mercury toxicity.[67]

Chromosome breakage in lymphocyte cultures from human subjects differing in amounts of fish they consumed was examined. The question was raised as to whether methylmercury in fish might cause chromosome abnormalities which in turn might be related to the fetal abnormalities associated with Minamata disease.[68]

MERCURY UPTAKE BY TISSUES

Three healthy male volunteers received an oral dose of methylmercury nitrate labeled with mercury-203. Samples of urine, feces, blood, and hair were collected at intervals up to 71 days after administration. Whole body counting of radioactivity was performed over five specified regions of the body. The main activity was localized in the liver, accounting for about 50% of the contents of the body. The head contained about 10% of the total body content, and the decline of mercury-203 from the head was less rapid than in the rest of the body. The main excretion occurred via the feces, but the urinary excretion increased with time to 30 days after the intake. Up to 0.12% of the dose per gram of hair was found as a maximum value after 40–50 days. There was a rapid uptake of mercury by the red blood cells. The biologic half-lives for the different regions of the body varied from 36–95 days.[69]

Rats were exposed to isotopically tagged mercuric chloride by intravenous injection at dose levels of 0.12 mg/kg and 1.2 mg/kg. The clearance of mercury from the blood was about 23% per hour for the low dose and about 5.8% per hour for the high dose. Initially, at both dose levels, about three-fourths of the mercury in the blood was associated with the red blood cells. At the lower dose level this persisted with time and the protein-bound mercury was associated with the α-globulins. At the higher dose level more of the mercury tended to be associated with the plasma with time and there was a shift in protein-bound mercury from the α-globulins to the albumin.[70] Essentially similar findings resulted in studies with mercuric nitrate, tagged with ^{203}Hg. It was noted that the *in vitro* distribution of mercury among plasma proteins was markedly different, being bound mostly to albumin.[71]

A model for mercury elimination was postulated consisting of four compartments: a long-term storage compartment, the kidneys; a short-term storage compartment, the liver; a tissue compartment, the rest of the body; and an excretion reservoir, in which both urinary and fecal mercury accumulate. When tested with rats injected with mercuric nitrate, it was found that the model was a reasonable representation of the elimination when the dose was low enough to preclude signs of acute mercury intoxication. It was suggested that medical surveillance of mercury workers would eliminate some uncertainties if both urinary and fecal mercury were monitored.[72] It has been observed that the elimination rate of mercury is dependent on the body burden. Thus the specific activity of radioactive mercury should be taken into account when the maximum body burden and maximum permissible concentrations are calculated.[73]

The ability of mercury vapor to be deposited in the lungs was clearly demonstrated by the larger amounts of mercury in the peripheral parts of the lung following inhalation as compared with injection. The conclusions reached were that mercury vapor penetrates to the alveoli; that a large proportion of mercury vapor diffuses into the blood; and that a smaller proportion is deposited in pulmonary tissues from which it is slowly eliminated with an initial half-life of about 5 hr.[74]

Animals exposed to mercury vapor had ten times as much mercury in their brain as animals receiving equivalent amounts of mercury by intravenous injection. It was possible to show that mercury vapor could be absorbed directly from the lungs into the blood and that this mercury was transported in the blood in a different mode than injected mercuric salts. It was suggested that the difference in uptake by the brain may in part be explained by the difference in transport mechanism.[75]

Radiolabeled methylmercury chloride was injected intravenously into rats at doses of either 1 or 10 mg mercury/kg. The clinical symptoms of poisoning by alkyl mercurial compounds have a characteristic lag period of 1 week after exposure. However it was found that the delay in symptoms could not be related to the biotransformation of methyl mercury since there was no evidence of an accumulation of inorganic mercury in brain tissue. Biotransformation was very important in the excretion with inorganic mercury being preferentially excreted in the feces. The distribution in the body, following injection of methyl mercury compounds, was strikingly different than the distribution observed after inorganic mercury. With time biotransformation leads to a higher proportion of inorganic mercury in the kidneys and a corresponding increase in the proportion of inorganic mercury in the urine.[76]

The distribution and biotransformation of inorganic mercury and organomercurial compounds have been reviewed. It was suggested that more needs to be known about the microdistribution of mercury in cells and tissues in order to pinpoint the specific biochemical lesions underlying mercury poisoning.[77,78]

Cadmium

Cadmium has been reported to cause proteinuria and kidney damage, emphysema, anosmia, hypertension, and early testicular atrophy. A recent review has discussed the biopathology of cadmium-induced proteinuria. One suggested mechanism for this proteinuria has involved the low-molecular weight protein metallothionein which is filtered through the glomerulus and reabsorbed by the tubular epithelial cells. This protein is unique in that it will bind up to 5.9% of cadmium. It has been suggested that this bound cadmium then poisons the mechanism of tubular reabsorption leading to proteinuria and ultimately killing the cell. An alternate mechanism has been suggested which is related to the finding that much of the urinary protein in cadmium intoxicated subjects is composed of fragments of immunoglobulins (γ-globulins). It was suggested that cadmium is a general poison of the peptidases which catabolize the fragments (light free chains) of the γ-globulins and possibly also of those catabolizing other plasma proteins. It follows that in cadmium poisoning the urine contains primarily the proteins or protein fragments which are not disposed of by the cadmium-inhibited enzymes.[79]

Six rabbits were given subcutaneous injections of cadmium chloride (0.25 mg cadmium/kg) 5 days a week for 24 weeks. After the last injection of cadmium the rabbits had changes in serum proteins and mild alterations in proximal tubular epithelial cells of the kidneys.[80] However at 7 months after exposure, although one rabbit had nephrosis, the other five rabbits had normal serum proteins and normal protein excretion. Signs of tubular epithelial regeneration were evident, although there were large amounts of cadmium in the renal cortex. It appeared that after exposure to cadmium had ceased, a considerable regression of tubular damage takes place and that large amounts of cadmium in the renal cortex of the rabbit can be tolerated without development of gross abnormalities.[81]

Chronic arterial hypertension has been produced in rats exposed to cadmium in their drinking water for 1–3 years. After exposure to 10 ppm cadmium, as acetate, in their drinking water for 2–4 months, rats showed a significant elevated mean blood pressure. An intraperitoneal injection of 1 mg cadmium acetate caused a rise in mean blood pressure within 5 min. It was also noted that the mean blood pressure of rats fed a commercial diet (about 0.6 μg cadmium/gm diet) was higher than in rats fed a special diet containing 0.1 μg cadmium/gm diet. Cadmium in rats did not sensitize the blood vessels to the action of norepinephrine and angiotensin. Thus it was suggested that the mechanism of cadmium-induced hypertension may differ fundamentally from that of other types of hypertension.[82,83]

The distribution and excretion of radiolabeled cadmium chloride was studied after subcutaneous injection in rats. The cadmium was predominantly concentrated in the liver, kidney, and pancreas. In spite of the high kidney concentration, only a small fraction of the dose appeared in the urine. The principal route of cadmium excretion was via the gastrointestinal tract. There was no information presented on possible enterohepatic circulation, and the rapid appearance of cadmium in the intestinal wall suggested that the mucosa of the intestine may play a role in the excretion.[84]

CADMIUM FUME INTOXICATION

Two workers were exposed to cadmium fumes when they cut a cadmium-coated bolt with an oxygen–propane torch in an underground vault without adequate ventilation. It was estimated that the maximum exposure was 38.6 mg cadmium/m^3; the time of the exposure was uncertain. Prominent symptoms were nausea, chills, cough, sore throat, and difficulty in breathing. One worker was treated at home for pneumonia from which he made a slow recovery over 3 months, but continued to complain of loss of appetite, exertional dyspnea, and increased thirst. The other worker was hospitalized with increasing respiratory difficulties. Two weeks after admission he died and autopsy revealed coronary arteriosclerosis with organizing thrombosis, massive infarction, and cor pulmonale. It was uncertain as to whether the cadmium fume exposure caused the fatal outcome and it was suggested that more detailed studies are needed on the effect of cadmium on the vascular system as well as on the methods of therapy.[85]

Beryllium

Guinea pigs were sensitized to beryllium to study the interaction of their epidermal tissue with beryllium. It was believed that the beryllium-binding with guinea pig epidermal constituents supported the mechanism of delayed allergic reaction for the toxic effects of beryllium. Alkaline phosphatase and nucleic acids of the epidermis were found to bind beryllium.[86]

A male and a female beagle dog were each exposed for 20 min. to the exhaust from firing a rocket motor. The exhaust products contained an average concentration of 115 mg total beryllium/m^3. Fifty percent of the total beryllium was in the form of beryllium oxide, 40% in the form of beryllium fluoride, and the remainder primarily beryllium chloride. The dogs lost weight during the first week after exposure, but remained clinically healthy until the time of necropsy 3 years later. There were no gross lesions evident, but small foci of granulomatous inflammation were scattered throughout the lungs of both dogs. Examination of the lungs with the electron microscope showed that beryllium particles were deposited in lysosomes in the cytoplasm of histiocytes in the interstitium of the septa. It was suggested that the lesions represented an early form of chronic beryllium disease.[87]

Nickel Carbonyl

Nickel carbonyl was found in the breath of rats after an intravenous injection indicating that some nickel carbonyl can pass across the pulmonary alveolus without decomposition to metallic nickel and carbon monoxide. The identity of the nickel carbonyl was established with four chromatographic columns and an electron capture detector. It was suggested that the method could serve as a practical method for monitoring the concentrations of nickel carbonyl in industrial atmospheres as well as providing a rapid and sensitive method for diagnosing nickel carbonyl poisoning in industrial workers.[88]

The ultrastructural and histochemical reactions of the pulmonary alveolar wall of the male rat to nickel carbonyl were determined after the intravenous injection of LD_{50} doses. This exposure caused acute pathologic reactions in the pulmonary parenchyma which affected all of the cellular elements of the alveolar septum. The most unusual alterations occurred in the membranous pneumocytes which developed changes suggesting they might be undergoing transformation into granular pneumocytes. There was no evidence that the membranous pneumocytes might undergo transformation to free alveolar cells or that they might develop any phagocytic activity. The most prominent and consistent ultrastructural abnormalities in liver cells were diffuse dilatation of endoplasmic reticulum with a paucity of related lesions in the mitochondria and other cytoplasmic structures.[89,90]

The metabolism of nickel carbonyl was investigated with isotopic labels on either the nickel (^{63}Ni) or the carbon (^{14}C). It was found that when the nickel carbonyl is dissociated, the carbon monoxide released becomes bound to hemoglobin and is eventually exhaled. The nickel is oxidized to divalent nickel and is transported in the blood plasma bound to albumin. It may bind with various tissue nucleic acids and proteins and is eventually excreted in the urine and feces.[91,92]

Arsenic Carcinogenicity?

There have been numerous attempts to verify the role of arsenic as a carcinogen in a variety of animals. A further attempt was made to see if arsenic trioxide at a concentration of 0.01% in the drinking water would influence the development of skin tumors produced by the application of methylcholanthrene to the skin of three strains of mice. There was no augmentation of papilloma production in Balb/C mice; in DBA mice the arsenic treatment appeared to augment papilloma production, but the effect was not statistically significant; and in C X C3H mice there was a statistically significant decrease in the number of papillomas with arsenic treatment. The experiment thus failed to demonstrate the carcinogenicity of arsenic, but in view of the complexities of species differences, cell line specificities, and interactions with other metals, it was suggested that it would be unwise to dismiss arsenic as a carcinogen.[93]

Arsine and Aluminum Ladders, Tanks

Three men entered a chemical evaporating tank containing aqueous sodium arsenite to inspect and repair a paddle. An aluminum ladder was used for descent into the tank instead of the usual wooden ladder. It was hypothesized that the aluminum reacted with the sodium arsenite to release arsine gas. The individual exposures of 2, 3, or 15 min were followed by massive hemolysis, anuria, and a peculiar bronze discoloration of the skin. Dimercaprol (BAL) and other supportive therapy was provided along with exchange blood transfusions and peritoneal dialysis during 4 weeks in intensive care.[94]

Three workers were exposed to arsine when they entered a tank via an aluminum ladder which rested in a moist mixture containing sodium arsenite. The symptoms were classical including early chills, abdominal cramps, nausea and vomiting followed by hematuria, anemia, jaundice, oliguria, and anuria. Vigorous therapy including exchange transfusions, BAL injections, peritoneal dialysis, and hemodialysis proved to be life saving.[95]

A 32-year-old truck driver was poisoned by arsine when he crawled into an aluminum tank to flush it out after it had contained sodium hydroxide and arsenic trioxide. He developed anuria within 24 hr after exposure and only returned to normal after more than 6 months with prolonged peritoneal dialysis and extracorporeal hemodialysis. Five renal biopsies showed acute tubular necrosis followed by rapid regeneration of tubular cells, which had a modified appearance.[96]

Five workmen were seriously poisoned while repairing a malfunctioning pump in a water well. The men were exposed to arsine gas which was released when a clean galvanized pipe placed into the water catalyzed a reaction between hydrochloric acid and sodium *m*-arsenite. One of the men died within half an hour, one suffered massive hemolysis and renal failure, and the other three had clinical evidence of poisoning without renal failure. The patient with renal failure was treated by exchange transfusion and then hemodialysis, resulting in good recovery. It was believed that this treatment was superior to peritoneal dialysis; dimercaprol failed to alter the course of the acute phase of arsine poisoning.[97]

Organotin Effects in Animals

Organotin compounds have been used as insecticides and as stabilizers in certain plastics. Mice were intubated by stomach tube with a single dose equivalent to 4 gm/kg of di-*n*-octyltin bis(2-ethylhexyl maleate) (DOT-EHM), di-*n*-octyltin bis(butyl maleate) (DOT-BM), and di-*n*-octyltin maleate (DOT-M). All of the mice were killed 24 hours after exposure for histopathologic study. The dose was in excess of the LD_{50} value for each compound. The major histologic findings were traces of fatty degeneration in liver cells and epithelial cells of the kidney tubules.[98]

Bis(tri-*n*-butyltin) oxide is used as a disinfectant because of its bacterial, fungicidal, insecticidal, and slimicidal properties. In tests on rats and rabbits it was found to be severely irritating to the skin and eyes, producing necrosis of the cornea and skin. It was estimated that the no-risk concentration of the compound for skin contact in industry was about 0.005–0.01%.[99,100]

Chromium Transport in Blood

Trivalent chromium is transported in blood bound to serum proteins. By immunoelectrophoretic analysis and autoradiography it was possible to identify siderophilin, one of the group of β-globulins, as the specific molecule to which most of the chromium is attached. It was suggested that this method should have wide applicability for studying the transport of other metals as well as other substances that are transported in a protein-bound form.[101]

Manganese Hazards

Rats were given an isotopically labeled dose of manganese maleate by intraperitoneal injection. Initially high activity appeared in the adrenals, pituitary, liver, and kidneys. A peak activity in the small and large intestine suggested the excretion of manganese into the feces. The cerebrum, cerebellum, and spinal cord, in contrast to other tissues, showed an increasing or steady activity from day 13 to day 34. It was suggested that the relatively high capacity of the central nervous system for manganese retention offers an explanation for the neurotoxic effects observed in chronic manganese poisoning.[102] A survey of 117 workers in Pennsylvania with manganese exposure in excess of the TLV level found that 7 of the workers had definite signs of manganese poisoning.[103] A group of workers in a factory manufacturing electrodes in Czechoslovakia were studied for the possibility of using manganese in feces as an indicator of exposure. There was too much variation to make the procedure useful for evaluation of individuals, but in general a level of 6 mg/100 gm of feces was taken to indicate industrial exposure to manganese. It was suggested that manganese in hair might be a useful diagnostic aid in manganese poisoning.[104]

Metal Fume Fever

A welder developed metal fume fever after an exposure to various metal oxides, nitrogen oxides, ozone, and heat. The worker experienced the typical symptoms of fever, nausea, malaise, and general fatigue. Serum zinc levels were elevated for as long as 12 days post exposure. It was suggested the measurement of the isoenzymes of serum lactic dehydrogenase (fraction no. 3 being the "pulmonary fraction") may be of value in the differential diagnosis of acute pulmonary symptoms.[105]

Radioactive Metals – Translocation Problems

Radioactive elements present special problems to the toxicologist over and beyond their specific radiation hazard, because specific elements undergoing radioactive decay are converted into new elements which may be handled biologically in quite a different manner from the parent element. Thus the toxic hazard of the daughter products are associated with possible translocation of the chemical in the body due to different biologic properties. An overexposure of twenty-five workers to plutonium oxide led to a comparative study of whole-body counting and urinary excretion of plutonium and its daughter, americium. It was found that the americium was excreted in the urine more readily than plutonium as shown by an α ratio of 1.5 to 1 in the urine as compared with a plutonium to americium ratio of 11 to 1 in the exposure material. Thus estimates of systemic body burden based on urinary excretion would have been approximately one-tenth the actual value established by whole-body counting.[106]

Rare Earths – Yttrium

Yttrium, the lowest molecular weight rare earth, is the second most abundant rare earth. Rats were exposed to aerosols containing different ratios of stable ^{89}Y and radioactive ^{88}Y. The distribution and excretion of ^{88}Y was markedly affected by the concentration of stable yttrium present. In all rats the highest concentrations of ^{88}Y were noted in skeleton, lung, and liver.[107]

Metals and New Phosphors

The development of color television has brought about the use of some new and exotic phosphors. One of these is a solid-state compound known as europium-activated yttrium orthovanadate. Workers exposed to this mixture of vanadium with rare earths have been studied closely and no permanent damage has been observed in any of the exposed men. It was noted that observations are continuing to ascertain if any delayed effects occur, such as was the case with beryllium phosphors.[108]

Metals and Reproduction and Interactions

Along with the increased interests in trace metals in the environment there has been concern for the role of trace metals in mammalian reproduction. Cadmium injected into pregnant hamsters was found to have a complex teratogenic interaction with the essential trace metal zinc.[109] When barely sublethal doses of cadmium sulfate or sodium arsenate were injected into the day 8 pregnant hamster, a high incidence of resorbed or malformed embryos resulted. Barely sublethal doses of sodium selenite, on the other hand, caused no change in the normal resorption rate, and furthermore, it caused a significant decrease in the teratogenic toxicity of both cadmium and arsenic when injected simultaneously.

Although these metals were injected at concentrations far in excess of the usual levels of environmental contaminations, it was suggested that such teratogenic responses be carefully evaluated pending determination of individual species sensitivity.[110]

Mice of four different inbred strains were given subcutaneous injections of cadmium-109 and zinc-65 together. Mice of two of the strains were known to be susceptible to cadmium-induced testicular necrobiosis and these mice exhibited a greater uptake of cadmium by the testes than did resistant mice. In all of the mice, the liver and kidney concentrated more cadmium than zinc, while in all other tissues, the uptake of zinc predominated that of cadmium. These studies lend further importance to the need for information about the complex interactions between nonessential metals and essential trace metals.[111]

Metals and Goldfish Behavior

Comet goldfish were conditioned to swim away from a lighted area in order to avoid an electric shock. They were then retested at 24-hour intervals during exposure to sodium arsenate, lead nitrate, mercuric chloride, or selenium dioxide. The lowest concentrations which gave significant impairment of the conditioned avoidance response were 0.1 ppm arsenic, 0.07 ppm lead, 0.003 ppm mercury and 0.25 ppm selenium. The LC_{50} based on a 7-day survival time was 32 ppm for arsenic, 110 ppm for lead, 0.8 ppm for mercury, and 12 ppm for selenium. Fifty ppm calcium carbonate was added to all solutions of metal compounds. When lead nitrate was tested alone, the LC_{50} value was 6.6 ppm, presumably because the formation of lead carbonate was avoided.[112]

Metals in Normal Blood

Blood from male donors at blood banks in nineteen different cities in the United States was analyzed for selenium, molybdenum, vanadium, copper, zinc, cadmium, and lead. The selenium concentrations ranged from 10–34 μg/100 ml whole blood with an overall mean value of 20.6 μg/100 ml. The limit of detection was 1 μg/100 ml. There was some evidence of a geographic variation, but no trend with the age of the donor. In over 75% of the samples the concentration of molybdenum was less than the detectable quantity of 0.5 μg/100 ml. When detected, the concentration of molybdenum was highly variable ranging up to 41 μg/100 ml. Over 90% of the samples had concentrations of vanadium less than the detectable quantity of 1 μg/100 ml; the highest concentration of vanadium encountered was 2 μg/100 ml. The overall mean value for the concentration of copper was 89(range 16–348) μg/100 ml of whole blood. The mean concentration of zinc was 530 (range 60–1987) μg/100 ml of whole blood. Less than one-half of the samples had more than the least detectable concentration of 0.5 μg cadmium/100 ml whole blood. The concentrations ranged up to 14 μg/100 ml and the overall mean for those samples with detectable quantities was 1.77 μg cadmium/100 ml whole blood. Nine of 243 samples had less than the detectable quantity of 1 μg lead/100 ml of whole blood. The values ranged up to 109 μg/100 ml and the overall mean for those samples with detectable quantities was 13 μg/100 ml whole blood.[113,114]

Metals – Hazardous or Nonhazardous in Air?

The hazards due to metals in the air, food and water, and soil were reviewed. Eight metals were considered to be essential to life or health of man: iron, zinc, copper, molybdenum, manganese, cobalt, selenium, and chromium. All of these were found to be present in soil, food, and water. All were present in air samples measured, except that cobalt was not found in nonurban air, and selenium was not measured but is probably present from coal and oil. It was suggested that the presence of these metals in air should cause no alarm from the public health viewpoint. There are seven additional metals which have low toxicities to cells and little or no adverse effects when fed orally to experimental animals in small doses for life. Titanium and vanadium were present in all air samples and vanadium was also present in motor vehicle exhaust. Barium was found only in urban air samples. Zirconium, niobium, aluminum, and strontium were not measured in these samples, but they are undoubtedly present. It was considered that these metals did not represent a health hazard under present conditions of use. Beryllium, antimony, bismuth, tin, lead, cadmium, and nickel were identified as metals with known toxicity. The first four, beryllium, antimony, bismuth, and tin were found rarely or infrequently in air, and it was considered that the largest real or potential public health hazards were related to lead, cadmium, and nickel. In addition arsenic, boron, germanium, and mercury were recognized as metals of variable toxicity and probably present in air from coal, oil, or fungicides. Of this latter group, only mercury was considered to be a potential hazard to health.[115-118]

MINERAL AND FIBROUS DUSTS

Cristobalite, Tridymite, and Quartz

Laboratory animals were exposed to flux-calcined diatomaceous earth of 61% cristobalite content which was dispersed as particles with a mean particle size of 0.7 μm. Dust exposure was at levels of 2 and 5 million particles per cubic foot (mppcf) as well as at intermittent levels of 50 mppcf; the 2 and 5 mppcf exposures were 6 hr/day, 5 days/week for 30 months. There was no fibrosis of the pulmonary parenchyma in any of the species exposed in this experiment, however there was no dust level which did not exhibit a histiocytic and giant cell infiltration in the pulmonary parenchyma. Dogs exposed to 5 mppcf developed a moderate number of hyalinized fibrotic nodules in hilar lymph nodes. It was suggested that the current TLV of 5 mppcf offered little if any safety factor.[119] Accordingly it has been recommended that the TLV for cristobalite be one-half the value calculated from the count formula for quartz, $300/(\%SiO_2 + 10)$.[9] For the specific mineral dust used in this study the calculated TLV would be $\frac{1}{2}[300/(61 + 10)]$ or 2.1 mppcf.

The mechanism by which silica dust interacts with cell membranes was studied in cultures of macrophages collected from the peritoneal cavity of guinea pigs. Silica dust (tridymite) was suspended in the medium of the cell cultures and various enzyme activities were measured over the course of several hours. The observations suggested that when silica is added to macrophage cultures *in vitro*, it interacts with the external cellular membrane and that the first cellular lesion occurs at this site. Further damage may then be inflicted by the liberation of lysosomal enzymes.[120]

Rats, hamsters, and guinea pigs were exposed to fine quartz dust at an average concentration of 280 mg/m^3 for 6 hr/day, 5 days/week for a total of 115 days. The outstanding feature was the paucity of classical silicotic nodules in the midst of large accumulations of quartz dust. In view of the very high concentrations it was expected that there might be a failure in the mechanism for alveolar clearance; however the failure to produce marked fibrotic response was unexpected. It was hypothesized that the findings might have been caused by the coating of the particles with a lipoproteinaceous material produced by the excessively high dust concentration.[121]

Dust Transport and Lung Disease

It is reasonable to assume that lungs with chronic obstructive disease may not be able to handle exposures to dust in a normal fashion, and that this may further complicate the lung disease. However in animals with papain-induced emphysema and subsequent exposure to quartz or coal dust there was less dust in the lungs than in animals with nonemphysematous lungs. It was suggested that the relatively smooth surfaces of the alveolar areas in the emphysematous lungs might have expedited the clearance of dust from these spaces.[122]

Rats were given intratracheal injections of particles of titanium dioxide, amorphous silica, or quartz in order to study the translocation of the dust from the lungs to the hilar lymph nodes. Intravenous injections of a low-virulent strain of bovine tubercle bacilli increased the retention of dust in the lungs and decreased the amount of dust in the hilar lymph nodes. It was reasoned that exposure to tubercle bacilli could affect the pulmonary lymph drainage and increase the risk of developing lung lesions after dust inhalation.[123]

Fibrous Dusts

When microquartz fibers, chrysotile asbestos, and tremolite talc dust were injected intratracheally into rats, they produced proliferative lesions which in time became permanent collagenized scars causing deformed bronchi and bronchioles. However other fibrous dusts, such as synthetic chrysotile, ceramic aluminum silicate, silicon carbide, glass, and brucite caused no permanent damage and did not destroy the anatomic integrity of the air spaces. It was noted that 4 days after the intratracheal injections there were foci of polyploid proliferative inflammation which disappeared in time. These lesions were not seen when the same dusts were inhaled in large concentrations and they are considered to be reversible changes related to the intratracheal injection technique and not the fibrous dusts.[124]

Glass dust may exist as fibers (a particle whose length is at least three times its diameter) or as nonfibrous particles or flakes. In addition the glass may be coated with resins or binders. Rats and hamsters were exposed to fibrous glass dust at an average concentration of 100 mg/m^3 for 6 hours/day, 5 days/week, for 24 months. The fibers had an average diameter of 0.5 μm and lengths which ranged from approximately 5–20 μm. The pulmonary response was characterized by small accumulations of macrophages without significant stromal change indicating a relative biological "inertness" in spite of the extremely high exposure.[125]

Byssinosis and Cotton Dusts

There has been a renewed emphasis on byssinosis in the textile industry in the United States. Several major conferences on the subject have been held. A grading system for byssinosis based upon forced expiratory volume was recommended at the Second International Conference on Respiratory Disease in Textile Workers held in Alicante, Spain in 1968. It was strongly urged that tests of pulmonary function be used to evaluate textile workers at risk of byssinosis.[126] The TLV value of 1 mg total cotton dust per cubic meter of air should be re-evaluated in light of considerations of respirable size and toxic qualities of specific dust components.

Asbestos Disease

The widespread use of asbestos combined with the long latent period before onset of asbestos disease has caused an increased concern about the assurance of safe working and living conditions. The research necessary to answer a series of pressing questions has been outlined. The areas of needed research include epidemiology, clinical and human pathology, animal experimentation and tissue culture, and analysis of chemical and physical characteristics.[127]

Attempts were made to induce cancer with asbestos deposited or implanted in rats by different methods, including deposition of asbestos in fluorocarbon-induced pulmonary infarcts, injection of asbestos-containing wax pellets into lung tissue, and the surgical insertion of fibrous glass pledgets saturated with asbestos onto the mesothelial and pericardial surfaces. In the first two methods no neoplasms resulted and it was suggested that this may be due to the small amount of asbestos present. Sarcomas of the pleura and pericardium developed in 74% of the rats in which the surface was covered with an asbestos-impregnated fibrous glass pad and it was suggested that this method may be effective for quantitative studies of the neoplastic response of the pleura to asbestos.[128]

Fibrous Dusts and Ferruginous Bodies

The term "ferruginous bodies" has been added to the vocabulary of health specialists concerned with dust diseases. For many years the encased fibers found in the lungs have been called asbestos bodies because it was known that asbestos fibers would remain in the lungs for a long period in this form. However it has now been established that other fibrous dusts lead to the identical type of body. They all have the characteristic of being visualized with an iron stain and thus have been given the name of ferruginous body. By means of electron diffraction and electron microscopic studies it was possible to show that ferruginous bodies obtained from human lungs did no' contain chrysotile, the most commonly used form of asbestos in the United States.[129–131]

IRRITANT GASES AND VAPORS

Sulfur Dioxide

ABSORPTION

Sulfur dioxide is considered to be a relatively soluble irritant gas and its rapid absorption on the moist surfaces of the pulmonary tree has been held responsible for its primary action on the upper airways. Recent experiments with radiolabeled sulfur dioxide have been able to compare the relative absorptive capacities of the mouth and nose of the dog when exposed to either 1 or 10 ppm sulfur dioxide. Nasal uptake of the incoming sulfur dioxide exceeded 99% at a flow of 3.5 liters/min and fell only a few per cent when the flow was

35 liters/min. In contrast uptake by the mouth averaged more than 95% at a flow of 3.5 liters/min, but fell to less than 50% when the flow was 35 liters/min. Thus a worker performing heavy labor with resulting high flow rates and obligatory mouth breathing is more likely to have an increased exposure of his lower airways to sulfur dioxide that is out of proportion to the associated increase in minute ventilation. It was interesting to note that some of the sulfur dioxide could be desorbed from the mucosa by flushing with ambient air. Thus the absorption of sulfur dioxide on the airways does not lead to an immediate irreversible reaction, but apparently leads in part to a physical solubility which is reversible.[132]

ENZYME INHIBITION

Concentrations of sulfur dioxide as low as 4 ppm bubbled for 1 hr through 60 ml of a solution containing bovine erythrocyte acetylcholinesterase caused a significant inhibition of enzyme activity. It was suggested that the bronchoconstriction induced by sulfur dioxide in animals and man might be mediated by the action of sulfur dioxide on acetylcholinesterase. Similar results were obtained with ozone at 0.3 ppm.[133]

LUNG DISEASE

It is generally believed that air pollution may play an important role in the pathogenesis of chronic obstructive pulmonary disease in man. Sulfur dioxide (650 ppm, 4 hr/day, 5 days/week for 19–74 days) did not produce an effect of significant importance on the development of papain-induced emphysema in Syrian hamsters. The hamster is remarkably tolerant to sulfur dioxide; in addition it was considered that, perhaps, sulfur dioxide must act in combination with other air pollutants in order for an adverse effect to become apparent.[134]

LONG TERM EFFECTS

Albino guinea pigs were exposed to sulfur dioxide at concentrations of 0.13, 1.01, and 5.72 ppm continuously for 12 months. All of the animals survived the exposure without adverse effect on their health as measured by body weight, growth, hematology, clinical chemistry, and pulmonary function tests. Microscopic examination of the liver revealed hepatocyte vacuolation in the guinea pigs exposed to 5.72 ppm, whereas the histopathologic examination of the lungs of these animals revealed a decrease in the incidence and severity of spontaneous pulmonary disease normally present in 1-year-old control guinea pigs. These findings are in contrast with increases in pulmonary flow resistance observed in guinea pigs and humans during short-term exposure. There appears to be the need for comparative studies with pulmonary function tests following both short and long-term exposure.[135]

LUNG FUNCTION

Miniature donkeys were exposed to concentrations of sulfur dioxide ranging from 26–713 ppm for a period of 30 min. Levels below 300 ppm did not produce impairment of bronchial clearance as measured by removal of ferric oxide particles tagged with ^{198}Au or ^{99m}Tc. At 300 ppm and above there was slowing in the rate of lower lung clearance and an abnormally large wave of activity in the upper lung with a lag in the onset of clearance. Although the donkeys exhibited signs of discomfort during exposures to all sulfur dioxide concentrations, impaired clearance occurred only when the sulfur dioxide produced severe coughing and mucus discharge from the nose.[136]

PARTICULATES AND LUNG FUNCTION

It has long been postulated that irritant gases which are primarily absorbed on the upper airways may have a greatly enhanced toxicity if they are absorbed on the surface of inert particles which are small enough to be deposited in the alveolar spaces. It has been reasoned that the particle enables the gas to reach more vulnerable lung surfaces and that it also concentrates the gas onto a small surface area. The synergistic action of sulfur dioxide and submicron particles of sodium chloride on the respiratory response by guinea pigs has provided sound support for the concept. However recent studies with humans exposed to low concentrations of sulfur dioxide and sulfur dioxide–aerosol mixtures have failed to demonstrate physiologic effects on measures of pulmonary mechanics. One of the ten healthy male volunteers was a possible "hyperreactor" to sulfur dioxide and it was suggested that in the future large numbers of normal subjects will need to be studied in order to locate individuals who might show effects of inhaled pollutants on pulmonary mechanics. It was also suggested that patients already affected by pulmonary disease might show positive results for irritant gas-aerosol synergism.[137]

Changes in pulmonary flow-resistance in unanesthetized guinea pigs were measured during exposures of various aerosols with and without the presence of sulfur dioxide. The aerosols included sodium chloride, potassium chloride, manganous chloride, ammonium thiocyanate, ferrous sulfate, sodium orthovanadate, activated carbon, spectrographic carbon, manganese dioxide, iron oxide fume, open hearth dust, fly ash, and triphenyl phosphate at concentrations ranging from 0.7–21 mg/m^3. The concentrations of sulfur dioxide ranged from 0.2–100 ppm. The soluble salts of manganese, ferrous iron, and vanadium, known to catalyze the oxidation of sulfur dioxide to sulfuric acid, potentiated the response to sulfur dioxide. In addition, sodium chloride, potassium chloride, and ammonium thiocyanate potentiated the response to sulfur dioxide in an order related to the solubility of sulfur dioxide in solutions of their salts. None of the insoluble aerosols were effective in potentiating the response to sulfur dioxide. It appeared that both solubility of sulfur dioxide in a droplet and oxidation to sulfuric acid play a major role in the observed potentiation of the response to sulfur dioxide by certain aerosols. It was suggested that the key to the role of sulfur dioxide in air pollution toxicology rested not with sulfur dioxide itself, but with its atmospheric chemistry.[138]

Laboratory rats were exposed to 1 ppm sulfur dioxide and graphite dust at a concentration of 1 mg/m^3 for 12 hr/day, 7 days per week for 4 months. It was not possible to show any gas-aerosol synergism as indicated by observations on body weight, hematocrit, histology of the respiratory structure, and the surface microflora at three sites of the respiratory system.[139]

PARTICULATES AND STANDARDS

The toxic effects of sulfur dioxide on plants, animals, and man have been extensively reviewed with particular reference to the setting of community air quality standards and their interactions with particulate matter.[140–146]

Nitrogen Dioxide

INTOXICATION

An industrial chemist was exposed to nitrous fumes while engaged in the process of reacting silver with nitric acid in a stainless steel retort. The patient experienced acute pulmonary edema with the classical, symptom-free lag of several hours intervening between exposure and onset. The patient was in the hospital for 7 days and returned to work 2 weeks later. On the first day back at work the patient experienced a recurrence which was similar in kind to the first, but more intense in degree, despite the fact that no further exposure to "nitrous fumes" had occurred. He was hospitalized for 28 days, but eventually made an apparent complete recovery. There was no evidence of bronchiolitis obliterans with permanent pulmonary disability which often develops with the typical biphasic pattern following nitrogen dioxide exposure. It was suggested that the corticosteroid therapy might have been responsible for the lack of pulmonary disability.[147]

LOW-LEVEL EFFECTS

Mice were exposed to nitrogen dioxide at 0.5 ppm for 6, 18 or 24 hr/day for up to 12 months. The general impression was of inflammation of the bronchioles with reduction of distal airway size and a concomitant expansion of alveoli. The number of expanded alveoli appeared to increase with exposure time.[148]

Rats exposed to nitrogen dioxide in air at concentrations of 0.8 or 2 ppm for 2 years showed some microscopic changes in the terminal bronchiolar epithelium, whereas levels of 10–25 ppm caused "emphysematous" lungs. It was pointed out that experiments with nitrogen dioxide alone may be misleading in that it is normally present with other materials, such as in tobacco smoke or automobile exhaust. It was considered that nitrogen dioxide may be one logical component in the complex etiology of emphysema.[149, 150]

PARTICULATES

Beagle dogs exposed to 26 ppm nitrogen dioxide plus iron oxide particles at a concentration of 0.9 mg/m^3 for 7 months showed a statistically significant higher average value for total pulmonary resistance. Diffusing capacity and pulmonary compliance were not significantly affected.[151]

Rats, hamsters, and guinea pigs were exposed to nitrogen dioxide at concentrations ranging from 22–74 ppm, for 2 hr/day, 5 days/week periods up to 1 year. Pneumoconiotic lungs which had been produced by intratracheal injections of quartz, coal, or blast furnace stack effluents were not aggravated by the nitrogen dioxide exposure. The pneumoconiosis did not have an aggravating effect on the naturally occurring emphysema of hamsters, nor on the papain-induced emphysema in rats, hamsters, or guinea pigs.[152]

VIRUSES AND BACTERIA

Germfree mice were exposed to 40 ppm nitrogen dioxide continuously for 6–8 weeks. They developed changes in the bronchial epithelium and showed resistance to inoculations of virus in a way similar to conventional (nongermfree) animals. It was suggested that the phenomenon of increased resistance to viral infection obviously needs further study.[153] On the other hand it has been well demonstrated that exposure to nitrogen dioxide has decreased the resistance to bacterial infections. Monkeys exposed to 50 ppm of nitrogen dioxide and challenged with *Klebsiella pneumoniae* died after the infectious challenge. Squirrel monkeys exposed to 10 or 35 ppm nitrogen dioxide did not die when challenged with bacterial infection, but did show a delayed clearance of the bacteria from the lungs.[154]

When mice were exposed to nitrogen dioxide continuously to 0.5 ppm for 3 months or longer, they demonstrated an increased susceptibility to airborne *Klebsiella pneumoniae.* Mice exposed intermittently for either 6 or 18 hr/day for 6 months also showed an increased mortality when challenged with viable bacteria. After 12 months exposure all of the groups of mice showed a reduced capacity to clear viable bacteria from the lung.[155]

It was also shown that squirrel monkeys were more susceptible to infection with influenza virus when challenged 24 hr before the onset of exposure to 10 ppm nitrogen dioxide.[156] It has been reported that nitrogen dioxide exposure inhibited the ability of rabbit alveolar macrophages to produce interferon and to develop resistance to viral infection. This effect was observed after the rabbits were exposed to 25 ppm of nitrogen dioxide for 3 hr.[157]

EMPHYSEMA MODEL

Rats exposed to 10 or 25 ppm of nitrogen dioxide for extended periods develop an enlarged thoracic cavity with the typical dorsal kyphosis which is characteristic of humans with emphysema of the lungs. It has been suggested that this may serve as a model for the study of emphysema in the experimental animal. Emphysema has also been produced in dogs after exposure to 25 ppm of nitrogen dioxide for 6 months.[158–160]

Ozone

LOW-LEVEL EXPOSURES

A group of ninety-nine male and female volunteer students at the University of Alberta were exposed to concentrations of ozone of 0.2–0.3 ppm for a period of 70 min. During the exposure the students were given psychologic tests to measure intelligence and anxiety. There was no noticeable effect on mental functioning during the exposure period.[161]

Very young mice ranging in age from 4 days to 8 weeks were exposed to ozone at concentrations from 0.6–1.3 ppm for 6–7 hr per day for 1 or 2 days. The 4-day-old mice were the most sensitive with the main target being the endothelial cells lining the small capillaries in the lungs.[162]

Rats, guinea pigs, monkeys, dogs, and mice were exposed to ozone continuously for 90 days at concentrations ranging from 0.26–3.0 mg/m^3. It was found that guinea pigs and rats were the most susceptible species, followed by squirrel monkeys; no dogs died in any of the exposures. The mice were used for measurement of spontaneous motor activity and a slow, progressive decline in activity was noted.[163]

PHAGOCYTE INHIBITION

Albino rabbits were exposed to ozone at a concentration of 5 ppm for 3 hr and then pulmonary cells were lavaged from the lungs at various time periods postexposure. This exposure produces an influx of heterophilic leukocytes into the lungs and a diminution of pulmonary alveolar macrophages. This phenomenon is evident immediately after 3 hr exposure, peaks about 6 hr later and is still present at a reduced level at 24 hr. In similar 3-hr exposures to lower concentrations of ozone, the rabbits were challenged with intratracheal injections of *Streptococcus* (Lancefield Group C) in order to test the effect of ozone exposure on phagocytic ability. The maximum depression in phagocytic activity was reached at a concentration of ozone of 4 ppm, and there was a slight decrease at the lowest concentration tested, 0.3 ppm. It was suggested that the decreased number of alveolar macrophages and depression in phagocytic activity may be the basis for the prolongation of intrapulmonary bacteria seen in mice following ozone exposure.[164]

The osmotic fragility of alveolar macrophages washed from the lungs of rabbits exposed to ozone was studied. Significant increases were found following single 3-hr exposures to 10 ppm or 7 daily 8-hr exposures to 2 ppm. Similar daily exposures to 0.5 ppm did not produce changes which were significantly different from controls.[165]

ENZYME INHIBITION

Mice were exposed to ozone at a concentration of 8 ppm for 4 hr and then examined for changes in the levels of acetylcholinesterase in their red blood cells. A marked reduction was found and since this enzyme is located solely on the red cell membrane and since there was no change in intracellular glutathione, it was suggested that the membrane is the most sensitive part of the red blood cell following ozone exposure. These findings support the conclusion that ozone has extrapulmonary effects.[166]

TUMORIGENESIS

Mice were exposed to 4.5 ppm of ozone for 2 hr every third day for 75 days. These almost lethal exposures produced edema, congestion, scattered hemorrhage, and early emphysematous changes. With prolonged exposure, inflammatory reactions ranging from acute bronchitis and bronchiolitis to bronchopneumonia resulted. Of the ninety-one ozone-exposed mice reaching autopsy, twenty-one (or 23%) exhibited cellular changes in the lungs of which six (or 6.6%) were pulmonary adenomas. In the control group of these tumor-resistant mice, only one of fifty-five (or 1.8%) demonstrated a mild hyperplasia of the bronchial epithelium of focal nature. It was stated that further elucidation of the relationship between ozone exposure and lung pathology is essential in order to understand the possible tumorigenic action of ozone on the general population.[167]

LUNG CELL CULTURES

Epithelial cells from human fetal lungs were cultured and exposed to 4 ppm ozone for 30 min. The ozone exposure caused an inhibition of cell proliferation. There were no microscopic changes observed following this exposure, but at higher concentrations (10, 15, and 20 ppm) it was believed that the formation of large blister-like vacuoles may be similar to changes observed following *in vivo* exposure to ozone.[168]

MECHANISM OF ACTION

There is evidence which implicates free radicals as the basic biochemical mechanism of ozone-induced cell damage. It has also been demonstrated that ozone concentrations present in urban areas can cause peroxidation of the lung lipids of laboratory mice. It was, therefore, speculated that photochemical air pollution might produce an acceleration of the aging process in the exposed human respiratory tract with relevance to chronic respiratory disease.[169]

Irritants – Lung Receptor Sites?

Deep-lung irritant gases, such as ozone, nitrogen dioxide, and phosgene produce pulmonary edema. It has been considered that these chemicals produce edema following a generalized contact with the alveolar surface dependent upon the individual conditions of exposure. It has been suggested that there may be specific receptor sites which when stimulated release neural or humoral impulses resulting in the spillage of fluid into the tissue and alveolus. However the microanatomic site of such potential edemagenic receptors has not been identified.[170]

Photochemical Smog – Lung Effects

A long-term study of laboratory animals exposed to the normally polluted atmosphere was conducted in Los Angeles County. Rabbits, guinea pigs, rats, and mice were maintained in ambient and filtered atmospheres in stations where air contaminants were monitored. The finding of pulmonary adenomas in mice was not always consistent, but in one experimental group of A/J mice there was a marked increase in the frequency of multiple tumors in mice 16 months and older when the ambient group was compared to controls. One group of C-57 mice tended to have an increased mortality during the first year of the lifespan of the male, but not the female, mice. Pulmonary function tests made on a routine basis were usually not significant, however tests made during severe episodes of photochemical smog indicated that the older guinea pigs had an increased pulmonary resistance of a substantial nature. Rabbits exposed for 2–3 years exhibited a reduced activity of serum glutamic oxalacetic transaminase, but not of serum alkaline phosphatase. Severe smog episodes caused ultrastructural alterations in the lung tissue of mice. It was found that continuous residence in the Los Angeles atmosphere conferred a detectable degree of resistance to the deleterious effect of synthetic photochemical smog. In the rats there was no discernible histologic effect upon lung tissue and no significant difference in life span or weight at death of rats living in ambient versus filtered atmosphere. However there was a greater overall incidence of advanced chronic nephritis at spontaneous death of the male rats exposed to ambient air.[171,172]

Auto Exhaust – Lung Effects

Beagle dogs were exposed for 18 months to one of the following atmospheres: control air, nonirradiated auto exhaust, irradiated auto exhaust, 0.5 ppm sulfur dioxide plus 100 $\mu g/m^3$ sulfuric acid mist, mixtures of the auto exhausts and sulfur dioxide atmospheres, 0.5–1.0 ppm nitrogen dioxide plus 0.2 ppm nitric oxide, and 1.5 to 2.0 ppm nitric oxide plus 0.2 ppm nitrogen dioxide. These exposures were not found to have any effect on the following pulmonary function tests: single-breath carbon monoxide diffusing capacity, dynamic pulmonary compliance, or total expiratory pulmonary resistance.[173]

PAN and Physical Activity

Mice were exposed to the photochemical air pollutant peroxyacetyl nitrate (PAN) at concentrations ranging from 2.8–8.6 ppm for a 6-hr period. All of the levels impaired the normal spontaneous activity of the mice as measured with a 6-inch revolving activity wheel. The concentration that would produce a 50% reduction in activity during 6 hr of exposure was calculated to be 4.4 ppm. This compares with a similar value of 0.23 ppm for ozone, 0.40 ppm for acrolein, and 16 ppm for nitrogen dioxide.[174]

Chlorine

Generally the bronchial and tracheal mucous membranes are so sensitive to chlorine that a man will not tolerate staying in concentrations which will cause alveolar edema. The exceptions to this generality are workers who become accustomed to the irritant effect and seem able to voluntarily reduce their respiratory tidal volume so as to tolerate concentrations which would drive another man out of the cloud of gas. Apparently such a boiler plant operator allowed himself to be in a chlorine cloud without a canister mask for about 30 min. He was given oxygen therapy and seemed to have recovered slightly, but died while being driven home by car. It was suggested that there needs to be particular alertness to the risk of pulmonary edema in exposure to irritant gases, especially in workers accustomed to minor degrees of exposure.[175]

Fluorine

Rats, mice, guinea pigs, rabbits, and dogs were exposed to various concentrations of fluorine gas for periods of 5, 15, 30, or 60 min. The LC_{50} values for these time periods were determined and were not significantly different between the four species, rat, mouse, guinea pig, and rabbit. Dyspnea, lethargy, red nose, and swollen eyes were observed at concentrations equivalent to one-half of the

LC_{50} values. At autopsy there was congestion, hemorrhage and atelectasis of the lungs. Exposure to concentrations at or below 100 ppm for 5 min, 70 ppm for 15 min, 55 ppm for 30 min, or 45 ppm for 60 min did not cause any apparent effects in the animals. Human volunteers experienced marked irritation of the eyes and nose at 100 ppm, but only slight irritation of the eyes at 25 ppm.[176]

Mice and rats were exposed to fluorine for 5, 15, 30, or 60 min at intervals ranging from 1 day to 1 week. Lungs, livers, and kidneys were the organs most affected. It was suggested that the repeated exposures might have provided some type of tolerance or protection as indicated by LC_{50} values and edema in the lungs.[177]

Other Irritant Vapors

The exposure of workers to toluene diisocyanate (TDI) in a factory producing polyurethane products was measured over an 18-month period. Periodic measurement of respiratory function of the workers was also made. Statistically significant decrements in ventilatory capacity were observed. Although the number of workers examined was relatively small, it was believed that the observations cast doubt on the safety of the current TLV. The concentrations of TDI as measured did not exceed 0.014 ppm on any occasion; the current TLV is 0.02 ppm as a ceiling limit.[178]

Certain alkyl 2-cyanoacrylate monomers are used as adhesives in many industrial operations. Fourteen individuals (five women and nine men) were exposed at a simulated workbench to vapor levels ranging from about 1–60 ppm. Odor was detected between 1 and 3 ppm, nose irritation was initiated near 3 ppm, and eye irritation at 5 ppm. Lacrimation and rhinorrhea occurred at concentrations of 20 ppm or higher. Concentrations of 50 and 60 ppm caused marked eye and nose irritation and resulted in a delayed, transient visual effect in two subjects.[179]

Animal experiments with methacrylonitrile indicated that the compound was highly hazardous when ingested, in contact with the skin, or when inhaled. The presence of cyanide ion in the blood and the effectiveness of the standard nitrite–thiosulfate therapy for cyanide intoxication suggested that the presence of cyanide in the blood may be a useful indicator of overexposure to methacrylonitrile vapors. It was noted that there was a delayed response to exposure which suggested that exposed humans must be observed carefully for a considerable time after exposure. Rats and dogs were exposed to concentrations ranging from 3.2–109.3 ppm, for periods of 7 hr/day, 5 days/week for a total of 91 days. The "no ill-effect" level of methacrylonitrile vapor for 91 days appeared to lie between 8.8 and 3.2 ppm for dogs. Human sensory response studies showed that 24 ppm for 1 min was irritating to four of eighteen subjects and 2 ppm for 10 min caused irritating effects which were of a transitory nature. All of the human experiments indicated that the vapor of methacrylonitrile has very poor warning properties. On the basis of these experiments it was recommended that humans should not be allowed to inhale more than 3 ppm for 8 hr/day, 5 days/week.[180]

Oxidants and Air-Quality Standards

The health effects of oxidants as they apply to community air quality standards with special emphasis on photochemical oxidants, ozone, and nitrogen dioxide have been extensively reviewed.[181-185]

Polymer Fume Fever

Three workers experienced the typical flu-like chills, mild chest pains, coughing, and fatigue related to polymer fume fever. The exposure resulted from polytetrafluoroethylene in the insulation of cables which were too close to an electric furnace. It was noted that smokers are predisposed to polymer fume fever and that smoking a cigarette after significant exposure to environmental polymer fume tends to precipitate an attack.[186]

Pyrolysis Products from Plastics

The toxicity of the pyrolysis products of plastics presents special problems in the description of the exposure system, the decomposition processes, as well as the toxic effects themselves. One of the most useful procedures is simply to give the weight and size of the plastic used and a careful description of the conditions surrounding the pyrolysis. However results may vary markedly with such conditions. It is necessary to know whether or not the plastic burned or only charred and whether excess oxygen was present. In addition it must be recognized that the conditions of exposure vary during the process of decomposition as opposed to the more classical inhalation experiment in which the concentration of air contaminants is maintained at a constant level.[187]

Plastic formulations containing polyvinyl chloride were pyrolyzed at approximately 600°C and the pyrolysis air stream was diluted with air or oxygen prior to exposure of rats. At high concentrations death was due to carbon monoxide. When sufficient oxygen was added to prevent deaths due to carbon monoxide then the pulmonary edema and interstitial hemorrhage due to hydrogen chloride and particulates became evident.[188]

Rats were exposed to the pyrolysis products of polychlorotrifluoroethylene (unspecified source) at varying concentrations for periods of 1 and 3 hr. There was only minimal mortality after 1-hr exposures at 375°C, but high mortality after 3-hr exposures at 375°C, or 1-hr exposures at 400°C. The mortality could not be accounted for by the measurable quantities of hydrolyzable fluorides and it was suggested that the particulate material may have accounted for the toxic action.[189]

Rats were exposed to the pyrolysis products from polytetrafluoroethylene (PTFE, sold under the trade name of Teflon 5 TFE-fluorocarbon resin). Small amounts of hydrolyzable fluorides were evolved at 400°C, but no mortality occurred until a temperature of 450°C was reached. This pyrolyzate contained hydrolyzable fluoride, tetrafluoroethylene, and hexafluoropropylene, but the component responsible for the toxicity was identified as a particulate material which may have had other toxicants adsorbed on it. When the temperature was raised to 480°C, the most toxic ingredient might have been octafluoroisobutylene. It was suggested that studies of the pyrolysis products of such compounds should specify completely the trade name of the specific product, since it was found that there were toxicologic differences between different types of polytetrafluoroethylene.[190] In a study with PTFE (unspecified source) at a temperature of 550°C, it was found that the principal toxic component was carbonyl fluoride. When the pyrolysis took place in a glass combustion tube, some silicon tetrafluoride was identified and appeared to contribute to the observed toxic effects.[191,192] Rats were exposed to products of the pyrolysis of PTFE at a concentration containing hydrolyzable fluoride equivalent to 50 ppm carbonyl fluoride for 1 hr/day for 5 days. The toxic effects were found to be cumulative and compatible with the descriptions of fluoride poisoning. There was an excellent correlation between the urinary fluoride and the inhibition of activity of succinic dehydrogenase in the kidney; this inhibition was reversible and had returned to normal by the eighteenth day after exposure.[193]

PESTICIDES

Intoxications

Malathion is one of the least toxic of the organophosphate insecticides, nevertheless poisonings continue to occur when sufficient Malathion is absorbed. A 63-year-old man intentionally ingested 60–90 gm of Malathion and died after 6 days in spite of hospitalization. In addition to the classical effects of enzyme inhibition, the patient developed cardiovascular changes which have only rarely been observed in man.[194]

A 4-year-old girl ingested an unknown quantity of 45% chlordane emulsifiable concentrate resulting in intermittent clonic convulsions, coordination loss, and increased excitability. The use of gastric lavage and parenteral phenobarbital was followed by disappearance of the symptoms and apparent complete recovery of health. The earliest serum chlordane concentration was 3.4 ppm and at 130 days postexposure the level was 0.138 ppm; the biologic half-life for chlordane in serum was estimated to be 88 days.[195]

The accidental ingestion of a nonlethal dose of technical chlordane by a 2-year-old child was reported. Fat samples taken from the buttock by needle biopsy showed that the concentration of chlordane in fat was still rising 1 week after ingestion. It was estimated that the half-life of chlordane in serum was 21 days, however this estimate is only based on two measurements, one at 8 days and one at 94 days after ingestion.[196]

Paraquat and diquat are dipyridilium weed killers which have created particularly serious problems because of their delayed action. They have been of special interest to toxicologists because of their somewhat unique ability to produce pulmonary edema following ingestion or injection. At least one case has been reported in which permanent damage of the cornea of the eye occurred.[197–203]

Carbofuran (trade name Furadan, Niagara Chemical Division of FMC Corporation) is a new carbamate insecticide with the chemical name of 2,3-dihydro-2,2-dimethyl-7-benzofuranylmethyl carbamate. It exerts its toxic action on the enzyme acetylcholinesterase directly without requiring metabolic alteration. It is in the class of highly toxic chemicals having an oral LD_{50} of about 11 mg/kg. Inhalation studies with rhesus monkeys led to the suggestion of a TLV of 0.25 mg/m^3. Several case histories have been presented illustrating the effect of carbofuran on the eye in addition to systemic intoxication.[204]

Human Exposures

Plastic strips releasing the insecticide Dichlorvos (DDVP) were used in hospital wards in a proportion of one strip to every 30 m^3 of room volume. The patients in the hospital wards included adults with liver disease, silicosis, acute bronchitis, lead poisoning, chronic bronchitis, and other diseases; women during labor and puerperium; and children with various illnesses. The airborne concentrations of DDVP varied from 0.04–0.28 mg/m^3, which are concentrations sufficient to kill fruit flies in 10–200 min. Patients exposed to concentrations above 0.1 mg/m^3 for several days showed a decrease in plasma cholinesterase ranging from 35–72%. Patients with liver diseases were more affected than patients with normal liver function. The red blood cell cholinesterase was unaffected in all patients. Since it is generally found that symptoms of intoxication by organophosphate insecticides occur when the red blood cell cholinesterase is reduced to 50% and plasma cholinesterase of 10% of preexposure levels, it was concluded that the use of strips containing DDVP for disinsection in hospitals seems to be clinically harmless.[205]

Twelve male adults received capsules containing Dieldrin daily for 18–24 months and then were observed over an 8-month postexposure period. The subjects remained in excellent health throughout the exposure. It was possible to establish significant relationships between the concentrations of Dieldrin in the blood and adipose tissue and the amount of Dieldrin ingested daily. There was a considerable variation in the biologic half-life among the subjects as indicated by the concentration of Dieldrin in the blood during the postexposure period and these observations were being continued in order to obtain a more precise estimate of the half-life of Dieldrin in man.[206]

A group of seventy-one men from a factory manufacturing pesticides participated in a study which included measurement of pesticides in plasma, fat, and urine. Eleven isomers or metabolites of six insecticides (DDT, DDD, BHC, Dieldrin, aldrin, and heptachlor) were detected in plasma, but not all were found in every sample. In fact, *p,p'*-DDE, the most significant metabolite of *p,p'*-DDT in humans, was the only compound found in every plasma, fat, and urine sample. The average concentration of Dieldrin in plasma was over nine times greater than that for the general population, while Dieldrin in the fat and urine was 19 and 30 times greater, respectively. No endrin was found in any sample although there had been some exposure to it. There was no significant relationship between the use of sick leave and the amount of Dieldrin in samples. Dieldrin in plasma was considered to be the best convenient way to monitor exposure in order to prevent excessive absorption. More information is necessary before urinary levels of Dieldrin can be relied upon for monitoring purposes.[207]

Synergists

It was found that pyrethrins, especially in combination with the synergist piperonyl butoxide, produced lesions in the liver of rats which were similar in character to changes caused by DDT. When DDT and pyrethrins were administered in combination, there were more than additive effects on liver pathology. Since pyrethrin aerosols have often been used for insect control in animal colonies which have been used to study the effects of DDT and other chlorinated hydrocarbon pesticides, it is possible that exposure of the animals to synergized pyrethrins may have influenced the extent of liver cell changes attributed to the compound under study. The same factor may also account for variable results in such studies.[208]

Insecticides and Their Potentiation

One of the relatively rare examples of a potentiation of toxic action caused by a mixture of chemicals has been that observed with various combinations of cholinesterase-inhibiting insecticides. This discovery led to the need to test new insecticides in combination with many others in order to establish safety of use. The theoretical combinations for such testing and the time required to test for potentiation of mortality and cholinesterase inhibition can be monumental. It has, therefore, been immensely satisfying to find a biochemical mechanism for a toxic interaction which not only can provide an understanding of the potentiated toxic action, but also offers the basis for biochemical assays which can predict potentiation without extensive and costly animal experiments. The mechanism of toxic action of organophosphate insecticides is the inhibition of acetylcholinesterase activity of nerve tissue. In addition several organophosphorus insecticides also inhibit tissue carboxyesterases which hydrolyze aliphatic esters. These compounds thus have the capability to potentiate Malathion which is detoxified by a carboxyesterase in mammalian tissue. Insecticides were administered to rats by intraperitoneal injection or by mixing with their diet. It was found that in order for a compound to potentiate other anticholinesterases, it must be more potent as a carboxyesterase inhibitor than a cholinesterase inhibitor and carboxyesterase activity must be the critical factor that controls the rate and extent of detoxication of the other insecticides.[209,210]

Dietary Stresses

The herbicide Chlorpropham was administered to rats which had been maintained on diets with varying quantities of protein. The oral LD_{50} was 10.4 gm/kg in rats on a normal (26%) protein diet, but was 1.2 gm/kg in rats previously fed no protein. It was suggested that these results may be due to the inhibition of formation of hepatic enzymes concerned with detoxification of the herbicide.[211]

Lindane (γ-benzene hexachloride) was given per-orally to male rats and the clinical signs and histopathologic findings were described in detail. The LD_{50} was determined to be 184 mg/kg when the rats were fed a synthetic diet with normal protein content, but was 95 mg/kg when the rats were fed a low-protein diet. The results suggested the possibility that lindane should be used as an insecticide with a greater caution in countries where the diet is deficient in protein.[212]

Rats were given subclinical doses of Parathion with simultaneous dietary stresses with low calcium and high fat. The high-fat diet itself caused an increase in red blood cell cholinesterase as compared with the low calcium diet alone. When parathion was administered the cholinesterase activities in rats on both diets dropped at the same rate. It was suggested that the influence of diet and other stresses on the reaction of experimental animals to organophosphorus pesticides needs further work.[213]

Reproduction

Mice were fed DDT at a concentration of 2.8–3.0 ppm (equivalent to 0.4–0.7 mg/kg/day) throughout six generations. There was no effect on reproductive performance, but an increased incidence of leukemia and malignant tumors was observed in the third and subsequent generations.[214]

When parathion, Diazinon, Tepa, apholate, Dichlorvos, and Malathion were administered intraperitoneally at doses producing toxicity in the dams, all but Malathion affected the fetuses by causing malformations, resorptions, or reducing the weight of the fetus and placenta. When the above compounds were administered in doses which were tolerated by the dams, the only effects seen were a greater number of resorptions with parathion, Diazinon, and Tepa. Metepa was used as a positive control in this study since it is highly teratogenic in Sherman strain rats.[215] Both Tepa and Metepa were found to produce mutagenic effects in mice.[216]

Carcinogenicity

Perhaps one of the most controversial reports of recent years has been a study of the tumorigenicity of 104 pesticides and 19 industrial chemicals which has become known popularly as the Bionetics study. The controversy has raged not only over the content of the report, but also over its introduction to the scientific community. The controversy led to initial publication of the report in the *Congressional Record–Senate* (May 1, 1969) prior to publication in a scientific journal. The report was a central issue in the Mrak Committee report to the Secretary of Health, Education, and Welfare.[217–219]

Dieldrin in Blood and Toxic Signs

Adult mongrel dogs of both sexes were given capsules containing Dieldrin once each day for periods up to 150 days. A good correlation was found between the level of Dieldrin in the blood and the clinical signs of intoxication. At concentrations averaging 0.37–0.39 ppm Dieldrin in whole blood there was a decreased food consumption. A significant decrease in body weight was seen at blood levels of 0.38–0.50 ppm. The first muscular spasms were seen at blood levels of 0.51–0.58 ppm, and the concentrations of 0.74–0.84 ppm occurred at the time of the first observed convulsions.[220]

Mirex Toxicity

The control of fire ants in the southeastern United States has involved use of the pesticide Mirex, which is dodecachlorooctahydro-1,3,4-metheno-2*H*-cyclobuta(cd)pentalene. Mirex was found to have a low order of acute toxicity, but a high chronicity factor compared to other chlorinated hydrocarbon pesticides. Effects on the liver included increased liver weight, enlarged liver cells, bile stasis, and an increase in smooth endoplasmic reticulum and formation of atypical lysosomes. Except for the bile stasis, the effects in the liver raised again the question of whether such changes are evidence of liver damage or rather the liver's normal adaptive change to the presence of a foreign chemical. It was found that Mirex passes through the placenta and is excreted in milk. Cataracts developed in pups, predominantly during the suckling period.[221]

HYDROCARBONS

Benzene

INTOXICATION

An employee of a benzol plant in a steel mill died when he went to attend an overflowing tank of "light oil." The light oil contained 67.7% benzene, 5.7% xylene, and 14.5% toluene. The employee had run to the area while a second employee who arrived at the scene without undue exertion was not affected, but found the first employee within 2 min, on the ground, apparently lifeless. Benzene was found in the blood, liver, and brain, but not at exceptionally high levels, and it was believed that the death was not due to anesthesia, the usual cause of acute benzene-induced mortality. It was suggested that the death may have been due to adrenalin and sympathin effect on the benzene-sensitized myocardium.[222]

CHROMOSOMAL ABERRATIONS

The lymphocytes from the blood of workers exposed to benzene were cultured for chromosome studies. Two of three populations of workers have shown increases in the number of cells with unstable chromosome aberrations.

There appeared to be some question as to the suitability of the control populations and it was suggested that environmental factors other than age might explain the results.[223]

Human Volunteer Exposures

Human volunteers were exposed for 8 hr to vapors of ethyl benzene, styrene, or α-methylstyrene. The concentrations of ethyl benzene ranged from 23–85 ppm, for styrene was 22 ppm, and for α-methylstyrene was not reported. No health damage was observed due to these exposures which are all below current TLV levels. With ethyl benzene it was found that 64% of that inhaled was retained by the respiratory tract and only traces could be found in the expired air. The main urinary metabolites were mandelic acid (64%) and phenylglyoxylic acid (25%) and methylphenylcarbinol (5%). With styrene it was found that 61% of that inhaled was retained by the respiratory tract and none was found in the expired air by the method used. Other studies[224] have shown that styrene can only be found in the expired air for a few hours after exposure. The main urinary metabolites were mandelic acid (85%) and phenylglyoxylic acid (10%). It was suggested that mandelic acid in urine may be used as an index of exposure to both ethyl benzene and styrene and the urinary concentrations recommended as biologic limits were 2000 and 1500 mg/liter (or 1.5 and 1 mg/mg creatinine) for ethyl benzene and styrene, respectively. It was found that atrolactic acid, the methyl derivative of mandelic acid, was the main urinary metabolite following inhalation of α-methylstyrene. However the data were too limited to propose a biologic threshold limit.[225]

Additional experimental human exposures have been conducted with styrene[224] and propylene glycol monomethyl ether.[226] Exposure to styrene vapors at a concentration of 376 ppm for a period of 1 hr caused unpleasant subjective symptoms and definite objective signs of impairment of neurologic function in the majority of subjects. Exposure to 200 ppm for 1 hr, 100 ppm for 2–7 hr and 50 ppm for 1 hr did not cause any subjective symptoms; odor was readily detected. Styrene could be measured in the expired air for a few hours after exposure. The measurement of hippuric acid in the urine was not a sensitive indicator of exposure.

Exposures to vapors of the monomethyl ether of propylene glycol at concentrations ranging from 50–250 ppm for periods up to 7 hr caused eye, nasal, and throat irritation. The ability to detect the very objectionable odor at 100 ppm was lost after 3 hr of exposure. Propylene glycol monomethyl ether is excreted in the breath for such a brief period that unless a breath sample can be obtained within 10–20 min following exposure, too little will be present for detection by available analytic techniques.

Polyphenyls

Rats were fed a mixture of polyphenyls in their daily diet for a period of 235 days. The mixture was given the trade name of Santowax OM by the Monsanto Chemical Corporation; it consisted or 4.7% biphenyl, 64.1% *o*-terphenyl, 25.1% *m*-terphenyl, and 6.1% *p*-terphenyl. The ingestion of 3 mg/kg/day produced no micropathologic changes; 31 mg/kg/day caused the accumulation of granules in the kidneys; and 350 mg/kg/day led to a pronounced degree of renal disease. In contrast a similar study with an irradiated moderator and coolant from a reactor showed only minimal renal injury, but marked liver changes. However this mixture consisted of less than 1% bi- and terphenyls; over 98% of the mixture was high-boiling materials consisting of triphenylenes, phenanthrenes, and alkylation, phenylation, and hydrogenation products of these and polyphenyl compounds. The radioactive components were ^{60}Co, ^{54}Mn, and ^{137}Cs.[227]

Mice were exposed to a terphenyl aerosol for several hours daily up to 8 days to simulate exposure to the concentration of the oil coolant at a reactor site. There was no evidence of lung injury seen by light microscope. With the electron microscope it was observed that many type 2 alveolar cells showed central vacuolation of mitochondria. This change was no longer evident after 6-weeks postexposure. The reactor coolant consisted of a mixture of *ortho-*, *meta-*, and *para*-terphenyls, with a small concentration of higher polyphenyls.[228]

Hydrocarbons – Toxicodynamics

One of the important principles of toxicology is that a chemical must be absorbed into the body and be transported to a site of toxic action where its effect will be proportional to its concentration at that site. In general it is not possible to measure exposure at the site of toxic action even when we know where that site is located. A series of hydrocarbons were studied in mice and rats to correlate the concentration of hydrocarbon in the brain with lethality after 2–4 hr of inhalation exposure. The sensitivity of gas–liquid chromatography for hydrocarbon analysis allowed such studies on small animals as well as an investigation of the distribution of butadiene and isobutylene in various sections of the central nervous system of the cat.[229]

HALOGENATED HYDROCARBONS

Accidental Poisonings

A female worker who operated a machine which removed paint from metal parts by washing in trichloroethylene was exposed to airborne trichloroethylene concentrations estimated to be 260–280 ppm. On examination she showed signs of narcosis and complained of nausea and feeling "drunk." She also had electrocardiographic abnormalities and an elevated level of urinary trichloroacetic acid.[230]

A chemist working with tetrabromoethane (acetylene tetrabromide) for 7.5 hr without benefit of local exhaust ventilation was hospitalized with near fatal liver injury. It was estimated that the exposure for most of this period was 1–2 ppm and for a single 10-min period was as high as 16 ppm. There were no factors which apparently could have caused the illness other than the exposure to tetrabromoethane and it was therefore suggested that man's reaction to this chemical is widely different from laboratory animals. The current TLV for tetrabromoethane is 1 ppm and it was recommended that until further information becomes available, workers exposed to concentrations approaching this value should be kept under medical surveillance.[231]

A radiator and metal tank repairman received a fatal dose of methyl chloroform while cleaning a metal tank with this solvent. Inadequate ventilation was used and a reconstruction of the operation indicated that concentrations may have been approximately 50,000 ppm. The worker, found unconscious, was not revived by artificial respiration. His blood contained 6 mg of methyl chloroform per 100 ml, and was negative for alcohol, barbiturate, and carbon monoxide.[232]

Ten case histories of workers exposed to methyl bromide used for fumigation, including four fatal cases, have been reviewed. The early symptoms were malaise, headache, visual disturbance, nausea, and vomiting. Pulmonary changes were those found with acute chemical pulmonary edema. Neurologic findings in fatal poisonings were principally clonic and tonic convulsions. Psychiatric disability was a problem in persons with prolonged disability following exposure. It was noted that the absolute mortality among persons exposed to methyl bromide is greater than for any other agricultural chemical used in the United States.[233]

Chlorinated Hydrocarbons – Breath Data

The interpretation of concentrations of chlorinated hydrocarbons in expired air involves four variables: the concentration of the exposure, the length of the exposure, the concentration in the breath after exposure, and the period of time after exposure. The visualization of the interrelationships of these variables requires an extensive series of elimination curves. A limited amount of such information has become available in recent years. In addition attempts have been made to show the relationship between such curves as obtained on animals and humans. It was felt that animal breath data could be reliably interpreted for human purposes.[234]

Human Volunteer Exposures

Experimental human exposures have been conducted with tetrachloroethylene,[235] trichloroethylene,[236] and methyl chloroform.[237] Exposures to tetrachloroethylene vapors at a concentration of 100 ppm for periods of 7 hr to a 5-day work week caused untoward subjective responses in 25% of the subjects and a small percentage exhibited early signs of central nervous system depression. These results indicated that there exists a range of individual susceptibility to this type of compound. Exposures to trichloroethylene vapor at concentrations of 100 and 200 ppm for periods of 1 hr to a 5-day work week produced only mild, inconsistent, untoward subjective responses at 200 ppm. The only troublesome response was a sensation of mild fatigue and sleepiness experienced by one-half of the subjects during their fourth and fifth consecutive days of exposure at 200 ppm. Measurements of the urinary metabolites, trichloroacetic acid and trichloroethanol, did not correlate well with exposure.

Exposure to methyl chloroform vapors at a concentration of 500 ppm for periods of 6.5–7 hr/day for 5 days caused subjective untoward responses which were mild, inconsistently present, and of doubtful clinical significance. The only adverse objective response was an abnormal modified Romberg test by two of the eleven subjects. This test requires the subject to balance on one foot with his eyes closed and his arms at his side. The two subjects had demonstrated difficulty performing this test before any exposure, but within 5–10 minutes after cessation of exposure they were able to perform a normal test. No abnormalities of organ function were revealed by clinical laboratory tests. Analyses of the breath of the subjects provided a means to establish a diagnosis of exposure.

Since trichloroethylene is a central nervous system depressant, it was hypothesized that low concentrations for short periods of exposure might significantly affect visual and motor skills and thus influence work performance. Eight male volunteers between the ages of 21 and 30 were given a battery of six standardized tests during a 2-hr exposure to either 0, 100, 300, or 1000 ppm trichloroethylene. There were no significant effects at the lower concentrations. At 1000 ppm there were adverse effects on performance in the depth perception, steadiness, and pegboard tests. There was no effect on flicker fusion, form perception, or code substitution tests. It was suggested that the standard of a peak concentration of 300 ppm for not more than 5 min in any 2-hr period offers a considerable safety factor with regard to the single criterion of "undesirable functional reactions that may have no discernible effects on health."[238]

Ten workmen exposed to vinyl chloride participated in on-the-job breath sampling conducted in association with careful environmental surveys of vinyl chloride concentrations. The data were evaluated based on elimination curves obtained from experimental human exposures. The results indicated that either continuous air monitoring or breath analysis is valid for estimating the worker's individual daily exposure to vinyl chloride. The experimental human exposures were to concentrations of either 50, 250, or 500 ppm as a time-weighted average for a 7.5-hr period including a 0.5 hr lunch period in an uncontaminated area. None of the men exposed could detect any odor at 50 ppm and a very slight odor was detectable by all four subjects entering the chamber at 250 ppm. The exposures had no noticeable effect on neurologic responses and there were no significant changes in psychomotor tests or clinical laboratory studies.[239]

Polychlorinated Biphenyls

The occupational hazards of the polychlorinated biphenyls have not given rise to serious problems, but the finding of widespread environmental contamination has been a bombshell. Which of the many industrial applications have contributed to this contamination is still unknown. In particular the problem has been an analyst's nightmare. Not only has there been confusion with DDT, but the large and variable numbers of polychlorinated biphenyls makes quantitation difficult to express. Three commercial preparations, Phenochlor DP6 (from Prodelec in France), Clophen A60 (from Bayer in Germany), and Aroclor 1260 (from Monsanto in the United States) were found to differ in their oral toxicity when administered at a dose of 400 ppm for 60 days to chickens.[240] It was found that the two most toxic preparations contained chlorinated dibenzofurans and chlorinated naphthalenes. The chlorinated dibenzofurans are highly toxic and are believed to have determined the toxicity of the commercial polychlorinated biphenyl preparations.[241]

Contamination of rice bran oil by leaking of chlorobiphenyls caused poisoning in 600 patients in Japan.[242] An experimental study of the long-term ingestion of a mixture of chlorinated biphenyls containing 48% chlorine was conducted in mice and monkeys. After 26 weeks of exposure the significant findings were enlargement of the liver and fatty degeneration of liver cells in light microscopy. Changes observed by electron microscopy consisted of an increase in the smooth endoplasmic reticulum and a reduction in the rough endoplasmic reticulum of liver cells. It was suggested that the proliferation of smooth endoplasmic reticulum may reflect adaptive phenomena for enhanced drug detoxification.[243]

Bromochloromethane – Pyrolysis Products

Bromotrifluoromethane and bromochloromethane are used for aircraft fires. A study of the toxicity of their pyrolysis products was made. The pyrolysis was conducted at 800°C in a hydrogen–oxygen flame and the identified products were hydrogen fluoride, hydrogen chloride, hydrogen bromide, and bromine; no phosgene or chlorine was detected. The pyrolysis products of bromochloromethane were more toxic than those of bromotrifluoromethane.[244]

Vinyl Bromide

Male rats were exposed to vinyl bromide at a concentration of 10,000 ppm for 7 hr/day, 5 days/week for 4 weeks. The exposure resulted in decreased body weight and physical activity. However there were no remarkable abnormal changes when rats, rabbits, and monkeys were exposed to either 250 or 500 ppm for 6 hr/day, 5 days/week for 6 months. It was suggested that the time-weighted average TLV should be 250 ppm with a ceiling limit of 500 ppm. However no official action has as yet been taken for setting a TLV for vinyl bromide.[245]

Trichloroethylene

Exposure of rats to 2500 or 3000 ppm of trichloroethylene for 30 min caused a depression in the rate of lever-pressing which produced stimulation of the hypothalamic area of the brain. Deprivation of water for 3 days did not markedly increase susceptibility to trichloroethylene. Previous studies had shown that deprivation of water had increased the susceptibility of rats and mice to the acute toxic effects of lead and antimony. It was suggested that the difference in results may be due to the difference in water–lipid solubility of these materials, or that the brain may be more adequately protected against dehydration than the tissues affected by lead and antimony.[246]

When the trichloroethylene exposures were given in conjunction with an antihistamine (Anahist) or a sedative (Equanil) there was no significant alteration in the results of the tests due to the drugs. However the combination of trichloroethylene and alcohol (bonded, bourbon whiskey) markedly augmented the adverse effects of both 300 and 1000 ppm trichloroethylene.[247]

CARBON MONOXIDE

Carbon Monoxide and Psychomotor Tests

The studies of the effect of carbon monoxide on humans as measured by psychologic tests have highlighted discussions of the basis for setting environmental standards. A major distinction is made between concentrations which will not cause structural damage to body tissues versus concentrations which will not produce a decrement in performance. It has been suggested that a logical approach would be to have multiple TLV's which could be geared to the level of alertness and neuromuscular function needed by a specific work situation. Continued research is needed to establish the relevance of neurophysiologic data to TLV setting.[248]

A group of eighteen healthy male volunteers were exposed to concentrations of carbon monoxide ranging from less than one to 1000 ppm for periods of ½–24 hours. The tests for effect of the exposures were those felt to be of practical significance in the performance of vocational endeavors and of automobile driving where significant impairment of visual or auditory acuity, coordination, reaction time, manual dexterity, or time estimation would be intolerable. It was found that an 8-hr exposure to 100 ppm or less produced no impairment of performance in the tests. In these experiments the carboxyhemoglobin saturation values went from preexposure levels of less than 1.5% up to 11–13%. Higher concentrations of carbon monoxide which produced carboxyhemoglobin levels of 15%–20%, resulted in delayed headaches, changes in the visual evoked response, and impairment of manual coordination.[249]

Workroom Standards

The Subcommittee on International Threshold Limits established by the Permanent Commission and International Association on Occupational Health recommended a TLV of 50 ppm for carbon monoxide. This value was recommended on the basis that such exposures will produce carboxyhemoglobin saturation values of about 8–10% in nonsmokers, and that well-defined objective symptoms may occur at 13% (vascular permeability) and 15% (enhancement of cholesterol uptake in the intima of blood vessels in rabbits). The committee felt that it would not be practical to include subjective symptoms which may be caused or greatly influenced by fatigue or be caused by conditions outside the workplace alone, e.g., tobacco smoking. Thus the finding of changes in flicker fusion frequency, time perception, and various psychomotor tasks at carboxyhemoglobin saturation levels of 2–5% were not utilized. It was considered that peak values, representing repeatable events during the work day, are not justified in the industrial situation. Emergency values of 400 ppm for 1 hr and 1000 ppm for 20 min were suggested, but these values should not apply in (*a*) heavy work, (*b*) work requiring precision, and (*c*) normal working conditions, and should thus occur as a rare event in an emergency or accidental situation.[250–253]

Air Quality Standards

The physiopathology of carbon monoxide as it relates to community air quality standards has been extensively reviewed.[254-256]

Interaction with Metals?

Rats were exposed to carbon monoxide at a concentration of 50 ppm, 5 hr/day, 5 days/week for periods up to 12 weeks. The livers of the rats were fractionated to isolate nuclear fragments, mitochondria, and the supernatant. Zinc, copper, cobalt, iron, and magnesium were analyzed by atomic absorption spectrophotometry in each fraction. Changes in concentrations of metals which were statistically significant were observed and it was considered that carbon monoxide at its TLV level exerted a trace metal effect. However the direction of the observed changes were quite variable, as were the changes with time of exposure, and it appears that further investigation is necessary.[257]

CHEMICAL CARCINOGENS

Pitch

A group of 144 workers in a fuel plant engaged in the fusion of coal dust and pitch with steam heat were examined for skin lesions and compared with 263 controls from a dermatologic clinic. The lesions in the pitch workers and controls were squamous keratoses (12 and 10%, respectively), chronic tar dermatosis (5 and 0%), squamous cell carcinoma (2.8 and 0.4%), pitch warts (10.4 and 2.7%), and acneiform lesions (93 and 31%). There were scrotal lesions in 13.5% of the pitch workers. It was indicated that pitch workers are at special risk and special precautions must be taken.[258]

Bladder Carcinogens

One of the most serious and striking toxicity problems in industry has been the bladder carcinogenesis of certain amino and nitro aromatic compounds. It has been observed that some patients with spontaneous bladder cancer excrete certain urinary metabolites of the essential amino acid tryptophan. Some of these metabolites are structurally similar to known exogenous human bladder carcinogens and in experimental animals they have produced cancers when surgically implanted into the bladder lumen. It was hypothesized that metabolites of tryptophan may be incomplete inducers of cancer requiring the presence of a vehicle, such as cholesterol, in the bladder lumen for the expression of bladder carcinogenicity. A clarification of the genesis of "spontaneous" bladder cancer would be a major assist in the investigation of "industrial" bladder cancer.[259]

Dimethylnitrosoamine

N,N-Dimethylnitrosoamine is a liver carcinogen that produces destruction of the endothelium of the central and sublobular veins in hepatic tissue. Since exposure to this compound in industrial and military applications may be accompanied by exposure to hydrazine sulfate, a study was conducted of their joint toxic action following large single doses to mice. It was found that no synergistic or antagonistic effect occurred with coincident exposure of mice to both dimethylnitrosoamine and hydrazine sulfate.[260]

Printing Inks

Mice were exposed to twenty-two printing inks by 15–21 weekly subcutaneous injections in a study of carcinogenic activity. With seventeen of the inks there was no observation of carcinogenic activity, but with the other five inks there were sarcomas at the site of injection in one of twenty mice for each ink. The five inks which produced local tumors were to be examined in more detail.[261]

MISCELLANEOUS COMPOUNDS OF SPECIAL INTEREST

Vinyl Acetate

A group of male workers who had an average length of service of 15.2 years in a vinyl acetate complex were studied. The average long-term exposure was estimated to range from 5–10 ppm with intermittent exposures near 50 ppm and potential acute exposures to 300 ppm. Clinical and medical studies revealed no evidence to suggest chronic effects at the determined long-term levels and there were no serious residual injury to acute exposure if treated promptly. Some individuals could detect the odor of vinyl acetate at levels of 0.4 ppm. No eye or upper respiratory tract irritation was evident at levels below 10 ppm. Levels of 21.6 ppm did produce eye and throat irritation in the majority of five subjects.[262]

Optical Brighteners

The soap and detergent industry has been faced with multiple problems as many ingredients of their formulations are questioned relative to both toxicologic and environmental hazards. Optical brighteners have been added to the list. Contact dermatitis in thirty patients was traced to the content of an optical brightener, Tinopal CH 3566, in washing powders. Since the compound has been used for several years and the dermatitis only observed recently, a question was raised as to whether some interaction with enzymes in the washing powders was having an effect.[263] Studies with mice exposed to optical brighteners and ultraviolet light showed a higher incidence of tumors than in mice exposed to ultraviolet light alone. Exposure to optical brighteners alone did not cause any tumors. These studies raise questions for research into the potential interactions of these chemicals and environmental factors.[264]

Surfactants

During the years 1950–1964 the major surfactant in detergent or cleaning formulations was tetrapropylene-derived alkylbenzene sulfonates. In mid-1965 this chemical was discontinued in favor of linear alkylate sulfonate and other surfactants more easily decomposed by bacterial action. The acute and chronic toxicity of surfactants, as well as their presence in the environment, has been reviewed. It was estimated that the amount of surfactant which might be ingested in the normal course of events was an order-of-magnitude figure of 1 mg/day for each person. It was inferred that there is a negligible degree of hazard associated with surfactants at environmental levels.[265]

Iodine

Rats were exposed to radioactive iodine (^{131}I) by either intravenous injections or 10-min inhalation periods of an aerosol (^{131}I incorporated onto nonradioactive CsCl particles). The general toxicity and metabolism were the same for both routes of administration. The doses administered resulted in initial body burdens ranging from 20–1200 μCi/kg. Fifty to sixty percent of the dose was excreted during the first 24 hr. The urinary excretion was about five times the fecal excretion. Most of the retained iodine was found in the thyroid and in the pelt. No effect on life span of the rats was observed. However the rats at the higher dose levels appeared to have a higher incidence of pituitary adenomas and thyroid tumors than did the controls and lower dose levels. It had been considered that the dose levels used in this study were safe "tracer" levels, but it was suggested that it may become necessary to accept radioiodine as a potential hazard at the levels used.[266]

Fluoride

The fluoride ion is a ubiquitous bone seeker with a variety of toxic effects. These have been reviewed under the headings of acute poisoning, kidney, thyroid, growth, chronic fluorosis, and dental caries. The subject of fluoride toxicology continues to be a much debated subject in communities considering fluoridation of the public water supply.[267,268]

Fluorine is an essential trace nutrient and produces toxic effects when ingested in excessive amounts (this statement is generally true for all essential nutrients). The toxicologic aspects of fluorine toxicosis in livestock has been reviewed. Procedures for alleviating fluorosis in animals have been recommended.[269]

p-Phenylenediamine

There is a TLV for *p*-phenylenediamine, but it is recognized that it will not protect workers who are already sensitized. Some of the substituted *p*-phenylenediamines are also sensitizers. Allergic contact dermatitis is caused by its *N*-substituted derivatives. One mechanism for skin sensitization is the binding of the chemical to skin proteins in order to form a complete antigen. It was found that 3-methyl-4-amino-*N,N*-diethylaniline, and 4-amino-*N,N*-diethlaniline do not form stable isolatable adducts with epidermal protein amino acids. Thus these allergenic *p*-phenylenediamines apparently differ from some other contact allergens.[270]

Carbon Disulfide and Alcoholics

The measurement of the iodine-azide reaction in urine has been recommended as a useful indicator of exposures to carbon disulfide in excess of 50 mg/m^3. It was found that dithiocarbamates in urine will interfere with the test. Dithiocarbamates will be present in the urine of workers who are undergoing disulfiram therapy for chronic alcoholism.[271]

Nitriles

The acute toxicity of thirty-six nitriles by various routes of administration in mice, rats, and rabbits and 137 cases of accidental intoxication resulting from occupational overexposure have been presented. The majority of the cases of accidental human intoxication were caused by acrylonitrile or *o*-phthalodinitrile.[272]

Monomethylhydrazine

The inhalation toxicity was determined in laboratory animals after single exposures to monomethylhydrazine for periods ranging from 15–240 min. Monomethylhydrazine has been of interest because of its use as a rocket fuel. It caused irritation of nose and eyes, emesis in large animals, ataxia, and convulsions; it was an active hemolytic agent, particularly in dogs. The 60-min LC_{50} values in ppm were 82 for squirrel monkeys, 96 for beagle dogs, 122 for mice, 162 for rhesus monkeys, and 244 for rats. The most consistent pathologic findings were changes in the kidneys, but pulmonary, hepatic, brain, and splenic changes were also observed.[273]

Rubber Additives

Rubber formulations include ingredients classified as elastomer, vulcanizer, primary accelerator, secondary accelerator, antioxidant, activator, processing aid, plasticizer, reinforcing agent, or blowing agent. Twenty-four of the most frequently used specific substances were listed and their toxic properties were reviewed. It was concluded that the compounds were not highly hazardous; none of them were considered to have carcinogenic activity. However it was pointed out that toxicity information for some of the compounds was scarce. In addition the chemical nature of the fumes from the curing process is not fully understood and the potential for toxic hazard needs further investigation.[274]

Aminoethanols

The 2-*N*-alkyl-substituted aminoethanols are widely used as curing agents, flotation agents, dispersants, and emulsifiers. Rats were exposed to 2-*N*-dibutyl-aminoethanol (DBAE) by ingestion and inhalation. The acute oral LD_{50} was 1.78 gm/kg for neutralized DBAE. The rats experienced periods of depression followed by tremors, incoordination, clonicotonic convulsions, and death. When DBAE was supplied to rats in their drinking water for 5 weeks, the only effect seen with a concentration of 1 gm/liter was that the growth curve remained consistently below that of the control group. At concentrations of 2 and 4 gm/liter in the drinking water there was a marked reduction in body weight initially which was not recovered. At the end of 5 weeks exposure no histopathologic changes were seen, but the males at 2 gm/liter and the males and females at 4 gm/liter showed increased kidney to body weight ratios. The kidney weights themselves were normal, but because of the decreased body weights, the ratios were elevated. This did not occur with the liver. Thus the liver weights varied directly with the body weight, while the kidneys either continued to grow during the body-weight-loss period or grew faster than normal during the recovery period.[275] Intraperitoneal doses have been shown to produce significant reductions in brain cholinesterase in rats.[276]

GENERAL TOPICS OF SPECIAL INTEREST

Perspectives from an Old Scout

Dr. Herbert E. Stokinger, the old scout, has made a comprehensive review and toxicologic evaluation of the potential human health hazards from man-made and natural environmental pollutants. His minimal list of substances which may reasonably be expected to exert an adverse effect on health now or in the foreseeable future includes carcinogens, asthmagens, asbestos, respiratory irritants, cadmium in food and beverages, selenium in food and water, nitrates in water, and "hard" waters. The disease states which may arise from these pollutants include accelerated aging, allergic asthma, cardiovascular disease, atherosclerotic heart disease, berylliosis, bronchitis, cancer of the gastrointestinal and respiratory tracts, dental caries, emphysema, mesotheliomas, methemoglobinemia infant death, and renal hypertension. Dr. Stokinger noted in particular that (*1*) pollutants in the air possess more potential for affecting health than do pollutants in water and food combined; (*2*) usually pollutants exert their effect through interaction with some other factor; (*3*) acceleration of aging is the dominant characteristic of the effect of many pollutants, and (*4*) the counteractants or natural antagonists of pollutants must be measured to avoid an overestimation of the health problem.[277]

Toxicologic Insignificance

Some years ago the TLV's for a couple of compounds were set at zero, and zero tolerances were established for a variety of food additives and pesticide residues. Such values had a practical legal significance and it was comforting to the toxicologist to feel that certain chemicals were not present. The analytic chemist has burst that bubble by developing techniques to measure minute quantities and has found toxic chemicals present in places which were not expected. Thus a zero concentration has lost all scientific validity that it might have had. In the case of TLV's the zero values were quickly removed and currently compounds of concern are placed in a special appendix with a notation that all contact should be prevented. In the case of food additives and pesticide residues it took much longer but eventually a regulation was promulgated requiring numerical values for all such safe limits. This, of course, gave rise to the practical aspects of the toxicity testing to be required on setting very low permissible levels. The same practical considerations regarding toxicity testing have been raised with respect to certain chemicals which are only present in the food supply at extremely low levels. It has been suggested that 0.1 ppm in the total diet of man be accepted as a toxicologically insignificant level of a food packaging component – with the exception of metals and pesticides. This type of reasoned approach to the toxicologic consideration of extremely small quantities of chemicals in the environment seems to be a practical solution for many problems facing both government and industry.[278–280] In particular the concept of zero tolerance as it applies to chemical carcinogens needs special attention.[281]

Dose–Response Relationships

The dose–response relationship is central to the quantitation of pharmacologic and toxicologic actions and it is essential to the setting of environmental standards for safe exposures to chemicals. The significant dimensions of this relationship have been explored as they apply to situations in industry. It was noted that the term dose basically applies to the quantity of a chemical agent which provides a stress at some site in the body, but that in practice it is often an exposure to external conditions which subsequently give rise to the stress. The response to a dose may consist of two components, one being the actual strain to the body giving rise to an immediate or a potential health effect, the other being a probability of responding related to the specific characteristics and susceptibilities of the population being tested. A statistically dependable dose–response relationship over a range of responses from essentially zero up to a level of disturbance which is clearly unacceptable provides the basis for setting a limit of tolerable exposure below which an unwanted level of disturbance will not occur in a group of exposed individuals with a frequency in excess of a stated probability.[282]

Biologic Classification System

An interesting classification system for air pollutants has been proposed which interrelates the route of entry with the site of toxic action. The routes of entry are the external surface, the respiratory tract, and the intestinal tract. The sites for either acute or chronic toxic action are the surface, the subsurface, or the systems of the body. Thus, categories include substances bringing about their effects by contact with body surface, substances that act on the membranes of the respiratory tract without significant accumulation or systemic absorption, substances that accumulate in pulmonary tissue but are not systemically absorbed, substances that are systemically absorbed by the lungs exerting their effects on extrapulmonary tissues, and substances settling out of the air and incorporated into food or water supplies. It was recognized that the system contains many oversimplifications; for example, several substances would appear in more than one place. Nevertheless such a system can provide a biologic meaning to consideration of parameters related to atmospheric pollution.[283]

Joint Toxic Action

Finney's mathematical model for additive joint toxicity, which yields the harmonic mean of the toxicity of the components, was tested with the determination of oral LD_{50} values of all possible pairs of twenty-seven industrial chemicals. The model satisfactorily predicted the toxicity of a large proportion of the pairs when the mixtures were made with equal volumes of the components, as well as when the mixtures contained the two chemicals in volumes directly proportional to their respective rat oral LD_{50} value.[284,285]

Physical Factors and Toxicity

Environmental factors, such as temperature and humidity, may alter the susceptibility of man and animals to toxic chemicals. Interest in the role of temperature in lead poisoning originated from the observations that most of the severe lead poisoning in children is observed during the summer months. Mice or rats injected with lead and exposed to an environmental temperature of 95°F began to die earlier, had a higher mortality, and a shorter average survival time than control animals at 72°F. If the animals were dehydrated, the mortality was increased at both temperatures. It was suggested that environmental temperature and humidity, by altering the physiologic functions of the host, may affect susceptibility to both exogenous and endogenous disease-producing agents and thus play a coincident role in such diseases.[286]

Interaction with Microsomal Enzymes

More and more information has become available about the role of the microsomal enzymes in the liver. These enzymes play a major role in the metabolism of many foreign compounds in the body and this enzyme activity is subject to both inhibition and stimulation. Some compounds will inhibit the enzyme activity at high concentrations, but will actually stimulate the induction of increased enzyme activity at low concentrations. Since the metabolic products resulting from the enzyme activity may, in some cases, be more toxic rather than less toxic than the parent compound, there is opportunity for toxic interaction. The toxicologist must be aware of the potential for many commonly used chemicals to thus interfere with his experiments. It was found that *m*-terphenyl stimulated liver hydroxylase activity, but the related compounds *o*-terphenyl, *p*-terphenyl, and triphenylmethane did not. Caffeine caused a depression in *N*-dealkylation in the liver of rats fed low doses over a 6-week period. Fasting for 48-hours also caused a depression in *N*-dealkylation activity. Mercuric chloride and dimethyl mercury caused marked depression of both hydroxylation and *N*-dealkylation.[287]

Mechanisms of Toxic Action

A major review of the research progress in the study of modes of action of toxic substances has been presented in a series of papers. This series will serve as an excellent introduction and up-dating to the processes used in order to understand the toxic action of a chemical.[288]

Mutagens

Since mutations can be induced by both radiation and chemical compounds, there has been an increased concern for the potential adverse effects due to chemical mutagens in the environment. Mutations are sudden changes in inherited material; they may (*1*) be lethal and give rise to dead offspring, (*2*) affect morphologic characteristics such as color, or (*3*) give rise to physical and mental defects. Mutations can be dominant expressing themselves in the first generation after induction, or they can be recessive requiring the same mutant gene in both parents for expression and thus possibly going through many generations before causing a defective offspring. The structural changes in inherited material are usually classified as chromosomal aberrations or point mutations. Chromosomal aberrations are drastic rearrangements of the base sequences in the DNA structure which often lead to retarded growth and many times to death with subsequent loss of the mutation. Point mutations are single base-pair changes in the DNA and are usually recessive and thus are not rapidly eliminated from the population which may be subject to hereditary disease. An Environmental Mutagen Information Center has been formed at the Biology Division, Oak Ridge National Laboratory, Oak Ridge, Tennessee.[289]

A major problem confronting the experimental toxicologist concerned with safety evaluation has been the methodology to establish the potential for certain chemicals to cause a mutagenic hazard. There have been several reviews of the commonly used testing procedures for mutagenicity.[289-291]

Range-Finding Toxicity Data

The seventh listing of range-finding toxicity data has been published by the Union Carbide-sponsored laboratory at Carnegie-Mellon University. This new list adds data on about 200 compounds. This continuing compilation has become an extremely useful reference for toxicologists. The toxicity information includes the single oral LD_{50} for rats, single penetration LD_{50} for rabbits, the maximum exposure time for a concentrated vapor producing no death, the mortality for specific concentrations and time of exposure, the irritation on the uncovered rabbit belly, and corneal injury in rabbits.[292]

Threshold Limit Values

TLV's are developed or validated with different types and kinds of data. It has been emphasized that in general the setting of a TLV for a new substance stems from basic data, including in particular, chronic animal inhalation toxicity data. Unfortunately these types of data are in the shortest supply. It has been suggested that this problem results from a combination of failure to develop long-term studies and failure to release the data to the open literature.[293]

The progress made in establishing international agreement for occupational exposure to toxic substances has been published. Agreement has been reached on twenty-four substances and a classification of biologic responses to airborne toxic chemicals has been compiled.[294]

The inhalation toxicity of 109 industrial chemicals was studied by exposing rats to known concentrations in air for periods of up to 6 hr/day, 5 days/week for up to 4 weeks. The observations are reported and a provisional maximum acceptable concentration in the working environment is suggested for each compound.[295]

Manpower Needs

The toxicologist has been recognized as one of the critical specialists needed to face the chemical challenge to man and his environment. The demand for toxicologists was found to be most visible as evidenced by advertisements with offers for positions. An invisible demand was apparent when various agencies were asked for information but could not supply it because there were not enough professionals to develop the required data or to validate and deliver the information.[296]

Governmental Advisory Center

The activities of the Advisory Center on Toxicology, under the auspices of the National Academy of Sciences, has been described. The charter of this Center restricts its services to seven federal agencies. The Center has been successful in obtaining information as yet unpublished from the chemical manufacturing industry for discreet usage. It was suggested that there is a general need for a repository and source of unpublished data. Some limited efforts in this direction have been made by the American Medical Association and the Industrial Health Foundation, Inc., but industry needs to supply information and financial support for these systems.[297]

General Toxicology Directory

A directory was compiled listing more than 750 organizations which can supply specific information relating to toxicology. Each organization is described according to its toxicology-related interests, its holdings, publications, and information services.[298]

REFERENCES

1. Wilson, R. H., McCormick, W. E., Tatum, C. F., and Creech, J. L., Occupational acroosteolysis. Report of 31 cases. *J. Amer. Med. Ass. 201*, 577 (1967).
2. Roush, G., Jr., Polycleaner's disease: Clinical aspects. Read before the 22nd Annual Meeting of the American Academy of Occupational Medicine, Cincinnati, Ohio, 1970.
3. Cook, W. A., Giever, P. M., Dinman, B. D., and Magnuson, H. J., Occupational exposures associated with acroosteolysis in vinyl chloride polymerization. Read before the American Industrial Hygiene Conference. Denver, Colorado, May 15, 1969.
4. Griffith, J. F., Weaver, J. E., Whitehouse, H. S., Poole, R. L., Newmann, E. A., and Nixon, G. A., Safety evaluation of enzyme detergents. Oral and cutaneous toxicity, irritancy and skin sensitization studies. *Food Cosmet. Toxicol. 7*, 581 (1969).
5. Flindt, M. L. H., Pulmonary disease due to inhalation of derivatives of *Bacillus subtilis* containing proteolytic enzyme. *Lancet I*, 1177 (1969).
6. Pepys, J., Hargreave, F. E., Longbottom, J. L., and Faux, J., Allergic reaction of the lungs to enzymes of *Bacillus subtilis. Lancet I*, 1181 (1969).
7. Newhouse, M. L., Tagg, B., Pocock, S. J., and McEwan, A. C., An epidemiological study of workers producing enzyme washing powders. *Lancet I*, 689 (1970).
8. McMurrain, K. D., Jr., Dermatologic and pulmonary responses in the manufacturing of detergent enzyme products. *J. Occup. Med. 12*, 416 (1970).
9. Committee on Threshold Limits, *Threshold Limit Values for 1970.* American Conference of Governmental Industrial Hygienists, Cincinnati, Ohio, 1970.
10. Van Duuren, B. L., Goldschmidt, B. M., Katz, C., Langseth, L., Mercado, G., and Sivak, A., Alpha-haloethers: A new type of alkylating carcinogen. *Arch. Environ. Health 16*, 472 (1968).
11. Friedman, L., Kunin, C. M., Nelson, N., Whittenberger, J. L., and Wilson, J. G., Panel on reproduction report on reproduction studies in the safety evaluation of food additives and pesticide residues. *Toxicol. Appl. Pharmacol. 16*, 264 (1970).
12. Shimkin, M. B., Environmental carcinogens. *Arch. Environ. Health 16*, 513 (1968).
13. Shimkin, M. B., Biologic tests of carcinogenic activity. *Arch. Environ. Health 16*, 522 (1968).
14. Saffiotti, U., Cefis, F., and Kolb, L. H., A method for the experimental induction of bronchogenic carcinoma. *Cancer Res. 28*, 104 (1968).

15. Spritzer, A. A., Watson, J. A., Auld, J. A., and Guetthoff, M. A., Pulmonary macrophage clearance: The hourly rates of transfer of pulmonary macrophages to the oropharynx of the rat. *Arch. Environ. Health 17,* 726 (1968).
16. Watson, J. A., Spritzer, A. A., Auld, J. A., and Guetthoff, M. A., Deposition and clearance following inhalation and intratracheal injection of particles. *Arch. Environ. Health 19*, 51 (1969).
17. Sanders, C. L., The distribution of inhaled plutonium-239 dioxide particles within pulmonary macrophages. *Arch. Environ. Health 18*, 904 (1969).
18. Holma, B., Scanning electron microscopic observation of particles deposited in the lung. *Arch. Environ. Health 18*, 330 (1969).
19. Lippman, M., and Albert, R. E., The effect of particle size on the regional deposition of inhaled aerosols in the human respiratory tract. *Amer. Ind. Hyg. Ass. J. 30*, 257 (1969).
20. Aerosol Technology Committee, American Industrial Hygiene Association, Guide for respirable mass sampling. *Amer. Ind. Hyg. Ass. J. 31*, 133 (1970).
21. Lippman, M., "Respirable" dust sampling. *Amer. Ind. Hyg. Ass. J. 31*, 138 (1970).
22. Morrow, P. E., Models for the study of particle retention and elimination in the lung. *In* "Inhalation Carcinogenesis" (M. G. Hanna, Jr., P. Nettesheim, and J. R. Gilbert, eds.), Symp. Ser. 18, U.S. At. Energy Comm., Div. Tech. Inform., Oak Ridge, Tennessee, 1970.
23. Kilburn, K. H., guest ed., Symposium on pulmonary responses to inhaled materials. An evaluation of model systems. *Arch. Intern. Med. 126*, 415 (1970).
24. Hinners, R. G., Burkart, J. K., and Punte, C. L., Animal inhalation exposure chambers. *Arch. Environ. Health 16*, 194 (1968).
25. Laskin, S., and Drew, R. T., An inexpensive portable inhalation chamber. *Amer. Ind. Hyg. Ass. J. 31*, 645 (1970).
26. P'an, A. Y. S., and Jegier, Z., A simple exposure chamber for gas inhalation experiments with small animals. *Amer. Ind. Hyg. Ass. J. 31*, 647 (1970).
27. Raabe, O. G., Generation and characterization of aerosols. *In* "Inhalation Carcinogenesis" (M. G. Hanna, Jr., P. Nettesheim, and J. R. Gilbert, eds.), Symp. Ser. 18, U.S. At. Energy Comm., Div. Tech. Inform., Oak Ridge, Tennessee, 1970.
28. Knott, M. J., and Malanchuk, M., Analysis of foreign aerosol produced in NO_2-rich atmospheres of animal exposure chambers. *Amer. Ind. Hyg. Ass. J. 30*, 147 (1969).
29. Beckley, J. H., Russell, T. J., and Rubin, L. F., Use of the rhesus monkey for predicting human response to eye irritants. *Toxicol. Appl. Pharmacol. 15*, 1 (1969).
30. Weil, C. S., Woodside, M. D., Bernard, J. R., and Carpenter, C. P., Relationship between single-peroral, one-week, and ninety-day rat feeding studies. *Toxicol. Appl. Pharmacol. 14,* 426 (1969).
31. Weil, C. S., and Carpenter, C. P., Abnormal values in control groups during repeated-dose toxicologic studies. *Toxicol. Appl. Pharmacol. 14*, 335 (1969).
32. Rieke, F. E., Lead intoxication in shipbuilding and shipscrapping 1941 to 1968. *Arch. Environ. Health 19,* 521 (1969).
33. Katchian, A., Lead poisoning orally in a painter, with hematemesis. *J. Occup. Med. 10*, 89 (1968).
34. Jovičić, B., Ulcer and gastritis in the professions exposed to lead. *Arch. Environ. Health 21*, 526 (1970).
35. Soliman, M. H. M., El-Sadik, Y. M., El-Kashlan, K. M., and El-Waseef, A., Biochemical studies on Egyptian workers exposed to lead. *Arch. Environ. Health 21*, 529 (1970).
36. Davis, J. R., Abrahams, R. H., Fishbein, W. I., and Fabrega, E. A., Urinary delta-aminolevulinic acid (ALA) levels in lead poisoning. II. Correlation of ALA values with clinical findings in 250 children with suspected lead ingestion. *Arch. Environ. Health 17*, 164 (1968).

37. Ornosky, M., Coproporphyrinuria and urine-lead findings: Fifteen years of experience. *Amer. Ind. Hyg. Ass. J. 29*, 228 (1968).
38. Selander, S., and Cramér, K., Interrelationships between lead in blood, lead in urine, and ALA in urine during lead work. *Brit. J. Ind. Med. 27*, 28 (1970).
39. Gibson, S. L. M., Mackenzie, J. C., and Goldberg, A., The diagnosis of industrial lead poisoning. *Brit. J. Ind. Med. 25*, 40 (1970).
40. Williams, M. K., King, E., and Walford, J., Method for estimating objectively the comparative merits of biological tests of lead exposure. *Brit. Med. J. 1*, 618 (1968).
41. Hernberg, S., and Nikkanen, J., Enzyme inhibition by lead under normal urban conditions. *Lancet I*, 63 (1970).
42. Hernberg, S., Nikkanen, J., Mellin, G., and Lilius, H., δ-Aminolevulinic acid dehydrase as a measure of lead exposure. *Arch. Environ. Health 21*, 140 (1970).
43. Ulmer, D. D., and Vallee, B. L., Effects of lead on biochemical systems. *In* "Trace Substances in Environmental Health" (D. D. Hemphill, ed.), Vol. II, pp. 7-27. Univ. Missouri Press, Columbia, Missouri, 1969.
44. Sandstead, H. H., Stant, E. G., Brill, A. B., Arias, L. I., and Terry, R. T., Lead intoxication and the thyroid. *Arch. Intern. Med. 123*, 632 (1969).
45. Sandstead, H. H., Michelakis, A. M., and Temple, T. E., Lead intoxication: Its effect on the renin-aldosterone response to sodium deprivation. *Arch. Environ. Health 20*, 356 (1970).
46. Schroeder, H. A., and Tipton, I. H., The human body burden of lead. *Arch. Environ. Health 17*, 965 (1968).
47. Barry, P. S. I., and Mossman, D. B., Lead concentrations in human tissues. *Brit. J. Ind. Med. 27*, 339 (1970).
48. Hine, C. H., Cavalli, R. D., and Beltran, S. M., Percutaneous absorption of lead from industrial lubricants. *J. Occup. Med. 11*, 568 (1969).
49. Goyer, R. A., Leonard, D. L., Moore, J. F., Rhyne, B., and Krigman, M. R., Lead dosage and the role of the intranuclear inclusion body. An experimental study. *Arch. Environ. Health 20*, 705 (1970).
50. Lessler, M. A., Cardona, E., Padilla, F., and Jensen, W. N., Effect of lead on reticulocyte respiratory activity. *J. Cell. Biol. 39*, 171a (1968).
51. Bingham, E., Pfitzer, E. A., Barkley, W., and Radford, E. P., Alveolar macrophages: Reduced number in rats after prolonged inhalation of lead sesquioxide. *Science 162*, 1297 (1968).
52. Bingham, E., Trace amounts of lead in the lung. *In* "Trace Substances in Environmental Health" (D. D. Hemphill, ed.), Vol. III, pp. 83-90. Univ. Missouri Press, Columbia, Missouri, 1970.
53. Schroeder, H. A., Mitchener, M., and Nason, A. P., Zirconium, niobium, antimony, vanadium and lead in rats: Life term studies. *J. Nutr. 100*, 59 (1970).
54. Selye, H., and Somogyi, A., Pulmonary calcergy induced by inhalation of irritating gases. *Arch. Environ. Health 16*, 827 (1968).
55. Stopps, G. J., Symposium on air quality criteria – lead. *J. Occup. Med. 10*, 550 (1968). Including discussions by R. J. M. Horton and T. J. Haley.
56. Kehoe, R. A., Toxicological appraisal of lead in relation to the tolerable concentration in the ambient air. *J. Air Pollut. Contr. Ass. 19*, 690 (1969).
57. Hammond, P. B., Lead poisoning. An old problem with a new dimension. *In* "Essays in Toxicology" (F. R. Blood, ed.), Vol. I, Chapt. 4, pp. 115-155. Academic Press, New York, 1969.
58. An American Chemical Society Symposium, Air quality and lead. *Environ. Sci. Technol. 4*, 217 (1970); *4*, 305 (1970).
59. El-Sadik, Y. M., and El-Dakhakhny, A., Effects of exposure of workers to mercury at a sodium hydroxide producing plant. *Amer. Ind. Hyg. Ass. J. 31*, 705 (1970).

60. West, I., and Lim, J., Mercury poisoning among workers in California's mercury mills. A preliminary report. *J. Occup. Med. 10*, 697 (1968).
61. Rentos, P. G., and Seliman, E. J., Relationship between environmental exposure to mercury and clinical observation. *Arch. Environ. Health 16*, 794 (1968).
62. Wada, O., Toyokawa, K., Suzuki, T., Suzuki, S., Yano, Y., and Nakao, K., Response to a low concentration of mercury vapor. Relation to human porphyrin metabolism. *Arch. Environ. Health 19*, 485 (1969).
63. Smith, R. G., Vorwald, A. J., Patil, L. S., and Mooney, T. F., Jr., Effects of exposure to mercury in the manufacture of chlorine. *Amer. Ind. Hyg. Ass. J. 31*, 687 (1970).
64. Miedler, L. J., and Forbes, J. D., Allergic contact dermatitis due to metallic mercury. *Arch. Environ. Health 17*, 960 (1968).
65. Joselow, M. M., Ruiz, R., and Goldwater, L. J., Absorption and excretion of mercury in man. XIV. Salivary excretion of mercury and its relationship to blood and urine mercury. *Arch. Environ. Health 17*, 35 (1968).
66. Joselow, M. M., Ruiz, R., and Goldwater, L. J., The use of salivary (parotid) fluid in biochemical monitoring. *Amer. Ind. Hyg. Ass. J. 30*, 77 (1969).
67. Taylor, W., Guirgis, H. A., and Stewart, W. K., Investigation of a population exposed to organomercurial seed dressings. *Arch. Environ. Health 19*, 505 (1969).
68. Skerfving, S., Hansson, K., and Lindsten, J., Chromosome breakage in humans exposed to methyl mercury through fish consumption. *Arch. Environ. Health 21*, 133 (1970).
69. Aberg, B., Ekman, L., Falk, R., Greitz, U., Persson, G., and Snihs, J.-O., Metabolism of methyl mercury (^{203}Hg) compounds in man. Excretion and distribution. *Arch. Environ. Health 19*, 478 (1969).
70. Cember, H., Gallagher, P., and Faulkner, A., Distribution of mercury among blood fractions and serum proteins. *Amer. Ind. Hyg. Ass. J. 29*, 233 (1968).
71. Farvar, M. A., and Cember, H., Difference between *in vitro* and *in vivo* distribution of mercury in blood proteins. *J. Occup. Med. 11*, 11 (1969).
72. Cember, H., A model for the kinetics of mercury elimination. *Amer. Ind. Hyg. Ass. J. 30*, 367 (1969).
73. Phillips, R., and Cember, H., The influence of body burden of radiomercury on radiation dose. *J. Occup. Med. 11*, 170 (1969).
74. Berlin, M. H., Nordberg, G. F., and Serenius, F., On the site and mechanism of mercury vapor resorption in the lung. *Arch. Environ. Health 18*, 42 (1969).
75. Berlin, M., Fazackerley, J., and Nordberg, G., The uptake of mercury in the brains of mammals exposed to mercury vapor and to mercuric salts. *Arch. Environ. Health 18*, 719 (1969).
76. Norseth, T., and Clarkson, T. W., Studies on the biotransformation of ^{203}Hg-labeled methyl mercury chloride in rats. *Arch. Environ. Health 21*, 717 (1970).
77. Miller, M. W., and Berg, G. G., eds., "Chemical Fallout. Current Research on Persistent Pesticides." Thomas, Springfield, Illinois, 1969.
78. Clarkson, T. W., Biochemical aspects of mercury poisoning. *J. Occup. Med. 10*, 351 (1968).
79. Vigliani, E. C., Yant Memorial Lecture: The biopathology of cadmium. *Amer. Ind. Hyg. Ass. J. 30*, 329 (1969).
80. Axelsson, B., Dahlgren, S. E., and Piscator, M., Renal lesions in the rabbit after long-term exposure to cadmium. *Arch. Environ. Health 17*, 24 (1968).
81. Piscator, M., and Axelsson, B., Serum proteins and kidney function after exposure to cadmium. *Arch. Environ. Health 21*, 604 (1970).
82. Schroeder, H. A., Baker, J. T., Hansen, N. M., Jr., Size, J. G., and Wise, R. A., Vascular

reactivity of rats altered by cadmium and a zinc chelate. *Arch. Environ. Health 21*, 609 (1970).

83. Kanisawa, M., and Schroeder, H. A., Renal arteriolar changes in hypertensive rats given cadmium in drinking water. *Exp. Mol. Pathol. 10*, 81 (1969).
84. Lucis, O. J., Lynk, M. E., and Lucis, R., Turnover of cadmium-109 in rats. *Arch. Environ. Health 18*, 307 (1969).
85. Zavon, M. R., and Meadows, C. D., Vascular sequelae to cadmium fume exposure. *Amer. Ind. Hyg. Ass. J. 31*, 180 (1970).
86. Belman, S., Beryllium binding of epidermal constituents. *J. Occup. Med. 11*, 175 (1969).
87. Robinson, F. R., Schaffner, F., and Trachtenberg, E., Ultrastructure of the lungs of dogs exposed to beryllium-containing dusts. *Arch. Environ. Health 17*, 193 (1968).
88. Sunderman, F. W., Jr., Roszel, N. O., and Clark, R. J., Gas chromatography of nickel carbonyl in blood and breath. *Arch. Environ. Health 16*, 836 (1968).
89. Hackett, R. L., and Sunderman, F. W., Jr., Pulmonary alveolar reaction to nickel carbonyl. Ultrastructural and histochemical studies. *Arch. Environ. Health 16*, 349 (1968).
90. Hackett, R. L., and Sunderman, F. W., Jr., Nickel carbonyl: Effects upon the ultrastructure of hepatic parenchymal cells. *Arch. Environ. Health 19*, 337 (1969).
91. Sunderman, F. W., Jr., and Selin, C. E., The metabolism of nickel-63 carbonyl. *Toxicol. Appl. Pharmacol. 12*, 207 (1968).
92. Kasprzak, K. S., and Sunderman, F. W., Jr., The metabolism of nickel carbonyl-^{14}C. *Toxicol. Appl. Pharmacol. 15*, 295 (1969).
93. Milner, J. E., The effect of ingested arsenic on methylcholanthrene-induced skin tumors in mice. *Arch. Environ. Health 18*, 7 (1969).
94. Levinsky, W. J., Smalley, R. V., Hillyer, P. N., and Shindler, R. L., Arsine hemolysis. *Arch. Environ. Health 20*, 436 (1970).
95. De Palma, A. E., Arsine intoxication in a chemical plant. Report of three cases. *J. Occup. Med. 11*, 582 (1969).
96. Muehrcke, R. C., and Pirani, C. L., Arsine-induced anuria. A correlative clinicopathological study with electron microscopic observations. *Ann. Intern. Med. 68*, 853 (1968).
97. Teitelbaum, D. T., and Kier, L. C., Arsine poisoning: Report of five cases in the petroleum industry and a discussion of the indications for exchange transfusion and hemodialysis. *Arch. Environ. Health 19*, 133 (1969).
98. Pelikan, Z., Černý, E., and Polster, M., Toxic effects of some di-*n*-octyltin compounds in white mice. *Food Cosmet. Toxicol. 8*, 655 (1970).
99. Pelikan, Z., Effects of *bis*(tri-*n*-butyltin)oxide on the eyes of rabbits. *Brit. J. Ind. Med. 26*, 165 (1969).
100. Pelikan, Z., and Černý, E. Toxic effects of "*bis*(tri-*n*-butyltin)oxide" (TBTO) on the skin of rats. *Berufsdermatosen 17*, 305 (1969).
101. Jett, R., Pierce, J. L., II, and Stemmer, K. L. S., Toxicity of alloys of ferrochromium. III. Transport of chromium (III) by rat serum protein studied by immunoelectrophoretic analysis and autoradiography. *Arch. Environ. Health 17*, 29 (1968).
102. Dastur, D. K., Manghani, D. K., Raghavendran, K. V., and Jeejeebhoy, K. N., Distribution and fate of Mn^{54} in the rat, with special reference to the C.N.S. *Quart. J. Exp. Physiol. 54*, 322 (1969).
103. Tanaka, S., and Lieben, J., Manganese poisoning and exposure in Pennsylvania. *Arch. Environ. Health 19*, 674 (1969).
104. Jindřichová, J., Usefulness of manganese determination in stools as an index of exposure. *Int. Arch. Gewerbepathol. Gewerbehyg. 25*, 347 (1969).

105. Fishburn, C. W., and Zenz, C., Metal fume fever: A report of a case. *J. Occup. Med. 11*, 142 (1969).
106. Hammond, S. E., Lagerquist, C. R., and Mann, J. R., Americium and plutonium urine excretion following acute inhalation exposures to high-fired oxides. *Amer. Ind. Hyg. Ass. J. 29*, 169 (1968).
107. Wenzel, W. J., Thomas, R. G., and McClellan, R. O., Effect of stable yttrium concentration on the distribution and excretion of inhaled radioyttrium in the rat. *Amer. Ind. Hyg. Ass. J. 30*, 630 (1969).
108. Tebrock, H. E., and Machle, W., Exposure to europium-activated yttrium orthovanadate: A cathodoluminescent phosphor. *J. Occup. Med. 10*, 692 (1968).
109. Ferm, V. H., and Carpenter, S. J., The relationship of cadmium and zinc in experimental mammalian teratogenesis. *Lab. Invest. 18*, 429 (1968).
110. Holmberg, R. E., and Ferm, V. H., Interrelationships of selenium, cadmium, and arsenic in mammalian teratogenesis. *Arch. Environ. Health 18*, 873 (1969).
111. Lucis, O. J., and Lucis, R., Distribution of cadmium-109 and zinc-65 in mice of inbred strains. *Arch. Environ. Health 19*, 334 (1969).
112. Weir, P. A., and Hine, C. H., Effects of various metals on behavior of conditioned goldfish. *Arch. Environ. Health 20*, 45 (1970).
113. Allaway, W. H., Kubota, J., Losee, F., and Roth, M., Selenium, molybdenum, and vanadium in human blood. *Arch. Environ. Health 16*, 342 (1968).
114. Kubota, J., Lazar, V. A., and Losee, F., Copper, zinc, cadmium, and lead in human blood from 19 locations in the United States. *Arch. Environ. Health 16*, 788 (1968).
115. Schroeder, H. A., Mitchener, M., Balassa, J. J., Kanisawa, M., and Nason, A. P., Zirconium, niobium, antimony, and fluorine in mice: Effects on growth, survival and tissue levels. *J. Nutr. 95*, 95 (1968).
116. Schroeder, H. A., Kanisawa, M., Frost, D. V., and Mitchener, M., Germanium, tin and arsenic in rats: Effects on growth, survival, pathological lesions and life span. *J. Nutr. 96*, 37 (1968).
117. Kanisawa, M., and Schroeder, H. A., Life term studies on the effect of trace elements on spontaneous tumors in mice and rats. *Cancer Res. 29*, 892 (1969).
118. Schroeder, H. A., A sensible look at air pollution by metals. *Arch. Environ. Health 21*, 798 (1970).
119. Wagner, W. D., Fraser, D. A., Wright, P. G., Dobrogorski, O. J., and Stokinger, H. E., Experimental evaluation of the threshold limit of cristobalite – calcined diatomaceous earth. *Amer. Ind. Hyg. Ass. J. 29*, 211 (1968).
120. Parazzi, E., Secchi, G. C., Pernis, B., and Vigliani, E., Studies on the cytotoxic action of silica dusts on macrophages *in vitro*. *Arch. Environ. Health 17*, 850 (1968).
121. Gross, P., and deTreville, R. T. P., Experimental "acute" silicosis. *Arch. Environ. Health 17*, 720 (1968).
122. Gross, P., and deTreville, R. T. P., Emphysema and pneumoconiosis: An experimental study on their interrelationship. *Arch. Environ. Health 18*, 340 (1969).
123. Göthe, C-J., and Swensson, A., Effect of BCG on lymphatic lung clearance of dusts with different fibrogenicity. An experimental study on rats. *Arch. Environ. Health 20*, 579 (1970).
124. Gross, P., deTreville, R. T. P., Cralley, L. J., Granquist, W. T., and Pundsack, F. L. The pulmonary response to fibrous dusts of diverse compositions. *Amer. Ind. Hyg. Ass. J. 31*, 125 (1970).
125. Gross, P., Kaschak, M., Tolker, E. B., Babyak, M. A., and deTreville, R. T. P., The pulmonary reaction to high concentrations of fibrous glass dust. A preliminary report. *Arch. Environ. Health 20*, 696 (1970).

126. Bouhuys, A., Gilson, J. C., and Schilling, R. S., Editorial. Byssinosis in the textile industry. *Arch. Environ. Health 21*, 475 (1970).
127. Cralley, L. J., Cooper, W. C., Lainhart, W. S., and Brown, M. C., Research on health effects of asbestos. *J. Occup. Med. 10*, 38 (1968).
128. Stanton, M. F., Blackwell, R., and Miller, E., Experimental pulmonary carcinogenesis with asbestos. *Amer. Ind. Hyg. Ass. J. 30*, 236 (1969).
129. Gross, P., deTreville, R. T. P., Cralley, L. J., and Davis, J. M. G., Pulmonary ferruginous bodies. Development in response to filamentous dusts and a method of isolation and concentration. *Arch. Pathol. 85*, 539 (1968).
130. Gaensler, E. A., and Addington, W. W., Current concepts: Asbestos or ferruginous bodies. *N. Engl. J. Med. 280*, 488 (1969).
131. Gross, P., deTreville, R. T. P., and Haller, M. N., Pulmonary ferruginous bodies in city dwellers. A study of their central fiber. *Arch. Environ. Health 19*, 186 (1969).
132. Frank, N. R., Yoder, R. E., Brain, J. D., and Yokoyama, E., SO_2 (^{35}S labeled) absorption by the nose and mouth under conditions of varying concentration and flow. *Arch. Environ. Health 18*, 315 (1969).
133. P'an, A. Y. S., and Jegier, Z., The effect of sulfur dioxide and ozone on acetylcholinesterase. *Arch. Environ. Health 21*, 498 (1970).
134. Goldring, I. P., Greenburg, L., Park S. S., and Ratner, I. M., Pulmonary effects of sulfur dioxide exposure in the syrian hamster. II. Combined with emphysema. *Arch. Environ. Health 21*, 32 (1970).
135. Alarie, Y., Ulrich, C. E., Busey, W. M., Swann, H. E., Jr., and MacFarland, H. N., Long-term continuous exposure of guinea pigs to sulfur dioxide. *Arch. Environ. Health 21*, 769 (1970).
136. Spiegelman, J. R., Hanson, G. D., Lazarus, A., Bennett, R. J., Lippmann, M., and Albert, R. E., Effect of acute sulfur dioxide exposure on bronchial clearance in the donkey. *Arch. Environ. Health 17*, 321 (1968).
137. Burton, G. C., Corn, M., Gee, J. B. L., Vasallo, C., and Thomas, A. P., Response of healthy men to inhaled low concentrations of gas–aerosol mixtures. *Arch. Environ. Health 18*, 681 (1969).
138. Amdur, M. O., and Underhill, D., The effect of various aerosols on the response of guinea pigs to sulfur dioxide. *Arch. Environ. Health 16*, 460 (1968).
139. Battigelli, M. C., Cole, H. M., Fraser, D. A., and Mah, R. A., Long-term effects of sulfur dioxide and graphite dust on rats. *Arch. Environ. Health 18*, 602 (1969).
140. Battigelli, M. C., Sulfur dioxide and acute effects of air pollution. *J. Occup. Med. 10*, 500 (1968). (Including discussions by J. A. Zapp and R. Frank.)
141. Daines, R. H., Sulfur dioxide and plant response. *J. Occup. Med. 10*, 516 (1968). (Including a discussion by F. A. Wood.)
142. Fraser, D. A., Battigelli, M. C., and Cole, H. M., Ciliary activity and pulmonary retention of inhaled dust in rats exposed to sulfur dioxide. *J. Air Pollut. Contr. Ass. 18*, 821 (1968).
143. "Air Quality Criteria for Sulfur Oxides." National Air Pollution Control Administration Publ. No. AP-50, Superintendent of Documents, Washington, D.C., Jan. 1969.
144. "Air Quality Criteria for Particulate Matter." National Air Pollution Control Administration Publ. No. AP-49, Superintendent of Documents, Washington, D.C., Jan. 1969.
145. Amdur, M. O., Toxicological appraisal of particulate matter, oxides of sulfur, and sulfuric acid. *J. Air Pollut. Contr. Ass. 19*, 638 (1969). (With discussion by J. W. Clayton, Jr.)
146. Amdur, M. O., and Underhill, D. W., Response of guinea pigs to a combination of sulfur dioxide and open hearth dust. *J. Air Pollut. Contr. Ass. 20*, 31 (1970).

147. Milne, J. E. H., Nitrogen dioxide inhalation and bronchiolitis obliterans. A review of the literature and report of a case. *J. Occup. Med. 11*, 538 (1969).
148. Blair, W. H., Henry, M. C., and Ehrlich, R., Chronic toxicity of nitrogen dioxide. II. Effect on histopathology of lung tissue. *Arch. Environ. Health 18*, 186 (1969).
149. Freeman, G., Stephens, R. J., Crane, S. C., and Furiosi, N. J., Lesion of the lung in rats continuously exposed to two parts per million of nitrogen dioxide. *Arch. Environ. Health 17*, 181 (1968).
150. Freeman, G., Crane, S. C., Stephens, R. J., and Furiosi, N. J., The subacute nitrogen dioxice-induced lesion of the rat lung. *Arch. Environ. Health 18*, 609 (1969).
151. Lewis, T. R., Campbell, K. I., and Vaughan, T. R., Jr., Effects on canine pulmonary function: Via induced NO_2 impairment, particulate interaction, and subsequent SO_X. *Arch. Environ. Health 18*, 596 (1969).
152. Gross, P., deTreville, R. T. P., Babyak, M. A., Kaschak, M., and Tolker, E. B., Experimental emphysema. Effect of chronic nitrogen dioxide exposure and papain on normal and pneumoconiotic lungs. *Arch. Environ. Health 16*, 51 (1968).
153. Buckley, R. D., and Loosli, C. G., Effects of nitrogen dioxide inhalation on germfree mouse lung. *Arch. Environ. Health 18*, 588 (1969).
154. Henry, M. C., Ehrlich, R., and Blair, W. H., Effect of nitrogen dioxide on resistance of squirrel monkeys to *Klebsiella pneumoniae* infection. *Arch. Environ. Health 18*, 580 (1969).
155. Ehrlich, R., and Henry, M. C., Chronic toxicity of nitrogen dioxide. I. Effect on resistance to bacterial pneumonia. *Arch. Environ. Health 17*, 860 (1968).
156. Henry, M. C., Findlay, J., Spangler, J., and Ehrlich, R., Chronic toxicity of NO_2 in squirrel monkeys. III. Effect on resistance to bacterial and viral infection. *Arch. Environ. Health 20*, 566 (1970).
157. Valand, S. B., Acton, J. D., and Myrvik, Q. N., Nitrogen dioxide inhibition of viral-induced resistance in alveolar monocytes. *Arch. Environ. Health 20*, 303 (1970).
158. Freeman, G., Crane, S. C., Stephens, R. J., and Furiosi, N. J., Environmental factors in emphysema and a model system with NO_2. *Yale J. Biol. Med. 40*, 566 (1968).
159. Freeman, G., Crane, S. C., Stephens, R. J., and Furiosi, N. J., Pathogenesis of the nitrogen dioxide-induced lesion in the rat lung: A review and presentation of new observations. *Amer. Rev. Resp. Dis. 98*, 429 (1968).
160. Riddick, J. H., Jr., Campbell, K. I., and Coffin, D. L., Histopathologic changes secondary to nitrogen dioxide exposure in dog lungs. *Amer. J. Clin. Pathol. 49*, 239 (1968) (abstract).
161. Hore, T., and Gibson, D., E., Ozone exposure and intelligence tests. *Arch. Environ. Health 17*, 77 (1968).
162. Bils, R. F., Ultrastructural alterations of alveolar tissue of mice. III. Ozone. *Arch. Environ. Health 20*, 468 (1970).
163. Jones, R. A., Jenkins, L. J., Jr., Coon, R. A., and Siegel, J., Effects of long-term continuous inhalation of ozone on experimental animals. *Toxicol. Appl. Pharmacol. 17*, 189 (1970).
164. Coffin. D. L., Gardner, D. E., Holzman, R. S., and Wolock, F. J., Influence of ozone on pulmonary cells. *Arch. Environ. Health 16*, 633 (1968).
165. Dowell, A. R., Lohrbauer, L. A., Hurst, D., and Lee, S. D., Rabbit alveolar macrophage damage caused by *in vivo* ozone inhalation. *Arch. Environ. Health 21*, 121 (1970).
166. Goldstein, B. D., Pearson, B., Lodi, C., Buckley, R. D., and Balchum, O. J., The effect of ozone on mouse blood *in vivo*. *Arch. Environ. Health 16*, 648 (1968).
167. Werthamer, S., Schwarz, L. H., Carr, J. J., and Soskind, L., Ozone-induced pulmonary

lesions. Severe epithelial changes following sublethal doses. *Arch. Environ. Health 20*, 16 (1970).
168. Pace, D. M., Landolt, P. A., and Aftonomos, B. T., Effects of ozone on cells *in vitro*. *Arch. Environ. Health 18*, 165 (1969).
169. Goldstein, B. D., Lodi, C., Collinson, C., and Balchum, O. J., Ozone and lipid peroxidation. *Arch. Environ. Health 18*, 631 (1969).
170. Gregory, A. R., Inhalation toxicology and lung edema receptor sites. *Amer. Ind. Hyg. Ass. J. 31*, 454 (1970).
171. Wayne, L. G., and Chambers, L. A., Biological effects of urban air pollution. V. A study of effects of Los Angeles atmosphere on laboratory rodents. *Arch. Environ. Health 16*, 871 (1968).
172. Gardner, M. B., Loosli, C. G., Hanes, B., Blackmore, W., and Teebken, D., Histopathologic findings in rats exposed to ambient and filtered air. *Arch. Environ. Health 19*, 637 (1969).
173. Vaughan, T. R., Jr., Jennelle, L. F., and Lewis, T. R., Long-term exposure to low levels of air pollutants. Effects on pulmonary function in the beagle. *Arch. Environ. Health 19*, 45 (1969).
174. Campbell, K. I., Emik, L. O., Clarke, G. L., and Plata, R. L., Inhalation toxicity of peroxyacetyl nitrate. *Arch. Environ. Health 20*, 22 (1970).
175. Dixon, W. M., and Drew, D., Fatal chlorine poisoning. *J. Occup. Med. 10*, 249 (1968).
176. Keplinger, M. L., and Suissa, L. W., Toxicity of fluorine short-term inhalation. *Amer. Ind. Hyg. Ass. J. 29*, 10 (1968).
177. Keplinger, M. L., Effects from repeated short-term inhalation of fluorine. *Toxicol. Appl. Pharmacol. 14*, 192 (1969).
178. Peters, J. M., Murphy, R. L. H., Pagnotto, L. D., and Whittenberger, J. L., Respiratory impairment in workers exposed to "safe" levels of toluene diisocyanate (TDI). *Arch. Environ. Health 20*, 364 (1970).
179. McGee, W. A., Oglesby, F. L., Raleigh, R. L., and Fassett, D. W., The determination of a sensory response to alkyl 2-cyanoacrylate vapor in air. *Amer. Ind. Hyg. Ass. J. 29*, 558 (1968).
180. Pozzani, U. C., Kinkead, E. R., and King, J. M., The mammalian toxicity of methacrylonitrile. *Amer. Ind. Hyg. Ass. J. 29*, 202 (1968).
181. Jaffe, L. S., Photochemical air pollutants and their effects on men and animals. II. Adverse effects. *Arch. Environ. Health 16*, 241 (1968).
182. Taylor, O. C., Effects of oxidant air pollutants. *J. Occup. Med. 10*, 485 (1968). (Including discussions by H. E. Heggestad and W. W. Heck.)
183. Tabershaw, I. R., Ottoboni, F., and Cooper, W. C., Oxidants: Air quality criteria based on health effects. *J. Occup. Med. 10*, 464 (1968). (Including discussions by D. V. Bates and D. L. Coffin.)
184. Stokinger, H. E., and Coffin, D. L., Biologic effects of air pollutants. *In* "Air Pollution" (A. C. Stern, ed.), 2nd ed., Vol. I, pp. 445-546. Academic Press, New York, 1968.
185. "Air Quality Criteria for Photochemical Oxidants." National Air Pollution Control Administration Publ. No. AP-63, Superintendent of Documents, Washington, D.C., March 1970.
186. Welti, D. W., and Hipp, M. J., Polymer fume fever: Possible relationship to smoking. *J. Occup. Med. 10*, 667 (1968).
187. MacFarland, H. N., The pyrolysis products of plastics – problems in defining their toxicity. *Amer. Ind. Hyg. Ass. J. 29*, 7 (1968).
188. Cornish, H. H., and Abar, E. L., Toxicity of pyrolysis products of vinyl plastics. *Arch.*

Environ. Health 19, 15 (1969).
189. Birnbaum, H. A., Scheel, L. D., and Coleman, W. E., The toxicology of the pyrolysis products of polychlorotrifluoroethylene. *Amer. Ind. Hyg. Ass. J. 29*, 61 (1968).
190. Waritz, R. S., and Kwon, B. K., The inhalation toxicity of pyrolysis products of polytetrafluoroethylene heated below 500 degrees centigrade. *Amer. Ind. Hyg. Ass. J. 29*, 19 (1968).
191. Coleman, W. E., Scheel, L. D., Kupel, R. E., and Larkin, R. L., The identification of toxic compounds in the pyrolysis products of polytetrafluoroethylene (PTFE). *Amer. Ind. Hyg. Ass. J. 29*, 33 (1968).
192. Scheel, L. D., Lane, W. C., and Coleman, W. E., The toxicity of polytetrafluoroethylene pyrolysis products – including carbonyl fluoride and a reaction product, silicon tetrafluoride. *Amer. Ind. Hyg. Ass. J. 29*, 41 (1968).
193. Scheel, L. D., McMillan, L., and Phipps, F. C., Biochemical changes associated with toxic exposures to polytetrafluoroethylene pyrolysis products. *Amer. Ind. Hyg. Ass. J. 29*, 49 (1968).
194. Namba, T., Greenfield, M., and Grob, D., Malathion poisoning: A fatal case with cardiac manifestations. *Arch. Environ. Health 21*, 533 (1970).
195. Aldrich, F. D., and Holmes, J. H., Acute chlordane intoxication in a child. *Arch. Environ. Health 19*, 129 (1969).
196. Curley, A., and Garrettson, L. K., Acute chlordane poisoning: Clinical and chemical studies. *Arch. Environ. Health 18*, 211 (1969).
197. Campbell, S., Death from paraquat in a child. *Lancet I*, 144 (1968).
198. Cant, J. S., and Lewis, D. R. H., Ocular damage due to paraquat and diquat. *Brit. Med. J. 2*, 224 (1968).
199. Fennelly, J. J., Gallagher, J. T., and Carroll, R. J., Paraquat poisoning in a pregnant woman. *Brit. Med. J. 3*, 722 (1968).
200. Oreopoulos, D. G., Soyannwo, M. A. O., Sinniah, R., Fenton, S. S. A., McGeown, M. G., and Bruce, J. H., Acute renal failure in case of paraquat poisoning. *Brit. Med. J. 1*, 749 (1968).
201. Clark, D. G., and Hurst, E. W., The toxicity of diquat. *Brit. J. Ind. Med. 27*, 51 (1970).
202. Joyce, M., Ocular damage caused by paraquat. *Brit. J. Ophthalmol. 53*, 688 (1969). From *Food Cosmet. Toxicol. 8*, 597 (1970).
203. Kimbrough, R. D., and Gaines, T. B., Toxicity of paraquat to rats and its effect on rat lungs. *Toxicol. Appl. Pharmacol. 17*, 679 (1970).
204. Tobin, J. S., Carbofuran: A new carbamate insecticide. *J. Occup. Med. 12*, 16 (1970).
205. Cavagna, G., Locati, G., and Vigliani, E. C., Clinical effects of exposure to DDVP (Vapona) insecticide in hospital wards. *Arch. Environ. Health 19*, 112 (1969).
206. Hunter, C. G., Robinson, J., and Roberts, M., Pharmacodynamics of Dieldrin (HEOD). Ingestion by human subjects for 18 to 24 months, and postexposure for eight months. *Arch. Environ. Health 18*, 12 (1969).
207. Hayes, W. J., Jr., and Curley, A., Storage and excretion of Dieldrin and related compounds. Effect of occupational exposure. *Arch. Environ. Health 16*, 155 (1968).
208. Kimbrough, R. D., Gaines, T. B., and Hayes, W. J., Jr., Combined effect of DDT, pyrethrum, and piperonyl butoxide on rat liver. *Arch. Environ. Health 16*, 333 (1968).
209. Murphy, S. D., and Cheever, K. L., Effect of feeding insecticides. Inhibition of carboxyesterase and cholinesterase activities in rats. *Arch. Environ. Health 17*, 749 (1968).
210. DuBois, K. P., Kinoshita, F. K., and Frawley, J. P., Quantitative measurement of

inhibition of aliesterases, acylamidase, and cholinesterase by EPN and Delnav®. *Toxicol. Appl. Pharmacol. 12*, 273 (1968).
211. Boyd, E. M., and Carsky, E., The acute oral toxicity of the herbicide Chlorpropham in albino rats. *Arch. Environ. Health 19*, 621 (1969).
212. Boyd, E. M., and Chen, C. P., Lindane toxicity and protein-deficient diet. *Arch. Environ. Health 17*, 156 (1968).
213. Kling, T. G., and Long, K. R., Blood cholinesterase in previously stressed animals subjected to Parathion. *J. Occup. Med. 11*, 82 (1969).
214. Tarjan, R., and Kemeny, T., Multigeneration studies on DDT in mice. *Food Cosmet. Toxicol. 7*, 215 (1969).
215. Kimbrough, R. D., and Gaines, T. B., Effect of organic phosphorus compounds and alkylating agents on the rat fetus. *Arch. Environ. Health 16*, 805 (1968).
216. Epstein, S. S., Arnold, E., Steinberg, K., Mackintosh, D., Shafner, H., and Bishop, Y., Mutagenic and antifertility effects of TEPA and METEPA in mice. *Toxicol. Appl. Pharmacol. 17*, 23 (1970).
217. Innes, J. R. M., Ulland, B. M., Valerio, M. G., Petrucelli, L., Fishbein, L., Hart, E. R., Pallotta, A. J., Bates, R. R., Falk, H. L., Gart, J. J., Klein, M., Mitchell, I., and Peters, J., Bioassay of pesticides and industrial chemicals for tumorigenicity in mice: A preliminary note. *J. Nat. Cancer Inst. 42*, 1101 (1969).
218. "Report of the Secretary's Commission on Pesticides and Their Relationship to Environmental Health." U.S. Department of Health, Education, and Welfare. Superintendent of Documents, Washington, D.C., Dec. 1969.
219. Weil, C. S., Selection of the valid number of sampling units and a consideration of their combination in toxicological studies involving reproduction, teratogenesis or carcinogenesis. *Food Cosmet. Toxicol. 8*, 177 (1970).
220. Keane, W. T., and Zavon, M. R., Validity of a critical blood level for prevention of Dieldrin intoxication. *Arch. Environ. Health 19*, 36 (1969).
221. Gaines, T. B., and Kimbrough, R. D., Oral toxicity of Mirex in adult and suckling rats. *Arch. Environ. Health 21*, 7 (1970).
222. Tauber, J., Instant benzol death. *J. Occup. Med. 12*, 520 (1970).
223. Tough, I. M., Smith, P. G., Court Brown, W. M., and Harnden, D. G., Chromosome studies on workers exposed to atmospheric benzene. The possible influence of age. *Eur. J. Cancer 6*, 49 (1970). From *Food Cosmet. Toxicol. 8*, 601 (1970).
224. Stewart, R. D., Dodd, H. C., Baretta, E. D., and Schaffer, A. W., Human exposure to styrene vapor. *Arch. Environ. Health 16*, 656 (1968).
225. Bardodej, Z., and Bardodejova, E., Biotransformation of ethyl benzene, styrene, and alpha-methylstyrene in man. *Amer. Ind. Hyg. Ass. J. 31*, 206 (1970).
226. Stewart, R. D., Baretta, E. D., Dodd, H. C., and Torkelson, T. R., Experimental human exposure to vapor of propylene glycol monomethyl ether. *Arch. Environ. Health 20*, 218 (1970).
227. Young, G., Petkau, A., and Hoogstraten, J., Chronic toxic effects of polyphenyl mixtures. *Amer. Ind. Hyg. Ass. J. 30*, 7 (1969).
228. Adamson, I. Y. R., Bowden, D. H., and Wyatt, J. P., The acute toxicity of reactor terphenyls on the lung. *Arch. Environ. Health 19*, 499 (1969).
229. Shugaev, B. B., Concentrations of hydrocarbons in tissues as a measure of toxicity. *Arch. Environ. Health 18*, 878 (1969).
230. Milby, T. H., Chronic trichloroethylene intoxication. *J. Occup. Med. 10*, 252 (1968).
231. Van Haaften, A. B., Acute tetrabromoethane (acetylene tetrabromide) intoxication in man. *Amer. Ind. Hyg. Ass. J. 30*, 251 (1969).
232. Hatfield, T. R., and Maykoski, R. T., A fatal methyl chloroform (trichloroethane)

poisoning. *Arch. Environ. Health 20*, 279 (1970).

233. Hine, C. H., Methyl bromide poisoning. A review of ten cases. *J. Occup. Med. 11*, 1 (1969).
234. Boettner, E. A., and Muranko, H. J., Animal breath data for estimating the exposure of humans to chlorinated hydrocarbons. *Amer. Ind. Hyg. Ass. J. 30*, 437 (1969).
235. Stewart, R. D., Baretta, E. D., Dodd, H. C., and Torkelson, T. R., Experimental human exposure to tetrachloroethylene. *Arch. Environ. Health 20*, 224 (1970).
236. Stewart, R. D., Dodd, H. C., Gay, H. H., and Erley, D. S., Experimental human exposure to trichloroethylene. *Arch. Environ. Health 20*, 64 (1970).
237. Stewart, R. D., Gay, H. H., Schaffer, A. W., Erley, D. S., and Rowe, V. K., Experimental human exposure to methyl chloroform vapor. *Arch. Environ. Health 19*, 467 (1969).
238. Vernon, R. J., and Ferguson, R. K., Effects of trichloroethylene on visual-motor performance. *Arch. Environ. Health 18*, 894 (1969).
239. Baretta, E. D., Stewart, R. D., and Mutchler, J. E., Monitoring exposures to vinyl chloride vapor: Breath analysis and continuous air sampling. *Amer. Ind. Hyg. Ass. J. 30*, 537 (1969).
240. Vos, J. G., and Koeman, J. H., Comparative toxicologic study with polychlorinated biphenyls in chickens with special reference to porphyria, edema formation, liver necrosis, and tissue residues. *Toxicol. Appl. Pharmacol. 17*, 656 (1970).
241. Vos, J. G., Koeman, J. H., van der Maas, H. L., ten Noever de Brauw, M. C., and de Vos, R. H., Identification and toxicological evaluation of chlorinated dibenzofuran and chlorinated naphthalene in two commercial polychlorinated biphenyls. *Food Cosmet. Toxicol. 8*, 625 (1970).
242. Kuratsune, M., *et al.*, An epidemiologic study on Yusho. *Fukuoka Acta Med. 60*, 513 (1969).
243. Nishizumi, M., Light and electron microscope study of chlorobiphenyl poisoning in mouse and monkey liver. *Arch. Environ. Health 21*, 620 (1970).
244. Haun, C. C., Vernot, E. H., Geiger, D. L., and McNerney, J. M., The inhalation toxicity of pyrolysis products of bromochloromethane (CH_2BrCl) and bromotrifluoromethane ($CBrF_3$). *Amer. Ind. Hyg. Ass. J. 30*, 551 (1969).
245. Leong, B. K. J., and Torkelson, T. R., Effects of repeated inhalation of vinyl bromide in laboratory animals with recommendations for industrial handling. *Amer. Ind. Hyg. Ass. J. 31*, 1 (1970).
246. Baetjer, A. M., Annau, Z., and Abbey, H., Water deprivation and trichloroethylene. Effect on hypothalamic self-stimulation. *Arch. Environ. Health 20*, 712 (1970).
247. Ferguson, R. K., and Vernon, R. J., Trichloroethylene in combination with CNS drugs. Effects on visual-motor tests. *Arch. Environ. Health 20*, 462 (1970).
248. Dinman, B. D., Editorial. Work demands and threshold limits. *Arch. Environ. Health 16*, 153 (1968).
249. Stewart, R. D., Peterson, J. E., Baretta, E. D., Bachand, R. T., Hosko, M. J., and Herrmann, A. A., Experimental human exposure to carbon monoxide. *Arch. Environ. Health 21*, 154 (1970).
250. Grut, A., Astrup, P., Challen, P. J. R., and Gerhardsson, G., Report of the Subcommittee on International Threshold Limits. Threshold limit values for carbon monoxide. *Arch. Environ. Health 21*, 542 (1970).
251. Astrup, P., and Pauli, H. G., eds., A comparison of prolonged exposure to carbon monoxide and hypoxia in man. Reports of a joint Danish–Swiss study. *Scand. J. Clin. Lab. Invest. Suppl.* 103, *22*, 3 (1968).
252. Astrup, P., Effects of hypoxia and of carbon monoxide exposures on experimental

atherosclerosis. *Ann. Int. Med. 71*, 426 (1969).
253. Wanstrup, J., Kjeldsen, K., and Astrup, P., Acceleration of spontaneous intimal–subintimal changes in rabbit aorta by a prolonged moderate carbon monoxide exposure. *Acta Path. Microbiol. Scand. 75*, 353 (1969).
254. Dinman, B. D., Pathophysiologic determinants of community air quality standards for carbon monoxide. *J. Occup. Med. 10*, 446 (1968). Including discussions by R. G. Smith and J. H. Schulte.
255. Goldsmith, J. R., and Landaw, S. A., Carbon monoxide and human health. *Science 162*, 1352 (1968).
256. "Air Quality Criteria for Carbon Monoxide." National Air Pollution Control Administration Publ. No. AP-62. Superintendent of Documents, Washington, D.C., March 1970.
257. Mazaleski, S. O., Coleman, R. L., Duncan, R. C., and Nau, C. A., Subcellular trace metal alterations in rats exposed to 50 ppm of carbon monoxide. *Amer. Ind. Hyg. Ass. J. 31*, 183 (1970).
258. Hodgson, G. A., and Whiteley, H. J., Personal susceptibility to pitch. *Brit. J. Ind. Med. 27*, 160 (1970).
259. Bryan, G. T., Role of tryptophan metabolites in urinary bladder cancer. *Amer. Ind. Hyg. Ass. J. 30*, 27 (1969).
260. Greenblatt, M., Raha, C., and Roe, C., Dimethylnitrosamine and hydrazine sulfate: An analysis of combined toxicity and pathology in mice. *Arch. Environ. Health 17*, 315 (1968).
261. Carter, R. L., Mitchley, B. C. V., and Roe, F. J. C., Preliminary survey of 22 printing inks for carcinogenic activity by the subcutaneous route in mice. *Food Cosmet. Toxicol. 7*, 53 (1969).
262. Deese, D. E., and Joyner, R. E., Vinyl acetate: A study of chronic human exposure. *Amer. Ind. Hyg. Ass. J. 30*, 449 (1969).
263. Osmundsen, P. E., Contact dermatitis due to an optical whitener in washing powders. *Brit. J. Dermatol. 81*, 799 (1969). From *Food Cosmet. Toxicol. 8*, 472 (1970).
264. Bingham, E., and Falk, H., Combined action of optical brighteners and ultraviolet light in the production of tumours. *Food. Cosmet. Toxicol. 8*, 173 (1970).
265. Swisher, R. D., Exposure levels and oral toxicity of surfactants. *Arch. Environ. Health 17*, 232 (1968).
266. Thomas, R. L., Scott, J. K., and Chiffelle, T. L., Metabolism and toxicity of inhaled and injected ^{131}I in the rat. *Amer. Ind. Hyg. Ass. J. 31*, 213 (1970).
267. Hodge, H. C., Highlights of fluoride toxicology. *J. Occup. Med. 10*, 273 (1968).
268. Hodge, H. C., and Smith, F. A., Fluorides and man. *Annu. Rev. Pharmacol. 8*, 395 (1968).
269. Shupe, J. L., Fluorine toxicosis and industry. *Amer. Ind. Hyg. Ass. J. 31*, 240 (1970).
270. Reynolds, R. C., Astill, B. D., and Fassett, D. W. Interaction of *N,N*-disubstituted *p*-phenylenediamines with guinea-pig epidermis *in vivo*. *Food Cosmet. Toxicol. 8*, 635 (1970).
271. Novak, L., Djuić, D., and Fridman, V., Specificity of the iodine–azide test for carbon disulfide exposure. I. Urinary excretion of sulfur compounds in persons treated with disulfiram. *Arch. Environ. Health 19*, 473 (1969).
272. Zeller, H., Hofmann, H. T., Thiess, A. M., and Hey, W. The toxicity of nitriles. (Animal experiment results and works medical experience of 15 years.) *Zentralbl. Arbeitsmed. Arbeitsschutz 19*, 225 (1969).
273. Haun, C. C., MacEwen, J. D., Vernot, E. H., and Eagan, G. F. Acute inhalation toxicity of monomethylhydrazine vapor. *Amer. Ind. Hyg. Ass. J. 31*, 667 (Nov. 1970).

274. Bourne, H. G., Yee, H. T., and Seferian, S., The toxicity of rubber additives. Findings from a survey of 140 plants in Ohio. *Arch. Environ. Health 16*, 700 (1968).
275. Cornish, H. H., Dambrauskas, T., and Beatty, L. D., Oral and inhalation toxicity of 2-*N*-dibutylaminoethanol. *Amer. Ind. Hyg. Ass. J. 30*, 46 (1969)
276. Hartung, R., and Cornish H., Cholinesterase inhibition in the acute toxicity of alkyl-substituted 2-aminoethanols. *Toxicol. Appl. Pharmacol. 12*, 486 (1968).
277. Stokinger, H. E., Cummings Memorial Lecture: The spectre of today's environmental pollution – USA brand: New perspectives from an old scout. *Amer. Ind. Hyg. Ass. J. 30*, 195 (1969).
278. Frawley, J. P., A reasoned approach to regulation based on toxicologic considerations. *Food Drug Cosmet. Law J. 23*, 260 (1968).
279. Oser, B. L., Much ado about safety. *Food Cosmet. Toxicol. 7*, 415 (1969).
280. Oser, B. L., Reflections on some scientific problems in regulatory compliance. *Ass. Food Drug Off. U.S. Quart. Bull. 34*, 129 (1970).
281. Weisburger, J. H., and Weisburger, E. K., Food additives and chemical carcinogens: On the concept of zero tolerance. *Food Cosmet. Toxicol. 6*, 235 (1968).
282. Hatch, T. F., Significant dimensions of the dose–response relationship. *Arch. Environ. Health 16*, 571 (1968).
283. Bartlett, D., Jr., and Carroll, R. E., Air quality parameters for epidemiologic studies. *Arch. Environ. Health 16*, 182 (1968).
284. Smyth, H. F., Jr., Weil, C. S., West, J. S., and Carpenter, C. P., An exploration of joint toxic action. II. Equitoxic versus equivolume mixtures. *Toxicol. Appl. Pharmacol. 17*, 498 (1970).
285. Smyth, H. F., Jr., Weil, C. S., West, J. S., and Carpenter, C. P., An exploration of joint toxic action: Twenty-seven industrial chemicals intubated in rats in all possible pairs. *Toxicol. Appl. Pharmacol. 14*, 340 (1969).
286. Baetjer, A. M., Role of environmental temperature and humidity in susceptibility to disease. *Arch. Environ. Health 16*, 565 (1968).
287. Cornish, H. H., Wilson, C. E., and Abar, E. L., Effect of foreign compounds on liver microsomal enzymes. *Amer. Ind. Hyg. Ass. J. 31*, 605 (1970).
288. Aldridge, W. N., ed., Mechanisms of toxicity. *Brit. Med. Bull. 25*, 219 (1969).
289. Malling, H. V., Chemical mutagens as a possible genetic hazard in human populations. *Amer. Ind. Hyg. Ass. J. 31*, 657 (1970).
290. Epstein, S. S., and Shafner, H., Chemical mutagens in the human environment. *Nature (London) 219*, 385 (1968).
291. Fishbein, L., Flamm, W. G., and Falk, H. L., "Chemical Mutagens. Environmental Effects on Biological Systems." Academic Press, New York, 1970.
292. Smyth, H. F., Jr., Carpenter, C. P., Weil, C. S., Pozzani, U. C., Striegel, J. A., and Nycum, J. S., Range-finding toxicity data: List VII. *Amer. Ind. Hyg. Ass. J. 30*, 470 (1969).
293. Stokinger, H. E., Current problems of setting occupational exposure standards. *Arch. Environ. Health 19*, 277 (1969).
294. Permissible levels of occupational exposure to airborne toxic substances. Sixth Report of the Joint ILO/WHO Committee on Occupational Health. *Tech. Rep. Ser. World. Health Org. 415*, 16 (1969).
295. Gage, J. C., The subacute inhalation toxicity of 109 industrial chemicals. *Brit. J. Ind. Med. 27*, 1 (1970).
296. Rosenblum, M., Manpower needs for toxicology. *Arch. Environ. Health 16*, 438 (1968).

297. Wands, R. C., The unpublished volumes of toxicology literature. *Amer. Ind. Hyg. Ass. J. 30*, 344 (1969).
298. National Referral Center for Science and Technology of the Library of Congress, "A Directory of Information Sources in the United States: General Toxicology." Superintendent of Documents, Government Printing Office, Washington, D.C., June 1969.

Noise

PAUL L. MICHAEL

INTRODUCTION

Noise pollution problems have received increasing attention in household, community, and occupational environments in recent years. Among the several reasons for increased interest in noise effects are (*1*) a greater awareness of the effects of noise on man, (*2*) the ever-increasing noise exposure levels, and (*3*) the recently enacted and impending legislation limiting noise exposures.

Much information on the effects of noise has been gained through research over the past few years, but it is obvious that many variables are not yet clearly defined. Effects that are generally understood today are:

1. Exposure to high level noise over a significant period of time can cause permanent damage to hearing.
2. Noise can mask speech communication, warning signals, or interfere with work performance.
3. Noise can disturb rest, relaxation, and sleep.

EFFECTS OF NOISE ON MAN: CRITERIA, GUIDELINES, STANDARDS, RULES, AND REGULATIONS

American Conference of Governmental Hygienists Threshold Limits for Noise

The American Conference of Governmental Industrial Hygienists (ACGIH) announced its intent to establish Threshold Limit Values for Noise at its annual meeting of 1968 and these limits were published later in 1969.[1] Limits were set for broad- and narrow-band noise which has "sound energy distributed more or less evenly throughout the eight octave bands with mid-frequencies from 63 to 8000 Hz," and for impulse or impact noises.

The limit set for broad-band noise is 90 decibels measured with an A-frequency weighting (dBA) for exposures of 4–8 hr/day; 95 dBA for 2–4 hr exposures; 100 dBA for 1–2 hr exposures; and 105 dBA for exposures less than 1 hr. The limit set for narrow-band noise, or pure tones, is 5 decibels lower than that for broad-band noises. A limit of 140 dB peak sound pressure is specified for impulse- or impact-type of noise.

The ACGIH limits further specifies that when the daily exposure to noise is composed of two or more periods at different levels, their individual exposures are to be added. That is, the total time of exposure to a specified noise level is designated as C, and T is the total time of exposure permitted at this level; hence, the fractions of allowable exposure times should be added in the following manner:

$$\frac{C_1}{T_1} + \frac{C_2}{T_2} + \ldots + \frac{C_n}{T_n}$$

The threshold limit is exceeded when the sum of these fractions totals more than unity.

Walsh-Healy Public Contracts Act

One of the most significant developments in noise pollution control during the last 3 years is the revision of the Walsh-Healy Public Contracts Act which became effective May 20, 1969.[2] The noise section of the revised Walsh-Healy Act, based for the most part on the ACGIH Threshold Limit Values, provides the first meaningful noise exposure control law enacted in the USA.

As in the ACGIH Threshold Limit Values, acceptable steady-state exposures are based on the total time of exposure to a specified level during a normal work day without regard to the number of occurrences of the noise exposure (see Table A.I). A peak pressure level limit of 140 dB is also specified in this Act for exposure to impulse or impact noises.

An important provision of the latest revision of the Walsh-Healy Act is that "when employees are subjected to sound exceeding those listed in Table I of this section, feasible administrative or engineering controls shall be utilized. If such controls fail to reduce sound levels within the levels of the table, personal protective equipment shall be provided and used to reduce sound levels within the levels of the table."* Department of Labor representatives have made it clear that these provisions mean that a reasonable effort is to be made at reducing the

*"Guidelines to the Department of Labor's Occupational Noise Standards for Federal Supply Contracts" (Bulletin 334) has been published recently by the Department of Labor to explain in detail what is expected in the way of threshold monitoring and personal protective equipment programs.

noise at the source by engineering means even though it is obvious that safe levels cannot be achieved in any practical way by these means in the near future. They hold to this position even though it is reasonably certain that personal ear protection equipment will have to be used, and that adequate protection could have been provided at the beginning by personal protective equipment.

This emphasis of engineering noise control means appears unreasonable if short-term results are considered; however, over the long term, quieter work areas will result only if engineering control efforts are required. Thus, the need for hearing conservation procedures should ultimately be reduced. Surely, if ear protectors were used in all cases where adequate protection is reasonably assured, there would be very little effort spent in lowering the noise levels produced by machinery.

The Department of Labor representatives have also made it clear that they expect hearing threshold monitoring for all persons in exposure levels above those listed in Appendix I even though personal protective equipment is provided. Threshold monitoring is the only way presently available to check the effectiveness of an ear protector program and to detect those persons who are very susceptible to noise-induced hearing loss; however, it should be realized that hearing testing is not a sensitive indicator of progressive hearing loss. Even the best industrial hearing measurement programs will not provide an accuracy in threshold measurements greater than about ±5 dB. Therefore, only losses greater than about 10 dB will be evident in most cases and these threshold shifts may not become obvious before several years of exposure. Temporary threshold shifts caused by colds, infections, etc., also confuse audiogram interpretation.

The value of hearing threshold monitoring should not be overlooked because of the problems listed above. However these problems emphasize the fact that every effort should also be made to assure that an effective hearing conservation program is underway wherever needed so that an evaluation by hearing testing is not relied upon exclusively.

As before, the provisions of the Walsh-Healey Act shall apply to all work areas where there is manufacturing or furnishing of materials, supplies, articles, and equipment in any amount exceeding $10,000 for any agency of the United States. Thus, this is the first meaningful ruling for noise exposure control in the United States that has specific limits and a base for widespread enforcement.

Guidelines for Noise Exposure Control

An Intersociety Committee published a paper entitled "Guidelines for Noise Exposure Control"[3] in various scientific journals in 1967 which was discussed in some detail by T. B. Bonney.[4] A second Intersociety Committee was formed in 1968 to review these Guidelines and the final draft of a revised document was released to participating associations in February 1970. If approved by these associations, this document should be published soon in their journals.

The revised guidelines draft differs from the original version mainly in that it now sets forth acceptable exposure levels as well as incidence of hearing impairment data in the general population and in selected populations by age groups and occupational noise exposure. The acceptable exposure levels cited for 8 hours or less are the same as the Walsh-Healey values shown in Appendix I; however, there is, in addition, a second set of acceptable exposure levels established as a function of the number of occurrences per day. This second set of acceptable levels has the effect of permitting higher exposure levels for any given duration over a 24-hr period than the Walsh-Healey limits when there is a large number of interruptions in noise exposure during the day, the acceptable levels increasing with the number of noise occurrences per day for any given total duration of exposure.

Proposed Health Standards for Underground Coal Mines

The Bureau of Mines, Department of the Interior, published a Proposed Rule Making entitled "Mandatory Health Standards for Underground Coal Mines" in the *Federal Register.*[5] The requirements of the noise portion of these proposed Health Standards are set forth in the form of acceptable exposures to noise in dBA as a function of the number of occurrences per day. Recent data[3] have shown that this method of evaluating noise exposures has merit for guidelines; however, its complexity in the present form will make it very difficult to enforce as a rule or regulation.

Federal, State, and Local Community Noise Activities

Community noise problems have received particular attention during the past 3 years in aircraft, industrial, and ground transportation areas. Efforts have continued toward the development of more meaningful standards, regulations, and methods of measurement. Most of these activities have been concentrated on annoyance[6-22] or interference with communication[23,24]; however, some are also related to the health, productivity, and well-being of persons exposed.[25-27]

The Federal Aviation Administration proposed the first Federal noise standards for new subsonic transport planes on January 6, 1969, which would reduce noise levels substantially below that of some noisy planes currently in use. Until this time, noise control at airports has been left to local authorities.

Growing concern about community noise around airports is shown by the increasing number of suits being filed over aircraft noise. The 1.4 billion dollar suit filed by 94,000 Inglewood residents against the city of Los Angeles is an example of recent protests against aircraft noise. The residents, who live near the

Los Angeles International Airport, claimed in their suit that the aircraft noise resulted in "nerve and emotional disturbances of a permanent nature."

Sonic boom problems continue to receive much attention in the press.[13-17, 28-30] Studies of the effects of sonic boom range from those on community response[13-15] to the effects on structural behavior.[28-30] Sonic boom and other noise-related problems, are, of course, major factors in the argument against the supersonic transport.

Ground transportation has also experienced increased activities in the development of standards and legislation in noise pollution on federal, state, and local levels. A precedent-setting ruling on the responsibility for transportation-related noise pollution was made by the New York Court of Appeals June 30, 1968, when it rejected the state's contention that traffic noise should be excluded as an element of damages because it is suffered generally by the public. Decision by the state's highest court upheld a $37,000 award to Ira and Dorothy Dennison whose land in the Lake George area was partially taken for the Luzerne Road–Lake George Interchange.

California, New York, and other states are moving toward a planned program to reduce noise levels on a broad front. Major effort in these programs will be to collect and interpret data on the nature and extent of community noise problems and their effects on health. In California, Dr. John M. Heslep, Chief of the Division of Environmental Health in the California State Board of Health, has indicated that the noise standards in their present Vehicle Code[6] "are far too high to result in any substantial reduction in vehicle noise" and that broad statewide legislation may be recommended to reduce noise levels. Several other states have used, or are in the process of using, the present inadequate California Vehicle Code as a model for their codes.[31]

Internationally, the Stockholm City Board of Health has successfully appealed a Swedish Board of Aviation decision to allow jet traffic at Stockholm's domestic airport, Bromma. The ruling is regarded as precedent-setting in that it places environmental factors above economic considerations.

In Japan, the Ministry of Health and Welfare is planning to establish monitoring stations for air and noise pollution in all Japanese cities with populations greater than 500,000 as a step toward a coordinated national program of nuisance abatement.

One of the most significant contributions to community noise assessment during the past 3 years is the latest International Organization for Standardization draft Recommendation No. 1996 Noise Assessment with Respect to Community Response (May 15, 1970)[32] which appears to be near approval. This proposed Recommendation sets forth a method for rating noises with respect to community response based on A-frequency weighting data that can be used as a basis on which limits for noises in various situations may be set by competent authorities.

Other Legislative Activities

The Federal Occupational Safety and Health Law was passed and signed on December 29, 1970, and became effective April 28, 1971.[212] The acceptable noise limits in this law are similar to those in the revised Walsh-Healey Act; however, the coverage of this law is much broader than that of the Walsh-Healey Act.

House Bill No. 2492 amending the Pennsylvania Occupational Disease Act (1939) was referred to the Committee on Labor Relations, the General Assembly of Pennsylvania, July 15, 1970. This amendment considers noise-induced hearing loss to be compensible as an occupational disease and outlines provisions for administrating claims.

Similarly, a bill concerning workman's compensation, and supplementing Chapter 15 of Title 34 of the Revised Statutes, State of New Jersey, was introduced to the New Jersey Assembly March 17, 1969. The provisions of this bill make occupational hearing loss compensible and outline means for administration of claims.

The State of California Division of Industrial Safety issued a revision of Group 6.1 Noise Control Safety Orders under Article 55, Standards for Occupational Noise Exposure July 1970. The revised Safety Orders now correspond closely with the provisions set forth in the revised Walsh-Healey Act.

Philadelphia will apply noise control provisions based roughly on the revised Walsh-Healey Act to all industry in the city.[33] The city health department will make 2500 mailings to local industries informing them of the new regulations and requesting conformance.

Numerous other rules and regulations on various aspects of noise pollution have been proposed on federal, state, and local levels during the past 3 years. Some of these proposals have shown considerable thought; however, most require extensive refinements before they can be classified as good legislation.

NOISE MEASUREMENT

Noise measurement instrumentation and survey techniques have changed very little in the past few years. Some new small portable instruments have been made available, but these devices provide essentially the same capability and accuracy of measurement as described by T. B. Bonney in Volume 1 of "Industrial Hygiene Highlights."[4] For the present state of knowledge on the effects of noise on man and for the rules and regulations based on this knowledge, presently available instruments are adequate if they are used in the prescribed manner and if they are maintained in calibration.[35-39]

At the present time, any sound level meter meeting the pertinent ANSI Specifications[34] should be satisfactory for monitoring steady-state noise levels for hearing conservation purposes if used properly. However, if noise control or personal ear protection programs are to be undertaken, or otherwise specified, octave- or narrow-band analyzers[40,41] may be required. Portable calibrators[42] are available that should be used daily with these instruments, and more thorough laboratory checks should be made periodically, perhaps once each year.

Noises with impulsive characteristics which have significant level variations in periods of time shorter than 0.2 sec cannot be measured accurately with a conventional sound level meter, and an impact-noise analyzer[43-45] or an oscilloscope must be employed to determine peak sound pressure levels which are specified in present hearing conservation criteria. Care must be taken when purchasing impact-noise instruments to be sure that they conform to the specifications of the Walsh-Healey Act if this is the purpose of the peak pressure measurements.

To summarize, instruments are available that will provide all of the information called for in present hearing conservation criteria and in most noise control applications. The importance of daily calibration of these instruments cannot be overemphasized.

HEARING MEASUREMENT

Standards

A new American National Standard Specifications for Audiometers, ANSI S3.6-1969[46] became effective September 1, 1970. These new specifications are much more complete and specify closer tolerance limits than the three specifications (Z24.5-1951, Z24.12-1952, and Z24.13-1953) which they replace.

The most significant change in the audiometer specifications, and the one that has caused so much controversy over the past 15 years while the new specifications were being written, is the new hearing threshold reference zero levels.[47] The sound pressure levels corresponding to the new threshold reference levels are lower (higher hearing sensitivity) at all test frequencies, the changes ranging from 14 dB at 500 Hz to 6 dB at 4000 Hz. The Scope and Purpose section of the new specifications clearly states that the new threshold levels should not be interpreted as altering previously implied physical sound pressure levels for specific purposes such as in laws and administrative rules and regulations relating to the impairment of hearing, to minimum requirements for employment, to audiometric screening levels in school systems, etc.

Wide-range, limited-range, and narrow-range audiometers are specified in the new specifications for pure-tone threshold measurement. The wide-range audiometer is "intended primarily for clinical and diagnostic purposes, or for the measurement of the hearing thresholds of children." The limited range audiometer is "intended for measuring the hearing threshold levels of adult populations such as those found in industry." The narrow-range audiometer "is more restricted than a limited range audiometer in its ranges of frequency and sound pressure levels."

The limited-range audiometer which is of primary interest to industry is intended for air conduction threshold measurements with test tones provided at least at 500, 1000, 2000, 3000, 4000, and 6000 Hz, with hearing levels from 10 dB to at least 70 dB referenced to the new (ISO 1963) threshold levels. Facilities for bone conduction measurements and masking may be omitted.

Other significant provisions which have been changed from the previous audiometer standard include:

1. The accuracy of sound pressure levels shall be ±3 dB at test frequencies from 250–3000 Hz, ±4 dB at 4000 Hz, and ±5 dB at all other test frequencies above or below this range.

2. The measured difference between two successive designations of hearing threshold level shall not differ from the dial-indicated difference by more than (*1*) three-tenths of the dial interval measured in decibels, or (*2*) 1 dB, whichever is larger.

3. The accuracy of test tone frequencies shall be ±3% of the indicated frequency for discrete-frequency audiometers.

4. The sound pressure level of any harmonic of the fundamental shall be at least 30 dB below the sound pressure level of the fundamental.

A separate standard specification is being written on test room requirements for hearing measurement which will correspond to the threshold levels specified in the new audiometer standard. These room specifications should be completed in the near future. It might be anticipated that the room requirements will specify levels about 5 dB quieter for industrial purposes than the present standard.[48]

Training Audiometric Technicians

Experience with many training courses over the past few years has shown that the 2½-day minimum training time set by the Intersociety Committee on Industrial Audiometric Technician Training[49] is adequate in most cases if the course is well planned and if the instructors are well prepared. However, it is obvious that an on-the-job check of his work, or a short refresher course would be very helpful after a technician has been trained and has had an opportunity to work in the field for a few months. This additional training period would enable the technician to discuss the many questions that will have arisen on the job, and it would provide an opportunity to check the technician's record-keeping and his overall measurement techniques.

It should be remembered that accuracy in the measurement of hearing thresholds requires a well-trained technician,[49] a quiet test environment,[48] and an accurately calibrated audiometer.[46] A weakness in any of these factors will result in inaccurate hearing measurement data.

NOISE CONTROL

Personal ear protective equipment in the form of plugs or muffs is, by far, the most common means for reducing noise exposures at this time. Considerable attention has been given to engineering means of noise control during the past few years, but much remains to be done in this area. The Department of Labor's long-term philosophy of emphasizing the need for noise reduction by engineering means through Walsh-Healey Act[2] enforcement should provide emphasis to further this work, but it is obvious that personal protective equipment will be a major factor in hearing conservation programs for some time.

Engineering Control

There have been many papers presented and published during the past 3 years on engineering means of noise control. The American Industrial Association, the Acoustical Society of America, the American Society of Safety Engineers, the National Safety Council, the American Machine and Tool Builders Association, and many others have had special sessions and many publications on noise control. In addition to scientific organizations, many suppliers of noise reduction equipment, universities, trade journals, the armed forces, and individuals have presented and published valuable information in this field. The many publications are too numerous to discuss here; however, some of these are referenced.[50-65,186-201]

The combined information from these and other references along with the present level of noise control technology provide the tools necessary to solve a very large majority of noise reduction problems. In most cases, however, the need for common sense and ingenuity in the application of present knowledge is of far greater importance than the need for new technology.

Personal Protective Equipment

Even though the long-term philosophy for noise reduction is now concentrated on engineering control means, it is obvious that personal protective equipment will retain a most important position in hearing conservation programs for a long time. Considerable effort continues in the development of new personal protective equipment and in the development of new performance[66] and evaluation standards.[67-68]

There have been no radically new muff-type personal protective designs produced in the last few years; however, new materials and improved individual protector models make possible somewhat improved and more consistent protection with these devices. The major United States suppliers of muff-type protective devices [Sellstrom Manufacturing Ltd., Montreal, Canada, is distributing muff-type protectors through Sellstrom Manufacturing Company, Palatine, Illinois 60067] now include American Optical Company,[69] Bausch & Lomb, Inc.,[70] David Clark Company, Inc.,[71] Flents Products Company,[72] Glendale Optical Company,[73] Mine Safety Appliances Company,[74] Welsh Manufacturing Company,[75] Willson Products Division, ESB,[76] and U.S. Safety Service Company.[77] Muff-type protectors from all of these major suppliers are of good quality, each having advantages and disadvantages for specified uses.

Insert-type protectors have also remained much the same as those described by T. B. Bonney 3 years ago.[4] A new molded insert-type protector has been introducted by Willson Products Division, ESB, since Mr. Bonney's report. This air-filled and flanged plug provides good protection and wearing comfort, particularly when used in a straight and round ear canal.

The major suppliers of insert-type protectors now include American Optical Company,[69] Flents Products Company,[72] Frontier Industrial Products,[79] Glendale Optical Company,[73] Mine Safety Appliances Company,[74] Rockford I. C. Webb, Inc.,[80] Safety Ear Protector Company,[81] Sigma Engineering Company,[78] Stayrite, Inc.,[82] Surgical Mechanical Research,[83] Welsh Manufacturing Company,[75] Willson Products Division, ESB,[76] and U. S. Safety Service Company.[77] The most recent additions to this list, Glendale Optical Company and Welsh Manufacturing Company, are supplying the V51-R insert protector design which is the same as that sold by American Optical Company and Mine Safety Appliances Company.

Protectors which are not classified as either insert or muff types include individually molded designs and others that provide an acoustic seal around the entrance to the ear canal. Mr. Bonney[4] has described the Sound Sentry[84] which makes use of a narrow headband to press two soft plastic, conical canal cups against the entrance to the ear canal. Individually molded devices which are seated in the concha of the external ear and extend into the ear canal without external support in a manner similar to a hearing aid earpiece have been available for several years. General Electric Company introduced a new version of the individually molded protector in 1970[85] which, because of the characteristics of the material used, can be molded and ready for use in about 15 min. The protection provided and the wearing comfort of the G. E. "Peacekeeper" are quite good if the protectors are carefully molded according to the instructions provided.

Each type of ear protector has distinct advantages and disadvantages[86] that should be considered when personal protective devices are chosen for a specific use. Briefly, the following characteristics might be considered:

INSERT AND CONCHA SEATED PROTECTORS (NO HEADBANDS)

Advantages:

1. They are small and can be easily carried.
2. They can be worn conveniently and effectively with other personally worn items such as glasses, various headgear, or hairstyles.
3. They are more comfortable to wear than muffs in hot environments.
4. They are convenient to wear where the head must be maneuvered in close quarters.
5. The cost of sized insert protectors is generally less than muff types (however, hand-formed and some molded insert-types may be more expensive than muff types).

Disadvantages:

1. Sized and molded insert-type protectors require more time and effort to provide an effective fit than for muff-type protectors.
2. The amount of protection provided by a good insert-type protector is generally less and more variable from one individual to another than that provided by a good muff-type protector.
3. Dirt may be inserted into the ear canal if the protectors are removed and reinserted with dirty hands.
4. Insert protectors are more difficult to see in the ear from a distance; hence, it is more difficult to police the wearing of insert-types than muff-type protectors.
5. Insert protectors can be worn only in healthy ear canals.

MUFF-TYPE PROTECTORS

Advantages:

1. The protection provided by a good muff-type protector is generally greater and less variable from one individual to another than that of insert-type protectors.
2. A single size of muff-type protector fits a very large percentage of heads.

3. The relatively large muff-type protector can be readily seen at a distance; thus, the wearing of these protectors is easily monitored.

4. The muff-type protector is accepted more readily by more people at the beginning of a hearing conservation program than sized insert protectors.

5. Muff-type protectors can be worn even with many minor ear infections.

6. Muff-type protectors are not misplaced or lost as easily as are insert types.

Disadvantages:

1. The muff-type protector is uncomfortable in hot environments.

2. Muff-type protectors are not as easily carried nor stored as insert types.

3. Muff-type protectors are not as compatible with other personally worn items such as glasses and headgear as are insert types.

4. Muff suspension force may be reduced by usage, or by deliberate bending, so that the protection provided will be substantially less than expected.

5. The relatively large size of the muff-type may not be acceptable where the head must be maneuvered in close quarters.

6. The muff-type protector is more expensive than most insert types.

It is doubtful that either the insert- or the muff-type protector alone can satisfy all needs in any sizable organization. The obvious advantages of each should be used wherever possible to make a hearing conservation program more acceptable.

Procedures for determining the effectiveness of ear protectors in hearing conservation programs are discussed in Appendix II.

INFORMATION SOURCES

Conferences

The number of conferences offering training on various noise-related subjects have increased significantly during the past 2 years. The American Industrial Hygiene Association, the American Conference of Governmental Industrial Hygienists, the American Society of Safety Engineers, the Industrial Medical Association, the American Medical Association, the Acoustical Society of America, the American Speech and Hearing Association, and other health-related associations and societies have had many papers presented on the broad topic of noise at their meetings and most of these papers are eventually published in their respective journals. Conferences on various aspects of noise pollution have also been sponsored by the National Safety Council, the National Council on Noise Abatement, universities, and various consulting groups.

A very successful and informative international Conference on Noise as a

Public Health Hazard was held in Washington, D.C., June 13–14, 1968.[87] This Conference sponsored by the U. S. Public Health Service and the American Speech and Hearing Association was organized in an effort to present the best evidence available bearing on the general question: To what extent is noise a public health hazard? Similar meetings have also been held in many foreign countries including Canada,[88] Austria,[89] and Finland.[90]

Short Courses

The number of short courses or seminars on noise lasting from a few days to 2 weeks have also increased significantly during the past 2 years. In addition to the established annual seminars held at Colby College[91] and The Pennsylvania State University[92] and the biennial seminar held at the Massachusetts Institute of Technology,[93] many other seminars have been sponsored by universities, colleges, equipment manufacturers, associations, societies, and consulting groups.

Publications

A number of worthwhile publications on noise have been made available since Mr. Bonney listed recommended noise publications in Volume I of "Industrial Hygiene Highlights" in 1968.[4]

BOOKS

The book "Noise and Man" by William Burns[94] (336 pp.) briefly covers the physical properties of sound and its measurements, the mechanism of hearing, health, and well-being of man. This book, written in England, contains material that is useful in all countries and is quite a bargain at about six dollars.

"The World of Sound," by Vernon M. Albers[95] (241 pp.), is another bargain at about six dollars for those interested in a basic introduction to the science of sound without the use of complicated mathematics. Dr. Albers discusses the nature of sound, units and reference quantities, sound propagation, transducers, measurement and analysis, music, room acoustics, speech and hearing, effects of noise on hearing, underwater acoustics, and careers in acoustics in this book.

"Acoustics-Room Design and Noise Control," by Michael Rettinger[96] (386 pp.), should serve as a valuable reference in room acoustics, noise reduction procedures, and room design for the practitioner of architectural acoustics or for those who have some training in acoustics. However, sections of this book may be beyond the scope of persons who are just getting into this field.

The Proceedings of the Conference on Noise as a Public Health Hazard[87] held in Washington, D.C., June 13–14, 1968 (384 pp.) is one of the best references on the state of the art on the general subject of noise hazards. Papers on all aspects of noise pollution by internationally recognized authors are included in this publication. These Proceedings are also quite a bargain at five dollars.

The second edition of the "American Industrial Hygiene Association Industrial Noise Manual"[97] was published in 1966 and a third edition is being prepared. This manual is highly recommended for those interested in the broad aspects of industrial noise problems. Topics include physics of sound, sound measurement, vibration, physiology of the ear, effects of noise on man, medical aspects, personal protection, engineering control, and legal aspects.

PAPERS AND OTHER SOURCES OF INFORMATION

An international quarterly journal introduced in 1968, *Applied Acoustics,*[65] will deal with a broad scope of noise problems. This English-based journal is intended generally for engineers, architects, industrial hygienists, and others who are interested in practical aspects of acoustics.

The IIT Research Institute is initiating a joint industrial program for the establishment of a Noise Information Service (NOISE).[211] The purpose of this service is to supply program sponsors with monthly bibliographies of published references on noise and pertinent related subjects, and to construct a data base for searches on specific topics.

In addition to those papers listed earlier, the following references covering various noise-related problems may be of interest:

1. Noise-induced hearing loss caused by steady-state noise exposures.[98–133]
2. Noise-induced hearing loss caused by intermittent exposure to steady-state noise.[134–136]
3. Noise-induced hearing loss resulting from exposure to airborne ultrasonic noises.[137–138]
4. Noise-induced hearing loss resulting from exposure to impulse or impact-type noise.[139–154]
5. Miscellaneous legislative activities.[155–163]
6. Community noise activities.[164–173]
7. Noise measurement.[174–176]
8. Hearing measurement.[177–185]
9. Noise control.[186–201]
10. General.[202–210]

APPENDIX I: WALSH-HEALEY PUBLIC CONTRACTS ACT*

Rules and Regulations

§50-204.10 OCCUPATIONAL NOISE EXPOSURE

(a) Protection against the effects of noise exposure shall be provided when the sound levels exceed those shown in Table A.I of this section when measured on the A scale of a standard sound level meter at slow response. When noise levels are determined by octave band analysis, the equivalent A-weighted sound level may be determined.

(b) When employees are subjected to sound exceeding those listed in Table A.I of this section, feasible administrative or engineering controls shall be utilized. If such controls fail to reduce sound levels within the levels of the table, personal protective equipment shall be provided and used to reduce sound levels within the levels of the table.

(c) If the variations in noise level involve maxima at intervals of 1 sec or less, it is to be considered continuous.

(d) In all cases where the sound levels exceed the values shown herein, a continuing, effective hearing conservation program shall be administered.

TABLE A.I
Permissible Noise Exposures[a]

Duration per day (hr)	*Sound level dBA, slow response*
8	90
6	92
4	95
3	97
2	100
1½	102
1	105
½	110
¼ or less	115

[a]When the daily noise exposure is composed of two or more periods of noise exposure of different levels, their combined effect should be considered, rather than the individual effect of each. If the sum of the following fractions: $C_1/T_1 + C_2/T_2 + \ldots + C_n/T_n$ exceeds unity, then the mixed exposure should be considered to exceed the limit value. C_n indicates the total time of exposure at specified noise level, and T_n indicates the total time of exposure permitted at that level.

*As reprinted from *Federal Register,* Vol. 34, No. 96, Tuesday, May 20, 1969. As revised by errata sheet dated July 15, 1969.

Exposure to impulsive or impact noise should not exceed 140 dB peak sound pressure level.

APPENDIX II

Two factors must be carefully considered when personal protection devices are evaluated for a hearing conservation program.

1. The ear protector's attenuation values used must be representative for all frequencies throughout the octave band, not just at a single frequency, or a small band, centered at the octave band center frequency. Unfortunately, details of protector performance throughout the octave band are not often available; thus, there is an obvious need to check the effectiveness of the ear protectors by periodic hearing tests.

2. The performance of most ear protectors varies considerably between wearers and between the way the protectors are worn. There are also significant differences in performance between some protectors of the same model. Standard deviations of 4–7 dB are commonly found in measurements of protector attenuation; therefore, if 95% confidence limits are used, the attenuation values will have a range of perhaps ±8 to ±14 dB. On this basis, the lowest rather than the mean attenuation values should be used when selecting an ear protector for a particular application.

TABLE A.II
Adjustment of a Linear Spectrum to an A-Frequency Weighted Spectrum

Frequency in Hz	*Adjustment to be added to linear spectrum values to obtain equivalent dBA levels in dB*
25	−44.8
32	−39.2
50	−30.2
63	−26.1
100	−19.1
125	−16.2
250	− 8.6
500	− 3.3
1000	− 0.8
1600	0
2000	+ 1.0
2500	+ 1.2
3150	+ 1.2
4000	+ 1.0
5000	+ 0.5
6300	− 0.2
8000	− 1.1

TABLE A.III
Table for Combining Decible Levels of Noises with Random Frequency Characteristics[a]

Sum (L_R) of dB levels L_1 and L_2	
Numerical difference between levels L_1 and L_2	L_3: Amount to be added to the higher of L_1 or L_2
0.0 to 0.1	3.0
0.2 to 0.3	2.9
0.4 to 0.5	2.8
0.6 to 0.7	2.7
0.8 to 0.9	2.6
1.0 to 1.2	2.5
1.3 to 1.4	2.4
1.5 to 1.6	2.3
1.7 to 1.9	2.2
2.0 to 2.1	2.1
2.2 to 2.4	2.0
2.5 to 2.7	1.9
2.8 to 3.0	1.8
3.1 to 3.3	1.7
3.4 to 3.6	1.6
3.7 to 4.0	1.5
4.1 to 4.3	1.4
4.4 to 4.7	1.3
4.8 to 5.1	1.2
5.2 to 5.6	1.1
5.7 to 6.1	1.0
6.2 to 6.6	0.9
6.7 to 7.2	0.8
7.3 to 7.9	0.7
8.0 to 8.6	0.6
8.7 to 9.6	0.5
9.7 to 10.7	0.4
10.8 to 12.2	0.3
12.3 to 14.5	0.2
14.6 to 19.3	0.1
19.4 to ∞	0.0

[a]Step 1. Determine the difference between the tow levels to be added (L_1 and L_2).
Step 2. Find the number (L_3) corresponding to this difference in the Table.
Step 3. Add the number (L_3) to the highest of L_1 and L_2 to obtain the resultant level ($L_R = L_1 + L_2$).

After the ear protector attenuation values have been determined, the following procedure may be used to predict the effectiveness of the protective devices when they are used in a hearing conservation program based on an A-weighted exposure level criterion:

Step 1. Take octave-band level measurements of the noise environment at the point of exposure.

Step 2. Subtract from the octave-band levels (in decibels) obtained in Step 1 the center-frequency adjustment values for the A-frequency weighting shown in Table A.II.

Step 3. Subtract from the A-weighted octave bands calculated in Step 2 the representative attenuation values provided by the protector for each corresponding octave band to obtain the A-weighted octave-band levels reaching the ear while wearing the ear protector.

Step 4. Calculate the equivalent steady-state A-weighted noise level reaching the ear while wearing the ear protector by adding the octave band levels as follows:

$$\text{Equivalent dBA} = 10 \log_{10} \left(\operatorname{antilog}_{10} \frac{L_{31}}{10} + \operatorname{antilog}_{10} \frac{L_{63}}{10} + \cdots + \operatorname{antilog}_{10} \frac{L_{8000}}{10} \right)$$

An alternate method for calculating equivalent A-weighted values from octave bands is shown in Table A.III.

REFERENCES

1. American Conference of Governmental Industrial Hygienists. Threshold Limit Values of Physical Agents Adopted by ACGIH for 1970.
2. Safety and Health Standards for Federal Supply Contracts (Walsh-Healey Public Contracts Act). U. S. Department of Labor *Fed. Regist. 34,* 7948 (1969).
3. Guidelines for Noise Exposure Control. *Amer. Ind. Hyg. Ass. J. 28,* 418 (1967); *ibid, Arch. Environ. Health 15,* 674 (1967); *ibid, J. Occup. Med. 9,* 571 (1967); *ibid, Amer. Ass. Ind. Nurses J. 16,* 17 (1968).
4. Cralley, L. V., Cralley, L. J., and Clayton, G. D., eds., "Industrial Hygiene Highlights." Industrial Hygiene Foundation of America, Pittsburgh, Pennsylvania, 1968.
5. Mandatory Health Standards for Underground Coal Mines. U. S. Department of the Interior. *Fed. Regist. 35,* 18671 (1970).
6. California Motor Vehicle Code, Divisions 11 and 12, Paragraphs 23130 and 27160 (1968). (Chapter 2, Title 13, California Administrative Code.)
7. "Toward a Quieter City." Report of Mayor's Task Force on Noise Control, New York City, 1970.
8. Apps. D., Automobile noise. *In* "Handbook of Noise Control" (C. M. Harris, ed.), Chapt. 31, p. 31. McGraw-Hill, New York, 1957.
9. "Objective Limits for Motor Vehicle Noise." Report No. 824, Bolt, Beranek and Newman, Inc., Van Nuys, California, Dec. 1962.
10. Campbell, R. S., A survey of passby noise from boats. *Sound Vibration 3*(9), 24 (1969).
11. Bender, E. K., and Heckl, M., Noise generated by subways aboveground and in stations. Report No. OST-DNA-70-1, Bolt, Beranek and Newman, Inc., Cambridge, Massachusetts, Jan. 1970.
12. Kryter, K. D., Psychological reactions to aircraft noise. *Science 151,* 1346 (1966).
13. Nixon, C. W., and Hubbard, H. H., Results of USAF NASA–FAA flight program to

study community responses to sonic booms in the greater St. Louis area. *NASA Tech. Note* TN D-2705 (1965).

14. Proceedings of the Sonic Boom Symposium. *J. Acoust. Soc. Amer. 39,* S1 (1966).
15. Kupferman, T., H. R. Bill 14608, Congressional Record, 8339 (April 21, 1966).
16. Aircraft Noise – Unrelenting, Unremitting, Intolerable. *Environ. Sci. Technol. 1,* 977 (1967).
17. Rosinger, G., Nixon, C. W., and Von Gierke, H. E., Quantification of the noisiness of "approaching" and "receding" sounds. *J. Acoust. Soc. Amer. 48,* 843 (1970).
18. Goodfriend, L. S., The wrong road to community noise regulation. *Sound Vibration 1*(2), 7 (1967).
19. Mehling, E. A., Community noise ordinances. *ASHRAE (Amer. Soc. Heat. Refrig. Air Cond. Eng.) J. 9,* 40 (1967).
20. Morse, K. M., Community noise – the industrial aspect. *Amer. Ind. Hyg. Ass. J. 29,* 368 (1968).
21. Donley, R., Community noise regulation. *Sound Vibration 3*(2), 12 (1969).
22. Bragdon, C. R., Community noise and the public interest. *Sound Vibration 3*(12), 16 (1969).
23. American National Standard Methods for the Calculation of the Articulation Index, ANSI S3.5-1969, American National Standards Institute, New York, 1969.
24. Webster, J. C., SIL – past, present and future. *Sound Vibration 3*(8), 22 (1969).
25. Cohen, A., Noise effects on health. Productivity and well-being. *Trans. N.Y. Acad. Sci. 30,* 910 (1968).
26. Anticaglia, J. R., and Cohen, A., Extra auditory effects of noise as a health hazard. *Amer. Ind. Hyg. Ass. J. 31,* 277 (May 1970).
27. Cohen, A., Anticaglia, J., and Jones, H., "Sociocusis"–hearing loss from non-occupational noise exposure. *Sound Vibration 4*(11), 12 (1970).
28. The Effects of Sonic Boom on Structural Behavior. SST Report No. 65-18, prepared for Federal Aviation Agency by J. Blume & Ass. (1965).
29. Sonic Boom Experiments at Edwards Air Force Base. By Stanford Research Institute, Contract AF 49 (638) - 1758 (July 1967).
30. Hubbard, H. H., Sonic booms. *Phys. Today 21,* 31 (1968).
31. Amendment of P. L. 58 ("The Vehicle Code"). House Bill No. 1660, The General Assembly of Pennsylvania (Oct. 6, 1969).
32. ISO, International Organization for Standardization. Draft Recommendation No. 1996. Noise Assessment with Respect to Community Response (May 15, 1970).
33. Technology Newsletter, *Chem. Week 106,* 67 (1970).
34. American National Standard Specification for General Purpose Sound Level Meters, S1.4-1961. American National Standards Institute, New York, 1961.
35. NMTBA Noise Measurement Techniques. National Machine Tool Builders Association, Washington, D.C., June 1970.
36. Ranz, J. R., A survey of noise measurement methods. *Mach. Des. 38,* 199 (1966).
37. Skode, F., Windscreening of outdoor microphones. *Tech. Rev. 1,* 3 (1966). B&K Instruments, Inc., Cleveland, Ohio.
38. Bauer, B. B., and Foster, E. J., Methodology for acoustical data gathering. *J. Audio Eng. Soc. 16,* 390 (1968).
39. Schneider, E. J., Microphone orientation in the sound field. *Sound Vibration 4*(2), 20 (1970).
40. American National Standard for Preferred Frequencies and Band Numbers for Acoustical Measurements, S1.6-1967. American National Stnadards Institute, New York, 1967.

41. American National Standard for Octave, Half-Octave, and Third-Octave Band Sets, S1.11-1966. American National Standards Institute, New York, 1961.
42. Sound Level Calibrator, Type 1562-A, General Radio Company, West Concord, Massachusetts.
43. "Acoustics Handbook." Hewlett-Packard Company, Palo Alto, California, 1968.
44. "Handbook of Noise Control." General Radio Company, West Concord, Massachusetts, 1967.
45. Broch, J. "Acoustic Noise Measurements," 2nd Ed. Bruel and Kjaer, Copenhagen, 1971.
46. American National Standard Specifications for Audiometers, ANSI S3.6-1969. American National Standards Institute, New York, 1969.
47. Michael, P. L., Standardization of normal hearing thresholds, *J. Occup. Med. 10,* 67 (1968).
48. American National Standard Criteria for Background Noise in Audiometer Rooms, ANSI S3.1-1960. American National Standards Institute, New York, 1960.
49. Intersociety Committee on Industrial Audiometric Technician Training, Guide for Training of Industrial Audiometric Technicians. *Amer. Ind. Hyg. Ass. J. 27,* 303 (1966).
50. Cerami, V. V., and Bishop, E. S., Control of duct generated noise. *Air Cond. Heat. Vent. 63*(9), 55 (1966).
51. Design for Quiet. *Mach. Des. 39,* 174 (1967).
52. Fenton, R. G., Reducing noise in cams. *Mach. Des. 38,* 187 (1966).
53. Quieting Noise Mathematically. Search 2 Published by General Motors Research Laboratory, Sept.–Oct. 1967.
54. Quiet Transit Wheel. *J. Acous. Soc. Amer. 41,* 537 (1967).
55. Uniroyal Rubber Mufflers. UNIROYAL Industrial Division, Dominion Rubber Co., Montreal, Quebec.
56. Fader, B., Practical designs for noise barriers based on lead. *Amer. Ind. Hyg. Ass. J. 27,* 520 (1966).
57. Hines, W. A., "Noise Control in Industry," p. 197, Business Applications Ltd., London, 1966.
58. Bishop, D. E., Reduction of aricraft noise measured in several school, motel and residential rooms. *J. Acoust. Soc. Amer. 39,* 907 (1966).
59. Lynch, C. J., Noise control. *Int. Sci. Technol. 32,* 32 (1966).
60. Rosenblith, W. A., Stevens, K. N., and Staff of Bolt, Beranek and Newman, Inc., "Handbook of Acoustic Noise Control," Vol. II, Noise and Man. USAF, WADC Tech. Rep. 52-204 (1953).
61. Use in Industry of Elasticity Measurements in Metals with the Help of Mechanical Vibrations. Notes on Applied Science No. 30, National Physical Laboratory. G. Bradfield (reviewed by W. P. Mason), *J. Acoust. Soc. Amer. 36,* 1752 (1964).
62. Jorgensen, R., "Fan Engineering." Buffalo Forge Company, Buffalo, New York (1961).
63. Geiger, P. H., "Noise Reduction Manual." Engineering Research Institute, University of Michigan (1953). Prepared under the direction of the Office of Naval Research, Undersea Warfare Branch.
64. Bolt, R. H., Beranek, L., and Newman, R., "Handbook of Noise Control," Volume I. Wright Air Development Center WADC Tech. Rep. 52-204 (Dec. 1952).
65. *Applied Acoustics.* Coordinating editor, Peter Lord. Elsevier Publishing Company. Annual subscription £7.10s. Reviewed in *Amer. Occup. Hyg. 2,* 269 (1968).
66. American National Standards Writing Group Z137 (Standard Practice for Hearing

Protection). American National Standards Institute, New York.
67. American National Standards Writing Group S3-52 (Method for the Measurement of Real-Ear and Physical Attenuation of Hearing Protectors). American National Standards Institute, New York.
68. Institute for Electrical and Electronics Engineers WG30.10 (Procedures for the Measurement of Circumaural Ear Protectors). Institute for Electrical and Electronics Engineers, New York.
69. American Optical Company, Safety Products Division, Southbridge, Massachusetts.
70. Bausch & Lomb, Rochester, New York.
71. David Clark Company, Ear Protector Department, Worcester, Massachusetts.
72. Flents Products Company, New York.
73. Glendale Optical Company, Woodbury, New York.
74. Mine Safety Appliances Company, Safety Products Division, Pittsburgh, Pennsylvania.
75. Welsh Manufacturing Company, Providence, Rhode Island.
76. Willson Products Division, ESB, Reading, Pennsylvania.
77. United States Safety Service Company, Kansas City, Missouri.
78. Sigma Engineering Company, North Hollywood, California.
79. Frontier Industrial Products, Los Angeles, California.
80. Billasholm products distributed by Rockford I. C. Webb, Inc. Rockford, Illinois.
81. Safety Ear Protector Company, Los Angeles, California.
82. Stayrite, Inc., Long Island City, New York.
83. Surgical Mechanical Research, Los Angeles, California.
84. H. E. Douglass Engineering Sales Company, Burbank, California.
85. General Electric Company, Medical Development Operation, Schenectady, New York.
86. Michael, P. L., Ear protectors – their usefulness and limitations. *Arch. Environ. Health 10,* 612 (1965).
87. Noise as a Public Health Hazard - Proceedings of the Conference. American Speech and Hearing Association. Report No. 4 (Feb. 1969). Available from Director, Public Information, ASHA, Washington, D.C.
88. Symposium on Aerodynamic Noise, 20-21 May 1968, Toronto, Canada. *J. Acoust. Soc. Amer. 42,* 1106 (1967).
89. Meeting of the Austrian Working Group for Noise Control, 4-6 September 1968. *J. Acous. Soc. Amer. 44,* 840 (1968).
90. Finnish Society for Noise Abatement. *J. Acoust. Soc. Amer. 39,* 985 (1966).
91. Institute on Occupational Hearing Loss, Colby College, Waterville, Maine.
92. Industrial Noise and Engineering Control Seminar. The Pennsylvania State University, University Park, Pennsylvania.
93. Noise and Vibration Reduction. Massachusetts Institute of Technology, Cambridge, Massachusetts.
94. Burns, W., "Noise and Man." Lippincott, Philadelphia, Pennsylvania, 1968.
95. Albers, V. M., "The World of Sound." A. S. Barnes and Company, Cranbury, New Jersey, 1970.
96. Rettinger, M., "Acoustics – Room Design and Noise Control" Chemical Publishing Company, London, 1968.
97. "Industrial Noise Manual." American Industrial Hygiene Association, Southfield, Michigan, 1966.
98. Baughn, W. L., Noise control – percent of population protected. *Int. Audiol. J. 5,* 331 (1966).
99. Nixon, J., and Glorig, A., Noise-induced permanent threshold shift at 2000 cps and 4000 cps. *J. Acoust. Soc. Amer. 33,* 904 (1961).

100. Cohen, A. Anticaglia, J. R., and Jones, H. H., Noise-induced hearing loss – exposures to steady-state noise. *Arch. Environ. Health 20,* 614 (1970).
101. Schneider, E. J., Mutchler, J., Hoyle, H., Ode, E., and Holder, B., The progression of hearing loss from industrial noise exposures. *Amer. Ind. Hyg. Ass. J. 31,* 368 (1970).
102. Robinson, D. W., The relationships between hearing loss and noise exposure. NPL Aero Rep. Ac 32, National Physical Laboratory, England, 1968.
103. Ward, W. D., Temporary threshold shift following monaural and binaural exposure. *J. Acoust. Soc. Amer. 38,* 121 (1965).
104. Ward, W. D., Temporary threshold shifts in males and females. *J. Acoust. Soc. Amer. 40,* 478 (1966).
105. Lawrence, M., Gonzales, G., and Hawkins, J. E., Some physical factors in noise-induced hearing loss. *Amer. Ind. Hyg. Ass. J. 28,* 425 (1967).
106. The Environmental Health Pilot for Chemical Industry Management, No. 6, Manufacturing Chemists Ass., Sept. 1967.
107. Hazardous Noise Exposure. Air Force Regulation No. 160-3, Washington, D.C. (Oct. 29, 1956).
108. Botsford, J. H., Simple method for identifying acceptable noise exposures. *J. Acoust. Soc. Amer. 42,* 810 (1967).
109. Flanagan, J. J., and Guttman, N., Estimating noise hazard with the sound-level meter. *J. Acoust. Soc. Amer. 36,* 1654 (1964).
110. Acceptable Noise Exposures. *Sound Vibration 1*(11), 4 (1967).
111. Karplus, H. B., and Bonvallet, G. L., A noise survey of manufacturing industries. *Amer. Ind. Hyg. Ass. Quart. 14,* 235 (1953).
112. Morse, K. M., The assessment of the work place – a prerequisite to the diagnosis of occupational chest diseases. *Amer. Ind. Hyg. Ass. J. 28,* 141 (1967).
113. Guides to Evaluation of Permanent Impairment - Ear, Nose, Throat and Related Structures. *J. Amer. Med. Ass. 177,* 489 (1961).
114. Cohen, A., and Baumann, K. C., Temporary hearing losses following exposure to pronounced single-frequency components in broadband noise. *J. Acoust. Soc. Amer. 36,* 1167 (1964).
115. Von Gierke, H. E., Effects of sonic boom on people: Review and outlook. *J. Acoust. Soc. Amer. 39,* S43 (1966).
116. Rintelman, W. F., and Borous, J. F., Noise-induced hearing loss and rock and roll music. *Arch. Otolaryngol. 88,* 377 (1968).
117. Flugrath, J. M., Modern day rock-and-roll music and damage-risk criteria. *J. Acoust. Soc. Amer. 45,* 704 (1969).
118. Lebo, C. P., and Oliphant, K., Music as a source of acoustic trauma. *Laryngoscope 78,* 1211 (1968).
119. Dey, F. L., Auditory fatigue and predicted permanent hearing defects from rock-and-roll music. *New Engl. J. Med. 282,* 467 (1970).
120. Anderson, J. H., The effect of high intensity "rock" music on teenage hearing thresholds. Master's thesis, North Dakota State Univ. Agriculture and Applied Science, Fargo, North Dakota, 1969.
121. Lipscomb, D. M., Ear damage from exposure to rock-and-roll music. *Arch. Otolaryngol. 90,* 29 (1969).
122. Ayley, J., Bartlett, B., Bedford, W., Gregory, W., and Hallum, G., Pilot study on the effects of "pop" group music on hearing. I. S. V. R. Memo. no. 266, Institute of Sound and Vibration, University of Southampton, Southampton, England, 1969.
123. Lipscomb, D. M., High intensity sounds in the recreational environment. *Clin. Pediat. 8,* 63 (1969).

124. Dougherty, J. D., and Welsh, O. L., Community noise and hearing loss. *N. Engl. J. Med. 275,* 759 (1966).
125. Tobias, J. V., Noise in light twin-engine aircraft. *Sound Vibration 3*, 16 (1969).
126. Somerville, G. W., and Kronoveter, K. J., Cockpit noise-induced hearing loss in pilots. Bureau of Occupational Safety and Health, Western Area Occupational Health Laboratory, Salt Lake City, Utah, 1968.
127. Tobias, J. V., Cockpit noise intensity: Fifteen single-engine light aircraft. *Aerosp. Med. 39,* 963 (1969).
128. Gottlieb, P., "Anomalous Hearing Loss and Jet Noise Exposure." Unpublished report available from the author (11339 Gladwin Street, Los Angeles, California 90049) (Sept. 1969).
129. Harris, J. D., Hearing-loss trend curves and the damage-risk criterion in diesel-engineroom personnel. *J. Acoust. Soc. Amer. 37,* 444 (1965).
130. Botsford, J. H., Predicting hearing impairment from A-weighted sound levels (abstract). *J. Acoust. Soc. Amer. 42,* 1151 (1967).
131. LaBenz, P., Cohen, A., and Pearson, B., A noise and hearing survey of earth-moving equipment operators. *Amer. Ind. Hyg. Ass. J. 28,* 117 (1967).
132. Ottoboni, F., and Milby, T. H., Occupational disease potentials in the heavy equipment operator. *Arch. Environ. Health 15*, 317 (1967).
133. Smith, P. E., Jr., A test for the susceptibility to noise-induced hearing loss. *Amer. Ind. Hyg. Ass. J. 30,* 245 (1969).
134. Kryter, K. D., Ward, W. D., Miller, J. D., and Eldredge, D. H., Hazardous exposure to intermittent and steady-state noise. *J. Acoust. Soc. Amer. 39,* 451 (1966).
135. Sataloff, J., Vassallo, L., and Menduke, H., Hearing loss from exposure to interrupted noise. *AMA Arch. Environ. Health 18*, 972 (1969).
136. Ward, W. D., Temporary threshold shift and damage-risk criteria for intermittent noise exposure. *J. Acoust. Soc. Amer. 48,* 561 (1970).
137. Parrack, H. O., Effect of airborne ultrasound on humans. *Int. Audiol. 5,* 294 (1966).
138. Acton, W. I., A criterion for the prediction of auditory and subjective effects due to air-borne noise from ultrasonic sources. *Ann. Occup. Hyg. 11,* 227 (1968).
139. Loeb, M., Fletcher, J. L., and Benson, R. W., Some preliminary studies of temporary threshold shift with an arc-discharge impulse-noise generator. *J. Acoust. Soc. Amer. 37,* 313 (1965).
140. Ward, W. D., Proposed damage risk criterion for impulse noise. NRC Committee on Hearing, Bioacoustics and Biomechanics, Report of Working Group 57, Washington, D.C., July 1968.
141. Ward, W. D., and Glorig, A., A case of firecracker-induced hearing loss. *Laryngoscope 71,* 1590 (1961).
142. Christiansen, A., and Rojskaer, C., Audio injuries from New Year's Eve fireworks. *Nord. Audio 13,* 33-40 (1964).
143. Taylor, G. D., and Williams, E., Acoustic trauma in the sports hunter. *Laryngoscope 76,* 863 (1966).
144. Acton, W. I., and Forrest, M. R., Hearing hazard from small-bore weapons. *J. Acoust. Soc. Amer. 44,* 817 (1968).
145. Hodge, D. C., and McCommons, R. B. Acoustical hazard of children's "toys." *J. Acoust. Soc. Amer. 40,* 911 (1966).
146. Gjaevenes, K., Measurements on the impulsive noise from crackers and firearms. *J. Acoust. Soc. Amer. 39,* 403 (1966).
147. Coles, R. R. A., and Rice, C. G., Audiotory hazards of sports guns. *Largyngoscope 76,* 1728 (1966).

148. Cohen, A., Kylin, B., and LaBenz, P., Temporary threshold shifts in hearing from exposure to combined impact/steady-state noise conditions. *J. Acoust. Soc. Amer. 40,* 1371 (1966).
149. Rice, C. G., and Coles, R. R. A., Impulsive noise studies and temporary threshold shift. *In* "Proceedings of the Fifth International Congress on Acoustics, 1965, Liege" (D. C. Commins, ed.) Vol. 1a, paper B67. Georges Thoni, Liege, 1965.
150. Hodge, D. C., and McCommons, R. B., Reliability of TTS from impulse noise exposure. *J. Acoust. Soc. Amer. 40,* 839 (1966).
151. Coles, R. R. A., and Rice, C. G., Towards a criterion for impulse noise in industry. *Ann. Occup. Hyg. 13,* 43 (1970).
152. Coles, R. R. A., Assessment of risk of hearing loss due to impulse noise. British Acoustical Society Meeting, March 23-25, 1970.
153. Rice, C. G., Deafness due to impulse noise. *Phil. Trans. Roy. Soc. London 263,* 279 (1968).
154. Forest, M. R., Ear protection and hearing in high-intensity impulsive noise. Meeting of the British Acoustical Society, 23-25 Mar. 1970.
155. Ross, E. M., State regulation of community noise (abstract). *J. Acoust. Soc. Amer. 47,* 54 (1970).
156. Little, R., State regulation of motor vehicle noise (abstract). *J. Acoust. Soc. Amer. 47,* 54 (1970).
157. Van Atta, F. A., Federal regulation of occupational noise (abstract). *J. Acoust. Soc. Amer. 47,* 54 (1970).
158. Foster, C. R., Federal regulation of transportation noise (abstract). *J. Acoust. Soc. Amer. 47,* 54 (1970).
159. Jones, H. H., State regulation of occupational noise (abstract). *J. Acoust. Soc. Amer. 47,* 54 (1970).
160. Schultz, T. J., Impact-noise recommendations for the FHA. *J. Acoust. Soc. Amer. 36,* 729 (1964).
161. Regulations Respecting Protection of Workers from Effects of Noise. Alta. Reg. 185/66, Provincial Board of Health, Edmonton, Alta., dated 1 June 1966. Entry into Force: 15 June 1966, *Alberta Gazette 62,* 403 (1966).
162. Sec. 25A, Article 101, Annotated Code of Maryland (1964 Replacement Volume), Workmen's Compensation, Approved April 14, 1967, Effective June 1, 1967.
163. National and International Aspects of Noise Legislation. *J. Acoust. Soc. Amer. 40,* 922 (1966).
164. Goodfriend, L. S., Community noise problems – origin and control. *Amer. Ind. Hyg. Ass. J. 30,* 607 (1969).
165. Kryter, K. D., Concepts of perceived noisiness, their implementation and application. *J. Acoust. Soc. Amer 43,* 344 (1968).
166. Ostergaard, P. B., and Donley, R., Background-noise levels in suburban communities. *J. Acoust. Soc. Amer. 36,* 409 (1964).
167. Buchta, E., Distributions of transportation and community noise. *J. Acoust. Soc. Amer. 47,* 60 (1970).
168. Donley, R., Measurement of community noise. *J. Acoust. Soc. Amer. 47,* 61 (1970).
169. Nelson, D. L., Gordon, C., and Galloway, W., Methodology for highway noise prediction. *J. Acoust. Soc. Amer. 47,* 111 (1970).
170. Berendt, R. D., Airborne, impact and structure-borne noise control in multifamily dwellings. U. S. Department of Housing and Urban Development, Washington, D.C., 152 pp. + 148 pp. test data (1967).
171. Young, R. W., Single-number criteria for room noise. *J. Acoust. Soc. Amer. 36,* 289 (1964).

172. Webster, J. C., Speech communications as limited by ambient noise. *J. Acoust. Soc. Amer. 37,* 692 (1965).
173. Botsford, J. H., Predicting speech interference and annoyance from A-weighted sound levels (abstract). *J. Acoust. Soc. Amer. 42,* 1151 (1967).
174. ISO Recommendation R495-1966 (E), General Requirements for the Preparation of Test Codes for Measuring the Noise Emitted by Machines. International Organization for Standardization, Aug. 1966.
175. Sound Rating of Outdoor Unitary Equipment, Standard 270, Air Conditioning and Refrigeration Institute, 1967.
176. Groff, G. C., Schreiner, J., and Bullock, C., Centrifugal fan sound power level prediction. [A condensed version is published in ASHRAE (Amer. Soc. Heat. Refrig. Air Cond. Eng.) *J. 9,* 71 (1967), ASHRAE Trans. 73, Part II (1967).]
177. Eagles, E. L., and Doerfler, L. G., Hearing in children: Acoustic environment and audiometer performance. *J. Speech Hear. Res. 4,* 149 (1961).
178. *Environ. Health Lett.* 6(17), 7 (1967).
179. Knight, J. J., Normal hearing threshold determined by manual and self-recording techniques. *J. Acoust. Soc. Amer. 39,* 1184 (1966).
180. Riley, E. C., Sterner, J. H., Fassett, D. W., and Sutton, W. L., Ten years experience with industrial audiometry. *Amer. Ind. Hyg. Ass. J. 22,* 151 (1961).
181. Glorig. A., "Noise and Your Ear." Grune and Stratton, New York, 1958.
182. Rosen, S., Plester, D., El-Mofty, A., and Rosen, H. V., High frequency audiometry in presbycusis: A comparative study of the Mabaan tribe in the Sudan with urban populations. *Arch. Otolaryngol. 79,* 1 (1964).
183. Rice, C. G., and Coles, R. R. A., Normal threshold of hearing for pure tones by earphone listening with a self-recording audiometric technique. *J. Acoust. Soc. Amer. 39,* 1185 (1966).
184. Whittle, L. S., and Delaney, M. E., Equivalent threshold sound-pressure levels for the TDH39/MX41-AR earphone. *J. Acoust. Soc. Amer. 38,* 1187 (1966).
185. "Guide for Industrial Audiometric Technicians." Employers Insurance of Wausau, Wausau, Wisconsin.
186. Goss, B. L., Electric motor noise: Control of noise at the source. *Amer. Ind. Hyg. Ass. J. 31,* 16 (1970).
187. Mills, R. O., Noise reduction in a textile weaving mill. *Amer. Ind. Hyg. Ass. J. 30,* 71 (1969).
188. Judd, S. H., and Spence, J. A. Noise control for electric motors. *Amer. Ind. Hyg. Ass. J. 30,* 588 (1969).
189. Torpey, P. J., Noise control of emergency power generating equipment. *Amer. Ind. Hyg. Ass. J. 30,* 596 (1969).
190. Lowson, M. V., Theoretical analysis of compressor noise. *J. Acoust. Soc. Amer. 47,* 371 (1970).
191. Prillwitz, H. Mechalkg, K., Karl-Heinz, and Seyfarth, B., Soundproof compressed air machine. Patent reviewed in *J. Acoust. Soc. Amer. 47,* 990 (1970).
192. Carlson, R. O., Ventilated sound-reducing enclosure for a teleprinter. Patent reviewed in *J. Acoust. Soc. Amer. 47,* 991 (1970).
193. Cunningham, J. M., Ventilated and soundproofed enclosure for printer. Patent reviewed in *J. Acoust. Soc. Amer. 47,* 989 (1970).
194. Jacobson, G. R., Air turning vane with removable closure for insertion of acoustical material. Patent reviewed in *J. Acoust. Soc. Amer. 47,* 990 (1970).
195. Amlott, N. J., and Karn, J. D., Air cleaner and silencer assembly. Patent reviewed in *J. Acoust. Soc. Amer. 46,* 72 (1969).
196. Thomas, D., Exhaust muffler. Patent reviewed in *J. Acoust. Soc. Amer. 46,* 1114 (1969).

197. Lowson, M. V., Reduction of compressor noise radiation. *J. Acoust. Soc. Amer. 43,* 37 (1968).
198. Lucht, R. F., and Scanlan, R. H., Method of axial testing for roller bearing noise qualities. *J. Acoust. Soc. Amer. 44,* 5 (1968).
199. Plummer, W. A., Pneumatic tool muffle. Patent reviewed in *J. Acoust. Soc. Amer. 37,* 774 (1965).
200. Reed, G. A., and Rosen, D., Acoustical cabinet for office business machines. Patent reviewed in *J. Acoust. Soc. Amer. 37,* 774 (1965).
201. Dahl, C. B., Suppression of objectionable noise in rotating machinery. Patent reviewed in *J. Acoust. Soc. Amer. 36,* 1242 (1964).
202. Industrial Noise Control and Hearing Conservation Program, International Brotherhood of Boilermakers, Iron Ship Builders, Blacksmiths, Forgers and Helpers, Brotherhood Building, Kansas City, Kansas, Jan. 1966.
203. Hearing Levels of Adults by Age and Sex, United States, 1960-1962. National Center for Health Statistics, U.S.P.H.S. Publ. No. 1000, Ser. 11, No. 11 (1965). Available from Superintendent of Documents, Govt. Printing Office, Washington, D.C.
204. "Foundry Noise Manual," 2nd Ed., American Foundrymen's Society, Des Plaines, Illinois, 1966.
205. "Industrial Noise, A Guide to Its Evaluation and Control," FS2.6/2:n 69. (1967). Available from Superintendent of Documents, Govt. Printing Office, Washington, D.C.
206. Bell, A., Noise, an occupational hazard and public nuisance. W.H.O. Publ. Health Papers 30, (1966). Obtainable through Columbia Univ. Press, International Documents Service, New York.
207. Stevens, S. S., Warshofsky, F., and the Editors of *Life,* "Sound and Hearing." Time, New York, 1965.
208. "Sound and Vibration," Acoustical Publications Inc., Cleveland, Ohio, 1967.
209. "Industrial Noise and Hearing Protection," Employers Insurance of Wausau, Wausau, Wisconsin.
210. A Study of Noise Induced Hearing Damage Risk for Operators of Farm and Construction Equipment. Tech. Rep. SAE Research Project R-4, Society of Automotive Engineers, New York, 1969.
211. Noise Information Service (NOISE), IIT Research Institute, Chicago, Illinois.
212. Occupational Safety and Health Standards (Williams-Steiger Occupational Safety and Health Act of 1970). U. S. Department of Labor, *Fed. Regist. 36,* 10518 (1971).

Nonionizing Radiation

DAVID H. SLINEY

GENERAL

Introduction

In considering biologic effects of electromagnetic radiation, a convenient division is often made between ionizing radiation (cosmic, γ, and x radiation) and nonionizing radiation (ultraviolet, visible, infrared, microwave, and radiofrequency radiation). Such a division is also convenient in industrial and environmental hygiene. The dividing line between any two adjacent regions in the electromagnetic spectrum is arbitrary since the regions are simply inventions for convenience by man, and nature provides no such divisions. Matelsky[1] suggests that an appropriate dividing line between ionizing and nonionizing radiation for our purposes would be based upon the minimum photon energy generally required to produce ionization in any of the principal atoms found in biologic tissue – an energy of approximately 12 eV which corresponds to a wavelength of 103 nm, which is in the vacuum ultraviolet. Vacuum ultraviolet radiation (wavelengths below ~200 nm), like soft x radiation, is of little practical importance as a health hazard since it is strongly absorbed in air. The dividing line between optical radiation and microwave radiation is also vague, but is generally placed at 0.1 or 1 mm. The extent of the visible spectrum (light) is given as 380–760 nm.[2,3] Some time could be spent discussing the merits of other dividing lines between adjacent regions within the nonionizing spectrum, but this would be wasteful since these divisions are only useful as a conceptual aid in this review. Hence divisions which have been chosen should be considered as quite arbitrary. Nature has not been so accommodating as to limit a given biologic effect to a precise band of wavelengths. For instance, given sufficient radiant energy, thermal injury (burns) of the skin could be produced at any wavelength although other effects, such as photochemical effects in the actinic ultraviolet, could mask the thermal injury unless the energy was delivered very rapidly.

Recent developments in optical technology, considered by many as revolutionary, have created a demand for a better understanding of the potential health hazards from high-intensity light sources. The invention of the laser at the beginning of the decade of the 1960's also spurred many improvements in other high intensity light sources such as electronic flashlamps and compact arc lamps. All of these devices possess a potential for chorioretinal injury, and biologic research has been performed to delineate the factors which must be known to evaluate this hazard. Many arc sources and also some lasers can produce hazardous levels of ultraviolet radiation. Recent biologic research has clarified the doses and action spectra of ultraviolet radiation which produce acute injury of the skin and eyes.

In 1960 the industrial and public use of microwave radiation was limited. Microwave applications which presented potential health hazards were limited essentially to high power radar equipment used by the military and high power microwave relay equipment used by the communications industry. Indeed, it was principally these two groups which developed the hazard criteria for microwave radiation at that time. The late 1960's showed a rapid increase in microwave power applications, i.e., industrial heating, drying and cooking applications, and even home microwave cooking. These developments which indicated an enormous potential increase in microwave exposure of the public prompted a renewed interest by 1970 in the biologic effects from microwaves. The exposure standards in this country dating to the 1950's and early 1960's were based on the premise that acute thermal effects presented the only significant hazard although other, perhaps "nonthermal," effects were recognized. Some authorities now question this premise and the subject is widely debated.

General Biologic Considerations

The biologic effects of nonionizing radiation vary considerably depending upon the spectral region and duration of exposure. Acute (single insult) effects have been studied extensively, while chronic effects – particularly in the ultraviolet and microwave regions – are now becoming the important area of interest in biologic research oriented toward more complete understanding of health hazards. Present controversies, where they exist, in establishing permissible exposure levels, generally center upon the uncertainties of what levels of chronic (i.e., daily) exposures could actually create adverse effects, and upon what precisely these effects are, if they do exist. The questions of cataractogenesis by chronic exposure to infrared or ultraviolet radiation is but one example of this problem.

Quantitative standards for exposure to radiation require a knowledge of the dose or dose rate to the biologic systems which cause an adverse effect. The correct quantity to define dose is energy per unit volume or energy per unit absorber mass (e.g., the rad used in the ionizing radiation region as the unit of absorbed dose). For dose rate, the appropriate quantity is power density. Such quantities have been used in microwave biologic research; however, these quantities are difficult to apply in the optical spectrum. In the ultraviolet, visible, and infrared regions (the optical region) of the electromagnetic spectrum, radiation is generally absorbed in a surface layer (i.e., the skin, the cornea of the eye, or the retina of the eye). Therefore, the optical radiation units of choice have been radiant exposure ($J \cdot cm^{-2}$ or $ergs \cdot cm^{-2}$) or irradiance ($W \cdot cm^{-2}$) incident upon the surface of interest; these quantities are also termed exposure dose and exposure dose rate, respectively. In some portions of the infrared and microwave regions of the electromagnetic spectrum, however, the radiant energy is absorbed in the deeper tissues and the use of power delivered or energy absorbed per unit volume is the more useful concept. Nevertheless, standards of personnel exposure to microwave radiation have always been expressed as free-field power density ($mW \cdot cm^{-2}$) or integrated power density ($J \cdot cm^{-2}$).

Several physical factors influence the potential hazard of radiation. Radiation incident upon tissue will cause an adverse effect dependent upon the spectral radiant energy absorbed in, reflected from, and transmitted through the tissue, and the spectral sensitivity of the particular tissue to the radiation. Inadequate appreciation of any of these factors has given rise to erroneous concepts of the relative biologic effectiveness of different types of radiation. A precise knowledge of the spectrum of a source of radiant energy is in some cases unnecessary and in other cases of great importance in defining the potential hazards to individuals exposed to the source. For example there is little difference in the relative hazards from ocular exposure to the radiation from a helium–neon laser (633 nm) and an argon laser (488 nm). However, the relative hazard to the skin from exposure to monochromatic radiation at only the 313 nm or only the 297 nm line of the low-pressure mercury lamp is a factor of more than fifty! These two examples illustrate the enormous importance of accurately knowing the spectral characteristics of the radiation as they relate to adverse effects upon tissue. The spectral sensitivity of the tissue is called the "action spectrum" for the specific effects.

Most bioeffects of concern from nonionizing radiation are generally thermal in nature. The principal nonthermal bioeffects of concern to health are ultraviolet photokeratitis, ultraviolet erythema, and certain athermal effects from microwaves. The eye is generally the organ most sensitive to injury, but the type of injury depends upon the region where most of the radiant energy is absorbed (Fig. 1). An adequate understanding of the health hazards from nonionizing radiation requires an appreciation for the extensive range of exposure levels

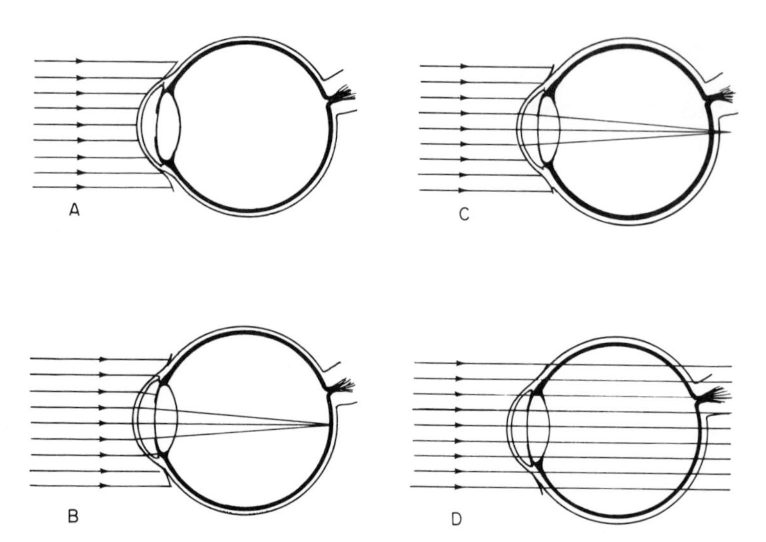

Fig. 1. Schematic diagram of the relative optical absorptive and refractive properties of the eye for: (A) ultraviolet and far-infrared radiation, (B) light, (C) near-infrared radiation, and (D) microwave radiation (frequencies < 10 GHz).

found in the work environment and the enormous range of exposure criteria depending upon the organ and the spectral region of interest. It is important to keep in mind these general considerations when reviewing the remainder of this chapter.

Nonionizing Radiation Bookshelf

I have often been asked to identify the principal source books which contain most of the information required to evaluate hazards from nonionizing radiation. A very brief core of reference books (journal articles excluded) would include one or several texts in each of the following areas: (*1*) optics, radiometry, and photometry[2–10]; (*2*) optical radiation biologic effects[11–23]; (*3*) microwave engineering and measurements[24–27]; and (*4*) microwave and *RF* biologic effects.[28–32] Several of the principal journals worthy of reviewing to keep abreast of new developments which have a high frequency of articles on this subject are *Non-Ionizing Radiation, Archives of Ophthalmology, American Journal of Ophthalmology, Investigative Ophthalmology, Vision Research, Acta Ophthalmologica, IEEE Transactions on Biomedical Engineering, IEEE Transactions on Microwave Theory and Techniques, Journal of Microwave Power, Journal of the Optical Society of America, Archives of Dermatology, Investigative Dermatology, Applied Optics, Laser Focus,* as well as the journals in the field of occupational and environmental health.

OPTICAL RADIATION

Terms and Units

To the uninitiated the terms and units used for calculations and measurements in radiometry and photometry may appear to present a confusing array of many synonomous terms for concepts often having subtle shades of meaning which may not seem worth trying to disentangle. Indeed it is unfortunate that so many photometric terms have been promulgated for each of several essentially comparable or synonomous concepts. For instance, luminance (brightness) may be measured in nits, stilbs, apostilbs, blondels, candelas-per-square-meter (or -per-square-centimeter, -millimeter, -inch, or -foot), millilamberts, foot-lamberts, scots, glims, microlamberts, lamberts, hefner-candles and more. Keeping clear the relationships of these units delights few people. Fortunately, on the radiometric side the corresponding units of $W \cdot cm^{-2} \cdot sr^{-1}$ or $W \cdot m^{-2} \cdot sr^{-1}$ are far more straightforward. The movement toward a single international (MKS) system of units – Système Internationale (SI) – and toward a common vocabulary through the International Commission on Illumination, or CIE (Commission Internationale de L'Eclairage)[2,33] are encouraging signs for the future. Table I lists most of the photometric and radiometric terms, CIE symbols, and SI units useful for the evaluation of optical radiation hazards. Some clarification of these terms are in order.

Note that the term "radiant energy" has units of joules and if used as a general descriptive term in place of "radiation," must be used carefully to avoid confusion. Likewise, the term "intensity" is often used loosely to mean either irradiance, radiant exposure, or illuminance and should probably be avoided to preclude confusion with "radiant intensity" or "luminous intensity" which describe the rate of energy emitted by a source per unit solid angle, or candlepower, respectively.

The terms "luminance" and "illuminance" are photometric terms related to the visual response of the eye of a "CIE Standard Observer" (a nonexistent individual created for mathematical simplicity) whereas "radiance" and "irradiance" are radiometric terms which are absolute physical units of radiation. Terms in the photometric system are valuable to illuminating engineers and industrial hygienists for describing levels of visible radiation in a human environmental situation (see "Industrial Hygiene Highlights,"[1] p. 154), whereas terms in the radiometric system are not dependent on the response of the eye. The conversion of light levels measured in one system to corresponding levels in the other system is not always easily accomplished since such a conversion would depend upon the spectral distribution of the radiant source, i.e., how much of the source energy is infrared and ultraviolet radiation not visible to the eye, and how much visible radiation weighted to the spectral

TABLE I

		Radiometric	
Term	*Symbol*	*Defining equation*	*SI units and abbreviation*
Radiant energy	Q_e	–	Joule (J)
Radiant energy density	W_e	$W_e = \frac{dQ_e}{dV}$	Joule per cubic meter ($J \cdot m^{-3}$)
Radiant power (radiant flux)	Φ_e, P	$\Phi_e = \frac{dQ_e}{dt}$	Watt (W)
Radiant exitance	M_e	$M_e = \frac{d\Phi_e}{dA} = \int L_e \cdot \cos\theta \cdot d\Omega$	Watt per square meter ($W \cdot m^{-2}$)
Irradiance or radiant flux density (dose rate in photobiology)	E_e	$E_e = \frac{d\Phi_e}{dA}$	Watt per square meter ($W \cdot m^{-2}$)
Radiant intensity	I_e	$I_e = \frac{d\Phi_e}{d\Omega}$	Watt per steradian ($W \cdot sr^{-1}$)
Radiance[c]	L_e	$L_e = \frac{d^2 \Phi_e}{d\Omega \cdot dA \cdot \cos\theta}$	Watt per steradian and per square meter ($W \cdot sr^{-1} \cdot m^{-2}$)
Radiant exposure (dose, in photobiology)	H_e	$H_e = \frac{dQ_e}{dA}$	Joule per square meter ($J \cdot m^{-2}$)
–	–	–	–
–	–	–	–
Radiant efficiency[d] (of a source)	η_e	$\eta_e = \frac{P}{P_i}$	Unitless
Optical density[e]	D_e	$D_e = -\log_{10} \tau_e$	Unitless
–	–	–	–

[a]The units may be altered to refer to narrow spectral bands in which case the term is preceded by the word spectral, and the unit is then per wavelength and the symbol has a subscript λ. For example, spectral irradiance H_λ has units of $W \cdot m^{-2} \cdot m^{-1}$ or more often, $W \cdot cm^{-2} \cdot nm^{-1}$.

[b]While the meter is the preferred unit of length, the centimeter is still the most commonly used unit of length for many of the above terms and the nm or μm are most commonly used to express wavelength.

Useful CIE Radiometric and Photometric Terms and Units[a, b]

Term	Symbol	Defining equation	SI units and abbreviation
		Photometric	
Quantity of light	Q_ν	$Q_\nu = \int \Phi_\nu dt$	Lumen-second (lm · s) (talbot)
Luminous energy density	W_ν	$W_\nu = \frac{dQ_\nu}{dV}$	Talbot per square meter ($lm \cdot m^{-3}$)
Luminous flux	Φ_ν	$\Phi_\nu = 680 \int \frac{d\Phi_e}{d\lambda} V(\lambda)\, d\lambda$	Lumen (lm)
Luminous exitance	M_ν	$M_\nu = \frac{d\Phi_\nu}{dA} = \int L_\nu \cdot \cos\theta \cdot d\Omega$	Lumen per square meter ($lm \cdot m^{-2}$)
Illuminance (luminous density)	E_ν	$E_\nu = \frac{d\Phi_\nu}{dA}$	Lumen per square meter ($lm \cdot m^{-2}$)
Luminous intensity (candlepower)	I_ν	$I_\nu = \frac{d\Phi_\nu}{dr}$	Lumen per steradian (lm · sr) or candela (cd)
Luminance[c]	L_ν	$L_\nu = \frac{d^2 \Phi_e}{dr \cdot dA \cdot \cos\theta}$	Candela per square meter ($cd \cdot m^{-1}$)
Light exposure	H_ν	$H_\nu = \frac{dQ_\nu}{dA} = \int E_\nu dt$	Lux-second (lx · s)
Luminous efficacy (of radiation)	K	$K = \frac{\Phi_\nu}{\Phi_e}$	Lumen per watt ($lm \cdot W^{-1}$)
Luminous efficiency (of a broad band radiation)	V(*)	$V(*) = \frac{K}{K_m} = \frac{K}{680}$	Unitless
Luminous efficacy[d] (of a source)	η_ν	$\eta_\nu = \frac{\Phi_\nu}{P_i}$	Lumen per watt ($lm \cdot W^{-1}$)
Optical density[e]	D_ν	$D_\nu = -\log_{10} \tau_\nu$	Unitless
Retinal illuminance in trolands	E_t	$E_t = L_\nu \cdot S_p$	Troland (td) = luminance of 1 $cd \cdot m^{-2}$ times pupil area in mm^2

[c] At the source, $L = \frac{dM}{d\Omega \cdot \cos\theta}$; at a receptor, $L = \frac{dE}{d\Omega \cdot \cos\theta}$

[d] P_i is electrical input power in watts.

[e] τ is the transmission.

sensitivity function of the eye is present. The conversion factor is the "luminous efficacy of radiation," not to be confused with the "luminous efficacy of a source."

The terms "radiance" and "luminance" are used to describe the rate at which light or radiant energy per unit area leaves a source per solid angle, i.e., the "brightness" of the source. Alternatively, the terms "irradiance" and "illuminance" are used to describe the level incident upon a surface or arriving at a given point in space, i.e., the level determined by a detector which measures the number of photons falling upon an area of the detector's surface but does not distinguish the direction from which the photons arrive.

The primary SI units listed in Table I are the principal units to be used in most instances; however, in some applications area units of square centimeters rather than square meters are accepted and are more convenient. Area units of square centimeters are used extensively in this review. Although radiometric units used in the United States follow the SI metric system almost exclusively, there are still two widely used nonmetric holdovers in the photometric system of units. These are the "footcandle" or "lumen per square foot" (1 fc = 10.764 lx) for illuminance and the "footlambert" ($1\ fL = 3.425\ cd \cdot m^{-2}$) for luminance. See "Industrial Hygiene Highlights,"[1] p. 159, for a further explanation of these units.

Several artificial units have been developed that are weighted to the "standard" erythema curve or to the germicidal curve – as photometric units are weighted to an approximate visual response of the eye. Such units include the EU, the E-viton, the Finsen, and the GU which are not in widespread use and are not used in this review.

Ultraviolet Radiation

The ultraviolet spectrum is often loosely divided into three or four regions. Physicists generally divide it differently than do dermatologists who use ultraviolet radiation for therapy. Table II illustrates four of the most commonly used schemes encountered in the literature. Since the same term is sometimes used to identify entirely different regions (e.g., far UV), and wavelength limits vary, it is always wise to determine each author's definition of the spectral regions in reading on this subject.

In considering the biologic effects of UV radiation one must be careful not to use absolute terms. For example, many an investigator has been puzzled by a biologic reaction to ultraviolet radiation because he considered a certain tissue layer or layers completely absorbing, yet biologic effects were produced in an underlying tissue. The ultimate explanation lay in the fact that a very small fraction of 1% of the incident radiation penetrated the "protective tissue" and affected the underlying tissue.[34-38] The photochemical effects of ultraviolet

TABLE II
Several Schemes for Dividing the Ultraviolet Spectrum

Physical #1	*Physical #2*	*Dermatologic (CIE)*	*Biologic*
Near UV (400–300 nm)	Near UV (400–300 nm)	UV-A (400–315 nm)	Near UV (400–380 to 315 nm)
Far UV (300–200 nm)	Middle UV (300–200 nm)	UV-B (315–280 nm)	Actinic UV (315–200 nm)
Vacuum or extreme UV (200 to 1 – 10 nm)	Far UV (200–100 nm)	UV-C (280–100 nm)	Vacuum UV (< 200 nm)
	Extreme UV (100 to 1 – 10 nm)		

radiation on the skin and eye are still not completely understood. The steep slopes of ultraviolet action spectra are quite impressive and should demonstrate the importance of not extrapolating data developed from one wavelength to another or assuming that narrower absorption peaks do not exist in the already familiar absorption bands. The inevitable development of a wavelength-tunable laser source in the ultraviolet region will surely revolutionize ultraviolet photobiology by making available a dramatic increase in spectral irradiance over that afforded by present ultraviolet sources. Present biologic studies often use xenon-arc or mercury-arc monochrometers having spectral bandwidths of several nanometers, whereas a laser could have a bandwidth less than a tenth of a nanometer. With the foregoing in mind when reviewing present biologic data we may examine the present available information.

ULTRAVIOLET EFFECTS ON THE SKIN

A sufficient exposure to actinic ultraviolet results in erythema. The action spectrum of ultraviolet erythema was investigated by several teams of physical scientists in the 1920's and early 1930's. Although the beneficial uses of ultraviolet as a tool in therapy was introduced by Finsen at the turn of the century, it was not until the 1920's that Hausser and Vahle at the Physics Laboratory of the Siemens Corporation in Germany carefully examined the influence of wavelength, exposure time, and exposure rate upon the nature, degree, and course of erythema.[39–40] They exposed the skin of several individuals to ultraviolet radiation from a mercury lamp passed through a double-quartz prism monochromator. Figures 2 and 3 illustrate some of the key findings of this classical study which are generally summarized by only one graph of a "standard" action spectra in most texts. Coblentz, Stair, and Hogue at the U. S. National Bureau of Standards[41] using a quartz-prism monochromator, and Luckiesh, Holladay, and Taylor of the General Electric Company[42] using a quartz-mercury arc lamp and filters found similar action spectra. All of these

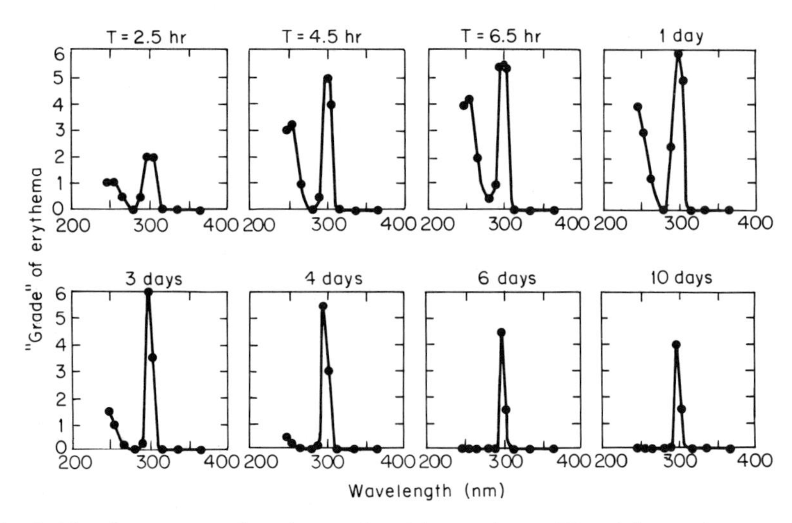

Fig. 2. The time course of moderate ultraviolet erythema. Ultraviolet erythema action spectra obtained by Hausser (1928) for eight different observation times.[39]

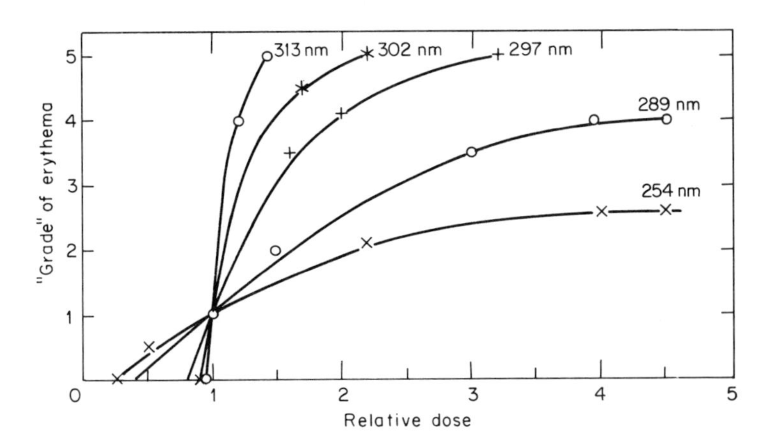

Fig. 3. The degree, or "grade," of erythema (skin reddening) for increasing dose obtained by Hausser.[39] He noted that the monochromatic UV-B radiation which penetrated more deeply into the skin than UV-C radiation was far more effective in producing serious erythema (as well as longer lasting tan). Hausser's grading system consisted of a comparison of skin redness with a logarithmic density scale obtained with red dye solutions. An erythema grade of one was above minimal erythema.

investigators explained the importance of noting the length of time after exposure for the various degrees of erythema to develop and the value of defining the action spectrum at a well-defined degree of redness and not at the "just perceptible" erythema (the MED). They all chose to work with exposure doses above those where the highly transitory erythema produced by UV-C played a significant role.

Recent investigators, in the field of dermatology, using xenon-arc monochromators[43-47] found the action spectra for this just-perceptible erythema to be quite different from the "classical" curve, which is not at all surprising. Figure 3 indicates that radiation at 254 nm and 289 nm is more effective in producing a very mild erythema (i.e., an erythema "grade" less than one) than the longer wavelengths at minimal doses. This is also evident in the results of the recent investigations (see "Industrial Hygiene Highlights,"[1] Fig. 3, p. 145). Other studies using a mercury lamp monochromator strongly substantiated the work of Hausser and Vahle and recent investigators.[48] Erythema thresholds vary significantly with skin pigmentation (over at least one order of magnitude). Erythema thresholds may be as great as a factor of ten or more for negro skin than for very light caucasian skin, with skin of intermediate pigmentation having thresholds in between. Clearly the action spectrum of interest depends upon the application. In applying an action spectrum to the development of hazard criteria in industry one must make a judgment of what exposure limits will not result in unwanted acute and chronic effects. This will be discussed further later. Erythema production is dependent only on the total dose and is independent of exposure rate and duration of exposure over a wide range of exposure times (~0.01 sec to several hours).[49-51] In other words there is reciprocity between exposure rate and exposure time.

Chronic exposure to ultraviolet radiation accelerates "skin aging" and it is now generally felt that such exposure increases the risk of developing skin cancer.[52-58] Ultraviolet radiation in the solar spectrum has long been identified with basal and squamous cell skin cancers.[13,59-62,35] Other types of skin cancer (such as malignant melanomas and sarcomas) do not appear to be related to UV exposure since their incidence is no greater in exposed areas such as the face, hands, and neck and no greater in lightly pigmented individuals. Although extensive animal studies have been conducted, few experts seem willing to suggest that the action spectrum for UV skin carcinogenesis is really known. Studies by Rusch, Kline and Bauman (see "Industrial Hygiene Highlights,"[1] Fig. 6, p. 149), which many feel are not sufficiently extensive, strongly indicate that UV-B is the spectrally effective band. Since UV-B penetrates more deeply into the skin than the rest of the actinic ultraviolet spectrum, it would be expected to be the most effective in affecting living tissue as is attested by its capability of producing the more severe grades of erythema. The actinic ultraviolet radiation from the sun, which penetrates the ozone layer in the upper atmosphere (air does not significantly attenuate actinic ultraviolet),[63] is limited to wavelengths above 295 nm. The UV-B which reaches the earth is dependent upon ground elevation and greatly dependent upon the slant path of the solar rays through the atmosphere (hence the time of day). The spectral irradiance of solar ultraviolet radiation at 300 nm is typically a factor of ten lower at ± 3 hours from zenith (solar noon).[35,64–66,367] Hence a sun bather who develops a mild erythema after 20 min of exposure to the summer noonday sun would have to sunbathe for several hours in the late afternoon or early morning to obtain the same effect. Several epidemiologic studies of skin cancer incidence reveal a very strong correlation with the solar UV-B levels found at given latitudes and ground elevations.[35,67–71,367]

A quantitative threshold for carcinogenesis by ultraviolet radiation appears to be very difficult to define. One of the principal investigators in this field, Dr. Harold F. Blum, concluded that carcinogenesis by ultraviolet radiation could be "considered as essentially nonthreshold." His studies of ultraviolet carcinogenesis in mouse skin indicated to him that there was a threshold dose at which recovery balanced carcinogenesis, but it was sufficiently low to preclude measurement because of the limited lifetime of his animals. He did conclude that UV carcinogenesis was in some way cumulative and "essentially irreversible."[13,53] The epidemiologic studies correlating solar ultraviolet exposure with skin cancer may shed some light on whether quantitative thresholds exist for human skin, but such studies could only place some quantitative limits on a "threshold value."

Some rare individuals are hypersensitive to irradiation from specific optical spectral bands and may develop skin reactions described as "photosensitivity"[12,72–74] following suberythemal exposure. Such photosensitivity may be due to an underlying systemic disease state, although more likely it results from ingestion or skin contact with chemical or therapeutic agents. In the latter case, the photosensitivity is elicited by those portions of the spectrum which correlate with the action spectrum of the chemical or therapeutic agent concerned.[75,366] In the industrial environment it would be highly unusual for the symptoms of photosensitization to be elicited solely by a limited emission spectrum from industrial light sources such as "black light"; sunlight usually will elicit or aggravate the skin response.[76,77]

ULTRAVIOLET RADIATION AND THE EYE

Regarding the visual spectral sensitivity of the human eye, the retina responds to near ultraviolet radiation as to light. However, the lens of the eye is a strong absorber of wavelengths shorter than 400 nm and the resultant spectral sensitivity for vision markedly decreases between 420–380 nm. Indeed an aphakic (an individual who has had a lens surgically removed) has a significant visual response to wavelengths down to 350 nm and perhaps to320 nm.[78–80] His visual acuity is, however, limited because of the eyes' strong chromatic aberrations in this end of the spectrum. As an interesting note, recent studies indicate that near-ultraviolet radiation from the sun may act upon animal lens constituents to increase attenuation by the lens in this spectral region.[81] A similar effect has been suggested as playing an etiologic role in the formation of a certain type of cataract in the human lens.[81-83]

Middle ultraviolet radiation (UV-B and UV-C) is absorbed in the cornea and conjunctiva, and in sufficient doses will cause keratoconjunctivitis. The action spectrum and threshold dosage of ultraviolet keratoconjunctivitis have been investigated by several groups.[84-89] General agreement may be found in the results of the different investigators if the differences in experimental techniques and instrumentation available is considered. Of the published studies those most relevant to human exposure were recently completed upon primate eyes and some human eyes by Pitts and his collaborators.[88,89] Using an arc monochromator they obtained data for 10 nm bands between 320 nm–200 nm. They found the peak of the photokeratitis action spectrum at the 265–275 nm band and a threshold at that wavelength of approximately 4 mJ · cm^{-2} for both human and primate eyes. This is somewhat different from the earlier studies of Cogan and Kinsey[84] who found a peak in the action spectrum at 288 nm using a mercury arc monochromator with wider wave bands (see "Industrial Hygiene Highlights,"[1] Fig. 6, p. 149). The reciprocity of irradiance and exposure times probably holds for time periods similar to those which hold for ultraviolet erythema of the skin. These studies did not reveal an action spectrum for conjunctivitis different from keratitis. The action spectrum of Pitts is given in Fig. 4 (solid-line histogram) with the approximate range of thresholds at each band (I). Unfortunately, this action spectrum when weighted against the ultraviolet spectrum of indirect daylight to which the eye is daily exposed would indicate that almost everyone would develop keratoconjunctivitis in a few hours while standing outdoors. This would indicate that the investigators may have had difficulty with stray light from the monochromator or other experimental difficulties in accurately determining the obviously extreme slope of the action spectrum in the 300–315 nm range. It also demonstrates that thresholds averaged over 10 nm intervals weighted against a source spectrum rapidly increasing in this region can lead to error since sufficiently narrow waveband intervals could not be used. Individuals do develop keratoconjunctivitis from daylight ultraviolet radiation but only after prolonged exposure to ultraviolet reflected from snow (an aspect of snow-blindness).[90] Snow is essentially the only material found in the natural environment with a high reflectance in the actinic ultraviolet spectral region; water reflects very little and transmits a large percentage. Nevertheless, there is little reason to doubt the shape of the curve or the threshold values at or near the 270 nm band. Accidental exposures to ultraviolet radiation from germicidal lamps[91] (which emit principally at 253.7 nm) indicate a human threshold for photokeratitis of approximately 10 mJ · cm^{-2} – in close agreement with the data of Pitts. Before leaving the subject of ultraviolet keratoconjunctivitis, it should be emphasized

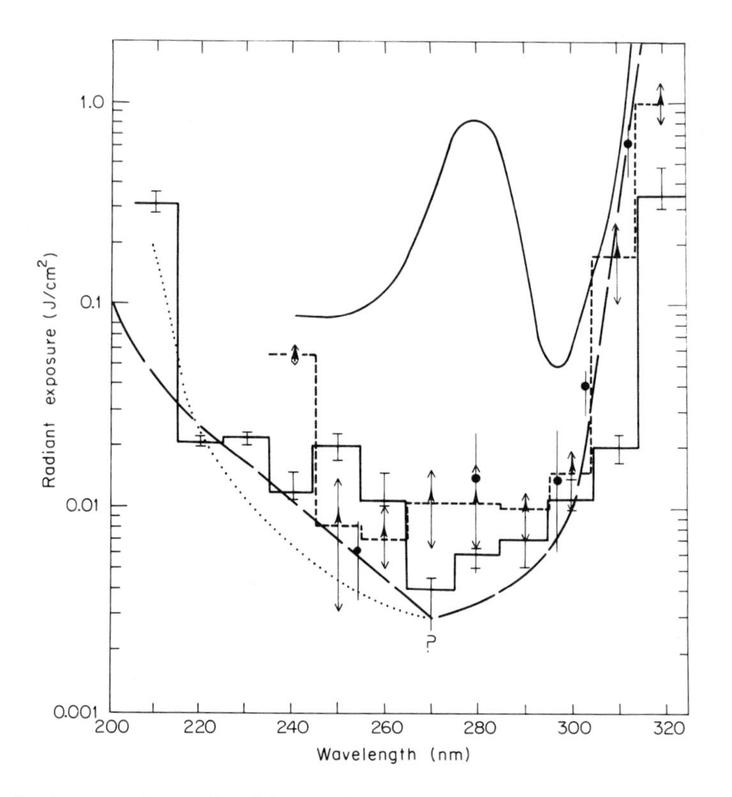

Fig. 4. A comparison of minimum thresholds for ultraviolet skin erythema and keratoconjunctivitis with "hazard envelopes" as a function of wavelength. The action spectrum for moderate skin erythema obtained by Coblentz *et al.*[41] is shown as a fine line near the top of the figure. The threshold data for minimal erythema obtained by Freeman *et al.*[47] and by Everett *et al.*[45,46] merge into one dotted-line histogram in this semi-logarithmic presentation; one standard deviation is shown by ↕ for each 10 nm band. The range of erythema thresholds obtained by Berger *et al.*[48] at several mercury lines are shown as ⧫ representing one standard deviation. Action spectrum for photokeratitis thresholds obtained by Pitts and co-workers[89] is presented as solid line histogram with approximate range of threshold data shown by I-bars. Bold, dashed curve is envelope used for tentative 1971 ACGIH TLV[99]; bold dotted line is an alternative envelope curve for hazard analysis. Note considerable safety factor between hazard envelope and "classical" erythema action spectrum (fine line). Action spectra are inverted from usual presentation in this figure.

that the susceptibility of the cornea and conjunctiva does not decrease with increased exposure to ultraviolet radiation, as does the susceptibility of the skin as a result of tanning and thickening of the stratum corneum.

GUIDELINES FOR OCCUPATIONAL EXPOSURE TO ULTRAVIOLET RADIATION

After reviewing the available biologic data for injury to the eye and skin from exposure to ultraviolet radiation, it is easy to understand why essentially no occupational exposure standards exist in this area. Add to this (*1*) the difficulty of performing accurate spectral irradiance measurements in this spectral region, (*2*) the limited number of optical radiation sources which emit significant actinic ultraviolet radiation, and (*3*) the ease of using protective eyewear and clothing which have enormous attenuation factors for actinic ultraviolet, and we see why exact quantitative exposure guidelines have normally not been required.

In one application – the use of germicidal low-pressure mercury lamps which emit principally at one wavelength (254 nm) – exposure limits have been promulgated. The Council on Physical Medicine of the American Medical Association proposed a limit of 0.1 $\mu W/cm^2$ for a 24-hr exposure and 0.5 $\mu W/cm^2$ for a 7-hr or shorter exposure to these lamps. These values were based upon a minimal erythema dose of 32 mJ/cm^2 delivered in a 15-min exposure at 254 nm.[92] Although a clearly evident safety factor (2.6 and 3.8) was applied by the Council in arriving at the preceding permissible irradiance values, a greater safety factor actually exists for the long exposure times of 7 and 24 hr since there is some loss of dose reciprocity for such long exposures. The dose of 0.5 $\mu W/cm^2$ for 7 hr is 13 mJ/cm^2 and the dose of 0.1 $\mu W/cm^2$ for 24 hr is 8.6 mJ/cm^2 which correspond well with the range of erythema threshold data given in Fig. 4.

Matelsky (p. 154)[1] recommended several values weighted against the action spectrum for erythema or photokeratitis as the case would apply. Such an approach requires an accurate determination of the spectral irradiance vs wavelength of the source, and this may be particularly difficult in the 300–315 nm spectrum for some sources. I – having tried this approach and encountered difficulties in evaluating some sources with sophisticated radiometric instruments – favor a somewhat simplified approach. If the threshold data for acute effects obtained from the recent studies of minimal erythema and minimal keratoconjunctivitis are combined in one graph (Fig. 4), an envelope curve can be drawn which does not vary significantly (in comparison with measurement errors and variations in individual response) from the collective threshold data. The shape of this envelope curve can more readily be fitted by a radiometric detector response with appropriate absorbing filter, than could an action spectrum with several dips and peaks. Since repeated exposure of the eye to potentially hazardous levels does not result in increasing the protective capabilities of the cornea (as does melanogenesis which increases the protective shield for the deeper skin tissue)[36,92] the preceding exposure guide is more readily applicable to the eye and should be considered as a limiting value for that organ. However, such a guide can only be a starting point for determining skin exposure since wide variations in actual thresholds exist among individuals and the threshold varies with exposure history for a given individual. Such a guideline would have a built-in safety factor for essentially all but very sensitive individuals. The magnitude of the safety factor would depend upon the spectrum of the source since at least two independent action spectra (e.g., the

300 nm vs the 254 nm bands for erythema production) may exist and would not be additive.[93-98] Sources such as the sun which have a rapidly increasing spectral irradiance in the 300–315 nm band would be difficult to accurately evaluate using this or any other exposure guideline. The guideline could be applied only with extreme caution to ultraviolet lasers since all the biologic data upon which it was based were obtained from relatively broad-band sources. Narrow absorption peaks of an appropriate chromophore if located at a laser wavelength could drastically change the action spectrum. Such an exposure guideline is clearly nonapplicable to photosensitive individuals since the action spectrum is likely to be significantly different and extend well into the UV-A or visible region. The Committee on Physical Agents of the ACGIH recently set a tentative TLV based upon this concept (Fig. 4 – dashed line); the TLV is given in Table III.[99] The formula required for determining permissible exposure time t from a broad-band ultraviolet source, for which the spectral irradiance is known, is as follows:

$$t = \frac{0.003 \mathrm{J/cm^2}}{\Sigma E_\lambda \cdot S_\lambda \cdot \Delta\lambda} \tag{1}$$

where E_λ is the spectral irradiance in $W \cdot cm^{-2} \cdot nm^{-1}$, S_λ is the relative spectral effectiveness (unitless), and Δ_λ is the band width in nanometers. Hopefully, the development of an inexpensive instrument will be favored by this type of TLV such that spectral irradiance measurements will not be required to evaluate an ultraviolet source.

TABLE III
Proposed ACGIH Threshold Limit Values for Ultraviolet Radiation

Wavelength (nm)	*TLV (mJ/cm²)*	*Relative spectral effectiveness S*
200	100	0.03
210	40	0.075
220	25	0.12
230	16	0.19
240	10	0.30
250	7.0	0.43
254	6.0	0.5
260	4.6	0.65
270	3.0	1.0
280	3.4	0.88
290	4.7	0.64
300	10	0.30
305	50	0.06
310	200	0.015
315	1000	0.003

Guidelines for limiting exposure of individuals to near ultraviolet (UV-A) radiation can be based on relatively little data, and are seldom required since few sources emit sufficient radiation limited to this spectral region to create any adverse biologic effects to normal individuals. Effects upon the skin are considered to be principally thermal, and guidelines applicable for skin exposure to visible and infrared radiation (100 mW/cm^2 for localized areas; 10 mW/cm^2 whole body) have been applied without problems (except of course for photosensitive individuals). Guidelines for ocular exposure are quite a different matter. The suggestion of near ultraviolet radiation playing a causitive role in cataractogenesis has not been sufficiently investigated.[78,81-83] Short-term (16-min) laser exposures to a rabbit eye at 325 nm have produced cataracts at a corneal irradiance of 0.85 W/cm^2 (10 mW in a 1.5-mm beam).[100] Bachem reported the experimental production of cataracts at 297 nm for doses less than previously reported.[86] What levels below this could create any lenticular opacities under chronic exposure conditions are unknown. There are no strong indications that the low levels of this radiation found in industrial environments present a problem, although it has been indicated as the causitive agent in glassblowers cataract in the past.[82,83] If concern exists in any situation, ultraviolet goggles, used by individuals working with "black light" to prevent the annoyance of lens fluorescence, provide considerable protection from UV-A. The ACGIH proposal for a TLV in the near-ultraviolet (UV-A) was adjusted in 1972 as follows: For exposure times greater than 1000 sec (~16 min) the TLV was 1 $mW \cdot cm^{-2}$. For exposure times less than 1000 sec the TLV was 1 $J \cdot cm^{-2}$.

Visible and Near-Infrared Radiation (400–1400 nm)

CHORIORETINAL BURNS

Retinal burns resulting in a loss of vision (scotoma) following observation of the sun have been described throughout history. Socrates in Plato's "Phaedo" discussed eclipse blindness (solar retinitis or eclipse scotoma) and suggested that the prevention was to observe the eclipse by viewing the sun's reflected image in water.[101] Man-made optical radiation sources, comparable to the sun in luminance, and capable of causing chorioretinal burns have been developed chiefly in the last century. The incidence of chorioretinal injuries from man-made sources is no doubt far less than the incidence of eclipse blindness. Until recently it was felt that chorioretinal burns would not occur from exposure to visible light in industrial operations.[102] Indeed, this is still largely true since the normal aversion to high brightness light sources (the blink reflex and movement of the head and eyes away from the source) provide adequate protection unless the exposure is hazardous within the duration of the blink reflex. However, the recent revolution in optical technology, forged principally by the invention of the laser, has meant a great increase in the use of high-intensity, high-radiance optical radiation sources. Many such sources have output parameters significantly different from those encountered in the past and may present serious chorioretinal burn hazards. In industry, besides lasers, one may encounter sources of continuous optical radiation, compact arc lamps (as in solar simulators), quartz-iodide-tungsten lamps, gas and vapor discharge tubes, electric welding units, and sources of pulsed optical radiation, such as flash lamps used in laser research and photochemical investigations, exploding wires, and superradiant light. These sources may be of concern when adequate protective measures are not being taken. While ultraviolet radiation from most of these devices must be considered – and may be the principal concern – the potential for chorioretinal injury should not be overlooked. As will be shown later, the evaluation of potential retinal hazards from extended sources can be more involved than for "point" sources such as intrabeam laser exposure.

STUDIES OF CHORIORETINAL INJURY

Retinal injury due to radiant energy was discussed at length long before the advent of the laser. Reports of accidentally and experimentally produced retinal injury from intense man-made radiant sources, such as electric arcs and the nuclear fireball form an extensive literature. Effects of viewing the sun, particularly during an eclipse, have been reported throughout history. As early as 1867, Czerny produced experimental retinal lesions in the rabbit[98] and several later investigators used experimental animals (usually the rabbit) to investigate such ocular damage. It was not until 1916 that a truly comprehensive study, both quantitative as well as qualitative, was published. This study, by Verhoeff and Bell[103] of the Massachusetts Eye and Ear Infirmary, described the role played by ultraviolet, visible, and infrared radiation in producing various ocular effects as well as chorioretinal burns and remains a classic in this field. A companion paper to the 1916 study was an extensive review by Walker[104] of the literature dating back to the Ancients – with 428 references. The review covered reports of eclipse blindness, observation of lightning bolts at close range, experiences of nineteenth century scientists who first worked with high-intensity arc lamps, and earlier experimental studies. More recent reviews of the subject may be found in a volume by Duke-Elder (1954)[98] and articles by Cogan (1950),[105] Kutscher (1946),[106] Newell (1964),[107] Bartleson (1968),[108] Clarke (1970),[109] and Sliney (1971).[110]

Verhoeff and Bell (1916) concluded that man-made sources of radiant energy would not be expected to present a retinal hazard in normal use. Only a few cases have been reported of injury following ocular exposure to arcs produced from electrical short circuits, welding arcs, sunlamps, and arc lamps. This record could change with the recent development of higher radiance sources.[111,112]

The development of the xenon-arc photocoagulator by Meyer-Schwickerath[113] showed the beneficial clinical applications of the arc lamp in ophthalmology and led to the laser photocoagulator. It also clearly demonstrated that xenon arcs viewed under appropriate conditions could produce retinal burns.

Studies of retinal injury conducted since the late 1950's using principally rabbits and monkeys – believed to be good surrogates for the human – have made use of both arc lamps and lasers.[109,110,114–149] These studies provide most of the information required to evaluate the hazards from lasers and other high-radiance sources. Many of these studies reviewed casually might appear to contradict one another; however, general trends and closer agreement become evident when it is realized that different experimental conditions – different image sizes, wavelengths, animals, and source distributions – have been used by different groups.

To summarize all of the results of these studies is difficult; however, several general points are important to understand when evaluating an optical radiation source:

1. Absorption of energy. Visible and near infrared radiation (~ 400–1400 nm) is transmitted through the ocular media (cornea, aqueous, lens, and vitreous) and is absorbed in significant doses principally in the retina where an image may be formed (Fig. 5). The radiation passes through the neural layers of the retina, with a small amount of light being absorbed by the visual pigments in the rods and cones to initiate the visual response, and the remaining energy absorbed in the retinal pigment epithelium and choroid. Since the retinal pigment epithelium is optically the most dense absorbing layer due to a high concentration of

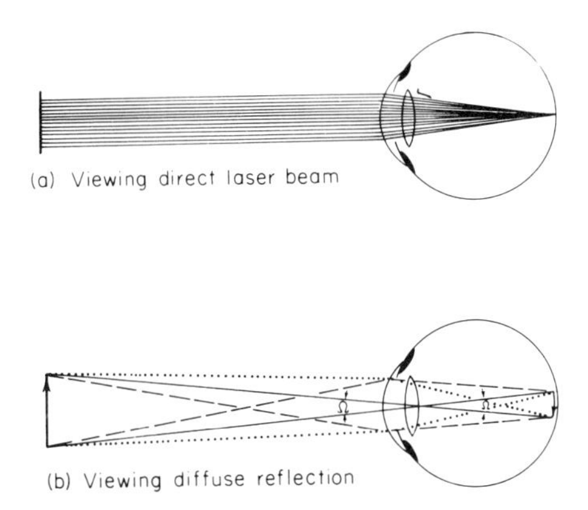

Fig. 5. A collimated beam of light (a) is brought to focus at the retina as a small "point" image, 10–20 μm diameter, by the relaxed normal eye. An extended source is imaged at the retina (b); the approximate size of the image is determined using the simple geometric relation that both the object and image subtend an angle Ω measured at the eye's nodal point (~1.7 cm in front of the retina).

melanin granules (layer of approximately 3–4 μm in thickness), the greatest temperature rise exists in this layer.[150]

2. Thermal injury. For accidental exposures from arc lamps, cw lasers, or the sun for durations of ~ 0.1 sec to ~ 100 sec, and for exposures from long-pulsed lasers or flash lamps (10 μsec to 10 msec) the mechanism of injury is largely thermal. Since this injury appears to result principally from protein denaturation and enzyme inactivation, the temperature-vs-time history of the retinal tissue during and following the insult must be considered. Several efforts to develop mathematical models for the light absorption, heat flow, and the rate-process injury mechanism within the complex structure of the retina have only been moderately successful. Nevertheless such attempts[150–169] are useful in gaining a broader understanding of the damage mechanism and provide the direction for future experimental studies. Such analyses of injury mechanisms make it possible to draw some general concepts. Since thermal injury is a rate-process, no single critical temperature exists above which injury will take place independent of exposure time.[109,150,169] Shorter exposures require higher temperatures for the same degree of retinal injury – at least for exposure times greater than ~ 1 μsec.[109] Since the very large, complex organic molecules absorbing the radiant energy would have broad spectral absorption bands, one would not expect the monochromatic nature of laser radiation to create any different effects than radiation from conventional sources, and indeed experimental evidence strongly supports this.[150] The coherence of laser radiation is also not considered to affect the hazard potential for chorioretinal injury. Hence chorioretinal injury from either a laser or a nonlaser source should not differ if image size (retinal irradiance distribution), exposure time, and wavelength are the same.

3. Image size. The tissue surrounding the absorption site can much more readily conduct away the absorbed heat for image sizes 10–50 μm diameter than it can for large image sizes (~ 1000 μm). Indeed retinal injury thresholds for the time domain of 0.1–10 sec show an enormous dependence upon image size (i.e., 1–10 W/cm^2 for a 1000 μm image; up to 1 kW/cm^2 for a 20 μm image). It is this dependence which explains the fact that momentary viewing of the sun (160 μm) by the unaided eye does not produce a retinal burn; however viewing the sun through a binocular or telescope, while not increasing the retinal irradiance, does greatly increase the image size, and a retinal burn is likely to result. This dependence is not so marked for much shorter exposure times, and some thermal models[151,163] would suggest that no such dependence should exist for exposure times shorter than approximately 1 msec. However, this is not borne out in animal experiments (Figs. 6 and 7). The retinal burn data for limiting-case, small images of 20–70 μm diameter are normally presented as total energy or power entering eye.[114,119,120,131,134–138] This is reasonable since the exact "image size" and the retinal exposure distribution cannot be accurately determined.

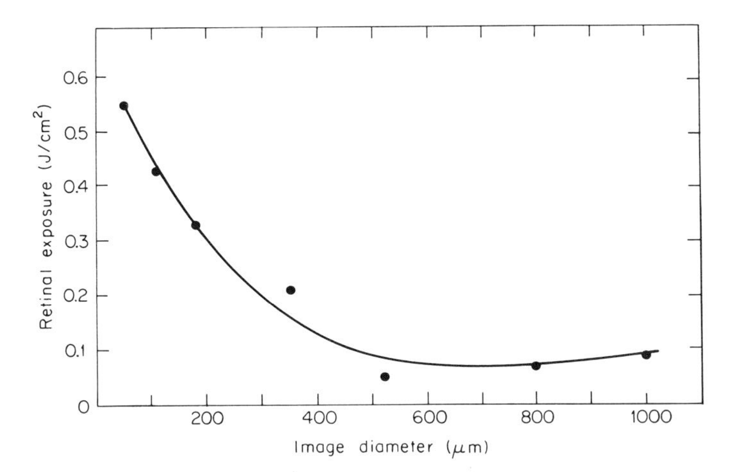

Fig. 6. Variation of the retinal injury threshold for a Q-switched ruby laser as a function of retinal image size for the rabbit eye, obtained by Ham and co-workers.[118]

4. Very short exposure times. For exposure times of Q-switched lasers (typically 5–50 nsec),[171] exploding wires, superradiant light (~ 2 nsec),[172] and mode-locked lasers (~ 10–100 psec)[171,173–175] exposure thresholds of injury are lowest.[109–110,138,150] Although it is believed that at Q-switched exposure times the injury mechanism is still largely thermal, the effect of acoustic transients due to the rapid heating and thermal expansion in the immediate vicinity of the absorption site[193] – each melanin granule – may play a role. For the still shorter exposure times, direct electric-field effects, Raman and Brillouin scattering, and multiphoton absorption could play a role in the damage mechanism.[115,150,155]

5. Location of retinal burns. Since the different regions of the retina (the fovea, the rest of the macula, and the peripheral retina) play different roles in vision, the functional loss of all or part of any one of these regions due to retinal injury varies in significance. As shown in Fig. 8, the greatest visual acuity exists only for central (foveal) vision, and the loss of this retinal area would dramatically reduce vision. In comparison, the loss of an area of similar size located in the peripheral retina could be subjectively unnoticed. Most studies indicate that retinal functional loss occurs at levels no less than one-half the ophthalmoscopic threshold.[134,178–182]

6. Very long exposure times. The human retina is normally subjected to irradiances below 10^{-4} W · cm^{-2} (see Fig. 9), except for occasional momentary exposure to the sun, arc lamps, photoflash lamps, incandescent filaments, and similar radiant sources. The retinal images resulting from viewing such sources are quite small (i.e., 160 μm for the sun), and exposure times are short: they are normally limited to the duration of the blink reflex (0.15–0.2 sec).[183] Since the natural aversion to bright light[184–192] normally limits further retinal

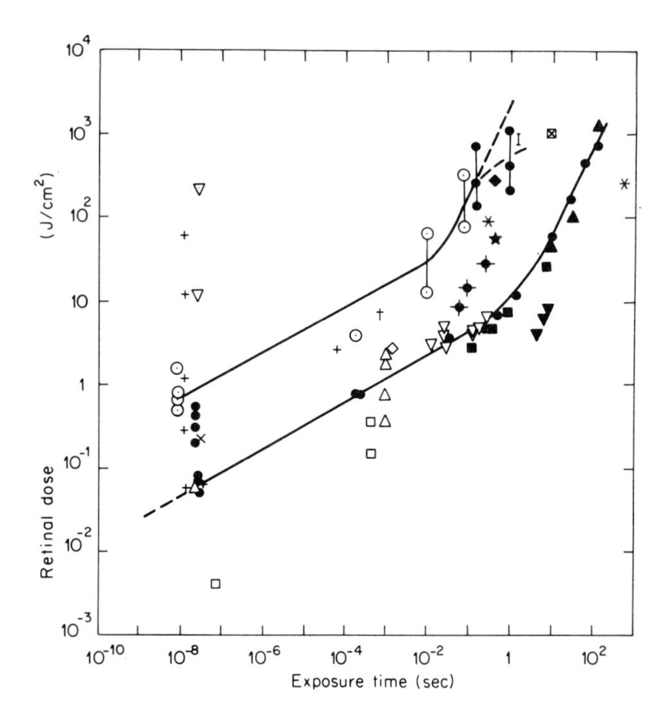

Fig. 7. Compilation of retinal injury thresholds obtained by several investigators. The upper curve passes through threshold data for small retinal image sizes (40–70 μm); the lower curve passes through data for 800 μm image size of Ham and co-workers.[118] Data points are identified as Ham and co-workers[118,114] (●); Bergqvist and co-workers[121] (+); Eccles and Flynn[170] (▲); Jones and McCartney[122] (◇); Jacobson and co-workers[125] (▽); Vassiliadis and co-workers[119, 120] (⊙); Kohtiao and co-workers[123] (□); Campbell and co-workers[126–128] (△); Zaret [129] (X); Demott and Davis[148] (■); Verhoeff and Bell[103] (*); Kohner and co-workers[149] (▼); Blabla and John[130] (◆); Ingram[124] (†); Bredemeyer and co-workers[145] (★); Lappin[136] (I); Davis and Mautner[134] (⊠). Thresholds for image sizes between 100 μm and 800 μm generally fall proportionately between the two solid lines. Recent data suggest that the upper curve may be much lower in the microsecond exposure range.[369] This finding effected a change in the ANSI Z-136 proposal in February 1972.

exposures above 10^{-4} W · cm^{-2}, few studies of retinal effects exist for the irradiance range of 10^{-4}–1 W · cm^{-2}. Studies in this range have generally centered on flashblindness effects following light exposures lasting up to 1 sec.[190–195] As flashblindness effects are related to bleaching of retinal visual pigments, photometric units must be used. The range of 10^{-4}–1 W · cm^{-2} generally falls within 10^{-2}–300 lm · cm^{-2} or 4×10^4 to 10^9 trolands. Staring into a cw laser at a level of approximately 1 μW · cm^{-2}, or viewing a periodically pulsed flashlamp or a high-radiance near-infrared source can occur without evoking a significant aversion response.

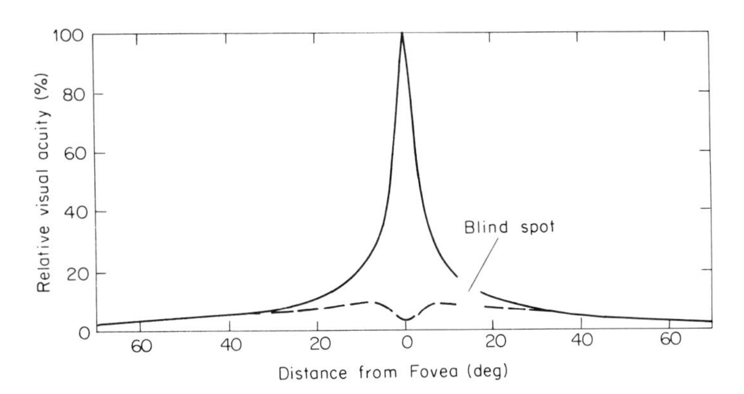

Fig. 8. Variation of visual acuity with angular distance from the fovea for scotopic (dashed line) and photopic (solid line) response. (Adapted from drawing in Walsh[176] and data of Randall *et al.*[177])

Exposure of large areas of the retina to moderately high-luminance light (10^{-5} to 10^{-4} W · cm^{-2} at the retina) for durations of one to several hours, have been investigated in experimental animals.[196–202] A thermally enhanced photochemical mechanism of injury or phototoxic effect appears to be possible under such conditions. Rats and nocturnal monkeys exposed for many hours to ordinary fluorescent lamps having a luminance of ~ 1 cd · cm^{-2} (3×10^{-5} W · cm^{-2} at the retina) developed irreversible retinal injury manifested as death of visual cells and degeneration of pigment epithelial cells, apparently due to a phototoxic reaction in the retina.[200] Some effects upon the outer segments of visual cells were shown to be reversible[199] if the photoreceptor cell body and adjacent pigment epithelium survived. Attempts to explain the exact mechanism of these effects have indicated a strong dependence upon an interruption of the diurnal light and dark cycle which normally assures retinal cell adaptation to light.[200] It is doubtful that these studies, conducted principally with nocturnal animals, can be related to any possible condition of human exposure. Only seldomly has it been suggested that continual exposure to high luminance levels in the natural environment (or work environment) could elicit significant effects in the human retina.[203,204]

One experimental study may have greater significance than the aforementioned. Exposure of the rhesus monkey retina to an irradiance of approximately 0.3 W · cm^{-2} for approximately 15 min from a clinical indirect ophthalmoscope (having a tungsten lamp) resulted in damage to the retinal pigment epithelium.[205] This damage threshold for a 4.5 mm diameter retinal image closely agrees with a 12-min exposure threshold of 0.42 W · cm^{-2} reported by Verhoeff and Bell in 1916 for a 3 mm diameter image for the rabbit. It has been suggested that the damage mechanism for these comparatively shorter exposure times of 10–30 min may be a thermally enhanced phototoxic effect.[165] Although the

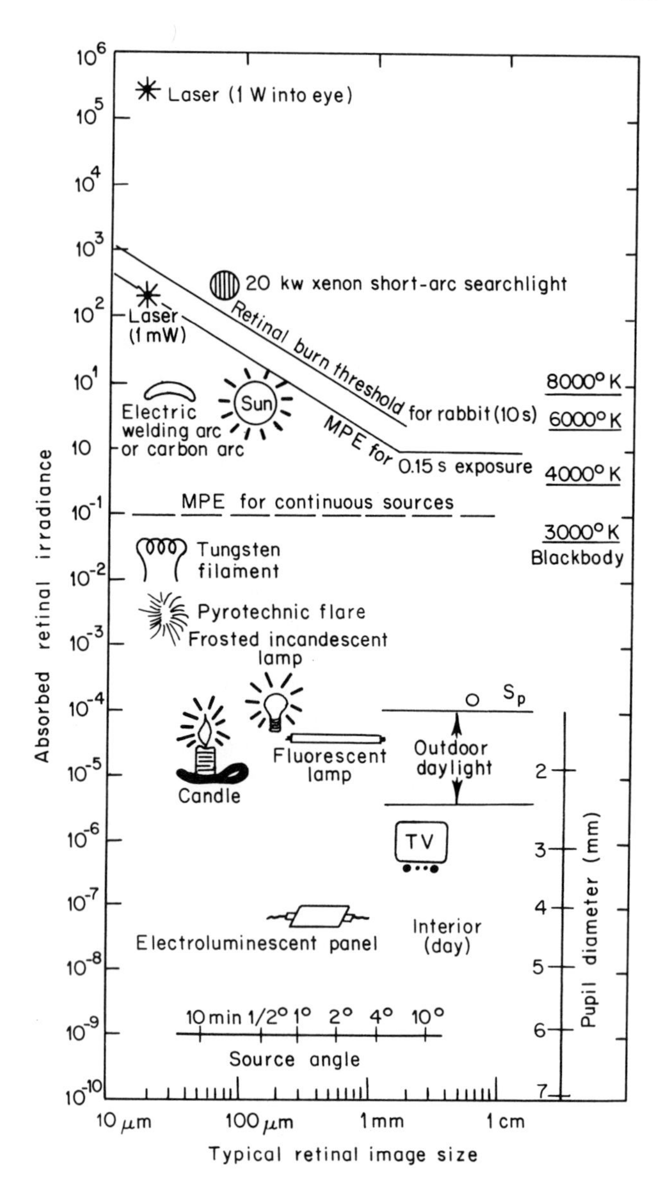

Fig. 9. The eye is exposed to light sources having radiances varying from $\sim 10^4 W \cdot cm^{-2} \cdot sr^{-1}$ to $\sim 10^{-6} W \cdot cm^{-2} \cdot sr^{-1}$ and less.

resulting retinal irradiance from viewing an incandescent tungsten filiament with a 2 mm pupil would normally not exceed $0.04\ W \cdot cm^{-2}$, a dilated (approximately 8 mm) pupil would permit the retinal irradiance to be as great as $0.6\ W \cdot cm^{-2}$ (see Fig. 9).

Monkeys exposed repeatedly to 10-Hz repetitively flashing xenon arc lamps for more than 17 min developed ophthalmoscopically visible retinal lesions at an average irradiance of $\sim 0.2\ W \cdot cm^{-1}$, a value which is well below that expected to be required for acute thermal injury.[206]

It is not clear why the threshold of injury for exposure durations of 12–15 min are so low since the corresponding calculated temperature elevations are less than 1°C. A thermally enhanced injury mechanism is occasionally suggested. Additionally, the "thermal equilibrium" that exists in the retina for these exposure durations is really only a condition of stabilized isotherms, which means that localized temperatures are significantly higher than the calculated average. These higher localized temperatures would exist at the pigment granules.[158]

Studies were conducted during and following World War II of the effects upon night vision and retinal sensitivity of prolonged exposure of individuals to bright outdoor environments. Lifeguards exposed to the high luminance environment (approximately $1\ cd \cdot cm^{-2}$) found at the seashore have showed both a short-term depression in photopic sensitivity and a marked long-term loss of scotopic (night) vision.[207–213] Repeated exposures of trained monkeys to high luminance blue light showed a "permanent" decrease in spectral sensitivity of the retina in this spectral range at retinal irradiances just above those experienced in a bright natural outdoor environment.[214] Exposure of the monkeys to narrow bands of wavelengths from the green to the red elicited a similar but not a lasting response. The depression of spectral sensitivity corresponded with the spectral absorption properties of each of the three cone photopigments (blue, green, and red). All of these effects required exposure of large areas of the retina. Prolonged erythropsia (red vision) in aphakics has also been reported following exposure to large-area, high-luminance sources such as snow fields.[215] The adverse effect produced by long-term exposure from repeatedly flashed electronic flash lamps[206] probably poses the only type of chronic retinal exposure hazard that could actually exist in any industrial environment.

To place the chorioretinal injury data in better perspective, Fig. 9 shows the retinal irradiance for many cw sources. It is reemphasized that several orders of magnitude in brightness (radiance or luminance) exist between sources which cause chorioretinal burns and those levels to which individuals are continuously exposed. The retinal irradiances shown are only approximate and assure minimal pupil sizes and some squinting for all the very high-luminance sources, except the xenon searchlight for which a 7-mm pupil was assumed (so as to apply to searchlight turn-on at night).

7. The optical properties of the eye play an important role in determining retinal injury. Such factors as the image quality, pupil size, spectral absorption and scattering by the cornea, aqueous, lens and vitreous, as well as the spectral reflectance of the fundus, and absorption and scattering in the various retinal layers must be known. These shall be considered separately.

8. Pupil size. The limiting aperture of the eye determines the amount of radiant energy entering the eye and thereby reaching the retina. The energy admitted is proportional to the area of the pupil. For the normal dark-adapted eye, pupil sizes range from 7–8 mm; for outdoor daylight the normal pupil size is 2–3 mm; momentary viewing of the sun causes the normal pupil to constrict to approximately 1.6 mm.[110,170] The ratio of areas between a 2 mm and 8 mm pupil is 4:64; hence a 2 mm pupil accepts one-sixteenth the light admitted by an 8 mm pupil. The angular subtense of the adapting field plays a role; hence, a light source of a given luminance causes a different pupil size dependent upon viewing distance (i.e., image area on the retina) and the luminance of the surrounding field. It is important to remember that pupil size for a given environment varies with age, emotional state, and other factors. The use of some medications creates abnormally large pupil sizes. Therefore, for a general population, the pupil size may vary greatly even for a bright environment.

9. Spectral transmission of the ocular media. The transmission of the ocular media between 300 and 1400 nm has been measured by several investigators.[110] Two studies are most often used: those of Geeraets and Berry[216] and those of Boettner and Wolter.[217–218] The measurement techniques differed. Figure 10 presents the results of these two studies. The highest transmission values (Geeraets and Berry) were obtained using intact enucleated human eyes and measuring total spectral transmission. Boettner and Wolter measured the spectral transmission of the cornea, aqueous, lens, and vitreous separately, which could account for the lower total transmission. They also measured "direct" transmission which limited the amount of scattered light accepted by the detector. The highest transmission values may be considered maximum even for large image sizes, while the lower values are probably more reasonable for smaller image sizes where forward scattering tends to limit the peak retinal illuminance. Unfortunately the experimental techniques used did not permit a clear indication of the comparable image sizes for each curve.

10. Spectral absorption by the retina and choroid. Geeraets and Berry also measured the fundus reflectance and the absorption in human retinal pigment epithelium and choroid. When these factors are taken into account, it is possible to multiply the absorption data by the transmission data for the ocular media to arrive at an estimate of the relative absorbed spectral dose in the retina plus

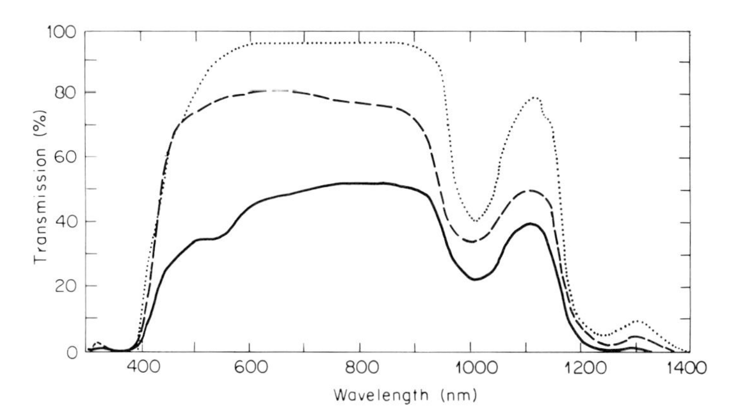

Fig. 10. Spectral transmission of the ocular media of the human eye. Upper curve was total transmission obtained by Geeraets and Berry,[216] using 28 intact enucleated eyes. Lower two curves by Boettner and Wolter[217] were obtained by combining separately measured transmission factors for the cornea, aqueous, lens, and vitreous for nine human eyes. Lowest curve is "direct" transmission obtained by eliminating forward scattered light, whereas middle curve was obtained by collecting total light transmitted.

choroid for the same spectral radiant exposure at the cornea. This should provide a relative spectral effectiveness curve for chorioretinal burns (at least for exposures times less than ~ 10 sec). Figure 11 provides two such curves using the maximum and minimum transmission values shown in Fig. 10. Comparison of actual threshold retinal burn data obtained at different wavelengths generally support the upper curve for large image sizes and the lower curve to some extent for smaller image sizes.

11. The optical image quality. For most extended sources the retinal image size can be calculated by geometric optics. As shown in Fig. 5, the angle subtended by an extended source defines the image size. Knowing the effective focal length of the relaxed normal eye (~ 1.7 cm), the retinal image size d_r can be calculated if the viewing distance r and the dimension of the light source D_L are known:

$$d_r = \frac{D_L f}{r} \tag{2}$$

Since this formula was derived (using similar triangles; see Fig. 5) by applying the approximation that the arc and the chord of a circle are approximately the same for small angles, the formula does not hold for large image angles (e.g., an error in d_r of ~ 5% for source sizes subtending an angle of 60 degrees). And by similar arguments, the solid angle Ω measured at the nodal point of the eye subtended by either the source or the image is the same (Fig. 5); hence the source area A_L and image area A_r are always proportional for small angles in the

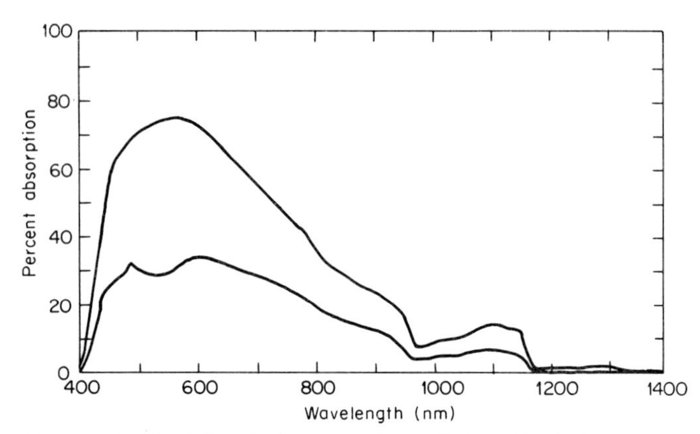

Fig. 11. Spectral absorbed dose in human retina and choroid relative to spectral corneal exposure as a function of wavelength. Retina and choroid spectral absorption values corrected for fundus reflection obtained by Geeraets and Berry[216] were multiplied by corresponding transmission spectral factors of the ocular media from upper and lower curves of Fig. 10.

ratio of the square of the viewing distance r to the square of the eye's focal length f. Again, a small error is introduced for large angles (e.g., an error in A_r of ~5% for a circular source area subtending an angle of 20 degrees). Hence the retinal irradiance and source radiance are likewise proportional. Indeed, by noting that the radiance L of the source can be defined from the irradiance at the cornea E_c for small angles as:

$$L = \frac{E_c}{\Omega} = E_c\left(\frac{r^2}{A_L}\right) = E_c\left(\frac{f^2}{A_r}\right) \tag{3}$$

and that the total power entering the eye (through the pupil of area A_c) that reaches the retina for a transmission T must be:

$$E_r \cdot A_r = T \cdot E_c \cdot A_c = T \cdot E_c \left[\frac{\pi \cdot d_p^{\,2}}{4}\right] \tag{4}$$

we obtain the quantitative relation of retinal irradiance to source radiance (or retinal illuminance to source luminance):

$$E_r = \frac{\pi\, d_p^{\,2} \cdot L \cdot T}{4 f^2} = 0.27\, d_p^{\,2} \cdot L \cdot T \tag{5}$$

for small source angles. Equation (5) is of great practical value since it permits one to define a permissible radiance from a permissible retinal irradiance or illuminance for a source of known radiance or luminance without having to be concerned with the viewing angle or viewing distance. Although the effective nodal point, hence f, varies slightly for the accommodated eye (viewing distances less than 6 m), the formula is not significantly affected even for very short viewing distances. The above derivation assumed a source of uniform radiance; however, since each source point has a corresponding image point, the retinal irradiance in an incremental area of the image is likewise related to the radiance of the corresponding source increment. As an aside, it may be pointed out that Eq. (5) is also useful in photographic radiometry. The ratio F of focal length f to the aperture d_p is known as the focal ratio or F-number of the camera setting, hence:

$$E_f \cdot t = H_f = \frac{\pi \cdot L \cdot t \cdot T}{4 F^2} \tag{6}$$

where t is the exposure time and H_f is the exposure at the film plane.

Equation (5) breaks down for very small image sizes (or for very small "hot spots" in an image) where the source or source element in question subtends an angle of less than ~ 10 min of arc. Figure 12 shows the approximate retinal irradiance profiles for circular sources subtending 1, 5, 10, and 20 min of arc. One arc-minute corresponds geometrically to approximately 5 μm at the retina. Note that for the same source radiance the image irradiance decreases as the

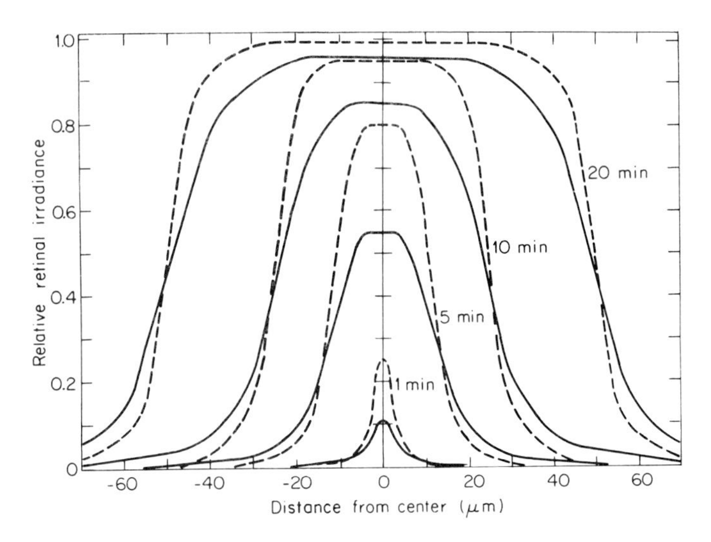

Fig. 12. Relative retinal image irradiance profiles for uniform disc sources subtending viewing angles of 1, 5, 10, and 20 arc-min corresponding geometrically to retinal image diameters of 5, 25, 50, and 100 μm respectively, obtained by Gubisch[219] for the normal human eye. As the source angle decreases appreciably below 20 arc-min, the peak retinal irradiance falls further below the normalized value of one, calculated by geometric optics. Dashed line profiles are for 3 mm pupil, solid line profiles are for 7 mm pupil.

image approaches a "point image." This decrease results from image blur due to diffraction of light at the pupil, due to the aberrations introduced by the cornea and lens, and due to scattering from the cornea and the rest of the ocular media. Since the effects of aberrations increase with increasing pupil size, greater blur, hence reduced peak retinal irradiance, is noted for larger pupil sizes. This effect is also evident in Fig. 13 which shows the optical gain (peak E_r/E_c) as a function of pupil size for a "point" source. This relation is useful in evaluating intrabeam viewing of a laser where the collimated light can produce a "point" image in the relaxed normal (emmetropic) eye. When a laser is directly viewed from within the beam or when an extended source is viewed at a great distance by the normal eye (i.e., the source angle is less than approximately one to four min of arc), the

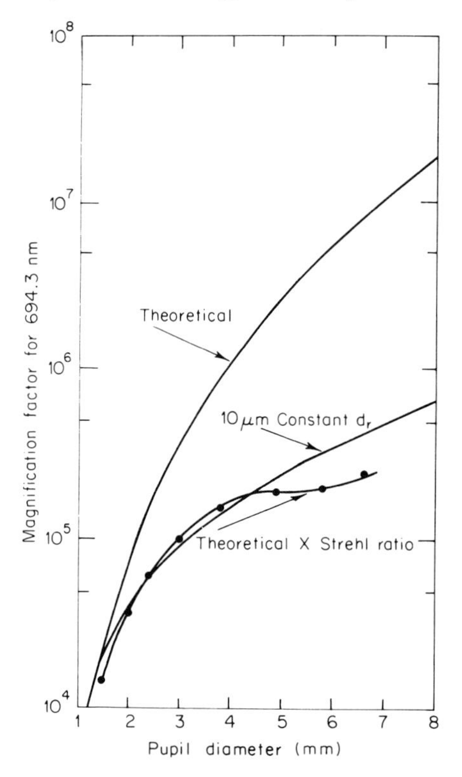

Fig. 13. Influence of pupil size on magnification factor of corneal to retinal irradiance for a point source viewed by the normal human eye. Theoretical curve was obtained using Airy formula for peak retinal intensity.[110] A second curve shows the magnification factor for a constant retinal image diameter of 10 μm. Final curve is believed to more accurately represent the actual magnification factor and was derived by multiplying the theoretical values by the Strehl ratios reported by Gubisch.[219] The Strehl ratios used may be high due to experimental difficulties of measuring low levels of forward scatter.[365]

retinal image profile is essentially the same regardless of the actual source angle. This fact requires us to determine the potential hazard by the corneal irradiance and not by the source radiance for intrabeam viewing of a laser. Investigators studying retinal injury using a collimated laser beam to produce a minimal image size also are limited to giving the total power entering the eye in presenting retinal injury thresholds.

HAZARD CRITERIA FOR RETINAL EXPOSURE

Having briefly discussed the principal factors which influence retinal hazards from lasers and other high-intensity light sources it is now easier to understand the several standards for personnel exposure which have been proposed for lasers emitting in the 0.4–1.4 μm spectral range. No standards are available for other high-brightness optical sources, but such sources can often be adequately evaluated using diffuse reflection standards for laser radiation which are based on viewing of an extended source.

More than twenty sets of exposure criteria have been proposed.[99,110,220–224] To limit the discussion, we shall consider only the most widely used. The earliest standards had large safety factors applied. Recent exposure standards generally incorporate a safety factor of ten. The most commonly used criteria presently in use are those of ACGIH and the military services (Table IV). The American National Standards Institute Committee Z136 was developing a standard at the time of this writing. The lower values given in Table IV were derived using retinal injury criteria which were well accepted for large (800 μm) image sizes and extrapolating to the minimal image size (10–20 μm).[225] The U.S. Air Force criteria were derived using retinal burn thresholds which were obtained for small image sizes and included safety factors less than ten.[222] The most dramatic difference between the USAF standard and the remaining standards is in the level for continuous wave (cw) lasers. The USAF standard is limited to visible cw lasers and assumes a momentary exposure of less than one sec, whereas the other standards assume continuous viewing, and some apply to near-infrared lasers as well.

The exposure criteria for intrabeam viewing of cw lasers is probably the single most important level applied in practical situations, since pulsed lasers are less commonly encountered and are generally enclosed in industrial applications. The wide variation in proposed permissible limits is therefore particularly unsettling. While most experts feel that an accidental intrabeam viewing of a visible cw laser at a level of 0.4 mW entering the eye (1 mW/cm^2 for a 7 mm pupil) would not result in a retinal burn since the exposure time would be limited to ~0.15 sec (the blink reflex), there is considerable disagreement as to a reasonable exposure limit for continuous viewing. Some feel that continued viewing at 1 mW/cm^2, even though dazzling, should be safe. Others point to recent investigations of

TABLE IV

Recommended Limits for Ocular Exposure to Laser Radiation (0.4–1.4 μm) for a 7 mm Pupil

A. Intrabeam Viewing of a Collimated Beam

Organization	*Principal data source*	*Type of laser*	*Continuous cw staring* ($W \cdot cm^{-2}$) ($t > 1$ sec)	*cw laser interrupted (1 msec to 1 sec)*	*Non-Q–switched long pulse* ($J \cdot cm^{-2}$) (20 μsec to 1 msec)	*Short Pulse (e.g., liquid dye laser)* ($J \cdot cm^{-2}$) (0.1–20 μsec)	*Q-switched pulse* ($J \cdot cm^{-2}$) (1–100 nsec)	*Ref.*
U. S. Departments of the Army and Navy (Feb. 1969)	MCV[a]	λ = 400 to 1400 nm	10^{-6}	($t < 0.1$ sec) 10^{-6} $J \cdot cm^{-2}$	($\sim$ 1 msec) 10^{-6}	–	(5–50 nsec) 10^{-7}	221
U. S. Department of the Air Force (September 1971)	SRI and USAF–SAM[a]	λ = 400 to 700 nm	–	10 to 500 msec 2.5 $mW \cdot cm^{-2}$ (2 to 10 msec) 5 $mW \cdot cm^{-2}$	(200 μsec to 2 msec) 10^{-5}	–	(10 to 100 nsec) 1.2×10^{-6}	222
		λ = 1060 nm	–	10 to 500 msec 6.2 $mW \cdot cm^{-2}$ (2 to 10 msec)	(200 μsec to 2 msec) 5×10^{-4}	–	(10 to 100 nsec) 6.2×10^{-6}	222
Ministry of Technology, United Kingdom (October 1969)	SRI[a]	Ruby	($t > 0.1$ sec) 4×10^{-7}	(occasional) 10^{-4}	(1 μsec to 0.1 sec) 10^{-6}	See applicable value	(1 nsec to 1 μsec) 3×10^{-5}	223
		Neodym.	2×10^{-6}	3×10^{-4}	3×10^{-6}	–	2×10^{-7}	223

		Argon or He-Ne	3×10^{-7}	(occasional) 3×10^{-5}	–	–	–	223
ACGIH (1971)	MCV[a]	Ruby	–	–	(1 μsec to 0.1 sec) 10^{-6}	See applicable value	(1 nsec to 1 μsec) 10^{-7}	99
		λ = 400 to 750 nm	10^{-5}	(t > 0.1 sec) 10^{-5} W · cm^{-2}	–	–	–	99
U. S. Department of Labor (29CFRA1518.54)	MCV and SRI[a]	He-Ne	10^{-6}	(incidental) 10^{-3} W · cm^{-2}	–	–	–	–
ANSI Z-136 Proposed Feb. 29, 1972[b] and ACGIH Proposed (1972)	MCV, SRI, USA–FA, and USAF–SAM	λ = 400 to 700 nm	(10 to 10^4 sec) 10 mJ · cm^{-2} (>10^4 sec) 1 μW · cm^{-2}	(20 μsec to 10 sec) 1.3 $t^{3/4}$ mJ · cm^{-2}	(20 μsec to 10 sec) 1.3 $t^{3/4}$ mJ · cm^{-2}	5×10^{-7}	5×10^{-7}	224
		λ = 1.06 to 1.4 μm	(10 to 10^4 sec) 50 mJ · cm^{-2} (>10^4 sec) 5 μW · cm^{-2}	(20 μsec to 10 sec) 6.5 $t^{3/4}$ mJ · cm^{-2}	(20 μsec to 10 sec) 6.5 $t^{3/4}$ mJ · cm^{-2}	2.5×10^{-6}	2.5×10^{-6}	224

[a] MCV, Medical College of Virginia; SRI, Stanford Research Institute; USA–FA, U. S. Army Frankford Arsenal; USAF–SAM, U. S. Air Force School of Aerospace Medicine.

[b] ANSI presented values between 0.7 and 1.06 μm varying with wavelength.

Table IV (Cont.)
B. Viewing a Diffuse Reflection[c]

Organization	*Type of exposure*	*Radiant exitance at diffuse surface*	*Surface radiance*	*Ref.*
U. S. Departments of the Army and Navy (Feb. 1969)	Continuous ($t > 0.1$ sec)	$2.5\ W \cdot cm^{-2}$	$\frac{2.5}{\pi}\ W \cdot cm^{-2} \cdot sr^{-1}$	221
	Long pulse ($t \sim 1$ msec)	$0.9\ J \cdot cm^{-2}$	$\frac{0.9}{\pi}\ J \cdot cm^{-2} \cdot sr^{-1}$	
	Q-switched (5–50 nsec)	$0.07\ J \cdot cm^{-2}$	$\frac{0.07}{\pi}\ J \cdot cm^{-2} \cdot sr^{-1}$	
ANSI Z-136 Proposed Feb. 29, 1972[d]	Long-term (10^4 sec)	$2 \cdot \pi\ mW \cdot cm^{-2}$	$2\ mW \cdot cm^{-2} \cdot sr^{-1}$	224
	Continuous (10–10^4 sec)	$20 \cdot \pi\ J \cdot cm^{-2}$	$20\ J \cdot cm^{-2} \cdot sr^{-1}$	
	Pulsed (1 nsec–10 sec)	$10 \cdot \sqrt[3]{t}\ J \cdot cm^{-2}$	$10\ \sqrt[3]{t}\ J \cdot cm^{-2} \cdot sr^{-1}$	
U. S. Department of Labor (29CFR1518.54)	Continuous	$2.5\ W \cdot cm^{-2}$	$\frac{2.5}{\pi}\ W \cdot cm^{-2} \cdot sr^{-1}$	–

[c] These levels, defined at the surface of a diffuse reflector, are applicable for extended sources. They are useful in defining hazard categories of lasers. The radiant exitance may be calculated – it is the product of the surface reflectance at the laser wavelength and the beam irradiance upon the diffuse surface.

[d] Tabulated values are for wavelengths between 0.4–0.7 μm only. Higher values apply for IR–A.

thermally enhanced "phototoxic" effects from chronic exposure to high-luminance sources, and decreasing retinal sensitivity from such exposure. Many practical considerations of natural avoidance behavior vs. image size,[184-192] or the reliance upon eye movements[226-229] must be considered to arrive at a reasonable solution to this problem. This author favors two exposure levels – one for chronic intentional exposure as occurs when individuals stare directly into the beam of a He-Ne alignment laser ($\sim 1\ \mu W/cm^2$) in certain construction site applications,[230] and another level ($\sim 1\ mW/cm^2$ for accidental exposure to cw lasers emitting between 0.4–0.75 μm) for which the blink reflex should provide protection.

Infrared Radiation

The infrared spectrum is considered to extend from 700–780 nm to approximately 1 mm. It is sometimes divided into specific regions by different specialists. For instance, the CIE committee which formulated the ultraviolet bands, UV-A, UV-B and UV-C, also defined three biologically significant infrared bands.[2] These bands, IR-A extending from 780–1400 nm, IR-B from 1.4–3 μm, and IR-C from 3 μm–1 mm, can be very useful to us in considering optical radiation biologic effects.

The principal adverse biologic effects attributed to infrared radiation are: infrared cataracts (also known as glassworker's or furnaceman cataracts), flash burns of both the skin and the cornea of the eye from very intense sources, and heat stress from less intense thermal radiation.

SKIN INJURY FROM VISIBLE AND INFRARED RADIATION

The imaging process of the eye makes that organ far more vulnerable to injury than the skin when exposed to very intense visible and near-infrared radiation. As a comparison, intrabeam viewing of a collimated laser beam may produce an exposure at the retina as much as 200,000 times the exposure incident upon the cornea. Therefore, hazardous radiant exposure of the skin for a pulsed laser may be as much as five orders of magnitude greater than levels hazardous to the eye. Hazardous levels for the skin and the eye are comparable in the infrared beyond 1.4 μm and in the near ultraviolet since the ocular media are opaque to UV-A, IR-B, and IR-C, and the injury mechanism is thermal.

For high energy pulsed lasers, skin injury thresholds generally range from $1–10\ J \cdot cm^{-2}$ for non-Q-switched lasers and $0.1–1\ J \cdot cm^{-2}$ for Q-switched lasers dependent upon skin pigmentation.[231-233] High power cw lasers operating in the visible and near-infrared spectrum have only recently been developed which are capable of producing significant burns to the skin in less than one second. As in the retina, the thresholds of injury are greater than

$1\ W \cdot cm^{-2}$, and are highly dependent upon the exposed skin area, particularly for areas less than one cm^2, and are dependent upon exposure time. The possibility of adverse effects from repeated or chronic laser irradiation of the skin has been suggested, although it is normally discounted.[232–234]

The reflectance of the skin varies significantly only in the visible and near-infrared spectrum as shown in Fig. 14. Not surprisingly, the spectral reflectance curve of the skin is somewhat similar in shape to the spectral irradiance curve for solar radiation. This permits the skin to reflect most of the solar radiation and yet have a thermal emissivity of nearly 1.0 in the far infrared to enhance its ability to rid the body of excess heat by thermal radiation. The human body has a black-body emission spectrum peaking at approximately 10 μm. This strong (almost total) absorption of IR-C, however, becomes a disadvantage to workers in industries where high levels of radiant heat from furnaces exist. Figure 14 also illustrates the reason for different indices for heat stress for outdoor and indoor work environments since the radiant heat load upon the body is different (see Review of Heat Exposures). If one could design a globe thermometer with a reflectance and emissivity similar to that of the skin, measurements to determine heat stress could be simplified for the radiant heat situation.

The warmth sensation by the skin resulting from absorption of radiant energy normally provides adequate warning to prevent thermal injury of the skin. The direct solar irradiance is approximately $40–70\ mW \cdot cm^{-2}$ in the northern

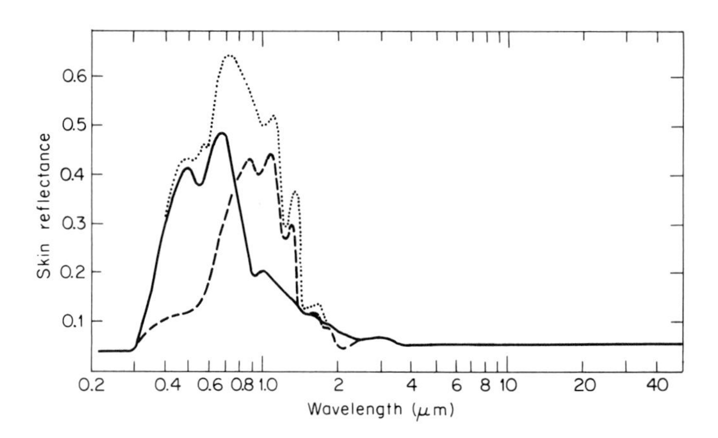

Fig. 14. Spectral reflectance of human skin. Dotted line and dashed line are data of Jacquez and co-workers[236,237] for individuals having a very fair complexion and for individuals (negro) having heavily pigmented skin. Spectral reflectance for moderately pigmented skin obtained by earlier investigators[235] and reconfirmed by Hardy and co-workers[238,239] is shown by the solid line. Significant differences in the IR-A spectral curves result from the use of different experimental techniques.

lattitudes and up to 100 mW · cm^{-2} in the tropics. These levels incident upon the skin produce a definite sensation of warmth. One could expect IR-C radiation to cause a similar sensation for a whole-body irradiance of only 10 mW · cm^{-2}, unless the deeper penetration of visible and IR-A radiation enhances the thermal sensation by the skin – which is unlikely.[240–244] For a CO_2 laser emitting at 10.6 μm, the beam size incident upon human skin must be nearly one centimeter or greater in diameter for 0.1 W · cm^{-2} to produce a definite sensation of warmth and at least 2–3 mm diameter for 1 W · cm^{-2} to produce this sensation, while 0.01 W · cm^{-2} would probably be sensed for whole-body exposure. This dependence upon the size of the irradiated area results from conduction of heat away from the absorbing area which limits surface temperature rise, the sensation being a function of temperature rise. Since the reflectance of the skin at a wavelength of 632.8 nm is nearly ten times greater than at a wavelength of 10.6 μm for a very lightly pigmented individual, the aforementioned irradiances would have to be increased by that factor to elicit the same sensation.

Flash burns of the skin or eyes from high-intensity optical radiation sources in industry are essentially nonexistent.[82] Studies to determine thermal injury thresholds for optical radiation, as from a nuclear fireball, required investigators to go to considerable lengths – using focused arcs and often blackened skin – to achieve threshold skin burns.[245–253] Approximately 0.2 W · cm^{-2} was required for an exposure time of 100 sec, 0.9 W · cm^{-2} for 10 sec, 7.5 W · cm^{-2} for 0.5 sec and 10^3 W · cm^{-2} (1 J · cm^{-2}) for 1 msec to produce second degree skin burns in blackened skin of the albino rat.[252–253] The 100 sec exposures produced a skin temperature elevation to 44–46°C. The skin temperature elevation required to elicit persistent pain is also ~45°C, whereas transient pain is noted at lower temperatures (38°–41° C).[242–244] The 0.5 sec exposure produced a temperature elevation to ~82°C. These levels are as much as a factor of ten lower than would apply to typical high-intensity sources which emit largely visible and IR-A radiation. Brownell and co-workers[254] found injury thresholds of two to three times greater than the aforementioned for the skin of the pig when irradiated with far-infrared radiation (10.6 μm) from a CO_2 laser for which skin absorption should be maximal. Again, the size of the irradiated area might account for this apparent discrepancy. The above levels are hardly likely to be encountered in the normal work environment, and if they are, considerations of heat stress require protective measures at lower levels.

INFRARED RADIATION AND THE EYE

The absorption properties in the near-infrared spectrum for the cornea, aqueous, lens, and vitreous are given in "Industrial Hygiene Highlights,"[1] Fig. 10, p. 162. These data show that the ocular media absorb an increasing amount of the radiant energy incident upon the cornea for increasing wavelengths in the near infrared (IR-A). For infrared wavelengths greater than 1.4 μm (IR-B and IR-C) the cornea and aqueous absorb essentially all incident radiation, and beyond 1.9 μm the cornea is the sole absorber. The absorbed energy may be conducted to interior structures of the eye and raise the temperature of that tissue as well as the cornea itself. Heating of the iris by absorption of visible and near-infrared radiation still appears to play the major role in the development of opacities in the crystalline lens.

Significant corneal doses of nearly $10\ J \cdot cm^{-2}$ are required to produce lenticular changes within the duration of the blink reflex (i.e., up to $100\ W \cdot cm^{-2}$).[257] The sensory nerve endings in both the cornea and iris are quite sensitive to small temperature elevations, and a temperature elevation to 47° corresponding to approximately $10\ W \cdot cm^{-2}$ absorbed in the cornea elicits a painful response in humans. Hence, it is generally considered that the blink reflex provides protection against infrared radiation up to levels in excess of those that cause flash burns of the skin.[255,256]

The most valuable quantitative information, however, applies to chronic exposure. Quantitative data for the production of lenticular opacities following exposure to infrared radiation are very limited. Radiant energy absorbed by the iris appears to play the principal role in raising the temperature of the crystalline lens for short exposure times,[82,258-260] hence one would expect the spectral absorptance and reflectance of the iris to determine the action spectrum for this effect. It is not clear whether a cataract which develops in a worker after many years of exposure to nearby radiant sources such as molten glass or metal can result from very moderate intraocular temperature rise during chronic exposure or from greater temperature elevations experienced during very occasional exposures to the sources at very close range.

Protein denaturation resulting from small temperature elevations may require extended periods of time. The temperature rises due to infrared radiation absorbed by the anterior structures of the eye have been quantitatively measured by the use of thermocouples implanted *in vivo*, and have been qualitatively determined by noting the aqueous flare (a measure of the dilation of blood vessels of the iris). As an example, optical radiation from a furnace operating at 1460°C when incident upon the human eye at an irradiance of $0.13\ W \cdot cm^{-2}$ created an aqueous temperature elevation from 36°C to 39°C in 30 sec.[82] Duke-Elder presents an excellent discussion of infrared effects upon the eye.[82] Since ocular temperature elevation is the parameter of interest, acute infrared exposures lasting up to a few minutes may be expressed in terms of corneal radiant exposure ($J \cdot cm^{-2}$), whereas chronic exposures must be described in terms of corneal irradiance ($W \cdot cm^{-2}$).

The infrared corneal dose–rate experienced out of doors in daylight is of the order of $10^{-3}\ W \cdot cm^{-2}$. As discussed in "Industrial Hygiene Highlights," p. 164, some workers exposed for 10–15 years to infrared irradiances of 0.08–0.4 $W \cdot cm^{-2}$ encountered in the glass and steel industries developed lenticular opacities.[1] Such irradiances, even if visible radiation, definitely produce a marked sensation of warmth, provided that a significant area of the skin is exposed (greater than a few square centimeters). A corneal infrared irradiance of 0.1 $W \cdot cm^{-2}$ is well below the level required to cause acute injury of the anterior ocular structures[261–266] and is often used as a guideline for occupational exposure to far-infrared CO_2 lasers (IR-C). However, safe chronic ocular exposure values particularly to IR-A, probably are of the order of 10^{-2} $W \cdot cm^{-2}$ for large, extended sources.

INFRARED EXPOSURE CRITERIA

Infrared exposure criteria are generally only needed for the evaluation of heat stress or for the evaluation of ocular hazards from chronic exposure conditions. Essentially no conventional source, including the nuclear fireball, is capable of producing flash burns of the cornea or acute lenticular injuries while not causing a retinal burn within the period of the blink reflex. Lasers are the exception to this rule. Infrared lasers such as the CO_2 laser (10.6 μm) or the erbium laser (1.54 or 1.64 μm) having cw output irradiances of the order of 10 $W \cdot cm^{-2}$ or greater could produce corneal lesions by delivering at least 0.5–1 $J \cdot cm^{-2}$ within the blink reflex. Exposure limits for infrared (IR-B and IR-C) lasers are 0.1 $W \cdot cm^{-2}$ for cw lasers, 0.01 $J \cdot cm^{-2}$ for Q-switched pulsed lasers, and 0.56 $t^{1/4}$ $J \cdot cm^{-2}$ for a pulsed exposure time t of 0.1 μsec to 10 sec based upon CO_2 laser studies. Studies with the erbium laser indicate however that 1.0 $J \cdot cm^{-2}$ may be permissible for some IR-B and IR-C pulsed lasers.[267,268] It is possible to estimate the threshold of acute thermal injury of the cornea at other IR-B and IR-C wavelengths using CO_2 laser injury data if greater accuracy in hazard analysis is essential. Such an estimate may be obtained by calculating the temperature rise in a thin layer of water, representing the cornea, which is a function of the extinction coefficient of water at that wavelength.[269] Such calculations indicate that thresholds vary by more than one order of magnitude.[368] The CO_2 laser criteria are considered conservative for any IR-B or IR-C acute exposure condition. For long-term chronic exposure to infrared optical sources the average ocular irradiance should be limited to ~10 $mW \cdot cm^{-2}$, although incidental exposures for several minutes up to 100 $mW \cdot cm^{-2}$ may be permitted.

Although an individual would not be expected to remain within a large diameter laser beam for an irradiance of 0.03–0.1 $W \cdot cm^{-2}$, the latter value is an acceptable level for skin exposure to an IR laser beam. Table V presents recommended limits for exposure to the skin in the visible and infrared spectrum.

TABLE V
Several Proposed Laser Guidelines for Skin Protection

Organization	*Spectral region*	*Continuous-wave* $(W \cdot cm^{-2})$	*Non-Q–switched* $(J \cdot cm^{-2})$	*Q–switched* $(J \cdot cm^{-2})$	*Ref.*
U. S. Departments of the Army and Navy (1969)	Vis, IR	0.1	0.1	0.01	221
U. S. Department of the Air Force (1971)	Vis, IR	0.1	0.1	0.01	222
Ministry of Technology, United Kingdom (1969)	UV-A, Vis, IR	0.1	0.1	0.1	223
ACGIH (1971)	Vis, IR	1.0	0.1	0.1	99
ANSI Z-136 Proposed February 29, 1972 and ACGIH Proposed (1972)	Vis, IR-A	0.2 ($t > 10$ sec)	$1.12\ t^{1/4}$ (100 nsec-10 sec)	20 $mJ \cdot cm^{-2}$	222
	IR-B, IR-C	0.1 ($t > 10$ sec)	$0.56\ t^{1/4}$ (100 nsec-10 sec)	10 $mJ \cdot cm^{-2}$	

OPTICAL RADIATION HAZARD CONTROLS

Since continuous visible sources elicit a normal aversion or pain response which can protect the eye and skin from injury, control measures have often been centered on using visual comfort as a hazard index and providing eye protection, baffles, etc., on this basis. The determination of shade number of welding goggles is but one example: since the ultraviolet and infrared corneal irradiances are reduced well below hazardous levels, the viewing safety may be based upon the visible radiation comfort alone.

Fortunately few arc sources are sufficiently large and sufficiently bright to be a retinal burn hazard under normal viewing conditions. Only when an arc or tungsten filament is greatly magnified, as in an optical projection system,[270] can hazardous irradiances be imaged on a sufficiently large area of the retina to cause a burn. Furthermore, individuals would normally not step into a projected beam at close range or view an arc with binoculars or a telescope. Conceivable accident situations require a hazardous exposure to be delivered within the period of the blink reflex. If an arc were initiated while an individual were located at very close viewing range, he could receive a retinal burn. Such situations require viewing distances of a few meters for all but the most powerful xenon searchlights or a few inches for a welding arc and most movie projection equipment[270,271] or movie lamps. It should be noted that if a searchlight or projector is hazardous, it is equally hazardous at viewing distances less than the flash distance.[270] Ultraviolet erythema and photokeratitis encountered in welding require the use of protective clothing, gloves, helmets, and screens. Engineering controls such as source enclosure are more commonly applied for ultraviolet sources other than the welding arc.

Aluminum, because of its exceedingly high reflectance throughout the optical spectrum is widely used as screening for the control of thermal radiation from furnaces and other infrared sources in the industrial environment. Enclosures such as baffled lamp housings are employed to control the potential hazards from laboratory arc sources, e.g., solar simulators, fadometers, spectroscopic equipment, and optical calibration sources. Optical projection systems such as movie projectors, searchlights, and spotlights are normally not a direct viewing hazard. However, the desire for smaller, more portable spotlights, movie lights, and searchlights have required the use of smaller, more efficient reflectors and more compact, brighter sources (xenon arcs and quartz-iodide tungsten lamps) which in a few cases can present a retinal burn hazard for short viewing distances. When highly collimated light beams are not required (e.g., movie lights) more diffuse reflectors and fresnel or diffusing lenses should be utilized to reduce the projected source radiance. Several hazard-reduction options are available to the designer of such equipment,[270] but if these are not sufficient to reduce the projected source radiance below 1–5 $W \cdot cm^{-2} \cdot sr^{-1}$, safety measures must be employed to prevent individuals from viewing the source at close range. Most electronic photoflash units are safe although marginally above exposure standards for extended sources.

Control of ultraviolet radiation is normally the most commonly encountered problem with arc sources. Koller,[3] and the Withrows,[272] and others have compiled a wealth of source data and optical transmission data useful in the design of source enclosures.[4,6,112,270,273-279] Essentially all "black lights" used in industry make use of glass filters and source envelopes which eliminate actinic as well as most visible radiant energy. Such black light sources become a problem only for photosensitive individuals. Figure 15 shows the optical transmission properties for several materials.

EYE PROTECTION

Eye protection filters for glass workers, steel and foundry workers, and welders were developed empirically; however optical transmission characteristics are now standardized as "shades" and specified for particular applications.[280-284] Although maximum transmittances for ultraviolet and infrared radiation are specified for each shade, the visual transmittance τ_V or visual optical density D_V defines the shade number, S# :

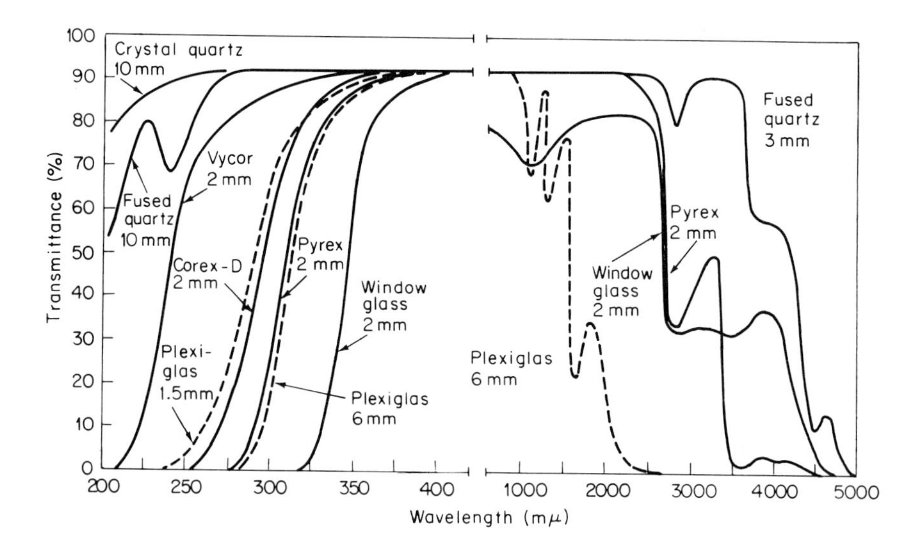

Fig. 15. Spectral transmission curves for several window materials. (From Withrow and Withrow,[272] with permission.)

$$S\# = 7/3\ D_V + 1 \qquad (7)$$

where

$$D_V = -\log_{10} \tau_V \qquad (8)$$

For instance a filter with a visual attenuation factor of 1000 (e.g., $D_V = 3$) has a shade number of 8. The standards for shade numbers were developed in the late 1920's at the National Bureau of Standards[280] and have since been adapted with slight modifications for transmission tolerances as a Federal Standard and as ANSI Standard Z2.1. As a note of caution, shade numbers should not be applied to laser eye protection since the radiometric optical density D_e must be specified at the laser wavelength. Since the luminance of acetelene flames are typically of the order of 10 cd · cm^{-2} a density of one, hence shade 3 or 4, is applicable. Electric welding arcs have luminances of the order of 10^4 to 10^5 cd · cm^{-2} and filter densities ranging from 4 to 5 corresponding to shades 10 to 13 are required for comfortable viewing. Likewise a shade of at least 13 is required to view the sun which has a luminance of approximately 10^5 cd · cm^{-2}. These densities are far in excess of those necessary to prevent retinal burns, but are required to reduce the luminance to 1 cd · cm^{-2} or less for viewing comfort. The user of the eye protection should therefore be permitted to choose the shade most desirable to him for his particular operation. Actinic ultraviolet radiation from welding arcs is effectively eliminated in all standard welding filters. High ambient light levels and the use of specialized filters to further attenuate intense spectral lines (e.g., class II, didymium welder's goggles) are the primary techniques to reduce glare in welding.

Some specialized filters do exist for particular applications. As an example, didymium glass strongly absorbs in the 575–595 nm band and has been used particularly by glass workers to reduce the intense sodium yellow spectral lines (589 and 589.6 nm) emitted from most molten glass.

Since the type of glass used in most eye protection filters have a short wavelength cutoff near 340 nm, actinic ultraviolet radiation is effectively eliminated. However, infrared from 1.4–4.6 μm is often transmitted to a varying extent.[281] Fortunately, electric welding arcs emit very little in this spectral region and the hot metal is the principal IR-B source. While this transmission might create a small corneal temperature rise, one must realize that all absorbing filters can become warm and radiate longer-wavelength infrared radiation. Reflective metal coatings are sometimes deposited on the front surface of an absorbing filter to reduce this problem. While this procedure normally has not been applied to welders filters, electrically conductive (EC) coatings have been deposited on glass substrates and used in windows of crane cabs in steel mills. Such coated filters have also been used with aluminized thermal protective suits.

Laser eye protection is designed to have the greatest visual transmission along with an adequate optical density at the laser wavelengths. A variety of eye protection filters are commercially available in spectacle frames or coverall goggle frames.[285–291] Absorbing glass filters are generally the most desirable since degradation is the least due to mechanical or optical damage of the filter material. For some applications where surface abrasion can be prevented, plastic filters are more desirable, principally because of their lighter weight. The optical density at appropriate laser wavelengths should be marked on the eye protection, since the use of goggles designed for one laser have been mistakenly used with another laser and could have resulted in ocular injury. Less technical markings (e.g., "use only with ruby laser") may also be desirable. Periodic visual inspection to insure the integrity of the eye protection is advisable. Periodic measurement of optical density of laser eye protective glass filters has not been shown to be necessary, and for densities greater than 5 is normally not even possible with present laboratory equipment.

LASER HAZARD CONTROLS

Most hazard controls are commonsense procedures designed to limit personnel from the beam path and limit the primary and any reflected beams from occupied areas. Procedures recommended in the early 1960's were limited to the laboratory environment[292–295] and were based on theoretical analyses of the laser hazard.[296] As the laser found widespread and varied applications, more specific procedures were developed.[110,225,230,297–302] In outlining control measures a common difficulty has been to adequately distinguish between different types of lasers which require different types and degrees of control.[303] Control measures developed for high-power laboratory ruby lasers have been too often mistakenly applied to the use of low-power helium-neon (He-Ne) lasers resulting in needless operational constraints. To reduce this difficulty several attempts have been made to classify lasers according to type, output parameters, and application.

For lasers operating in the 0.4–1.4 μm spectral region a useful distinction can be made between those which can produce hazardous diffuse reflections and those which cannot. The probability of viewing a diffuse reflection, if hazardous, from a laser beam is enormously greater than viewing the direct or specularly reflected beam itself. In the latter case, the normally small diameter beam must intercept the eye's pupil. Figure 16A shows a typical laboratory environment where the beam, and therefore the hazard, is confined to well-defined horizontal paths, whereas, if the same beam were sufficiently intense to produce a hazardous diffuse reflection from the backstop, the hazardous viewing area could extend throughout the room (Fig. 16B). Although an individual viewing the diffuse reflection from across the room would have a smaller retinal image than if viewing it at closer range, the retinal irradiance would remain the same regardless of viewing distance or viewing angle as long as the diffuse source

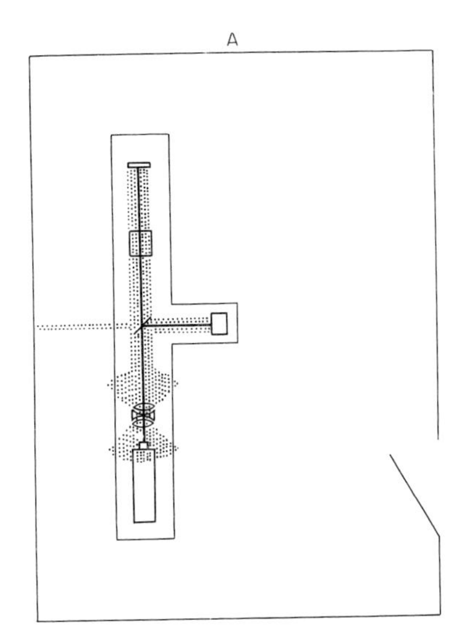

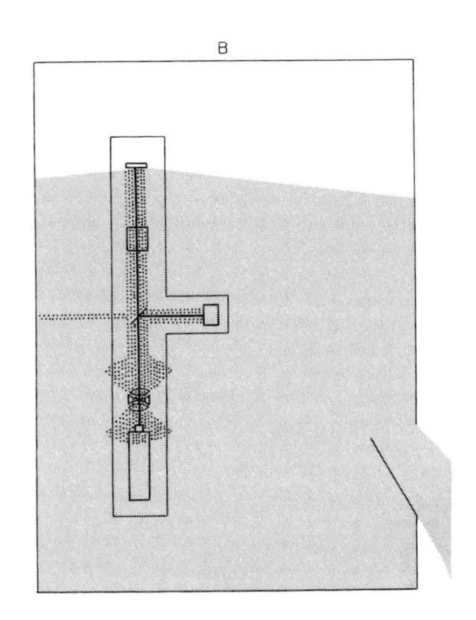

Fig. 16. Potentially hazardous locations (shaded areas) for an individual viewing reflections of a laser beam differ significantly between a laser capable of producing only hazardous specular reflections (A), and a laser capable of producing hazardous diffuse reflections (B).

subtended an angle greater than ~3 mrad. Of course, retinal injury thresholds vary with image size, but for simplicity of analysis this variation is normally neglected and a single radiance or diffuse radiant exitance criteria is applied as a standard.[110] Visible and IR-A lasers capable of producing a fire hazard also are capable of producing hazardous diffuse reflections.

At least four clearly distinct hazard categories for visible and IR-A lasers are therefore evident: high-power lasers, which can produce hazardous diffuse reflections and may also present a fire hazard; medium-power lasers, which cannot produce hazardous diffuse reflections but which do present an intrabeam viewing hazard; low-power visible lasers (power less than 1 mW) which are safe for momentary viewing; and nonhazardous lasers, or exempt lasers, which emit less than the recommended limits for intrabeam viewing. These categories may be further divided, if desired. Preliminary hazard analysis is reduced to laser classification by calculating or measuring the output or focused-beam irradiance (or radiant exposure).

Hazard protection from high-power lasers generally centers on engineering controls such as locating the laser in a closed facility or within an opaque housing. The latter approach has been followed with most industrial laser cutting, drilling, and welding equipment. To insure against accidental personnel exposure from opening the enclosure or entering the closed room, interlocks are installed which disable the laser in such an event. Remote TV monitoring or interlocked viewing optics are sometimes used to view the target. In laboratory situations where personnel work within the closed installation, eye protection becomes mandatory. Some high-power lasers are employed outdoors for atmospheric and astrophysical studies. In these situations the beam path can be limited to restricted air space. Optical and/or radar surveillance of the region of the beam path may be necessary, and elevation and azimuth traverse stops are sometimes installed. Continuous wave lasers operating above 0.5 W require an elevation of potential fire hazards.

Hazard controls for medium-power lasers are less stringent and engineering controls are less common. Since diffuse reflections do not present a problem, controls center upon terminating the beam paths and limiting hazardous specular reflections. If beam enclosure is possible, it should be encouraged. Eye protection is required only if the direct beam and specular reflections cannot be avoided. The use of a diffuse matte to locate the beam is advised. The education of individuals in the operating environment, particularly the operator, has been shown to be the single most important hazard control measure. Labels should be placed on the equipment warning individuals not to stare into the direct beam. Output beam irradiances of low power He-Ne lasers used in construction alignment should be limited to 1 $mW \cdot cm^{-2}$ where feasible. Unfortunately most laser operations of this type in sunlight appear to require a beam power of approximately 5 mW and irradiances above 5 $mW \cdot cm^{-2}$; shading the target and the use of retroreflective targets can be helpful. Shutters on the exit port of low power lasers have been shown to be useful when the beam is only required periodically. Limits on beam traverse are often required for outdoor laser operations.

Hazard controls for low-power lasers are limited to the use of precautions against staring into the beam. A caution label should be attached to the laser.

Three factors enter into any hazard analysis: (*1*) the type of laser and potential hazards of associated equipment, (*2*) the environment, and (*3*) potentially exposed individuals. Since many combinations are possible, numerous, rigid laser safety regulations should be avoided.

Optical Radiation Measurements and Calculations

Instruments designed specifically for hazard evaluation of a variety of optical radiation sources are not presently available. Such an instrument will probably not be manufactured for some time since the cost would not be justified and exposure criteria have not yet been standardized. Several types of commercially available thermal radiometers, brightness meters, ultraviolet and visible irradiance meters, and illuminance meters can be utilized to evaluate specific sources.[6; 7, 9, 304–309]

The measurement of laser beam pulsed radiant exposure is most readily accomplished by using a calibrated biplanar vacuum photodiode as a detector with a high-frequency oscilloscope or sensitive electrometer for readout. Such equipment can be quite expensive. Fortunately, to determine if a pulsed laser can produce a hazardous diffuse reflection, fully developed, black photographic print paper or similar thermally sensitive surfaces may be used.[306] The down-range radiant exposure H (or irradiance) of an unfocused circular beam can be adequately estimated at a range r by calculation. If the output energy Q, the beam diameter, a, and the divergence ϕ, are known the following expression may be used:

$$H = \frac{1.27\,Q \cdot e^{-\mu r}}{(a + r \cdot \phi)^2} \tag{9}$$

where μ is the atmospheric attenuation coefficient (typically $10^{-6}-10^{-7}\ cm^{-1}$). The term $e^{-\mu r}$ may be neglected in most cases, and the beam diameter may be neglected when the term $r \cdot \phi$ has a considerably larger value. Beam scintillation from atmospheric turbulence limit these calculations to estimations.[202,310,311]

Beam irradiance from a cw laser may be measured with a sensitive illuminance meter if the beam is larger than the detector aperture. The instrument must be calibrated at the laser wavelength. For a He-Ne laser emitting at 632.8 nm, an irradiance of $10^{-6}\ W \cdot cm^{-2}$ is equal to 0.17 ft cd and $10^{-3}\ W \cdot cm^{-2}$ is 170 ft cd.

Present commercially available ultraviolet meters are adequate except where a significant level of IR-A and visible light is present. The development of a single, simplified ultraviolet hazard spectrum (Fig. 4) should encourage the development of a direct-reading hazard meter. Such an instrument could make use of a "solar-blind" detector with a low-quality silica-glass window.

One can achieve reasonably accurate measurements of many visible sources using less expensive luminance spot meters and illuminance meters. "Standard" spectral data for the source from the literature and the use of a few selected filters may enable the user to translate his photometric measurements to radiometric values if necessary.

Recently developed thermal detectors, constructed with a very thin, vacuum-deposited thermopile surface, have sufficient sensitivity and sufficiently short time responses to be of practical value in optical measurements. Although they are still rather expensive because they use sophisticated ambient-compensating features, these instruments can measure irradiances below $10^{-3}\ W \cdot cm^{-2}$ from the ultraviolet to the far-infrared. Many pitfalls exist in radiometry and photometry and the user of any of the aforementioned instruments should consult one or several of the standard texts on this subject.[3,6,7,10,307]

Considering the present cost of equipment, one is forced to reconsider the need for hazard evaluation measurements and look for alternative techniques. Fortunately, most high-intensity light sources and lasers have fairly consistent output parameters. Because of this consistency and the uncertainties in present exposure criteria, there is seldom a need for periodic monitoring of a source. Quite often a source can be determined to have a radiant output either far exceeding or greatly below the present exposure standards. Thermally sensitive paper is often all that is required to classify a laser. Manufacturers of arc lamp sources and lasers generally provide radiant output data which is often sufficient to calculate the radiant exposure or irradiance at a point of interest with an accuracy adequate for hazard analysis. Beam irradiance and radiance calculations for searchlight and other projector sources are quite difficult, however, and do require an understanding of optical projector theory.[270] Some illuminance measurements, source specifications, and photography of such extended sources are sometimes adequate for hazard analysis. Proper installation of germicidal lamps sometimes can be accomplished without the need for measurements.

MICROWAVE RADIATION HAZARDS

The microwave region of the electromagnetic spectrum is generally considered to extend (at least for hazard considerations) from either 0.1–100 GHz (300 cm to 0.3 cm) or 0.3–300 GHz (100 cm to 1 mm). In either case, the wavelength or frequency range is 1000-fold. This range of wavelengths is enormous in comparison to the optical bands (UV-A through UV-C – a wavelength range of two; visible light – factor of two; IR-A – factor of two, etc.) With this in mind it is not surprising that the biologic effects vary considerably both qualitatively and quantitatively over this range of frequencies. The very short wavelength microwaves are absorbed at the skin surface as is far-infrared radiation.[1] Longer wavelength radiation ($>$ 1 cm) penetrates into deeper tissue. Although further investigations of microwave biologic effects have been reported in the past few years, the summary provided in "Industrial Hygiene Highlights"[1] remains essentially unchanged.

Units of $mW \cdot cm^{-2}$ or $\mu W \cdot cm^{-2}$ are generally used to define microwave flux, although field strengths in $V \cdot m^{-1}$ are sometimes used ($194 V \cdot m^{-1} = 10$ $mW \cdot cm^{-2}$). In some biologic studies absorbed dose is given $mW \cdot gm^{-1}$ or $mW \cdot cm^{-3}$. Since hazard evaluation requires free-field measurements, hazard criteria are given as free-field power densities in $mW \cdot cm^{-2}$ or $\mu W \cdot cm^{-2}$.

Microwave Exposure Criteria

The increased interest in microwave radiation hazard criteria prompts many to ask the basis of these criteria. In the early 1950's the interest of the military services and the communications industry in hazards to health from radar and microwave communication-relay equipment prompted the establishment of the first guidelines for permissible exposure. In 1953 Schwan proposed to the U. S. Navy that an individual receiving whole-body irradiation from one direction could tolerate 10 $mW \cdot cm^{-2}$.[312] He calculated that this would correspond to a total dose rate to the body of 100 W or approximately 1 $mW \cdot cm^{-3}$. The basal metabolic heat load is just under 100 W for an adult individual. Experimental and theoretical studies carried out during the 1950's[28-32,312-319] showed that this level was approximately a factor of ten below most acute thermal effects demonstrated in animals. Extrapolation of animal data to man could only be approximate since animal size and shape played a great role in determining absorption effects. Several organizations in the late 1950's adopted the 10 $mW \cdot cm^{-2}$ level. It was later felt that the single criterion of 10 $mW \cdot cm^{-2}$ was unnecessarily conservative for the short exposure times of several minutes. Following brief exposures, the body could readily dissipate the added thermal load created by the microwave exposure. One modification, a single integrated dose of 1 $mW \cdot hr \cdot cm^{-2}$ for exposure times less than 6 min proposed in 1956 by Schwan and Li[315,320] was later adopted by the American National Standards Institute[321] (see Fig. 17) and Canadian Standards Institute. Another approach proposed by Palmisano[322] was adopted by the U. S. Army and Air Force[323] and later by the French Ministry of Armies.[324]

A technique has been proposed[325] for lowering the continuous exposure criterion for use in hot environments which, although mentioned in the ANSI C95 standard, was not clarified. The United Kingdom and German standards retained the value of 10 $mW \cdot cm^{-2}$ for all exposure times. The American Conference of Governmental Industrial Hygienists in 1970 proposed a single step in their proposed TLV from 10 $mW \cdot cm^{-2}$ for long exposure times to 25 $mW \cdot cm^{-2}$ for exposure times less than 10 min. The Soviet Union and eastern Europe have far lower standards, and are based upon a cumulative daily dose concept (Fig. 17). The basis for the lower standards presupposes "nonthermal" or microthermal effects upon the central nervous system[29,31,32,326-328] at levels as low as 0.1 $mW \cdot cm^{-2}$. Besides referencing animal studies, the architects of these standards pointed to their studies with microwave workers who complained of headaches, eye strains, nausea, and similar disturbances when exposed to microwave radiation. A safety factor of ten was applied by the Soviets in their standard development.[29,31] While authorities in the United States, who established the present criterion, recognized that CNS effects were possible and did exist,[323,329] most did not feel that such effects could occur at such low levels, or if they could, would still not justify exposure criteria much lower than 10 $mW \cdot cm^{-2}$. Most behavioral and neural-effect studies with animals, which are difficult to perform, do not indicate significant effects at levels much lower than 10 $mW \cdot cm^{-2}$.[329-335]

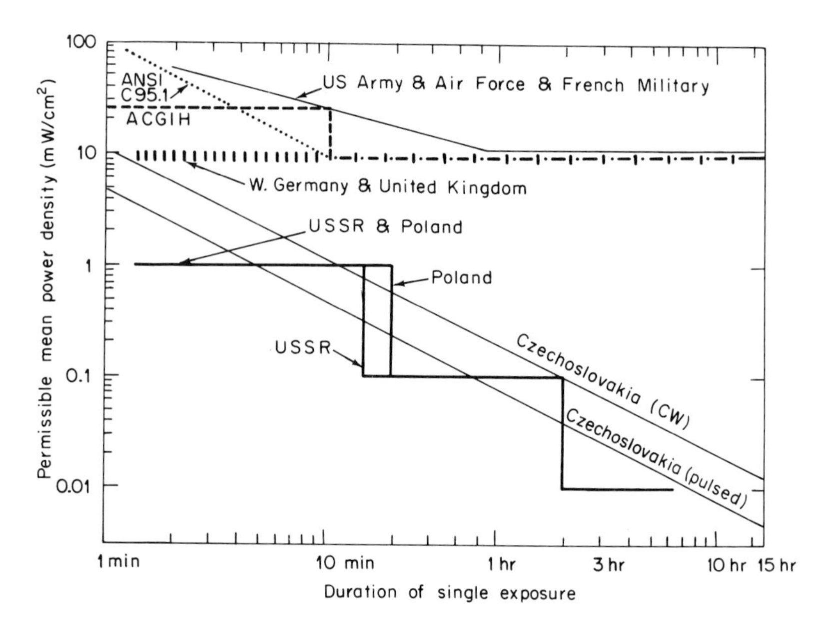

Fig. 17. A comparison of recommended exposure limits for occupational exposure to microwave radiation. The U. S. Army,[323] U. S. Air Force[323] and French Military[324] (solid line) permit exposure to power densities above continuous limit of 10 mW · cm^{-2} for periods less than 1 hr in any 1-hr period. ACGIH[99] (dotted line) and ANSI (dashed line)[321] permit exposure to power densities in excess of 10 mW · cm^{-2} for period less than 6 min within any 1-hr period. A 1960 Bell System Standard,[346] and West Germany[324] and the United Kingdom,[324] consider 10 mW · cm^{-2} as a ceiling value; and the earlier Bell System Standard considered 1 mW · cm^{-2} to 10 mW · cm^{-2} only permissible for occasional exposure.[346] The Soviet Union,[29] Czechoslovakia,[31] and Poland[324] have criteria which tend to follow a concept of a total daily dose (up to ~0.1 J · cm^{-2}).

Much of the present Soviet work seems to be full of speculation of the possible roles played by microwaves in the natural environment[32] and this outlook has no doubt influenced the setting of low standards.

It should be stated, however, that few workers in the West are probably actually exposed to levels approaching 10 mW · cm^{-2} for even brief periods. In fact, Palmisano of the U. S. Army Environmental Hygiene Agency, who has had a vast experience in evaluating personnel health hazards from Army radars throughout the world, has indicated that seldom are individuals exposed to levels for sufficient periods to exceed even the Soviet standards.[336] Hence we are unable to point to a large body of individuals who have been continuously exposed without ill effects to significant levels of microwave radiation for extended time periods.

The enormous increase in the use of the microwave cooking oven by the general public has prompted a renewed effort in this country to conduct laboratory investigations which could substantiate the existence of adverse effects below 10 mW · cm^{-2}. Several of the major Soviet and Czechoslovakian works in this field have recently been translated.[29,31,32] Microwave bioeffects research should be watched closely by the health specialist. Studies of particular interest are those concerned with cataractogenesis from chronic low-level exposure[336-339] and those studies directed at clarifying reports of adverse cellular effects reported during cell division and specialization,[340] and frequency-specific effects upon the central nervous system.[341,342]

Radio Frequencies below 300 MHz

Biologic effects for frequencies below 300 MHz (0.3 GHz) have not been extensively investigated and have been thought by many to be essentially nonexistent. Schwan has spoken in terms of a current density tolerance in an individual of approximately 1 mA · cm^{-3} as a basis for establishing field strength exposure standards for different field configurations.[343] Of the Western exposure criteria only the ANSI C95 standard specifies the 10 mW · cm^{-2} value much below 300 MHz – to 10 MHz. The USSR has exposure criteria of 20 V · m^{-1} for the 100 kHz to 1.5 MHz band, 20 V · m^{-1} for the 1.5–30 MHz band and 5 V · m^{-1} for the 30–300 MHz band. In Great Britain, Rogers has recommended a safe level of 10^3 V · m^{-1} between 1 and 30 MHz.[344]

Microwave Measurements

From 1959–1968 essentially no changes were made in microwave hazard monitoring equipment.[305,345-347] However, since 1968 several new instruments have been introduced, designed principally for measuring radiation leakage from microwave ovens. In using any of these instruments one must be careful to know the frequency dependence of the instrument and the limitations of its use in a perturbed field or very near the source.[348-354] It should be noted that exposure criteria, in particular the value 10 mW · cm^{-2}, were developed assuming "far-field" conditions, and its applicability to complex fields can be questioned.

Microwave Hazard Controls

The control of microwave radiation hazards from radar equipment centers on limiting beam elevation and traverse to preclude exposure of occupied areas.[323,355-357] The greatest difficulties have been encountered with the use of radar and microwave communications equipment aboard ship where occupied space is already greatly confined.[357]

Since line-of-sight, troposcatter, and ground-to-satellite communication transmitting antennas must be located and directed above any obstacles, the microwave beam paths for such equipment normally cannot be occupied. Some secondary lobes in the transmitted beam from high power equipment must, however, be evaluated. Protective suits and eyewear may be required for antenna maintenance personnel.[355] Microwave eye protection filters which have been developed use either metallic coated glass[29] or glass imbedded with a thin wire mesh.[357] Such eyewear is of little value however unless the user also wears protective head covering.

Industrial microwave equipment,[358] such as ink or paint dryers, cooking equipment, and plywood manufacturing equipment are normally designed to limit the leakage of potentially hazardous levels into occupied areas.

The increasing use of small microwave cooking ovens in food service facilities[359] and in the home has prompted several studies of radiation leakage from such equipment.[360-364] These studies pointed to two types of hazardous conditions: first, the leakage which normally occurs in all microwave ovens along the door periphery could exceed threshold limits due to poor design or misuse; second, hazardous conditions can occur if the oven door safety interlocks fail with the door open, resulting in hazardous levels of radiation in front of the oven.

Taylor of the U. S. Army Environmental Hygiene Agency, after evaluating hundreds of ovens stated that he was firmly convinced that a properly accomplished visual inspection was sufficient to evaluate ovens.[364] Although the actual level of microwave radiation emanating from an oven can only be determined with suitable instrumentation, he stated that it was possible to detect those ovens that could be "excessive leakers." He defined "excessive leakers" as those leaking greater than 20 $mW \cdot cm^{-2}$. Early detection of a potential "excessive leaker" could thereby prevent unnecessary exposure of personnel to extremely hazardous levels of microwave radiation by removal of the oven from service prior to total degeneration of the oven safety or control features.

Taylor listed a six-step procedure to check for faults causing improper seating of the door seal which were the principal cause of leakage from microwave ovens. Figure 18 shows the points to look for in evaluating an oven as recommended by Taylor. His six-step checklist of typical problem areas which should be considered when evaluating an oven is as follows:

1. Safety interlocks. There should be a minimum of two safety interlocks to prevent oven operation when the door is open. These interlocks should be adjusted so that as soon as the door latch is released, the oven will cease operation. This should be the first test conducted on the oven.

2. Door. A warped or misaligned door may result in improper closure permitting excessive leakage between the door seal and oven frame.

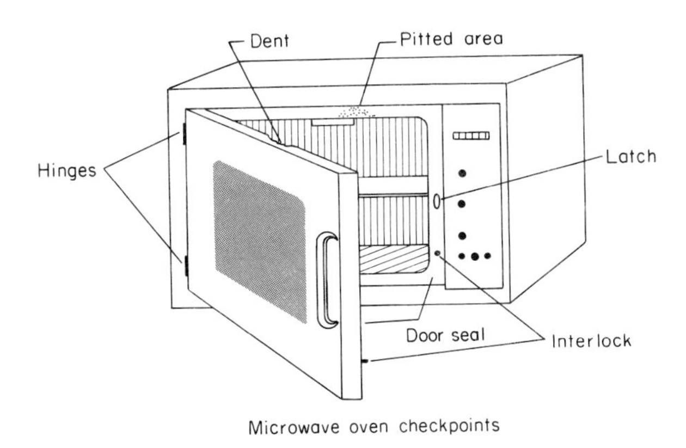

Fig. 18. Microwave oven checkpoints. Sites to inspect which could indicate presence (or tendency toward development) of excessive microwave radiation leakage, according to Taylor.[364]

3. Door seal. Leakage can occur from areas where the door seal plate has been scarred, dented, or improperly seated. One major cause of damage to the door seal is inadequate cleaning, since the presence of grease or food particles may permit arcing which could lead to pitting and burnt spots.

4. Hinges If the door hinges are loose or damaged, or if screws are missing from the hinge area, then misalignment of the door may occur and permit leakage.

5. Latch. The door latch should be properly adjusted. If the oven operates with the door partially open, it is possible that a loose latch could be at fault. Inspecting the latch area is also a good means to determine door misalignment.

6. Operation. The oven should not be operated with the cavity empty; for testing purposes a small container of water should be placed in the cavity. Metallic cooking vessels should not be used; reflections from such objects could increase the leakage. Cleanliness of the oven is necessary both for sanitary conditions and to aid in preventing the development of excessive leakage.

Taylor felt that evaluation of an oven should be done as often as possible, but at least once a month.

OUTLOOK FOR THE FUTURE

With the advance of technology we may expect expanding applications for high-brightness optical radiation sources and for high-power microwave sources. The industrial hygienist and the health specialists will be required to keep abreast of these developments and the associated potential health hazards.

Reasonable guidelines for acute exposure to most nonionizing radiation sources now exist, and with the expanding scope of both state and federal occupational health and safety codes, many of these guidelines may become mandatory. A knowledge of the underlying physical principles of these guidelines as well as a knowledge of natural exposure conditions of the individual in his environment should aid the reader in attaining the proper perspective required to evaluate new sources of nonionizing radiation.

Insufficient knowledge is presently available to adequately evaluate the potential health hazard of several sources: lasers which emit ultraviolet radiation, intense sources of near-infrared radiation, mode-locked lasers, very high-power pulsed microwave sources, and repetitively pulsed lasers and flashlamps. It is hoped that future biologic research will fill in these gaps.

REFERENCES

1. Matelsky, I., The non-ionizing radiations. "Industrial Hygiene Highlights," Vol. I, pp. 140-178. Industrial Hygiene Foundation of America, Pittsburgh, Pennsylvania, 1968.
2. Commission Internationale de L'Éclairage (International Commission on Illumination), "International Lighting Vocabulary," 3rd Ed., Publ. CIE No. 17 (E-1.1). CIE, Paris, 1970.
3. Koller, L. R., "Ultraviolet Radiation," 2nd. Ed.. Wiley, New York, 1965.
4. Illuminating Engineering Society (New York), "I. E. S. Lighting Handbook," 4th Ed. Waverly Press, Baltimore, 1966.
5. Born, M., and Wolf, E., "Principles of Optics," 4th Ed. Pergamon, Oxford, 1970.
6. Bauer, G., "Measurement of Optical Radiations." Focal Press, New York, 1965.
7. Heard, H. G., "Laser Parameters Measurement Handbook." Wiley, New York, 1968.
8. Jenkins, F. A., and White, H. E., "Fundamentals of Optics," 3rd Ed. McGraw-Hill, New York, 1957.
9. Ross, D., "Lasers, Light Amplifiers and Oscillators." Academic Press, New York, 1969.
10. Walsh, J. W. T., "Photometry." Dover, New York, 1958.
11. Adler, F. H., "Physiology of the Eye," 4th Ed. Mosby, St. Louis, Missouri, 1965.
12. Blum, H. F., "Photodynamic Action and Diseases Caused by Light." Hafner, New York, 1964.
13. Blum, H. F., "Carcinogenesis by Ultraviolet Light." Princeton Univ. Press, Princeton, New Jersey, 1959.
14a. Duke-Elder, S., "Text-Book of Ophthalmology," Vol. VI, Injuries. Mosby, St. Louis, Missouri, 1954.
14b. Urbach, F., ed., "The Biologic Effects of Ultraviolet Radiation." Pergamon, Oxford, 1969.
15. "Electronic Product Radiation and the Health Physicist," BRH/DEP 70-26. U. S. Department of Health, Education, and Welfare, Bureau of Radiological Health, Rockville, Maryland, Oct. 1970.
16. Giese, A. C., ed., "Photophysiology," Vols. I–IV. Academic Press, New York, 1964 and 1968.
17. Graham, C. H., "Vision and Visual Perception." Wiley, New York, 1965.
18. Herzfeld, C. M., and Hardy, J. D., eds., "Temperature, Its Measurement and Control

in Science and Industry," Vol. 3. Reinhold, New York, 1963.
19. Hollaender, A., "Radiation Biology," Vol. 2, Ultraviolet and Related Radiations and Vol. 3, Visible and Near-Visible Light. McGraw-Hill, New York, 1955 and 1956.
20. Le Grand, Y., "Form and Space Vision." Indiana Univ. Press, Bloomington, Indiana, 1967.
21. Licht, S. H., ed., "Therapeutic Electricity and Ultraviolet Radiation." Waverly Press, Baltimore, Maryland, 1967.
22. Straatsma, B. R., Hall, M. O., Allen, R. A., and Crescitelli, F., eds., "The Retina." Univ. Calif. Press, Los Angeles, California, 1969.
23. Wolbarsht, M. L., ed., "Laser Applications in Medicine and Biology." Plenum, New York, 1971.
24. Blake, L. V., "Antennas." Wiley, New York, 1966.
25. Jasik, H., "Antenna Engineering Handbook," McGraw-Hill, New York, 1961.
26. Okress, E. C., ed., "Microwave Power Engineering," 2 Vols. Academic Press, New York, 1968.
27. Reich, H. J., Ordnung, P. F., Krauss, H. L., and Skalnik, J. G., "Microwave Theory and Techniques." Van Nostrand, New York, 1953 or Boston Tech. Publ., Boston, Massachusetts, 1965.
28. Cleary, S. F., ed., "Biological Effects and Health Implications of Microwave Radiation BRH/DBE 70-2." U. S. Department of Health, Education, and Welfare, Rockville, Maryland, June 1970.
29. Gordon, Z. V., "Biological Effect of Microwaves in Occupational Hygiene." Published for the National Aeronautics and Space Administration and the National Science Foundation, Washington, D.C. by the Israel Program for Scientific Translations, Jerusalem, 1970.
30. Peyton, M. F., ed., "Proceedings of the Fourth Annual Tri-Service Conference on the Biological Effects of Microwave Radiation," Vol. I. Plenum, New York, 1961.
31. Marha, K., Musil, J., and Tuha, H., "Electromagnetic Fields and the Life Environment." San Francisco Press, San Francisco, California, 1971.
32. Presman, A. S., "Electromagnetic Fields and Life." Plenum, New York, 1970.
33. Mayer-Arendt, J. R., Radiometry and photometry: Units and conversion factors. *Appl. Opt. 7*, 2081 (1968).
34. Everett, M. A., Yeagers, E., Sayre, R. M., and Olson, R. L., Penetration of epidermis by ultraviolet rays. *Photochem. Photobiol. 5*, 533 (1966).
35. Daniels, F., Jr., Ultraviolet radiation and dermatology. *In* "Therapeutic Electricity and Ultraviolet Radiation" (S. H. Licht, ed.), pp. 379–402. Waverly Press, Baltimore, Maryland, 1967.
36. Johnson, B. E., Daniels, F., Jr., and Magnus, I. A., Response of human skin to ultraviolet light. *In* "Photophysiology, Current Topics, (A. C. Giese, ed.), Vol. IV, pp. 139-202. Academic Press, New York, 1968.
37. Daniels, F., Jr., Optics of the skin as related to ultraviolet radiation. *In* "The Biologic Effects of Ultraviolet Radiation" (F. Urbach, ed.), pp. 151–157. Pergamon, Oxford, 1969.
38. Pathak, M. A., and Stratton, K., Effects of ultraviolet and visible radiation and the production of free radicals in skin. "The Biologic Effects of Ultraviolet Radiation" (F. Urbach, ed.), pp. 207–221. Pergamon, Oxford, 1969.
39. Hausser, K. W., Influence of wavelength in radiation biology. *Strahlentherapie 28*, 25 (1928).
40. Hausser, K. W., and Vahle, W., Sunburn and suntanning. *Wiss. Veröff. Siemens Konzern 6(1)*. 101 (1927). Translated in "The Biologic Effects of Ultraviolet

Radiation" (F. Urbach, ed.), Pergamon, Oxford, 1969.

41. Coblentz, W. W., Stair, R., and Hogue, J. M., The spectral erythemic reaction of the human skin to ultra-violet radiation. *Proc. Nat. Acad. Sci. (U. S.) 17*, 401 (1931).
42. Luckiesh, M., Holladay, L. L., and Taylor, A. H., Reaction of untanned human skin to ultraviolet radiation. *J. Opt. Soc. Amer. 20*, 423 (1930).
43. Sayre, R. M., Straka, E. R., Anglin, J. H., Jr., and Everett, M. A., A high intensity ultraviolet light monochromator. *J. Invest. Dermatol. 45(3)*, 190 (1965).
44. Berger, D., Magnus, I., Rottier, P. B., Sayre, R. M., and Freeman, R. G., Design and construction of high-intensity monochromators. *In* "The Biologic Effects of Ultraviolet Radiation" (F. Urbach, ed.), pp. 125–138. Pergamon, Oxford, 1969.
45. Everett, M. A., Olsen, R. L., and Sayre, R. M., Ultraviolet erythema. *Arch. Dermatol. 92*, 713 (1965).
46. Everett, M. A., Sayre, R. M., and Olson, R. L., Physiologic response of human skin to ultraviolet light. *In* "The Biologic Effects of Ultraviolet Radiation" (F. Urbach, ed.), pp. 181–186. Pergamon, Oxford, 1969.
47. Freeman, R. G., Owens, D. W., Knox, J. M., and Hudson, H. T., Relative energy requirements for an erythemal response of skin to monochromatic wave lengths of ultraviolet present in the solar spectrum. *J. Invest. Dermatol. 47*, 586 (1966).
48. Berger, D., Urbach, F., and Davies, R. E., The action spectrum of erythema induced by ultraviolet radiation. Preliminary report. *In* "XIII Congressus Internationalis Dermatologiae – Munchen 1967" (W. Jadassohn and C. G. Schirren, eds.), pp. 1112-1117. Springer-Verlag, Berlin and New York, 1968.
49. Wiskemann, A., Light, inflammation and light pigmentation. *Strahlentherapie* 117, 608 (1962).
50. Schmidt, K., Comparison of intermittent and continuous UV irradiation in induction of skin erythema. *Strahlentherapie* 121, 383 (1963).
51. Schmidt, K., On the skin erythema effect of UV flashes. *Strahlentherapie* 124, 127 (1964).
52. Blum, H. F., Hyperplasia induced by ultraviolet light: Possible relationship to cancer induction. *In* "The Biologic Effects of Ultraviolet Radiation" (F. Urbach, ed.), pp. 83–89. Pergamon, Oxford, 1969.
53. Blum, H. F., Quantitative aspects of cancer induction by ultraviolet light: Including a revised model. *In* "The Biologic Effects of Ultraviolet Radiation" (F. Urbach, ed.), pp. 543–549. Pergamon, Oxford, 1969.
54. Hueper, W. C., Environmental and occupational cancer hazards. *Clin. Pharmacol. Ther. 3*, 776 (1962).
55. Hueper, W. C., Environmental carcinogenesis in man and animals. *Ann. N.Y. Acad. Sci. 108*, 963 (1963).
56. Caron, G. A., Xeroderma pigmentosum, a clue to carcinogenesis. *Arch. Dermatol. 98*, 548 (1968).
57. Freeman, R. G., Hudson, H. T., and Carnes, R., Ultraviolet wavelength factors in solar radiation and skin cancer. *Int. J. Dermatol. 9*, 232 (1970).
58. Tan, E. M., Freeman, R. G., and Stoughton, R. B., Action spectrum of ultraviolet light-induced damage to nuclear DNA *in vivo*. *J. Invest. Dermatol. 55*, 439 (1970).
59. Brodkin, R., Kopf, A. W., and Andrade, R., Basal-cell epithelioma and elastosis: A comparison of distribution. *In* "The Biologic Effects of Ultraviolet Radiation" (F. Urbach, ed.), pp. 581–618. Pergamon, Oxford, 1969.
60. Epstein, J. H., and Fukuyama, K., Ultraviolet light carcinogenesis. *In* "The Biologic Effects of Ultraviolet Radiation" (F. Urbach, ed.), pp. 551–568. Pergamon, Oxford, 1969.

61. Jakac, D., The importance of light injury in the development, localization and frequency of skin cancer. *Hautarzt* 19, 157 (1968).
62. Forbes, P. D., and Urbach, F., Vascular and neoplastic changes in mice following ultraviolet radiation. *In* "The Biologic Effects of Ultraviolet Radiation" (F. Urbach, ed.), pp. 279–289. Pergamon, Oxford, 1969.
63. London, J., The depletion of ultraviolet radiation by atmospheric ozone. *In* "The Biologic Effects of Ultraviolet Radiation" (F. Urbach, ed.), pp. 335–339. Pergamon, Oxford, 1969.
64. Bener, P., Spectral intensity of natural ultraviolet radiation and its dependence on various parameters. *In* "The Biologic Effects of Ultraviolet Radiation" (F. Urbach, ed.), pp. 351-358. Pergamon, Oxford, 1969.
65. Schulze, R., and Grafe, K., Consideration of sky ultraviolet radiation in the measurement of solar ultraviolet radiation. *In* "The Biologic Effects of Ultraviolet Radiation" (F. Urbach, ed.), pp. 359–373. Pergamon, Oxford, 1969.
66. Buettner, K. J. K., The effects of natural sunlight on human skin. *In* "The Biologic Effects of Ultraviolet Radiation" (F. Urbach, ed.), pp. 237–249. Pergamon, Oxford, 1969.
67. Schulze, R., On the radiation climate of the earth. *Strahlentherapie 119*, 321 (1962).
68. Lotmar, R., UV radiometry using the resist procedure. II. Dependence of UV intensity upon altitude and local conditions; dosage scheme. *Strahlentherapie* 137, 238 (1969).
69. Urbach, F., Geographic pathology of skin cancer. *In* "The Biologic Effects of Ultraviolet Radiation" (F. Urbach, ed.), pp. 635–650. Pergamon, Oxford, 1969.
70. Gordon, D., and Silverstone, H., Deaths from skin cancer in Queensland, Australia. *In* "The Biologic Effects of Ultraviolet Radiation" (F. Urbach, ed.), pp. 625–634. Pergamon, Oxford, 1969.
71. Robertson, D. F., Correlation of observed ultraviolet exposure and skin-cancer incidence in the population in Queensland and New Guinea. *In* "The Biologic Effects of Ultraviolet Radiation" (F. Urbach, ed.), pp. 619–623. Pergamon, Oxford, 1969.
72. Pathak, M. A., Basic aspects of cutaneous photosensitization. *In* "The Biologic Effects of Ultraviolet Radiation" (F. Urbach, ed.), pp. 489–511. Pergamon, Oxford, 1969.
73. Burchkhardt, W., Photoallergy and phototoxicity due to drugs. *In* "The Biologic Effects of Ultraviolet Radiation" (F. Urbach, ed.), pp. 527–532. Pergamon, Oxford, 1969.
74. Musajo, L, and Rodighiero, G., The mechanism of action of the skin-photosensitizing furocoumarins. *In* "The Biologic Effects of Ultraviolet Radiation" (F. Urbach, ed.), pp. 57–59. Pergamon, Oxford, 1969.
75. Wiskemann, A., Abnormal action spectra. *In* "The Biologic Effects of Ultraviolet Radiation" (F. Urbach, ed.), pp. 197–205. Pergamon, Oxford, 1969.
76. Auerbach, R., and Weinstein, G. D., Occupational ultraviolet light and skin disease. *Arch. Dermatol.* 87, 691 (1963).
77. Birmingham, D. J., Key, M. M., Tubich, G. E., and Perone, V. B., Phototoxic bullae among celery harvesters. *Arch. Dermatol. 83*, 127 (1961).
78. Ogilvie, J. C., Ultraviolet radiation and vision. *Arch. Ophthalmol. 50*, 748 (1953).
79. Wald, G., Human vision and the spectrum. *Science 101*, 653 (1945).
80. Wald, G., Alleged effects of the near ultraviolet on human vision. *J. Opt. Soc. Amer. 42*, 171 (1952).
81. Zigman, S., Eye lens color: Formation and function. *Science 171*, 807 (1971).
82. Duke-Elder, S., "Textbook of Ophthalmology, Injuries," Vol. VI, pp. 6443-6523.

Mosby, St. Louis, Missouri, 1954.
83. Clark, J. H., The effect of ultraviolet radiation on lens protein in the presence of salts and the relation of radiation to industrial cataract. *Amer. J. Physiol. 113*, 539 (1935).
84. Cogan, D. G., and Kinsey, V. E., Action spectrum of keratitis produced by ultraviolet radiation. *Arch. Ophthalmol. 35,* 670 (1946).
85. Friedenwald, J. S., Buschke, W., Crowell, J., and Hollaender, A., Effects of ultraviolet irradiation on the corneal epithelium. *J. Cell. Comp. Physiol. 32*, 161 (1948).
86. Bachem, A., Ophthalmic ultraviolet action spectra. *Amer. J. Ophthalmol. 4*, 969 (1965).
87. Pitts, D. G., and Kay, K. R., The photo-ophthalmic threshold for the rabbit. *Amer. J. Optom. Arch. Amer. Acad. Optom. 46,* 561 (1969).
88. Pitts, D. G., A comparative study of the effects of ultraviolet radiation on the eye. *Amer. J. Optom. Arch. Amer. Acad. Optom. 47* 535 (1970).
89. Pitts, D. G., and Tredici, T. J., The effects of ultraviolet on the eye. *Amer. Ind. Hyg. Ass. J. 32,* 235 (1971).
90. Hedblom, E. E., Snowscape eye protection. *Arch. Environ. Health 2*, 685 (1961).
91. Leach, W. M., "Biological Aspects of Ultraviolet Radiation, A Review of Hazards BRH/DBE 70-3." U. S. Public Health Service, Bureau of Radiological Health, Rockville, Maryland, Sept. 1970.
92. Council on Physical Medicine. Acceptance of ultraviolet lamps for disinfecting purposes. *J. Amer. Med. Ass. 137*, 1600 (1948).
93. Szabo, E., and Hokay, I., Effect of ultraviolet light on the epidermis. *Acta Morphol. Acad. Sci. Hung. 15,* 71 (1967).
94. Everett, M. A., Waltermire, J. A., Olson, R., and Sayre, R., Modification of ultra-violet erythema by epidermal stripping. *Nature (London) 205*, 812 (1965).
95. Findlay, G. H., An automatic fractionator for light dosage on the skin. Its application to the polychromatic minimal erythema dose. *Brit. J. Dermatol. 79*, 148 (1967).
96. Van der Leun, J. C., Delayed pigmentation and UV erythema. *In* "Recent Progress in Photobiology" (E. J. Bowen, ed.), pp. 387–388. Academic Press, New York, 1965.
97. Van der Leun, J. C., Theory of ultraviolet erythema. *Photochem. Photobiol. 4*, 453 (1965).
98. Van der Leun, J. C., Delayed pigmentation and ultraviolet erythema. *Photochem. Photobiol. 4*, 459 (1965).
99. American Conference of Governmental Industrial Hygienists. "Threshold Limit Values of Physical Agents Adopted by ACGIH for 1971." ACGIH, Cincinnati, Ohio, 1971.
100. MacKeen, D., Fine, S., Aaron, A., and Fine, B. S., Preventable Hazard at UV wavelengths. *Laser Focus 7(4)*, 29 (1971).
101. Plato, "Phaedo, The Dialogues of Plato," p. 242. Translated by B. Jowett, Encyclopaedia Britannica, Chicago, Illinois, 1952.
102. Kuhns, H. S., "Eyes and Industry," 2nd Ed., pp. 211–227. Mosby, St. Louis, Missouri, 1950.
103. Verhoeff, F. H., and Bell, L., The pathological effects of radiant energy on the eye: An experimental investigation. *Proc. Amer. Acad. Arts Sci. 51*, 629, 819 (1916).
104. Walker, C. B., Systematic review of the literature relating to the effects of radiant energy upon the eye. *Proc. Amer. Acad. Arts Sci. 51*, 760 (1916).
105. Cogan, D. G., Lesions of the eye from radiant energy. *J. Amer. Med. Ass. 142*, 145 (1950).
106. Kutscher, C. F., Ocular effects of radiant energy. *Ind. Med.* 15, 311 (1946).

107. Newell, F. W., Radiant energy and the eye. *In* "Industrial and Traumatic Ophthalmology" (A. H. Keeny, H. S. Kuhn, R. MacDonald, F. W. Newell, J. F. Novak, R. W. Ryan, and L. E. Zimmerman, eds.), pp. 158–187. Mosby, St. Louis, 1964.
108. Bartleson, C. J., Retinal burns from intense light sources. *Amer. Ind. Hyg. Ass. J. 29*, 415 (1968).
109. Clarke, A. M., Ocular hazards from lasers and other optical sources. *Crit. Rev. Environ. Contr. 1*, 307 (1970).
110. Sliney, D. H., The development of laser safety criteria – biological considerations. *In* "Laser Applications in Medicine and Biology" (M. L. Wolbarsht, ed.), pp. 163–238. Plenum, New York, 1971.
111. McLanahan, D., Beyond the candle or lamps from A to Z, *Electro-Opt. Sys. Des. 3(2)*, 18 (1971).
112. Goncz, J. H., and Newell, P. B., Spectra of pulsed and continuous xenon discharges. *J. Opt. Soc. Amer. 56*, 87 (1966).
113. Meyer-Schwickerath, G., Lichtkoagulation: Eine Methode zur behandlugund Verhutung der Netzautablosung. *Von Graefe's Arch. Ophthalmol. 156*, 2 (1954).
114. Ham, W. T., Jr., Geeraets, W. J., Mueller, H. A., Williams, R. C., Clarke, A. M., and Cleary, S. F., Retinal burn thresholds for the helium-neon laser in the rhesus monkey. *Arch. Ophthalmol. 84*, 797 (1970).
115. Ham, W. T., Jr., Clarke, A. M., Geeraets, W. J., Cleary, S. F., Mueller, H. A., and Williams, R. C., The eye problem in laser safety. *Arch. Environ. Health 20*, 156 (1970).
116. Ham, W. T., Jr., Laser hazards and control in industry. *Ind. Hyg. Found. Trans. Bull. 41*, 36 (1967).
117. Clarke, A. M., Ham, W. T., Jr., Geeraets, W. J., Williams, R. C., and Mueller, H. A., Laser effects on the eye. *Arch. Environ. Health 18*, 424 (1969).
118. Ham, W. T., Geeraets, W. J., Williams, R. C., Guerry, D., and Mueller, H. A., Laser radiation protection. *In* "Proceedings of the First International Congress of Radiation Protection," pp. 933–943. Pergamon, Oxford, 1968.
119. Vassilliadis, A., Zweng, H. C., Peppers, N. A., Peabody, R. R., and Honey, R. C., Thresholds of laser eye hazards. *Arch. Environ. Health 20*, 161 (1970).
120. Zweng, H. C., Vassiliadis, A., Peppers, N. A., Peabody, R. R., and Honey, R. C., Thresholds of laser eye hazards. *Ind. Hyg. Found. Trans. Bull. 42*, 75 (1968).
121. Bergqvist, T., Kleman, B., and Tengroth, B., Retinal lesions produced by Q-switched lasers. *Acta Ophthalmol. 44*, 853 (1966).
122. Jones, A. E., and McCartney, A. J., Ruby laser effects on the monkey eye. *Invest. Ophthalmol. 5*, 474 (1966).
123. Kohtiao, A., Resnick, I., Newton, J., and Schwell, H., Threshold lesions in rabbit retinas exposed to pulsed ruby laser radiation. *Amer. J. Ophthalmol. 62*, 664 (1966).
124. Ingram, H. V., A laser ophthalmoscope for retinal phototherapy. *Brit. Med. J. 5438*, 823 (1965).
125. Jacobson, J. H., Najac, H. W., Cooper, B., "Investigation of the Effects of Ruby Laser Radiation on Ocular Tissue." New York Eye and Ear Infirmary, New York College, Report R-1815 of U. S. Army, Frankford Arsenal, Philadelphia, Pennsylvania (1966) (AD 638 917).
126. Campbell, C. J., Rittler, M. C., and Swope, C. H., The ocular effects produced by experimental lasers. IV. The argon laser. *Amer. J. Ophthalmol. 67*, 671 (1969).
127. Campbell, C. J., Rittler, M. C., Swope, C. H., and Koester, C. J., Ocular effects produced by experimental lasers, I. Q-switched ruby laser. *Amer. J. Ophthalmol. 66*, 459 (1968).
128. Campbell, C. J., Rittler, M. C., Innis, R. E., and Shiner, W. H., Ocular effects

produced by experimental lasers, III. Neodymium laser. *Amer. J. Ophthalmol. 66*, 614 (1968).

129. Zaret, M. M., Ocular exposure to Q-switched laser irradiation. Rep. AFAL-TR-65-279. Wright-Patterson AFB, Fairborn, Ohio, USAF Avionics Laboratory (1966) (AD483970).
130. Blabla, J., and John, J., The saturation effect in retina measured by means of He-Ne laser. *Amer. J. Ophthalmol. 62*, 659 (1966).
131. Bresnick, G. H., Frisch, G. D., Powell, J. O., Landers, M. B., Holst, G. C., and Dallas, A. G., Ocular effects of argon laser radiation I. Retinal damage threshold studies. *Invest. Ophthalmol. 9*, 901 (1970).
132. Powell, J. O., Bresnick, G. H., Yanoff, M., Frisch, G. D., and Chester, J. E., Ocular effects of argon laser radiation. II. Histopathology of chorioretinal lesions. *Amer. J. Ophthalmol. 71*, 1267 (1971).
133. Yanoff, M., Landers, M. B., and Bresnick, G. H., Technique for flat-mount retinal pigment epithelial preparations. *Ann. Ophthalmol. 2*, 475 (1970).
134. Davis, T. P., and Mautner, W. J., "Helium-Neon Laser Effects on the Eye, Report C106-59223." EG&G Inc., Santa Monica Division, Los Angeles, California, April 1969.
135. Lappin, P. W., and Coogan, P. S., Relative sensitivity of various areas of the retina to laser radiation. *Arch. Ophthalmol. 84*, 350 (1970).
136. Lappin, P. W., Ocular damage thresholds for the helium-neon laser. *Arch. Environ. Health 20*, 177 (1970).
137. Lappin, P. W., and Coogan, P. S., Histologic evaluation of ophthalmoscopically subvisible retinal laser exposures. *Invest. Ophthalmol. 9*, 537 (1970).
138. Smart, D., Manson, N., Marshall, J., and Mellerio, J., New ocular hazard of mode locking in CW lasers. *Nature (London) 227*, 1149 (1970).
139. Leibowitz, H. M., and Peacock, G. R., The retinal pigment epithelium. Radiation thresholds associated with the Q-switched ruby laser. *Arch. Ophthalmol. 82*, 332 (1969).
140. L'Esperance, F. A., Jr., The ocular histopathologic effect of krypton and argon laser radiation. *Amer. J. Ophthalmol. 68*, 263 (1969).
141. L'Esperance, F. A., Jr., and Kelly, G. R., The threshold of the retina to damage by argon laser radiation. *Arch. Ophthalmol. 81*, 583 (1969).
142. L'Esperance, F. A., Jr., Clinical photocoagulation with the frequency-doubled neodymium yttrium-aluminum-garnet laser. *Amer. J. Ophthalmol. 71*, 631 (1971).
143. Rosan, R. C., Flocks, M., Vassiliadis, A., Rose, H. W., Peabody, R. R., and Hammond, A., Pathology of monkey retina following irradiation with an argon laser. *Arch. Ophthalmol. 81*, 84 (1969).
144. Geeraets, W. J., Retinal injury by ruby and neodymium laser. *Acta Ophthalmol. 45*, 846 (1967).
145. Bredemeyer, H. G., Wiegmann, O. A., Bredemeyer, A., and Blackwell, H. R., "Radiation Thresholds for Chorioretinal Burns." Institute for Research in Vision and Department of Ophthalmology, Ohio State University, Columbus, Ohio, Air Force Technical Documentary Rep. No. AMRL-TDR-63-71 (July 1963) (AD 416 652).
146. Allen, R. G., Bruce, W. R., Kay, K. R., Morrison, L. K., Neish, R. A., Polaski, C. A., and Richards, R. A., "Research on Ocular Effects Produced by Thermal Radiation," Final Rep. AF41 (609)–3099. Brooks AFB, San Antonio, Texas (1967) (AD 659 146).
147. Jacobson, J. H., Cooper, B., and Najac, H. W., "Effects of Thermal Energy on Retinal Function," Rep. AMRL-TDR-62-96. Aerospace Medical Division, Air Force Systems

Command, Wright-Patterson AFB, Fairborn, Ohio (1962) (AD 290 808).

148. DeMott, D. W., and Davis, T. P., Irradiance thresholds for chorioretinal lesions. *Arch. Ophthalmol. 62*, 653 (1959).
149. Kohner, E. M., Dollery, C. T., Henkind, P., Paterson, J. W., and Ramalho, P. S., Retinal vascular changes following exposure to high-intensity light. *Amer. J. Ophthalmol. 63*, 1748 (1967).
150. Vassiliadis, A., Ocular damage from laser radiation. *In* "Laser Applications in Medicine and Biology" (M. L. Wolbarsht, ed.), pp. 125–162. Plenum, New York, 1971.
151. Vos, J. J., A theory of retinal burns. *Bull. Math. Biophys. 24*, 115 (1962).
152. Wray, J. L., "Model for Prediction of Retinal Burns," Rep. DASA-1282. Defense Atomic Support Agency, Washington, D.C. (1962) (AD 277 363).
153. Vos, J. J., Heat-damage to the retina by lasers and photocoagulators. *Ophthalmologica 151*, 652 (1966).
154. Felstead, E. B., and Cobbold, R. S., Analog solution of laser retinal coagulation. *Med. Electron. Biol. Eng. 3*, 145 (1965).
155. Hayes, R., and Wolbarsht, M. L., Thermal model for retinal damage induced by pulsed lasers. *Aerosp. Med. 39*, 474 (1968).
156. Peacock, G. R., "Surface Temperature as a Parameter in Estimating Laser Injury Thresholds," Rep. No. 733. U. S. Army Medical Research Laboratory, Fort Knox, Kentucky, 1967.
157. Hansen, W. P., and Fine, S., Melanin granule models for pulsed laser induced retinal injury. *Appl. Opt. 7*, 155 (1968).
158. Hayes, J. R., and Wolbarsht, M. L., Models in pathology – mechanisms of action of laser energy with biological tissues. *In* "Laser Applications in Medicine and Biology" (M. L. Wolbarsht, ed.), pp. 255–274. Plenum, New York, 1971.
159. Fahs, J. H., "A Model for the Study of Retinal Damage Due to Laser Radiation," Rep. No. 3678. U. S. Army Materiel Command, Picatinny Arsenal, Dover, New Jersey (1968) (AD 668 906).
160. Ridgeway, D., Three dimensional steady-state temperature distribution about cylinders and discs. *Bull. Math. Biophys. 30*, 701 (1968).
161. Makous, W. L., and Gould, J. D., Effects of lasers on the human eye. *IBM J. Res. Develop. 12*, 257 (1968).
162. Clarke, A. M., Geeraets, W. J., and Ham, W. T., Jr., An equilibrium thermal model for retinal injury. *Appl. Opt. 8*, 1051 (1969).
163. Mainster, M. A., White, T. J., Tips, J. H., and Wilson, P. W., Retinal-temperature increases produced by intense light sources. *J. Opt. Soc. Amer. 60*, 264 (1970).
164. Mainster, M. A., White, T. J., and Allen, R. G., Spectral dependence of retinal damage produced by intense light sources. *J. Opt. Soc. Amer. 60*, 848 (1970).
165. Mainster, M. A., Destructive light adaptation. *Ann. Ophthalmol. 2*, 44 (1970).
166. Mainster, M. A., White, T. J., Tips, J. H., and Wilson, P. W., Transient thermal behavior in biological systems. *Bull. Math. Biophys. 32*, 303 (1970).
167. White, T. J., Mainster, M. A., Tips, J. H., and Wilson, P. W., Chorioretinal thermal behavior. *Bull. Math. Biophys. 32*, 315 (1970).
168. White, T. J., Mainster, M. A., Wilson, P. W., and Tips, J. H., Chorioretinal temperature increases from solar observation. *Bull. Math. Biophys. 33*, 1 (1971).
169. Hu, C. L., and Barnes, F. S., The thermal-chemical damage in biological material under laser irradiation. *IEEE Trans. Biomed. Eng. 17*, 220 (1970).
170. Eccles, J. C., and Flynn, A. J., Experimental photo-retinitis. *Med. J. Aust. 1*, 339 (1944).

171. Rockwell, R. J., Jr., Developments in laser instrumentation and calibration. *Arch. Environ. Health 20*, 149 (1970).
172. Brewster, J. L., Stimulated emission from CdS at ultra-high current density electron beam pumping. *Appl. Phys. Lett. 13*, 385 (1968).
173. Rentzepis, P. M., Ultrafast processes. *Science* 169, 239 (1970).
174. Duguay, M. A., and Hansen, J. W., Compression of pulses from a mode-locked He–Ne laser. *Appl. Phys. Lett. 14*, 14–15 (1969).
175. DeMaria, A. J., Glenn, W. H., and Mack, M. E., Ultrafast laser pulses. *Phys. Today 24(7)*, 19 (1971).
176. Walsh, J. W. T., "Photometry," 3rd Ed. Dover, New York, 1958.
177. Randall, H. G., Brown, D. J., and Sloan, L. L., Peripheral visual acuity. *Arch. Ophthalmol. 75*, 500 (1966).
178. Vos, J. J., Ham, W. T., and Geeraets, W. J., What is the functional damage threshold for retinal burn? "Loss of Vision from High Intensity Light," pp. 39–53. Advisory Group for Aerospace Research and Development (NATO), Paris (1966), AGARD-CP11.
179. Wolbarsht, M. L., Decrement in visual acuity from laser lesions in the fovea. *Aerosp. Med. 37*, 1250 (1966).
180. Nicholson, A. N., and Allwood, M. J., Laser lesions: Changes in retinal excitability. *Nature (London) 210*, 637 (1966).
181. Geeraets, W. J., Retinal injury from laser and light exposure. *In* "Laser Eye Effects" (H. G. Sperling, ed.), pp. 20–56. Armed Forces – National Research Council Committee on Vision, Washington, D.C., 1968, (AD 667 494).
182. Farrer, D. N., Graham, E. S., Ham, W. T., Jr., Geeraets, W. J., Williams, R. C., Mueller, H. A., Cleary, S. F., and Clarke, A. M., The effect of threshold macular lesions and subthreshold macular exposures on visual acuity in the Rhesus monkey. *Amer. Ind. Hyg. Ass. J. 31*, 198 (1970).
183. Gerathewohl, S. J., and Strughold, H., Motoric responses of the eyes when exposed to light flashes of high intensities and short duration. *J. Aviat. Med. 24*, 200 (1953).
184. Bourassa, C. M., and Wirtschafter, J. D., Mechanism of binocular increase of discomfort to high luminance. *Nature (London) 212*, 1503 (1966).
185. Craik, K. J. W., On the effects of looking at the sun. *In* "The Nature of Psychology" (S. L. Sherwood, ed.), pp. 98–101. Cambridge Univ. Press, London and New York, 1966.
186. Doesschate, J., and Alpern, M., Effect of photo-excitation of the two retinas on pupil size. *J. Neurophysiol. 30*, 562 (1967).
187. Fugate, J. M., Physiological basis for discomfort glare. *Amer. J. Optomol. 34*, 377 (1957).
188. Guth, S. K., Brightness relationships for comfortable seeing. *J. Opt. Soc. Amer. 41*, 235 (1951).
189. Lebensohn, J. E., Photophobia: Mechanism and implications. *Amer. J. Ophthalmol. 34*, 1294 (1951).
190. Lele, P. P., and Weddell, G., The relationship between neurohistology and corneal sensibility. *Brain 79*, 119 (1956).
191. Mathis, W., and Bourassa, C. M., Fusion and nonfusion as factors in aversion to high luminance. *Vision Res. 8*, 1501 (1968).
192. Wirtschafter, J. D., and Bourassa, C. M., Binocular facilitation of discomfort with high luminances. *Arch. Ophthalmol. 75*, 683 (1966).
193. Cleary, S. F., and Hamrick, R. E., Laser-induced acoustic transients in the mammalian eye. *J. Acoust. Soc. Amer. 46*, 1037 (1969).

194. Davies, J. M., and Randolph, D. T., eds., "Proceedings of the U. S. Army Natick Laboratories Flash Blindness Symposium," 264 pp. Armed Forces–National Research Council Committee on Vision, Washington, D.C., 1967.
195. Weale, R. A., Photochemistry and vision. *In* "Photophysiology" (A. C. Giese, ed.), Vol. IV, pp. 1–45. Academic Press, New York, 1968.
196. Noell, W. K., Walker, V. S., Kang, B. S., and Berman, S., Retinal damage by light in rats. *Invest. Ophthalmol. 5*, 450 (1966).
197. Gorn, R. A., and Kuwabara, T., Retinal damage by visible light, a physiologic study. *Arch. Ophthalmol. 77*, 115 (1967).
198. Kuwabara, T., and Gorn, R. A., Retinal damage by visible light, an electron microscopic study. *Arch. Ophthalmol. 79*, 69 (1968).
199. Kuwabara, T., Retinal recovery from exposure to light. *Amer. J. Ophthalmol. 70*, 187 (1970).
200. Noell, W. K., and Albrecht, R., Irreversible effects of visible light on the retina: Role of vitamin A. *Science 172*, 76 (1971).
201. Noell, W. K., Delmelle, M. C., and Albrecht, R., Vitamin A deficiency effect on retina: Dependence on light. *Science 172*, 72 (1971).
202. Ts'o, M. O., Fine, B. S., and Zimmerman, L. E., Photic maculopathy produced by the indirect ophthalmoscope. *Amer. J. Ophthalmol. 73*, 686(1972).
203. Smith, H. E., Actinic macular retinal pigment degeneration. *U. S. Naval Med. Bull. 42*, 675 (1944).
204. Medvedovskaya, Ts. P., Data on the condition of the eye in workers at a glass factory. *Hyg. Sanit. 35*, 445 (1970).
205. Friedman, E., and Kuwabara, T., The retinal pigment epithelium IV. The damaging effects of radiant energy. *Arch. Ophthalmol. 80*, 265 (1968).
206. Vassiliadis, A., Rosan, R. C., Hayes, J., and Zweng, H. C., "Investigation of Retinal Hazard Due to Pulsed Xenon Lamp Radiation." Stanford Research Institute, Menlo Park, California, 1970.
207. Livingston, P. C., The study of sun glare in Iraq. *Brit. J. Ophthalmol. 6*, 577 (1932).
208. Clark, B., Johnson, M. L., and Dreher, R. E., The effect of sunlight on dark adaptation. *Amer. J. Ophthalmol. 29*, 828 (1946).
209. Hecht, S., Hendley, C. D., Ross, S., and Richmond, P. M., The effect of exposure to sunlight on night vision. *Amer. J. Ophthalmol. 31*, 1573 (1948).
210. Clark, B., Johnson, M. L., and Dreher, R. E., The effect of sunlight on dark adaptation. *Amer. J. Ophthalmol. 29*, 828 (1946).
211. Peckham, R. H., The protection and maintenance of night vision for military personnel. *Amer. J. Ophthalmol. 30*, 1588 (1947).
212. MacDonald, P. R., Evaluation of night vision. *Amer. J. Ophthalmol. 32*, 1535 (1949).
213. Peckham, R. H., and Harley, R. D., The effect of sunglasses in protecting retinal sensitivity. *Amer. J. Ophthalmol. 34*, 1499 (1951).
214. Harwerth, R. S., and Sperling, H. G., Prolonged color blindness induced by intense spectral lights in rhesus monkeys. *Science 174*, 520 (1971).
215. MacDonald, J. E., and Fordon, L., Erythropsia and light toxicity thresholds. Presented at the meeting of the Association for Research in Vision and Ophthalmology, Sarasota, Florida, 28 Apr. 1971.
216. Geeraets, W. J., and Berry, E. R., Ocular spectral characteristics as related to hazards from lasers and other light sources. *Amer. J. Ophthalmol. 66*, 15 (1968).
217. Boettner. E. A., and Wolter, J. R., Transmission of the ocular media. *Invest. Ophthalmol. 1*, 776 (1962).
218. Boettner, E. A., "Spectral Transmission of the Eye." Final Rep. AF 41(609)-2996.

University of Michigan, Ann Arbor, Michigan (1967) (AD663 246).
219. Gubisch, R. W., Optical performance of the human eye. *J. Opt. Soc. Amer. 57*, 407 (1967).
220. Powell, C. H., Bell, H. E., Rose, V. E., Goldman, L., and Wilkinson, T. K., Current status of lasers threshold guides. *Amer. Ind. Hyg. Ass. J. 31*, 485 (1970).
221. "Control of Hazards to Health from Laser Radiation," TB MED 279, NAVMED P-5052-35. U. S. Department of the Army and U. S. Department of the Navy, Washington D.C., Feb. 1969.
222. "Laser Health Hazards Control," AFM 161-8. U. S. Department of the Air Force, Washington, D.C., April 1969 and September 1971.
223. Weston, B. A., "Laser Systems–Code of Practice." Ministry of Technology Safety Services Organization, Station Square, Kent, Oct. 1969.
224. "The Safe Use of Lasers." Proposed Standard Z-136. American National Standards Institute, New York, Feb. 1972.
225. Sliney, D. H., and Palmisano, W. A., The evaluation of laser hazards. *Amer. Ind. Hyg. Ass. J. 29*, 425 (1968).
226. Carpenter, J. A., Lehmiller, D. J., and Tredici, T. J., U. S. Air Force permissible exposure levels for laser irradiation. *Arch. Environ. Health 20*, 171 (1970).
227. Ditchburn, R. W., and Ginsborg, B. L., Involuntary eye movements during fixation. *J. Physiol. (London) 119*, 1 (1953).
228. Yarbus, A. L., "Eye Movements and Vision." Translated from the Russian by B. Haigh. Plenum, New York, 1967.
229. Wheeless, L. L., Cohen, G. H., and Boynton, R. M., Luminance as a parameter of the eye-movement control system. *J. Opt. Soc. Amer. 57*, 394 (1967).
230. Sliney, D. H., The amazing laser. "Transactions of the 56th National Safety Congress," Chicago, National Safety Council *8*, 38 (1968).
231. Parr, W. H., "Skin Lesion Threshold Values for Laser Radiation as Compared with Safety Standards," Rep. No. 813. U. S. Army Medical Research Laboratory, Fort Knox, Kentucky (Feb. 1969) (AD 688 871).
232. Kuhns, J. G., Hayes, J., Stein, M., and Helwig, E. B., Laser injury in skin. *Lab. Invest. 17*, 1 (July 1967) (AD 657 880).
233. Goldman, L., The skin. *Arch. Environ. Health 18*, 434 (1969).
234. Goldman, L., Rockwell, R. J., Jr., and Richfield, D., Long-term laser exposure of a senile freckle. *Arch. Environ. Health 22*, 401 (1971).
235. Büttner, K., On thermal radiation and the reflection properties of human skin. *Strahlentherapie 58*, 345 (1937).
236. Jacquez, J. A., Huss, J., McKeehan, W., Dimitroff, J. M., and Kuppenheim, H. F., Spectral reflectance of human skin in the region 0.7–2.6 μm. *J. Appl. Physiol. 8*, 297 (1955).
237. Jacquez, J. A., Kuppenheim, H. F., Dimitroff, J. M., McKeehan, W., and Huss, J., Spectral reflectance of human skin in the region 235–700 nm. *J. Appl. Physiol 8*, 212 (1955).
238. Hardy, J. D., Hammel, H. T., and Murgatroyd, D., Spectral transmittance and reflectance of excised human skin. *J. Appl. Physiol. 9*, 257 (1956).
239. Clark, C., Vinegar, R., and Hardy, J., Goniometric spectrometer for the measurement of diffuse reflectance and transmittance of skin in the infra-red spectral region. *J. Opt. Soc. Amer. 43*, 993 (1953).
240. Hardy, J. D., and Oppel, T. W., Studies in temperature sensation. III. The sensitivity of the body to heat and the spatial summation of the end organ responses. *J. Clin. Invest. 16*, 533 (1937).

241. Lele, P. P., Weddell, G., and Williams, C. W., The relationship between heat transfer, skin temperature, and cutaneous sensibility. *J. Physiol. (London) 126*, 206 (1954).
242. Hendler, E., Hardy, J. D., and Murgatroyd, D., Skin heating and temperature sensation produced by infrared and microwave irradiation. *In* "Temperature, Its Measurement and Control in Science and Industry" (C. M. Herzfeld and J. D. Hardy, eds.), Vol. 3, pp. 211–230. Reinhold, New York, 1963.
243. Webb, P., Pain limited heat exposures. *In* "Temperature, Its Measurement and Control in Science and Industry" (C. M. Herzfeld and J. D. Hardy, eds.), Vol. 3, pp. 245–250. Reinhold, New York, 1963.
244. Hardy, J. D., Pain following step increase in skin temperature. *In* "The Skin Senses" (D. R. Kenshalo, ed.), pp. 444–456. Thomas, Springfield, Illinois, 1968.
245. Henriques, F. C., Jr., Studies of thermal injuries. *Amer. J. Pathol. 23*, 489 (1947).
246. Henriques, F. C., Jr., and Moritz, A. R., Studies of thermal injury. I. The conduction of heat to and through skin and the temperatures attained therein. A theoretical and an experimental investigation. *Amer. J. Pathol. 23*, 531 (1947).
247. Moritz, A. R., and Henriques, F. C., Jr., Studies of thermal injury. II. The relative importance of time and surface temperature in the causation of cutaneous burns. *Amer. J. Pathol. 23*, 695 (1947).
248. Buettner, K., Effects of extreme heat and cold on human skin. Numerical analysis and pilot experiments on penetrating flash radiation effects. *J. Appl. Physiol. 5*, 207 (1952).
249. Sheline, G. E., Alpen, E. L., Kuhl, P. R., and Ahokas, A. J., Effects of high intensity radiant energy on skin. I. Type of injury and its relation to energy delivery rate. *Arch. Pathol. 55*, 265 (1953).
250. Schmidt, F. H., Williams, R. C., Ham, W. T., Brooks, J. W., and Evans, E. I., Experimental production of flash burns. *Surgery 36*, 1163 (1954).
251. Davis, T. P., The heating of skin by radiant energy. *In* "Temperature, Its Measurement and Control in Science and Industry" (C. M. Herzfeld and J. D. Hardy, eds.), Vol. 3, pp. 149–169. Reinhold, New York, 1963.
252. Derksen, W. L., Monahan, T. I., and deLhery, G. P., The temperatures associated with radiant energy skin burns. *In* "Temperature, Its Measurement and Control in Science and Industry" (C. M. Herzfeld and J. D. Hardy, eds.), Vol. 3, pp. 171–175. Reinhold, New York, 1963.
253. Derksen, W. L., Bracciaventi, J., and Mixter, G., Jr., "Burns to Skin by Millisecond Light Pulses." DASA Rep. No. 1532. U. S. Naval Applied Science Laboratory, Brooklyn, New York (July 1964) (AD 607 388).
254. Brownell, A. S., Parr, W. H., and Hysell, D. K., Skin and carbon dioxide laser radiation. *Arch. Environ. Health 18*, 437 (1969).
255. Kenshalo, D. R., Comparison of thermal sensitivity of the forehead, lip, conjunctiva, and cornea. *J. Appl. Physiol. 15*, 987 (1960).
256. Dawson, W. W., The thermal excitation of afferent neurones in the mammalian cornea and iris. *In* "Temperature, Its Measurement and Control in Science and Industry" (C. M. Herzfeld and J. D. Hardy, eds.), Vol. 3, pp. 199–210. Reinhold, New York, 1963.
257. Jacobson, J. H., Cooper, B., Najac, H. W., and Kohtiao, A., "The Effects of Thermal Energy on Anterior Ocular Tissues." AMRL-TDR-63-53, 6570th Aerospace Med. Res. Lab., Wright-Patterson Air Force Base, Fairborn, Ohio (June 1963) (AD 412 730).
258. Goldmann, H., König, H., and Mäder, F., The transmission of the lens of the eye for infrared. *Ophthalmologica 120*, 198 (1950).

259. MacDonald, J. E., and Light, A., Photocoagulation of iris and retina. *Arch. Ophthalmol. 60*, 384 (1958).
260. Langley, R. K., Mortimer, C. B., and McCulloch, C., The experimental production of cataracts by exposure to heat and light. *Arch. Ophthalmol. 63*, 473 (1960).
261. Fine, B. S., Fine, S., Feigen, L., and MacKeen, D., Corneal injury threshold to carbon dioxide laser irradiation. *Amer. J. Ophthalmol. 66*, 1 (1968).
262. Fine, B. S., Berkow, J. W., and Fine, S., Corneal calcification. *Science 162*, 129 (1968).
263. Peabody, R. R., Rose, H., Zweng, H. C., Peppers, N. A., and Vassiliadis, A., Threshold damage from CO_2 lasers. *Arch. Ophthalmol. 82*, 105 (1969).
264. Peppers, N. A., Vassiliadis, A., Dedrick, K. G., Chang, H., Peabody, R. R., Rose, H., and Zweng, H. C., Corneal damage thresholds for CO_2 laser radiation. *Appl. Opt. 8*, 377 (1969).
265. Gullberg, K., Hartmann, B., Kock, E., and Tengroth, B., Carbon dioxide laser hazards to the eye. *Nature (London)* 215, 857 (1967).
266. Liebowitz, H. M., and Peacock, G. R., Corneal injury produced by carbon dioxide laser radiation. *Arch. Ophthalmol. 81*, 713 (1969).
267. Lund, D. J., Bresnick, G. H., Landers, M. B., Powell, J. O., Chester, J. E., and Carver, C., Ocular hazards of the Q-switched erbium laser. *Invest. Ophthalmol. 9*, 463 (1970).
268. Bresnick, G. H., Lund, D. J., Landers, M. B., Powell, J. O., Chester, J. E., and Carver, C., Ocular hazards of Q-switched erbium laser. "Electronic Product Radiation and the Health Physicist BRH/DEP 70-26," pp. 305–316. U. S. Department of Health, Education, and Welfare, Bureau of Radiological Health, Rockville, Maryland, Oct. 1970.
269. Centeno, M. V., The refractive index of liquid water in the near infra-red spectrum. *J. Opt. Soc. Amer. 31*, 244 (1941).
270. Sliney, D. H. and Freasier, B. C., The evaluation of optical radiation hazards. *Appl. Opt. 11,* in press (1972).
271. Joffe, T. M., Retinal injury due to radiant energy from a motion picture projector. *Sov. Vestn. Oftal'mol. 9*, 882 (1936).
272. Withrow, R. B., and Withrow, A. P., Generation, control, and measurement of visible and near-visible radiant energy. "Radiation Biology" (A. Hollaender, ed.), Vol. III, pp. 125–258. McGraw-Hill, New York, 1956.
273. Douda, B. E., and Bair, E. J., Visible radiation from illuminating-flare flames. II. Formation of the sodium resonance continuum. *J. Opt. Soc. Amer. 60*, 1257 (1970).
274. Baum, W. A., and Dunkelman, L., Ultraviolet radiation of the high pressure xenon arc. *J. Opt. Soc. Amer. 40*, 782 (1950).
275. Daniels, F., Jr., Plastic materials for ultraviolet protection. *Arch. Dermatol. 84*, 392 (1961).
276. Ferry, J. J., Ultraviolet emission during inert-arc welding. *Amer. Ind. Hyg. Quart. 15*, 73 (1954).
277. Kinsey, V. E., Cogan, D. G., and Drinker, P., Measuring flash from arc welding. *J. Amer. Med. Ass. 123*, 403 (1943).
278. "The Spectral-Transmissive Properties of Plastics for Use in Eye Protection." American National Standards Institute, New York, 1955.
279. Council on Physical Medicine, Eye discomfort caused by improperly shielded black light ultraviolet lamps. *J. Amer. Med. Ass. 131*, 287 (1946).
280. Coblentz, W. W., and Stair, R., Correlation of the shade numbers and densities of eye-protective glasses. *J. Opt. Soc. Amer. 20*, 624 (1930).

281. Stair, R., Spectral-transmissive properties and use of eye-protective glasses. *Nat. Bur. Stand.* Circ. 471, U. S. Department of Commerce, Washington, D.C., Feb. 1948.
282. "Safety Code for Head, Eye, and Respiratory Protection." ANS Z2.1, American Standards Institute, New York, 1959.
283. Rutgers, G. A. W., Protective glasses for welding. *Doc. Ophthalmol. 4*, 320 (1950).
284. Bauer, G. H., Hubner, J., and Sutter, E., Measurement of light scattered by eye protection filters. *Appl. Opt.* 7, 325 (1968).
285. Schreibeis, W. J., Laser eye protection goggles, based on manufacturer's information. *Amer. Ind. Hyg. Ass. J. 29*, 504 (1968).
286. Straub, H. W., Protection of the human eye from laser radiation. *Ann. N.Y. Acad. Sci. 122*, 773 (1965).
287. Straub, H. W., Laser eye protection in the USA. *Berufsgenossensch. Z. Unfallversicherung Betriebssicherheit 3*, 83 (1970).
288. Swope, C. H., Design considerations for laser eye protection. *Arch. Environ. Health 20*, 184 (1970).
289. Swope, C. H., and Koester, C. J., Eye protection against lasers. *Appl. Opt. 4*, 523 (1965).
290. Swope, C. H., The eye – protection. *Arch. Environ. Health 18*, 428 (1969).
291. Sherr, A. E., Tucker, R. J., and Greenwood, R. A., New plastics absorb at laser wavelengths. *Laser Focus 5 (17)*, 46 (1969).
292. Solon, L. R., Occupational safety with laser beams. *Arch. Environ. Health 6*, 414 (1963).
293. Sykos, M., Safety Considerations of lasers. *J. Amer. Soc. Safety Eng. 8*, 14 (1962).
294. Goldman, L., and Hornby, P., The design of a medical laser laboratory. *Arch. Environ. Health 10*, 493 (1965).
295. Daniels, R. G., and Goldstein, B., Lasers and masers – health hazards and their control. *Fed. Proc., Fed. Amer. Soc. Exp. Biol. 24*, S-27 (1965).
296. Solon, L. R., Aronson, R., and Gould, G., Physiological implications of laser beams. *Science 134*, 1506 (1961).
297. Fine, S., Implementation of procedures and techniques for safe operation of lasers. "Proceedings of the First Conference on Laser Safety" (G. W. Flint, ed.), pp. 97–109. Martin Company, Orlando, Florida, 1966.
298. Stone, H. A. G., Safety with lasers. *Ordnance 50*, 315 (1965).
299. Worden, F. X., Roberts, W. C., and Dunn, J. P., Safety with the laser. *Nat. Safety News 94(4)*, 20 (1966).
300. Kaufman, J. C., Protect your sight from laser light. *Microwaves 5(4)*, 38 (1966).
301. White, D. F., The control of the hazards from lasers. *Ann. Occup. Hyg., Laser Safety Suppl.* 55 (1967).
302. Sliney, D. H., Evaluating health hazards from military lasers. *J. Amer. Med. Ass. 214*, 1047 (1970).
303. Honey, R. C., Hammer safety. *J. Occup. Med. 10*, 245 (1968).
304. Honey, R. C., Laser instrumentation and dosimetry. "Electronic Product Radiation and the Health Physicist BRH/DEP 70-26," pp. 263–272. U. S. Department of Health, Education, and Welfare, Bureau of Radiological Health, Rockville, Maryland, Oct. 1970.
305. Fanney, J. H., Jr., and Powell, C. H., Field measurement of ultraviolet, infrared, and microwave energies. *Amer. Ind. Hyg. Ass. J. 28*, 335 (1967).
306. Sliney, D. H., Bason, F. C., and Freasier, B. C., The measurement of ultraviolet, visible and infrared radiation. *Amer. Ind. Hyg. Ass. J.* 32, 45 (1971).
307. Garbuny, M., "Optical Physics." Academic Press, New York, 1965.

308. Stair, R., Measurement of natural ultraviolet radiation – historical and general introduction. "The Biologic Effects of Ultraviolet Radiation" (F. Urbach, ed.), pp. 377–390. Pergamon, Oxford, 1969.
309. Rockwell, R. J., Jr., Developments in laser instrumentation and calibration. *Arch. Environ. Health 20*, 149 (1970).
310. Deitz, P. H., Probability analysis of ocular damage due to laser radiation through the atmosphere. *Appl. Opt. 8*, 371 (1969).
311. Ochs, R. G., and Lawrence, R. S., Saturation of laser beam scintillation under conditions of strong turbulence. *J. Opt. Soc. Amer. 59*, 226 (1969).
312. Deichmann, W. B., and Stephens, F. H., Microwave radiation of 10mW/cm² and factors that influence biological effects at various power densities. *Ind. Med. Surg. 30*, 221 (1961).
313. Schwan, H. P., and Li, K., Capacity and conductivity of body tissues at ultrahigh frequencies. *Proc. IRE 41*, 1735 (1953).
314. Schwan, H. P., and Li, K., The mechanism of absorption of ultrahigh frequency electromagnetic energy in tissues, as related to the problem of tolerance dosage. *IRE Trans. Med. Electron. PGME-4*, 45 (1956).
315. Schwan, H. P., and Li, K., Hazards due to total body irradiation by radar. *Proc. IRE 44*, 1572 (1956).
316. Schwan, H. P., Microwave biophysics. "Microwave Power Engineering" (E. C. Okress, ed.), pp. 213–244. Academic Press, New York, 1968.
317. Carpenter, R. L., Experimental microwave cataract: A review. 'Biological Effects and Health Implications of Microwave Radiation BRH/DBE 70-2" (S. F. Cleary, ed.), pp. 76–81. U. S. Department of Health, Education, and Welfare, Rockville, Maryland, June 1970.
318. Hirsch, F. G., Microwave cataracts – a case report reevaluated. "Electronic Product Radiation and the Health Physicist BRH/DEP 70-26," pp. 111–140. U. S. Department of Health, Education, and Welfare, Bureau of Radiological Health, Rockville, Maryland, Oct. 1970.
319. Knauf, G. M., The biological effects of microwave radiation on Air Force personnel. *Arch. Ind. Health 17*, 48 (1958).
320. Hoeft, L. O., Microwave heating: A study of the critical exposure variables for man and experimental animals. *Aerosp. Med. 36*, 621 (1965).
321. "Safety Level of Electromagnetic Radiation with Respect to Personnel, ANSI-C95." American National Standards Institute, New York, 1966.
322. Palmisano, W. A., and Peczenik, A., Some considerations of microwave exposure criteria. *Mil. Med. 131*, 611 (1966).
323. U. S. Department of the Army and U. S. Department of the Air Force, "Control of Hazards to Health from Microwave Radiation" (TB-MED-270/AFM-161-7). Washington, D.C., 1965.
324. Swanson, J. R., Rose, V. E., and Powell, C. H., A review of international microwave exposure guides. *Amer. Ind. Hyg. Ass. J. 31*, 623 (1970).
325. Mumford, W. W., Heat stress due to R. F. radiation. *Proc. IEEE 57*, 171 (1969) and "Biological Effects and Health Implications of Microwave Radiation BRH/DBE 70-2" (S. F. Cleary, ed.), pp. 21–34. U. S. Department of Health, Education, and Welfare, Rockville, Maryland, June 1970.
326. Marha, K., Microwave radiation safety standards in eastern Europe. *IEEE Trans. Microwave Theory Tech. MTT-19*, 165 (1971).
327. Dodge, C. H., Clinical and hygienic aspects of exposure to electromagnetic fields. "Biological Effects and Health Implications of Microwave Radiation BRH/DBE

70-2" (S. F. Cleary, ed.), pp. 140–149. U. S. Department of Health, Education, and Welfare, Rockville, Maryland, June 1970.

328. Healer, J., Review of studies of people occupationally exposed to radio-frequency radiations. "Biological Effects and Health Implications of Microwave Radiation BRH/DBE 70-2" (S. F. Cleary, ed.), pp. 90–97. U. S. Department of Health, Education, and Welfare, Rockville, Maryland, June 1970.
329. Frey, A. H., Effects of microwave and radio frequency energy on the central nervous system. "Biological Effects and Health Implications of Microwave Radiation BRH/DBE 70-2" (S. F. Cleary, ed.), pp. 134–139. U. S. Department of Health, Education, and Welfare, Rockville, Maryland, June 1970.
330. Justesen, D. R., and King, N. W., Behavioral effects of low level microwave irradiation in the closed space situation. "Biological Effects and Health Implications of Microwave Radiation BRH/DBE 70-2" (S. F. Cleary, ed.), pp. 154–179. U. S. Department of Health, Education, and Welfare, Rockville, Maryland, June 1970.
331. Korbel, S. F., Behavioral effects of low intensity UHF radiation. "Biological Effects and Health Implications of Microwave Radiation BRH/DBE 70-2" (S. F. Cleary, ed.), pp. 180–184. U. S. Department of Health, Education, and Welfare, Rockville, Maryland, June 1970.
332. McAfee, R. D., Analeptic effect of microwave irradiation on experimental animals. *IEEE Trans. Microwave Theory Tech. MTT-19*, 251 (1971).
333. McAfee, R. D., The neural and hormonal response to microwave stimulation of peripheral nerves. "Biological Effects and Health Implications of Microwave Radiation BRH/DBE 70-2" (S. F. Cleary, ed.), pp. 150–153. U. S. Department of Health, Education, and Welfare, Rockville, Maryland, June 1970.
334. Frey, A. H., Biological function as influenced by low-power modulated RF energy. *IEEE Trans. Microwave Theory Tech. MTT-19*, 153 (1971).
335. Michaelson, S. M., Microwave hazards evaluation: Concepts and criteria. *J. Microwave Power 4*, 115 (1969).
336. Palmisano, W. A., Personal communication (1971).
337. Baillie, H. D., Thermal and nonthermal cataractogenesis by microwaves. "Biological Effects and Health Implications of Microwave Radiation BRH/DBE 70-2" (S. F. Cleary, ed.), pp. 59–65. U. S. Department of Health, Education, and Welfare, Rockville, Maryland, June 1970.
338. Baillie, H. D., Heaton, A. G., and Pal, D. K., The dissipation of microwaves as heat in the eye. "Biological Effects and Health Implications of Microwave Radiation BRH/DBE 70-2" (S. F. Cleary, ed.), pp. 85–89. U. S. Department of Health, Education, and Welfare, Rockville, Maryland, June 1970.
339. Neidlinger, R. W., Microwave cataract. *IEEE Trans. Microwave Theory Tech. MTT-19*, 250 (1971).
340. Carpenter, R. L., and Livstone, E. M., Evidence for nonthermal effects of microwave radiation: Abnormal development of irradiated insect pupae. *IEEE Trans. Microwave Theory Tech. MTT-19*, 173 (1971).
341. Shapiro, A. R., Lutomirski, R. F., and Yura, H. T., Induced fields and heating within a cranial structure irradiated by an electromagnetic plane wave. *IEEE Trans. Microwave Theory Tech. MTT-19*, 187 (1971).
342. Guy, A. W., Analyses of electromagnetic fields induced in biological tissues by thermographic studies on equivalent phantom models. *IEEE Trans. Microwave Theory Tech. MTT-19*, 205 (1971).
343. Schwan, H. P., Interaction of microwave and radio frequency radiation with biological systems. *IEEE Trans. Microwave Theory Tech. MTT-19*, 146 (1971).

344. Rogers, S. J., Radio frequency radiation hazards to personnel at frequencies below 30 MHz. "Biological Effects and Health Implications of Microwave Radiation BRH/DBE 70-2" (S. F. Cleary, ed.), pp. 222–232. U. S. Department of Health, Education, and Welfare, Rockville, Maryland, June 1970.

345. Morgan, W. E. Microwave radiation hazards. *Arch. Ind. Health 21*, 90, 570, 573 (1960).

346. Weiss, M. M., and Mumford, W. W., Microwave radiation hazards. *Health Phys. 5*, 160 (1961).

347. Palmisano, W. A., and Sliney, D. H., Instrumentation and methods used in microwave hazard analysis. Presented at the American Industrial Hygiene Conference, Chicago, May 4, 1967.

348. Silver, S., Microwave aperture antennas and diffraction theory. *J. Opt. Soc. Amer. 52*, 131 (1962).

349. Wacker, P. F., and Bowman, R. R., Quantifying hazardous electromagnetic fields: Scientific basis and practical considerations. *IEEE Trans. Microwave Theory Tech. MTT-19*, 178 (1971).

350. Bowman, R. R., Quantifying hazardous electromagnetic microwave fields: Practical considerations. "Biological Effects and Health Implications of Microwave Radiation BRH/DBE 70-2" (S. F. Cleary, ed.), pp. 204–209. U. S. Department of Health, Education, and Welfare, Rockville, Maryland, June 1970.

351. Coats, G. I., Nelson, C. B., and Underwood, R. G., The dipole/slot radiation pattern and its use in understanding microwave leakage and survey techniques. "Electronic Product Radiation and the Health Physicist BRH/DEP 70-26," pp. 141–158. U. S. Department of Health, Education, and Welfare, Bureau of Radiological Health, Rockville, Maryland, Oct. 1970.

352. Aslan, E. E., Electromagnetic radiation meter. *IEEE Trans. Microwave Theory Tech. MTT-19*, 249 (1971).

353. Moore, R. L., Smith, S. W., Cloke, R. L., and Brown, D. G., A comparison of microwave detection instruments. "Electronic Product Radiation and the Health Physicist BRH/DEP 70-26," pp. 423–430. U. S. Department of Health, Education, and Welfare, Bureau of Radiological Health, Rockville, Maryland, Oct. 1970.

354. Crapuchettes, P. W., Microwave leakage instrumentation. "Biological Effects and Health Implications of Microwave Radiation BRH/DBE 70-2" (S. F. Cleary, ed.), pp. 210–216. U. S. Department of Health, Education, and Welfare, Rockville, Maryland, June 1970.

355. Thompson, R. L., Microwave hazards surveillance and control. "Electronic Product Radiation and the Health Physicist BRH/DEP 70-26," pp. 463-464. U. S. Department of Health, Education, and Welfare, Bureau of Radiological Health, Rockville, Maryland, Oct. 1970.

356. Lamaster, F. S., Equipment Surveys for RF Radiation Hazards. "Electronic Product Radiation and the Health Physicist BRH/DEP 70-26," pp. 420-430. U. S. Department of Health, Education, and Welfare, Bureau of Radiological Health, Rockville, Maryland, Oct. 1970.

357. Glaser, Z. R., and Heimer, G. M., Determination and elimination of hazardous microwave fields aboard naval ships. *IEEE Trans. Microwave Theory Tech. MTT-19*, 232 (1971).

358. "Survey of Selected Industrial Applications of Microwave Energy BRH/DEP 70-10." U. S. Department of Health, Education, and Welfare, Bureau of Radiological Health, Rockville, Maryland, May 1970.

359. Breysse, P. A., Microwave cookers. *J. Environ. Health 29(2)*, 144 (1966).

360. Britain, R. G., and McConeghy, D. J., The impact of the electronic product radiation control program. "Electronic Product Radiation and the Health Physicist BRH/DEP 70-26," pp. 40–50. U. S. Department of Health, Education, and Welfare, Bureau of Radiological Health, Rockville, Maryland, Oct. 1970.
361. Gilbert, H., A study of microwave radiation leakage from microwave ovens. *Amer. Ind. Hyg. Ass. J. 31(6)*, 772 (1970).
362. Eden, W. M., Microwave oven repair: Hazard evaluation. "Electronic Product Radiation and the Health Physicist BRH/DEP 70-26," pp. 159–172. U. S. Department of Health, Education, and Welfare, Bureau of Radiological Health, Rockville, Maryland, Oct. 1970.
363. Voss, W. A. G., Microwave hazard control in design. "Biological Effects and Health Implications of Microwave Radiation BRH/DBE 70-2" (S. F. Cleary, ed.), pp. 217–221. U. S. Department of Health, Education, and Welfare, Rockville, Maryland, June 1970.
364. Taylor, J. T., The evaluation of hazards from microwave ovens (to be published).
365. Peppers, N. A., and Honey, R. C., Private communication.
366. McCullough, E. C., Qualitative and quantitative features of the clear day terrestrial solar ultraviolet radiation environment. *Phys. Med. Biol. 15(4)*, 723, Oct. 1970.
367. MacDonald, E. J., The epidemiology of skin cancer. *J. Invest. Dermatol. 32*, 379 (1959).
368. Peacock, G. R., Private communication (1971).
369. Skeen, H. S., Private communication (1972).

Ionizing Radiation

HARRY F. SCHULTE

Problems presented by the increasing use of radiation and radioactive materials are of concern to everyone, including the industrial hygienist. Public interest in the subject is high and, because of the complexity of the subject, misunderstandings frequently arise at the public level. Honest differences of opinion probably are no greater in this than in any other field of public health, but they seem to create greater public concern and perplexity. Partly as a result of this interest, developments in the field are rapid and the output of technical information is immense. Certainly, *Health Physics,* the journal of the Health Physics Society, is by far the major source of publications in the field, although by no means the only one. This journal published nearly 4400 pages of technical information over the past 3 years. It should be emphasized that the industrial hygienist is only concerned with part of the field of health physics, although the boundaries of his interests will vary greatly, depending on his employer and his responsibilities. In this review, an effort has been made to focus on developments which directly involve the working environment.

Major sources of information were reviewed in "Industrial Hygiene Highlights"[169] and there have been no important changes. A new international journal, *Aerosol Science*, has appeared which will include radioactive aerosols, among the many within its scope. The English translations of *Staub* (German) and *Hygiene and Sanitation* (Russian) contain frequent articles dealing with radioactive materials. *Nuclear Safety*, published by the U.S. Atomic Energy Commission (USAEC), publishes frequent articles of interest, mostly literature reviews on specific topics. *Radiological Health Data*, published by the Bureau of Radiological Health, presents much specific information and quantitative data, although most of it is not directly related to the work of the industrial hygienist.

Just prior to the period covered in this review, the International Radiation

Protection Association (IRPA) was formed and held its first congress in Rome. The Proceedings of that congress are now available and the IRPA Constitution has been published.[1] The second IRPA Congress was held in Brighton, England, in 1970, and the published abstracts indicate the high level of interest among many nations.[2] Another source of foreign information, particularly in radiobiology, is the series of translations from the Russian published under the general title, "The Toxicology of Radioactive Substances," of which five volumes have appeared so far.[3] For popular expositions of topics related to atomic energy, the series of booklets on "Understanding the Atom," published by the U. S. Atomic Energy Commission, are truly superb.[4] Each deals with a specific topic and is written by an authority in the field. While intended for popular consumption, they contain enough technical information to make them very useful to anyone except a specialist in the particular field covered.

Again, in dealing with this large volume of available material, I have tried to treat as many topics and subtopics as possible, in view of the diversity of interests among industrial hygienists. As a result each topic is covered very briefly, but the references are rather complete to permit the reader to pursue his own special interests in depth. While the reviewer's own interests do influence the treatment of topics, the material covered is a fair representation of the areas of new developments.

APPLICATIONS OF RADIATION AND RADIOACTIVE MATERIALS

While this reviewer is not aware of any remarkably new applications of these materials, processes involving radiation continue to grow and their use has proliferated many branches of industry. A simple example, which is close to the industrial hygienist, is the use of radioactive foils as an ionization source in gas chromatograph detectors. The use of nuclear energy for the generation of electrical power is increasing steadily and new reactor concepts are being introduced. The fast breeder reactor is now under development in several places. This uses uranium and plutonium for fuel and converts the uranium into plutonium as it supplies heat for power generation. Conventional reactors work almost entirely by the fission of uranium-235, which comprises only 0.7% of natural uranium. By proper placement of natural uranium to absorb fast neutrons from fission, uranium-238 can be converted to plutonium which, itself, can be used as a fuel in the fission reaction. In this system of "breeding," all of the natural uranium can be used to produce heat. A liquid metal, usually sodium, is used as a coolant and heat-transfer medium in this type of reactor.

Partly because of the work on the fast breeder and its important potential, interest in plutonium has increased greatly. Plutonium health hazards are similar to those of radium, and new techniques have had to be developed to work with larger quantities of this element. The plutonium used in power production is largely plutonium-239, although other isotopes are also formed and their concentration increases with the length of time the plutonium remains in the reactor. One isotope, plutonium-238, is finding increasing use as an isotope power source, ranging from powering the heart pacemaker to serving as a portable power source on the moon. This isotope has a specific activity more than 250 times as high as plutonium-239, and hence it is more hazardous and requires tighter controls. The quantities of neutrons and γ rays produced by this material also are greater, requiring shielding in addition to enclosure for certain operations. The neutron emission can be reduced drastically by complete removal of traces of impurities and such purification is a necessary prelude to the use of plutonium-238 in such applications as the pacemaker and the artificial heart.

Natural thorium also can be converted to the fissionable isotope uranium-233 in a suitable breeder reactor, but this process has not been studied or developed greatly as yet. It is certain to become important some day and more knowledge of thorium health effects will be required. Most power reactors presently in use are the pressurized water type, but the boiling water reactor is being used in many new installations. Advanced reactor types, such as gas-cooled reactors, are being studied; the gas-cooled reactor has been the standard power unit in the United Kingdom for many years. More distant is the use of the fusion of light elements as a source of heat and power, but work is continuing steadily on this in many places. At present, it seems to offer essentially unlimited power with little pollution potential. Technologic problems are extremely difficult in this field but progress is being made.

The peaceful uses of nuclear energy are many and are distributed throughout a variety of fields. A booklet on this subject, issued by the USAEC, outlined many of these applications and gives the reader some appreciation of the possibilities.[5] The Plowshare Program is concerned with the peaceful applications of nuclear explosions, such as the stimulation of natural gas flow, by shattering underground rock structures. Many projects are under study in this program, but it has also generated much controversy, as might be expected.

URANIUM MINING

In the past 3 years, there have been many advances made in the control of the radiation hazard in underground uranium mines. Nevertheless, there are still gaps in our knowledge and even more in the application of knowledge in this field. From the standpoint of making knowledge available to a wide audience, the special issue of *Health Physics* dedicated to Duncan Holaday is outstanding.[6] This issue is wholly devoted to this subject and contains many valuable papers. However, this is not the sole source of information and the number of publications was extremely large during this period. For those wishing to acquaint themselves with the history and background of this problem, the lead article by Holaday in the special issue is an excellent source.[7] Later a contribution from Czechoslovakia added information relating to the Joachimsthal mines.[8] Much technical information is given by Evans in a review of the physics aspect of the background.[9] Related to the latter is a plea by Morken that the working level concept is defective and not a satisfactory index of exposure or hazard.[10] This criticism has largely been ignored and practically all studies are based on measurement of working level. The difficulties in selecting a safe working level and the range of possible values has been discussed by Tompkins of the Federal Radiation Council.[11]

It was noted 3 years ago that a decision on the working level standard for mine operation was dependent on receipt of additional epidemiologic data.[12] Much of these data are still not available and the decision on lowering the standard below 12 working level months annually has been deferred until July 1, 1971. Data published by Lundin give some indication of biologic effects at relatively low exposure levels, but this interpretation has been disputed by others.[13] Lundin's work does bring out the fact that smoking is an extremely important factor in the production of lung cancer when added to that of radiation exposure. Two British papers on exposure to radon daughters and incidence of lung cancer among hematite miners are important and point to the need for more studies among nonuranium miners.[14,15], The subject of the epidemiology of lung cancer among uranium miners is covered in detail in the review on Epidemiology in this volume.

Attempts to calculate the magnitude of the radiation dose to a critical cell layer from radon daughter deposition in the respiratory tract are still receiving study, although only one publication on this phase was noted in the open literature.[16] This type of calculation probably has little application in deriving acceptable levels of daughter concentrations in the mines, but is very helpful in understanding the mechanisms involved and in pointing to factors to be controlled in reducing exposure. Information that may be useful in such a calculation was found in the work of Raabe who investigated the probable particle size distribution from adsorption of the daughters on aerosols[17,18] and in the total respiratory deposition measurements made by workers at Colorado State University[19] and the New York Health and Safety Laboratory.[20]

Major emphasis in estimating exposure of miners has been on air sampling. Since the original Kusnetz and Tsivoglou methods[168] of measuring working level by counting a sample after a fixed sampling period followed by fixed waiting periods, there have been many significant improvements and variations.

Some of these involve utilizing total counts rather than count rate measurements, others use an α spectrometer, some involve single counts and others follow the decay curves.[21-26] While there are still controversies about these various details, and improvements are valuable, the investigator does have the means at hand now to make good estimates of the working level at any one time and place. Since it is the radon which produces the daughters, and its concentration cannot be ignored as an exposure source, there is still interest in improving methods of measuring the concentration of this gas. Use of a portable ionization chamber has been described[27] and a method of using a charge to collect the daughters from a trapped filtered sample of gas has been incorporated into a usable instrument.[28] An older idea of drawing air through two filters spaced some distance apart has been studied and made practicable for use in mines.[29]

Such air sampling merely gives results at a single location at a single time. In order to estimate the exposure of each miner, an enormous number of individual air concentrations must be measured and these results must be related to the movements of the individual miner. A great deal of effort is now being devoted to this and much could be gained if a satisfactory device could be built to give the total daily integrated exposure at a given place or, even better, if a dosemeter device could be produced which the miner could wear. A type of film badge, utilizing a plastic plate in which α tracks are made visible by etching, has been described and tested in mines.[30,31] Most other devices are based on drawing air through a collector located close to a radiation-sensitive material, like lithium fluoride, which serves as a thermoluminescent detector.[32] A number of these devices have been tested and none of them are quite satisfactory as yet.[33] The measurement of unattached radioactive atoms is still of interest, although no additional data have been published on work in this country. A new method of making this measurement has been described from the United Kingdom[34] and some actual measurements in mines have been reported in Japan.[35] Although widely variable results have been reported for this measurement, it is hard to see how its significance can really be appraised at the present stage. Other factors of unknown significance, but presently presumed unimportant in mine exposures, are the concentration of long-lived daughters and the intensity of direct γ radiation.

Another approach to the measurement of total individual exposure has been to measure the concentration of one of the longer-lived radon daughters in the body. Most studies have concerned the measurement of lead-210, although polonium-210 also has been looked at. Using material obtained from autopsies, several investigators have analyzed bones for lead-210 and tried to relate these to exposures in terms of working level months as obtained from other measurements.[36-38] There seems little doubt from these data that there is a correlation, although it is not an exact one. One of the factors complicating the relationship is that all lead-210 in the body does not originate from the decay of inhaled radon daughters. Inhaled radon itself is carried by the bloodstream where it decays to lead. Lead is also inhaled directly from the decay of daughter products in air and lead-210 occurs in the mine dust itself.[39-42] In any case, this approach seems very promising, especially where it can be applied *in vivo.* A sensitive method for lead-210 in urine has been reported[43] and urine assay for both lead-210 and polonium-210 has been investigated in uranium miners with somewhat discouraging results.[44] A more promising method seems to be the use of body or *in vivo* counters for the direct measurement of lead-210 in bone, usually the skull.[45]

While epidemiologic and dosimetry studies continue, a great deal of useful effort is going into the reduction of exposures in the mines. It has been observed that falling barometric pressure increases the release of radon into the mine as expected.[46] A theoretical study, followed by a practical demonstration, showed that working levels can be significantly reduced by rapid mixing of the air within the working volume.[47,48] Workers at the Bureau of Mines have studied this problem and, in a booklet, have reported on the effects of pressuring, sealing, dilution, filtering, radon removal, and mine planning.[49] The most widely used method at present is dilution ventilation, but this is limited by the high air velocities required in tunnels. Careful planning of mine layout seems likely to be the most helpful technique in new mines. Respirators find some limited application and these have been the object of at least one study.[50] The Bureau of Mines has established a new amendment to Approval Schedule 21B for such respirators.[51] A powered filter respirator also has been developed for use in uranium mines.[52] Much additional work is going on in this field, but the results have not been published as yet. Application of all methods of control is producing a gradual lowering of exposure levels in most mines, but the limits which can be achieved vary from one mine to another. Complete reliance on dilution ventilation alone will be impossible in many mines, and individual exposures must be restricted or new or additional control measures introduced. The New Mexico State Health Department has leased an inactive uranium mine in which experimental work on the evaluation and control of hazards can be done.

Two problems have arisen as outgrowths of the uranium mine problem. Tailings piles from uranium ore mills contain radium and other uranium daughters. Considerable concern has been expressed over the possibility that radon emitted from the piles will be hazardous to the nearby communities. A portable instrument for collecting and evaluating samples from this source has been developed and results, to date, indicate that this hazard is minimal.[53,54] However, in Grand Junction, Colorado, one tailings pile was used as a source of fill material in and around buildings. An intensive study of more than 4000 buildings is now going on to evaluate the possible hazard and to develop appropriate action, if needed.

ESTIMATION OF EXPOSURE TO EXTERNAL RADIATION

This subject includes exposure to radiation from sources external to the body such as x rays, γ rays, and particulate radiation, principally, electrons, and neutrons. Under special circumstances, protons and heavier particles also are to be evaluated as external sources but, because of their low penetrating power, α particles are practically never considered a source of external radiation. Judging by the number of publications, this specialty is certainly one of the most active in the whole field of radiation protection, reflecting the heavy emphasis on evaluation and records-keeping in the health physics profession. The needs and problems involved in personnel radiation dosimetry were discussed in a symposium on the subject in 1967, and a summary has been published in *Health Physics.*[55] The introduction of thermoluminescent detectors (TLD) and the radiophotoluminescent (RPL) detectors several years ago has caused a real change as these replace film badges in many applications. The principles of these detectors were discussed in the previous review in "Industrial Hygiene Highlights."[169] Since that time there have been further advances, especially in the investigation of new substances useful as detectors. Beryllium oxide,[56] strontium fluoride,[57] and aluminum oxide[58] were found useful and many others were screened. Calcium fluoride, calcium sulfate, and lithium fluoride have been incorporated in plastic to make a product capable of fabrication into a wide variety of useful forms.[59] Bibliographies on both TLD[60,61] and RPL[62] have been published. Those on TLD include 678 references and that on RPL include 150, which are in addition to the previous list that contained 253 references.[63] This topic is too large and diverse for a summary and the reader must refer to the literature. Unfortunately, there has been no recent critical review of the subject. Among newer developments are the use of TLD for high dose-rate dosimetry,[64] for electrons,[65] and for personnel neutron dosimetry.[66] Many installations have switched to TLD in place of films, and descriptions of the complete systems have been described for the National Reactor Testing Station[67] and at Hanford.[68,69]

Significant, but somewhat fewer, results have been published on work with glass dosimeters using RPL. A system which is receiving increasing attention is thermally stimulated exoelectron emission (TSEE). This is similar to the TLD except that instead of measuring the light emitted when the dosimeter material is heated, emitted electrons are measured. The electrons may be measured with a Geiger tube or similar device. Emission of electrons may be stimulated by light (optical), as well as by heat, or ultraviolet light may be emitted following optical stimulation.[70] Such detectors can be made very sensitive; they respond differently to particulate radiations of varying linear energy transfer, and they can be made very thin for microdosimetry.[71-75]

Estimation of high level exposures received as a result of accidents is important, both in planning therapy and in documenting exposure for comparison with clinical observations. A wide variety of systems has been used, including chemical and TLD dosimeters,[76] electron spin resonance in teeth and hair,[77] and direct counting of phosphorus-32 in hair.[78]

Measurement of exposure to neutrons or to high energy particles around accelerators, or in outer space, are extremely difficult because of the complexity of the interactions of such particles with matter, such as tissue. Such measurements usually are outside the field of the industrial hygienist. A manual on neutron protection is in preparation by the National Council on Radiation Protection and Measurements,[79] but even this will require specialized knowledge for its use.Detection of the presence of neutrons, or estimation of their flux intensity (monitoring) as distinguished from dose measurement, is closer to the industrial hygienist's field. Here, there is still a need for more satisfactory instrumentation.

ESTIMATION OF RADIATION EXPOSURE FROM INTERNAL SOURCES

Radioactive materials may be taken into the body by inhalation, ingestion, or skin penetration through wounds or the intact skin. Their biologic effect then depends on their physical and radioactive properties and also on their residence time in the body, site of deposition, and rate and path of translocation to various organs. Thus, any attempt to predict biologic effects from environmental and personnel measurements requires a detailed knowledge of the radiobiology of the substance absorbed. There are three general methods of estimating such exposure, (*a*) direct measurement of radiation from the material while present in the body, (*b*) measurement in excreta, and (*c*) measurement of concentrations in the environment. All of these are complex and based on many assumptions which are difficult to confirm. In most cases, the assumptions used necessarily apply to an average or "standard" man and numbers used may diverge considerably from those correct for the individual being studied.

At first sight, it may seem that direct measurement involves no assumptions and it is true that this method probably is the most accurate where it can be applied. In this method, an instrument external to the body measures the amount of radiation being emitted from the body and so gives results which can reveal how much radioactive material is deposited in the body. For strong γ emitters and where a single isotope has been deposited the results are excellent, although there are still assumptions about the precise location in the body and the effect of shielding by body tissues. Crystal counters, which have largely replaced liquid scintillation counters for *in vivo* counting, can be used to localize the deposition site and by energy discrimination can provide a γ-ray spectrum of the radioactivity. This permits separate identification and estimation of several separate isotopes.

The greatest need and the most significant advances have been made in the detection and analysis of emitters of low energy γ rays, such as plutonium in lungs. Use of a thin cesium iodide crystal coupled to a thick sodium iodide crystal was reported in 1967.[80] Since then advances in this technique, which involves pulse shaping, have improved the sensitivity and reduced backgrounds and minimum detection levels of 4 nCi are being reported for plutonium-239.[81] Proportional counters also have improved and in the United Kingdom and Japan a detection level of 8 nCi has been reported.[82-88] Because of the difficulty of measuring the 17 keV x ray from plutonium-239, the 60 keV radiation from americium-241 is frequently measured, if the ratio of the two isotopes in the parent material is known. A recent study on implanted material indicates that translocation from the implant site causes changes in the ratios of these isotopes.[85] Thus, an error might be introduced in estimating plutonium in organs resulting from a contaminated wound. For pure β emitters like strontium-90, measurement of bremsstrahlung can be used to detect body burdens of about 30 nCi.[86] Simplified whole-body counters for more penetrating radiation, using shielded chairs[87,88] and a rotating unit for localization of deposition, have been found useful.[89]

Whole-body counters are bulky and expensive and their use is limited to certain materials and conditions. Most estimation of exposure to internal sources is done by the analysis of excreta, particularly urine. This involves the use of some type of mathematical model, if the urinary excretion rate is to be related to the amount of material resident in specific tissues and organs.[90,91] Such models must include assumptions about rates of transfer from respiratory tract to blood by various routes and rates of transfer from one organ to another and from blood to kidney to urine. Specific data on these, where available, have been published by the ICRP[92] and this compendium is useful but illustrative of the difficulties and gaps in required information. It is especially hazardous to estimate lung exposures from urinary excretion data, particularly for insoluble materials. Analysis of feces is more useful as an indication of such exposure but is seldom of quantitative value. The use of lung and whole-body counters have revealed the serious errors of such calculations from excretion data.[93-97] However, urine assay is still very valuable as an indication of exposure and, as such, it will continue to be used widely. In the use of urine assay, particularly for long-lived α emitters, it is necessary to use extreme caution in the interpretation of results since apparent excretion rates may be influenced by many factors such as metabolic changes, contamination, sample collection methods, and fluid intake.[98]

Accumulation of long-lived emitters in the body is usually the result of a series of acute exposures rather than continuous chronic exposure, since most operations involving these materials are carried out in almost completely closed systems. Analysis of tissue samples available at autopsy is the only way in which present methods of estimating deposition can be checked. Some data of this nature already have been published and a U. S. Transuranium Registry has been organized and funded to see that such information is obtained and made available.[95,99,100] It is apparent that most interest has centered on the long-lived α emitters, particularly plutonium-239 and, more recently, plutonium-238. Data are still needed on any human exposure cases, such as the 27-day effective chest half-time reported for seven persons exposed to ruthenium-103.[101] Obviously, this field is closely related to radiobiology and results on animal data are reported in that section.

The use of urine assay for the quantitative assessment of exposure to tritium oxide always has been considered the most successful application of this technique. Since the tritium is presumably in the body water and this is identical to the water in the urine, the urine assay is effectively a measure of a sample of the body. Recently, some concern has been expressed over the possibility of tritium entering into organic molecules where it may remain longer than the normal turnover time of body water. Measurements on deer living near a source of small amounts of tritium contamination suggest that there may, indeed, be some possibility of this, although the deviations are quite small.[102] This also should show up as a long component at the end of an excretion curve following tritium exposure. Little data have been published on this and the question still remains open and a matter of concern.

The estimation of exposure by means of environmental measurements, particularly the analysis of air samples, is almost universally used and such data usually are combined with urine assay data to produce the best estimate of exposure. However, air sampling is more closely related to monitoring or the determination of the probable upper limits of inhalation of radioactive materials. It also is closely related to the evaluation of the control system designed to prevent the release of material into the air. For these reasons, this topic is covered in the section on Hazards Evaluation. It should be remembered that the primary value of any measure or estimation of exposure, external or internal, is as a check on the control system.

HAZARD EVALUATION

Evaluation of radiation hazards is closely related to exposure estimation as previously discussed. This section deals largely with measurements in the working environment rather than laboratory studies. Measurements in or on people comprise a part, but only a part, of the evaluation process. Frequently in a single installation, a wide variety of exposure types is encountered and sometimes these are combined in a single operation. Thus, in processing the transplutonium elements operators are exposed to α, β, γ, and neutron radiation, including both internal and external exposures to materials emitting these. Interest in such materials continues to grow and the problems involved have been described in reports on two installations.[103,104] One of these materials is californium-252 which emits neutrons by spontaneous fission. This isotope then can be fabricated into a compact neutron source, useful in many applications, including therapy. Formulas for dose rate, as a function of distance and attenuation by absorbers, have been worked out for users.[105]

External radiation problems continue to arise, some from old operations or some from those which are new or newly studied. x Rays have been used for a long time for analyzing materials for various elements and this application is increasing with thousands of units in use in this country. A survey of experiences with the units indicates that overexposures and injuries continue to happen.[106-108] In fact, such operations constitute a major source of acute radiation injuries as distinct from overexposures. Very high-intensity pulsed x-ray machines have been in use in research organizations for nearly 10 years, and now they are being used in industry to take flash x-ray photographs. These present unique problems in measuring radiation intensity in the working area and in the design of shielding and other protection.[109] Television receivers have been noted as an x-ray source but this is largely a problem of consumer exposure; however, large television projectors may be a serious source of radiation to their operators.[110] Evaluation of this hazard is straightforward but its presence may be overlooked. Another easily neglected source of radiation comes from adhesive tapes frequently used to hold objects being irradiated for neutron activation. The presence of trace materials can cause emission of appreciable induced radiation and varies considerably among various brands and types of tape.[111]

Combinations of various types of exposure are not uncommon even in relatively simple operations. As tritium has been substituted for radium in luminous dial painting, the degree of hazard from this operation has decreased considerably. However, it has not disappeared entirely and relatively few reports are available on studies of this operation. In one published report, it was noted that the most serious route of tritium into the operator is via the intact skin and that no satisfactory protective gloves are presently available.[112] Another new application of radioactive materials is as ionization sources in detectors in chromatographs. Some of these emit appreciable radiation when removed from the apparatus for repair, while others lose radioactive material into the work area.[113] A long used method of evaluating certain hazards from radium sources is leak-testing but there are various methods of performing this test, all not equally reliable. A comparative study of eight test methods showed the charcoal method to be best and only four of the methods had the required sensitivity.[114]

For the evaluation of internal hazards, air sampling is the principal technique and, where possible, it is combined with urine assay or whole-body counting.

While this seems perfectly straightforward and much like evaluating chemical hazards, it has its own difficulties. Since the effects of inhaled radioactive materials are only apparent after long periods and since such effects are never trivial or reversible, a high degree of reliability is required. It is seldom sufficient to collect occasional air samples using a survey technique and, hence, samplers must run continuously at appropriate locations, if indeed they are required at all. The International Commission on Radiological Protection has issued a guide on monitoring the workplace which gives some suggestions on where air sampling is actually required.[115] This guide suggests that surface monitoring may be sufficient in many areas.

Surface monitoring is another technique of hazard evaluation which is seldom discussed in the literature since it appears so simple. Reports on the significance of surface monitoring and on correlation of its results with other techniques would be welcome and valuable. By means of a simple (or complex) instrument for detecting the presence of radioactive contamination on floors, furniture or other surfaces, a great deal can be learned concerning the potentiality of an internal hazard. Except in the case of strong γ-emitting material, the mere presence of radioactive material on a surface does not present a hazard. However, the unexpected detection of such material is an indication of a failure of control methods and the material may have deposited from air or can be resuspended into the air. Thus, there is a close connection between airborne material and surface deposits. Surface monitoring usually is simple and, in the complete absence of surface contamination, an airborne hazard is unlikely. Where very hazardous material is handled, both air and surface monitoring is required.

Plutonium is an α emitter and any instrument for detecting its presence, based on this α-ray emission, must be held very close to the surface or the plutonium will not be noticed. The presence of a film of oil or water will completely block the α radiation and the deposit will escape detection. If in addition to detection it is desired to estimate the quantity present, accurate measurement of the α radiation for this purpose is nearly impossible. This was the situation that confronted the surveyors following the crash of a bomb-carrying Air Force plane near Thule AFB in Greenland in January 1968. Plutonium was scattered on the ice in a mixture of gasoline, snow, and ice. Fortunately, a newly developed meter capable of detecting low energy x rays was made available when the use of α survey meters proved impracticable under the difficult circumstances.[116] Plutonium and americium emit small quantities of such radiation, as do most α emitters, and this radiation will penetrate oil, water, and such materials. This type of meter has since proven useful in many applications.

Where air sampling is used for routine hazard evaluation, many samplers are often used and the amount of data requiring evaluation is enormous. Computers and automation have been used in an attempt to cope with this and several systems have been described.[117,118] These do not solve the basic problem since they only assist in collecting, processing, and storing data and do not give much aid in the interpretation of the assembled data. For this, more than concentration measurements are required. Workers in the United Kingdom atomic energy establishments have been particularly advanced in studying all phases of the problem including particle size, presence of dominant particles, data from personal samplers, and other measurements. Reports of their work are covered in descriptions of the air sampling programs at various locations[119] and in several papers describing special techniques.[120-123] Descriptions of air sampling systems and techniques of other countries and installations are described in proceedings of a symposium on Assessment of Airborne Radioactivity convened by the International Atomic Energy Agency in 1967.[118]

Some very sophisticated sampling devices have been developed.[124] Nevertheless, developments in this field are somewhat disappointing in view of the intense interest prevailing 3 years ago. Most emphasis is still on hardware and not enough on the interpretation of results or on defining the information required. It seems reasonable to expect that careful statistical treatment of data could lead to the collection of fewer but more significant samples. The large amounts of accumulated data are a challenge to the industrial hygienist since in no other field than radiation protection is such a mass of material available. The results of developing a method of handling these data also should be of great value in the evaluation of airborne chemical hazards such as coal, silica, lead, beryllium, and others. Possibly the air pollution evaluation people who face a similar task could be of assistance.

Tritium is a separate problem since it is gaseous rather than particulate. Available air monitoring instruments are mostly flow-through ionization chambers and are not specific for tritium. By means of filters, ion traps, and compensating ion chambers, most interfering sources, except radioactive gases can be eliminated. High sensitivity can be obtained, if necessary, by the use of large ion chambers. There have been no outstanding developments in this area. Radioiodine is still sampled by means of adsorption on charcoal with few recent improvements reported.

HANDLING OF RADIOACTIVE WASTES

Liquid and solid wastes are created whenever radioactive materials are produced or processed and the amount of radioactivity per pound of waste varies widely. While numerous methods of handling, storage, and disposal are used in dealing with these wastes, nothing can be done to increase the rate at which the radioactivity diminishes. These rates vary from half times of minutes to thousands of years and, thus, the problems created are varied. While the prospects of disposing of plutonium wastes having a half life of 25,000 years seems incredibly difficult even to consider, it should not be forgotten that such stable materials as mercury and arsenic have infinite lifetimes and we are beginning to face this implication. Reactors of all sorts generate enormous quantities of highly active wastes, and the future of nuclear power may be dependent on the solution of the waste disposal problem. Storage of highly active wastes in excavated salt formations deep underground is the system presently being favored.[125] These are naturally dry, stable, unaffected by radiation, and located in areas of little seismic activity. They also are capable of dissipating the heat emitted as a result of the radioactivity.

Gaseous or airborne wastes are more likely to concern the industrial hygienist who deals with similar air cleaning processes for nonradioactive materials. For particulates, filtration is still used almost exclusively. High efficiency air filters are not a recent development for this purpose, but there has been increased attention given to assuring the integrity of filter systems. Testing of filter systems "in-place" after installation and at periodic intervals is a necessity in systems where efficiencies of 99.9% and higher are a requirement. Leaks in and around filters must be detected and repaired, especially where plutonium is handled. Filter banks in series are now being used to a degree, almost unheard of a few years ago. Partly, this resulted from a disastrous fire in a plutonium facility which directed attention to the possibilities inherent in the loss of a single filter bank. Methods of protecting filters from fire, heat, and shock are under study. Tighter restrictions on emission limits are forcing attention on containment, recycling, total enclosure, and modification of processes to reduce airborne wastes. Recirculation of room air, which is anathema to the industrial hygienist, may be reconsidered when the air cleaning system consists of double or triple filtration with 99.9% efficiency at each filter stage.

Gases such as tritium, krypton, and xenon still pose serious problems of removal, and research is being done on these problems. In 1968, the International Atomic Energy Agency sponsored a symposium, Treatment of Airborne Radioactive Wastes, and the proceedings of this symposium are now available as a source of much detailed information.[126] The U. S. Atomic Energy Commission and the Harvard Air Cleaning Laboratory sponsor periodic air cleaning conferences and proceedings of these also are available.[127] The high-efficiency filter remains the mainstay of air cleaning in nuclear installations and much experience and data on such systems have been accumulated.[128] For radioiodine, activated charcoal filters usually are used and their efficiencies are tested with radioiodine. A recent paper suggests that the efficiency decreases at low iodine concentrations and, hence, filters should be tested using the same total iodine concentration that they are expected to encounter in service.[129] A manual on the "Management of Radioactive Wastes at Nuclear Power Plants," based on worldwide experience, has been published by the International Atomic Energy Agency.[130]

RADIOBIOLOGY

Radiobiology is a very large field and bears much the same relation to radiation protection as toxicology and pharmacology do to protection against chemical agents. As in the latter case, only a small part of the field of radiobiology is of concern to the industrial hygienist. The greatest need is for information on tissue distributions and effects following inhalation, and there are comparatively few such studies reported, but those that are available provide excellent information. A second need is for information on effects following wounds which may be simulated by subcutaneous administration. Studies following intravenous injection must be applied cautiously in relation to actual exposure situations or standards setting.

A considerable variety of nuclides have been subject to radiobiological investigation. Europium was studied following a single intravenous injection in rats, revealing a comparatively long residence time in bone.[131] Radioiodine inhaled by humans was completely absorbed, with 40% going to the thyroid.[132] Silver-110m was administered to several mammalian species by oral, intravenous and intraperitoneal routes and was largely excreted in the feces.[133] Cesium-137 and strontium-90, given intraperitoneally to rats, showed altered distribution patterns at high dosages.[134] Zirconium-95 and niobium-95, given to rats by oral and intravenous routes, showed gastrointestinal absorption higher than expected and possible effects on the fetus at MPC levels.[135] Promethium-143, given intravenously to humans, showed a very long residence time in the body.[136] Cerium-144, inhaled by mice, was excreted principally in the feces and insoluble forms remained almost entirely in the lung, while soluble forms translocated to the liver and skeleton.[137] A study of radioyttrium distribution in organs, following inhalation of various mixtures of stable and radioyttrium, showed that the distribution was affected by the relative amount of stable yttrium used.[138] A similar effect was noted with radiomercury.[139]

A large number of persons are exposed to uranium dusts under conditions where considerable masses of material can be inhaled. Fortunately, excellent studies continue to be made on this material. Fifteen chronically exposed workers in an industrial plant were studied by whole-body counting and excretion analysis. The data suggest a maximum permissible lung burden of 30 mg that corresponds approximately to a daily urinary excretion of 20–50 μg.[140] The report of results from a five-year inhalation study of uranium oxide with monkeys, rats, and dogs yields a wealth of valuable data which cannot be summarized but suggest that long-term continuous exposure, even at MPC levels, can produce radiation effects in the respiratory tract.[141] The subject of uranium uptake from the gastrointestinal tract is still active and a recent study confirms the revised figure adopted by the ICRP.[142] The fact that insoluble substances deposited in the lung by inhalation are carried to the lymph nodes has been recognized in toxicology for many years. This fact is repeatedly confirmed for radioactive materials and quantitative information is being accumulated.[141,143-145]

Little additional data on the inhalation of plutonium have been published despite the continuing need for it. One study with rats clearly demonstrated clearance of inhaled plutonium oxide by macrophages.[146] Highly active plutonium oxide beads of about 200 μm, important because of their use in isotope power devices, are receiving study.[147] A preliminary model of the possible tumorigenicity of such particles indicates large errors can be introduced, using the usual dose averaging method for estimating tissue dose.[148] The problem of "hot particles" is a very old one and more studies of this sort are needed. Thorium, which has great potentialities as a source of nuclear fuel, has been studied very little recently by radiobiologists. Russian work in this field, using dogs, rabbits, and rats has recently been published in English. Their studies support the concern over the long retention time of thorium dioxide and the serious biologic changes leading to tumor production long after deposition.[149]

Little additional information of use to the industrial hygienist is available with respect to radiobiologic effects of external radiation sources. The fact that such radiation produces detectable chromosome aberrations is being studied as a possible early indication of radiation damage.[150] It can be useful in detecting moderate or high exposures to γ rays and neutrons.

MEDICAL ASPECTS

No attempt is made to review the medical literature in relation to the effects, uses and other aspects of ionizing radiation. There have been no outstanding changes in the methods of treating cases of radiation exposure, but more physicians are now trained in the field. Even in an installation where high levels of radiation can be encountered, the occupational physician finds that almost all of his work deals with nonradiation problems,[151] unless he also directs the preventive programs of health physics and radiation protection. In such installations, training and preparation for the possibility of serious radiation accidents are required but, fortunately, seldom used. Where radiation is only one hazard among many others, there is apt to be more stress on its importance than can be justified in a balanced program. To quote from a physician of a large university, "There are valid reasons for a physical examination program in the practice of modern preventive medicine but there is no specific reason why radiation exposure, when there is good environmental control, should be used as the excuse for initiating and continuing such a program."[152] The International Atomic Energy Agency has prepared a manual on "Medical Supervision of Radiation Workers."[153] It is a useful handbook for physicians in a nuclear energy industry, but only a very small fraction of the manual is devoted to medical aspects. Even there the topics covered relate to conditions that would be encountered extremely rarely.

Fortunately, there have been relatively few major radiation accidents recently where serious clinical cases demand the best in medical care and also add to the knowledge of effects of severe overexposure. Nor have there been any reports of drugs or chemicals useful in mitigating the serious course of illness from radiation exposure. Chelating agents are still used in certain cases for the accidental intake of radioactive materials. DTPA administered to dogs following inhalation of the oxides of cerium-144 and praseodynium-144 reduced the amount retained in the body as compared with controls.[154] A study on rats given plutonium oxide and DTPA intraperitoneally showed diminished phagocytic action with DTPA, suggesting a possibly adverse effect on lung clearance if DTPA is used following inhalation exposure.[155] A study of DTPA effects following implantation of plutonium and americium oxides subcutaneously in dogs revealed a reduction in body retention of the americium but little effect on the plutonium.[156] Further work is still needed in this field to assist the physician in planning therapy.

An alternative procedure in inhalation cases is suggested by experiments on rats[146] and on dogs[157] using lavage or saline washing of the lungs following inhalation of plutonium oxide dust. The animals tolerated the treatment well and the authors suggest its possible therapeutic use, although no one has applied it to humans as yet.

The physician is often called on to testify in workmen's compensation cases on the question of whether radiation was the cause of a disease of the claimant. Sagan, after reviewing all the complicating factors involved in such a judgment concludes, "that often no witness, or group of witnesses, no matter how sagacious, can offer any but highly arbitrary judgments."[158] Medicolegal problems in radiation cases remain very difficult to handle.

INSTRUMENTATION

Instruments used in detecting and controlling ionizing radiation are extremely varied and many of them are built by electronic specialists closely associated with the health physicist. Many of these are built to supply a special need such as monitoring a waste stream or checking the air around a specific operation. Most instruments have already been discussed in previous sections of this review. There are no specifically new developments in instrumentation but only gradual changes and improvements. There is increasing use of solid-state, semiconductor detectors since these lend themselves well to energy discrimination with the consequent screening out of unwanted background. It is interesting to note that improvements are still being made in the simple Geiger tube instrument which is hand-carried and used for the measurement of γ-ray intensity.[159] For x rays, a liquid-filled tissue equivalent ionization chamber has been developed which has many applications.[160]

Instrumentation has always been a highly developed aspect of radiation protection, and maintenance is a continual need. In estimating the costs of the protection program, this maintenance factor must not be overlooked. The detection and measurement of radiation lends itself so well to amplification, energy discrimination, alarm systems, and all forms of automation that there is a real temptation to overdo the use of these devices in the interests of "saving manpower." The result is often having trained men waiting for equipment repairs or doing it themselves. This can be very costly.

CONTROL METHODS

In contrast to other branches of industrial hygiene, the literature dealing with radiation protection contains relatively little on control measures and equipment. Design of shielding, glove boxes, hot cells, ventilation, remote handling equipment, etc., generally is considered the responsibility of other professions. In large measure, this is because of the complexity involved. The field of ionization chambers, Geiger counters, spectrometers, scintillometers, and similar electronic equipment is intricate enough and far removed from remote manipulators, exhaust blowers, and periscopes. Shielding is often designed by specialists in this aspect of the radiation protection field. Many health physicists actually do considerable work in shielding design using penetration tables from various sources but such work is not usually reported in the literature since it seems relatively simple. Occasionally, the existing tables are more complex than is desirable for shielding to serve in simple installation and then measurements are made and reported on common materials of construction.[161] Shielding is a serious problem around all types of high energy accelerators and the complexity of the calculations involved is well illustrated in one of the very few articles on this subject in *Health Physics.*[162]

A common method of working with γ-emitting isotopes is behind a shielding wall with an open top. This is an economical design, but consideration must be given of possible exposure from radiation scattered by air or other materials above the cell. The necessary calculations to predict this scatter are given in an article by Birchall who, significantly, is from a department of mechanical engineering.[163] The sole article on ventilation was an interesting description of a system used to permit repair of a reactor vessel without the uses of bulky protective equipment.[164] The International Atomic Energy Agency has published a "Manual on Safety Aspects of the Design and Equipment of Hot Laboratories," which contains descriptions of many important features of such equipment.[165]

STANDARDS, REGULATIONS, AND GUIDES

If one judges solely by the publicity in the public press, it would seem that the matter of radiation standards was almost the only area of activity in this field. Most of this controversy has not directly involved the industrial hygienist since it deals with exposure of the public rather than that in the workplace. Nevertheless, the industrial hygienist is interested and affected. This is an important topic but much too lengthy to even present here. It should be recognized that the establishment of a permissible level for exposure of the public is not solely a scientific procedure. This does not mean that it is unscientific but only that value judgments enter which are correctly based on more concerns than the scientific. The scientists can supply information on the effects or risks of various levels of radiation exposure, at least where such information is obtainable. Scientists and economists may estimate the values of the benefits from the use of radiation requiring various levels of exposure. Ultimately only the public can decide where the risk is commensurate with the gains. For this, education and debate is necessary and, hence, it is regrettable that much of the debate has descended to a level of slander and counterslander. It hardly seems appropriate, at this time, to choose sides when the evidence is still so vague and in need of strengthening. Another example of the serious problems of benefit versus risk is given by Morgan in a discussion of medical use of radiation.[166]

In the previous volume, the various standards-making agencies were listed and their activities noted. In the past 3 years, the National Council on Radiation Protection has published a number of useful handbooks on radiation protection in medicine, veterinary medicine, and dentistry. Most recently the recommendations of the Council on Basic Radiation Protection Criteria have appeared. These latter make no changes in most of the numerical standards but it is an excellent review and updating of the entire subject.[167] The International Commission on Radiological Protection also has been active and those publications of particular interest to the industrial hygienist have been cited elsewhere in this chapter. The International Commission on Radiation Units and Measurements has issued a number of important reports giving basic data necessary for making radiation measurements.

The Radiation Control for Health and Safety Act of 1968 was passed, which is designed to protect the public from unnecessary exposure to harmful radiation from "electronic products." This is consumer protection legislation and affects the industrial hygienist only if he also is concerned with product liability or as a member of the public. The law covers all forms of both ionizing and nonionizing radiation, including sound and has had the effect of stimulating interest in developing standards for nonionizing radiation. The recently organized Environmental Protection Administration includes part of the former Bureau of Radiological Health, the Federal Radiation Council, and some standards-making personnel from the Atomic Energy Commission. This agency has the responsibility for setting federal standards for radiation protection and, undoubtedly, will have considerable impact on the development of regulations.

Activity in the field of ionizing radiation of interest to the industrial hygienist is too great to summarize in any single review or even a single book. It is hoped that this review will stimulate interest in further reading and that the bibliography may be a helpful introduction.

REFERENCES

1. Constitution of the International Radiation Protection Ass. (IRPA). *Health Phys. 14*, 59 (1968).
2. Second International Congress of the International Radiation Protection Ass., Abstracts of Papers. *Health Phys. 19*, 67 (1970).
3. Letavet, A. A., and Kurlyandskaya, E. B., eds., "The Toxicology of Radioactive Substances." Pergamon, Oxford, 1970.
4. "Understanding the Atom" Series, Div. of Tech. Inform. U. S. Atomic Energy Comm., Washington, D.C. A single copy of any one booklet may be obtained from USAEC, Oak Ridge, Tennessee.
5. "Peaceful Uses of Nuclear Energy." Speeches by Glenn T. Seaborg, U. S. Atomic Energy Comm., 1970. Available from Div. Tech. Inform. Extension, USAEC, Oak Ridge, Tennessee, 1970.
6. Special Issue of Health Physics Dedicated to Duncan Holaday. *Health Phys. 16*, 545 (1969).
7. Holaday, D. A., History of the exposure of miners to radon. *Health Phys. 16*, 547 (1969).
8. Behounek, F., History of the exposure of miners to radon. *Health Phys. 19*, 56 (1970).
9. Evans, R. D., Engineers guide to the elementary behavior of radon daughters. *Health Phys. 17*, 229 (1969).
10. Morken, D. A., The relation of lung dose rate to working level. *Health Phys. 16*, 796 (1969).
11. Tompkins, P. C., Problems involved in setting basic guidance for radiological protection, *Amer. J. Pub. Health. 59*, 305 (1969).
12. "Radiation Exposure of Uranium Miners, A Report of an Advisory Committee from

the Division of Medical Sciences, National Academy of Sciences, August 1968." Published by the Federal Radiation Council, Washington, D.C.

13. Lundin, F. E., Lloyd, J. W., Smith, E. M., Archer, V. E., and Holaday, D. A., Mortality of uranium miners in relation to radiation exposure, hard rock mining and cigarette smoking, 1950 through Sept. 1967. *Health Phys. 16*, 571 (1969).
14. Boyd, J. T., Doll, R., Faulds, J. S., and Leiper, J., Cancer of the lung in iron ore (haematite) miners. *Brit. J. Ind. Med. 27*, 97 (1970).
15. Duggan, M. J., Soilleux, P. J., Strong, J. C., and Howell, D. M., The exposure of United Kingdom miners to radon. *Brit. J. Ind. Med. 27*, 106 (1970).
16. Parker, H. M., The dilemma of lung dosimetry. *Health Phys. 16*, 553 (1969).
17. Raabe, O. G., The adsorption of radon daughters to some polydisperse submicron polystyrene aerosols. *Health Phys. 14*, 397 (1968).
18. Raabe, O. G., Concerning the interactions that occur between radon decay products and aerosols. *Health Phys. 17*, 177 (1969).
19. Holleman, D. F., Martz, D. E., and Schiager, K. J., Total respiratory deposition of radon daughters from inhalation of uranium mine atmospheres. *Health Phys. 17*, 187 (1969).
20. George, A., and Breslin, A., Deposition of radon daughters in humans exposed to uranium mine atmospheres. *Health Phys. 17*, 115 (1969).
21. Loysen, P., Errors in measurement of working level. *Health Phys. 16*, 629 (1969).
22. Harley, N. H., and Pasternack, B. S., The rapid estimation of radon daughter working levels when daughter equilibrium is unknown. *Health Phys. 17*, 109 (1969).
23. Martz, D. E., Holleman, D. F., McCurdy, D. E., and Schiager, K. J., Analysis of atmospheric concentrations of RaA, RaB and RaC by alpha spectroscopy. *Health Phys. 17*, 131 (1969).
24. Raabe, O. G., and Wrenn, M. E., Analysis of the activity of radon daughter samples by weighted least squares. *Health Phys. 17*, 593 (1969).
25. Thomas, J. W., Modification of the Tsivoglou method for radon daughters in air. *Health Phys. 19*, 691 (1970).
26. Rolle, R., Improved radon daughter monitoring procedure. *Amer. Ind. Hyg. Ass. J. 30*, 153 (1969).
27. Waters, J. R., and Howard, B. Y., Calibration of instrument for measurement of radon concentrations in mine atmospheres. *Health Phys. 16*, 657 (1969).
28. Costa-Ribeiro, C., Thomas, J., Drew, R. T., Wrenn, M. E., and Eisenbud, M., A radon detector suitable for personnel or area monitoring. *Health Phys. 17*, 193 (1969).
29. Thomas, J. W., and LeClare, P. C., A study of the two-filter method for radon-222. *Health Phys. 18*, 113 (1970).
30. Rock, R. L., Lovett, D. B., and Nelson, S. C., Radon-daughter exposure measurement with track etch films. *Health Phys. 16*, 617 (1969).
31. Lovett, D. B., Track etch detectors for alpha exposure estimation. *Health Phys. 16*, 623 (1969).
32. McCurdy, D. E., Schiager, K. J., and Flack, E. D., Thermoluminescent dosimetry for personal monitoring of uranium miners. *Health Phys. 17*, 415 (1969).
33. White, O., An evaluation of six radon dosimeters. USAEC Health and Safety Lab. Rep. 69-23A (1969).
34. Duggan, M. J., and Howell, D. M., The measurement of the unattached fraction of airborne RaA. *Health Phys. 17*, 423 (1969).
35. Fusamura, N., Kurosawa, R., and Maruyama, M., Determination of f-value in uranium mine air. *In* "Assessment of Airborne Radioactivity." Int. Atomic Energy Agency, Vienna, 1968.

36. Black, S. C., Archer, V. E., Dixon, W. C., and Saccomanno, G., Correlation of radiation exposure and lead-210 in uranium miners. *Health Phys. 14*, 81 (1968).
37. Blanchard, R. L., Archer, V. E., and Saccomanno, G., Blood and skeletal levels of lead-210–polonium-210 as a measure of exposure to inhaled radon daughter products. *Health Phys. 16*, 585 (1969).
38. Fisher, H. L., A model for estimating the inhalation exposure to radon-222 and daughter products from the accumulated lead-210 body burden. *Health Phys. 16*, 597 (1969).
39. Holtzman, R. B., Sources of lead-210 in uranium miners. *Health Phys. 18*, 105 (1970).
40. Raabe, O. G., Concerning the relationship of lead-210 and inhalation exposure to radon-222. *Health Phys. 18*, 733 (1970).
41. Fisher, H. L., Relationship of lead-210 and short-lived radon-222 daughter products. *Health Phys. 19*, 697 (1970).
42. Blanchard, R. L., Radon-222 daughter concentrations in uranium mine atmospheres. *Nature (London) 223*, 287 (1969).
43. Cohen, N., and Kneip, T. J., A method for the analysis of lead-210 in the urine of uranium miners. *Health Phys. 17*, 125 (1969).
44. Gotchy, R. L., and Schiager, K. J., Bioassay methods for estimating current exposures to short-lived radon progeny. *Health Phys. 17*, 199 (1969).
45. Eisenbud, M., Laurer, G. R., Rosen, J. C., Cohen, N., Thomas, J., and Hazle, A. J., *In vivo* measurement of lead-210 as an indicator of cumulative radon daughter exposure in uranium miners. *Health Phys. 16*, 637 (1969).
46. Pohl-Rüling, J., and Pohl, E., The radon-222 concentration in the atmospheres of mines as a function of the barometric pressure. *Health Phys. 16*, 579 (1969).
47. Wrenn, M. E., Rosen, J. C., and Van Pelt, W. R., Steady state solutions for the diffusion equations of radon-222 daughters. *Health Phys. 16*, 647 (1969).
48. Wrenn, M. E., Eisenbud, M., Costa-Ribeiro, C., Hazle, A. J., and Siek, R. D., Reduction of radon daughter concentrations in mines by rapid mixing without makeup air. *Health Phys. 17*, 405 (1969).
49. Rock, R. L., and Walker, D. K., "Controlling Employee Exposure to Alpha Radiation in Underground Uranium Mines." Vol. 1, U. S. Bureau of Mines, 1970.
50. Martz, D. E., and Schiager, K. J., Protection against radon progeny inhalation using filter type respirators. *Health Phys. 17*, 219 (1969).
51. *Fed. Regist. 34*, 9617 (1969).
52. Burgess, W. A., and Shapiro, J., Protection from the daughter products of radon through the use of a powered air-purifying respirator. *Health Phys. 15*, 115 (1968).
53. Sill, C. W., An integrating air sampler for the determination of radon-222. *Health Phys. 16*, 371 (1969).
54. Shearer, S. D., and Sill, C. W., Evaluation of atmospheric radon in the vicinity of uranium mill tailings. *Health Phys. 17*, 77 (1969).
55. Ziemer, P. L., Personal radiation dosimetry. *Health Phys. 14*, 1 (1968).
56. Tochilin, E., Goldstein, N., and Miller, W. G., Beryllium oxide as a thermoluminescent dosimeter. *Health Phys. 16*, 1 (1969).
57. Jones, J. L., and Martin, J. A., Strontium fluoride as a thermoluminescent dosimeter. *Health Phys. 16*, 790 (1969).
58. McDougall, R. S., and Rudin, S., Thermoluminescent dosimetry of aluminum oxide. *Health Phys. 19*, 281 (1970).
59. Berstein, I. A., Bjarngard, B. E., and Jones, D., On the use of phosphor-Teflon thermoluminescent dosimeters in health physics. *Health Phys. 14*, 33 (1968).
60. Lin, F. M., and Cameron, J. R., A bibliography of thermoluminescent dosimetry

(letter). *Health Phys. 14*, 495 (1968).
61. Spurny, I., Additional bibliography of thermoluminescent dosimetry. *Health Phys. 17*, 349 (1969).
62. Becker, K., Radiophotoluminescence dosimetry, Bibliography II. *Health Phys. 17*, 631 (1969).
63. Becker, K., Radiophotoluminescence dosimetry, Bibliography. *Health Phys. 12*, 1367 (1966).
64. Goldstein, N., Tochilin, E., and Miller, W. G., Millirad and megarad dosimetry with LiF (letter). *Health Phys. 14*, 159 (1968).
65. Ehrlich, M., and Placious, R. C., Thermoluminescence response of CaF_2 : Mn in polytetrafluoroethylene to electrons. *Health Phys. 15*, 341 (1968).
66. Korba, A., and Hay, J. E., A thermoluminescent personnel neutron dosimeter. *Health Phys. 18*, 581 (1970).
67. Cusimano, J. P., and Cipperley, F. V., Personnel dosimetry using thermoluminescent dosimeters. *Health Phys. 14*, 339 (1968).
68. Kocher, L. F., Kathren, R. L., and Endres, G. W. R., Thermoluminescence personnel dosimetry at Hanford. *Health Phys. 18*, 311 (1970).
69. Endres, G. W. R., Kathren, R. L., and Kocher, L. F., Thermoluminescence personnel dosimetry at Hanford. II. *Health Phys. 18,* 665 (1970).
70. Rhyner, C. R., and Miller, W. G., Radiation dosimetry by optically stimulated luminescence of BeO. *Health Phys. 18*, 681 (1970).
71. Becker, K., and Robinson, E. M., Integrating dosimetry by thermally stimulated exoelectron (after) emission. *Health Phys. 15*, 463 (1968).
72. Becker, K., Thermally-stimulated exoelectron emission (TSEE) as a method for dose measurements using lithium fluoride. *Health Phys. 16*, 527 (1969).
73. Robinson, E. M., and Oberhofer, M., A sensitive ceramic BeO-TSEE dosimeter. *Health Phys. 18*, 434 (1970).
74. Becker, K., Cheka, J. S., and Oberhofer, M., Thermally-stimulated exoelectron emission, thermoluminescence, and impurities in LiF and BeO. *Health Phys. 19*, 391 (1970).
75. Becker, K., Principles of TSEE dosimetry. *At. Energy Rev. 8*, 173 (1970).
76. Duffy, T. L., and Kasper, R. B., Studies of gamma dosimetry systems used for nuclear accident dosimetry. *Health Phys. 14*, 45 (1968).
77. Brady, J. M., Aarestad, N. O., and Swartz, H. M., *In vivo* dosimetry by electron spin resonance spectroscopy. *Health Phys. 15*, 43 (1968).
78. Hankins, D. E., Direct counting of hair samples for phosphorus-32 activation, *Health Phys. 17*, 740 (1970).
79. National Council on Radiation Protection and Measurements. "Protection Against Neutron Radiation." NCRP Rep. No. 38 (1971).
80. Laurer, G. R., and Eisenbud, M., *In vivo* measurements of nuclides emitting soft penetrating radiations. *In* "Diagnosis and Treatment of Deposited Radionuclides" (H. Kornberg and W. Norwood, eds.). Excerpta Med. Found., Amsterdam, 1968.
81. Dean, P. N., Ide, H. M., and Langham, W. H., External measurement of plutonium lung burdens. Paper presented at the Health Physics Society meeting, June 1970.
82. Ramsden, D., The measurement of plutonium-239 *in vivo*. *Health Phys. 16*, 145 (1969).
83. Taylor, B. T., A proportional counter for low level measurement of plutonium-239 in lungs. *Health Phys. 17*, 59 (1969).
84. Tomitani, T., and Tanaka, E., Large area proportional counter for assessment of plutonium lung burden. *Health Phys. 18*, 195 (1970).

85. Johnson, L. J., Watters, R. L., Lagerquist, C. R., and Hammond, S. E., Relative distribution of plutonium and americium following experimental PuO_2 implants. *Health Phys. 19*, 743 (1970).
86. Dudley, R. A., and Ben Haim, A., Assay of skeletally-deposited strontium-90 in humans by measurement of bremsstrahlung. *Health Phys. 14*, 449 (1968).
87. Chabra, A. S., A whole body counter for routine monitoring. *Health Phys. 16*, 719 (1969).
88. Masse, F. X., and Bolton, M. M., Experience with a low-cost chair-type detector system for the determination of radioactive body burdens of M.I.T. radiation workers. *Health Phys. 19*, 27 (1970).
89. Anderson, J. I., Parker, D., and Olson, D. G., A whole body counter with rotating detectors. *Health Phys. 16*, 709 (1969).
90. Dyson, E. D., and Beach, S. A., The movement of inhaled material from the respiratory tract to blood. *Health Phys. 15*, 385 (1968).
91. Nelson, I. C., Urinary excretion of plutonium deposited in the lung. *Health Phys. 17*, 514 (1969).
92. International Commission on Radiological Protection, "Report of Committee IV on Evaluation of Radiation Doses to Body Tissues from Internal Contamination Due to Occupational Exposure." IRCP Publ. No. 10, Pergamon, Oxford, 1968.
93. West, C. M., and Scott, L. M., Uranium cases showing long chest burden retention – an updating. *Health Phys. 17*, 781 (1969).
94. Ramsden, D., Bains, M. E. D., and Fraser, D. C., *In vivo* and bioassay results from two contrasting cases of plutonium-239 inhalation. *Health Phys. 19*, 9 (1970).
95. Lagerquist, C. R., Bokowski, D. L., Hammond, S. E., and Hylton, D. B., Plutonium content of several internal organs following occupational exposure. *Amer. Ind. Hyg. Ass. J. 30*, 417 (1969).
96. Hammond, S. E., Lagerquist, C. R., and Mann, J. R., Americium and plutonium urine excretion following acute inhalation exposures to high-fired oxides. *Amer. Ind. Hyg. Ass. J. 29*, 169 (1968).
97. Bains, M. E. D., and Rowbury, P. W. J., The biological excretion pattern from a person involved in the inhalation of a mixture of enriched uranium oxide and lead metal powders. *Health Phys. 16*, 449 (1969).
98. Moss, W. D., Campbell, E. E., Schulte, H. F., and Tietjen, G. L., A study of the variations found in plutonium urinary data. *Health Phys. 17*, 571 (1969).
99. Newton, C. E., Larson, H. V., Heid, K. R., Nelson, I. C., Fuqua, P. A., Norwood, W. D., Marks, S., and Mahony, T. D., Tissue analysis for plutonium at autopsy. *In* "Diagnosis and Treatment of Deposited Radionuclides" (H. Kornberg and W. Norwood, eds.). Excerpta Med. Found., Amsterdam, 1968.
100. Bruner, H. D., A plutonium registry. *In* "Diagnosis and Treatment of Deposited Radionuclides" (H. Kornberg and W. Norwood, eds.). Excerpta Med. Found., Amsterdam, 1968.
101. Pusch, W. M., Determination of effective half-life of ruthenium-103 in man after inhalation. *Health Phys. 15*, 515 (1968).
102. Evans, A. G., New dose estimates from chronic tritium exposures. *Health Phys. 16*, 57 (1969).
103. Moyer, R. A., Savannah River experience with transplutonium elements. *Health Phys. 15,* 133 (1968).
104. Denham, D. H., Health physics considerations in processing transplutonium elements. *Health Phys. 16*, 475 (1969).
105. Wright, C. N., Radiation protection for safe handling of californium-252 sources. *Health Phys. 15*, 466 (1968).

106. Lindell, B., Occupational hazards in x-ray analytical work. *Health Phys. 15*, 481 (1968).
107. Lubenau, J. O., Davis, J. S., McDonald, D. J., and Gerusky, T. M., Analytical x-ray hazards: A continuing problem. *Health Phys. 16*, 739 (1969).
108. Matthews, J. D., Accidental extremity exposures from analytical x-ray beams. *Health Phys. 18*, 75 (1970).
109. Paschal, L., Flash x-ray machines. *Amer. Ind. Hyg. Ass. J. 31*, 109 (1970).
110. Matthews, J. D., Radiation emission from a television projection system. *Health Phys. 18*, 541 (1970).
111. Neely, G. W., Thermal neutron activation of adhesive tapes. *Health Phys. 18*, 285 (1970).
112. Moghessi, A. A., Toerber, E. D., Regnier, J. E., Carter, M. W., and Posey, C. D., Health physics aspects of tritium luminous dial painting. *Health Phys. 18*, 255 (1970).
113. Howley, J. R., Robbins, C., and Brown, J. M., Health physics considerations in the use of radioactive foils for gas chromatography detectors. *Health Phys. 18*, 76 (1970).
114. Morris, J. O., Menker, D. F., and Dauer, M., A comparison of leak test procedures for sealed radium sources. *Amer. Ind. Hyg. Ass. J. 29*, 279 (1968).
115. International Commission on Radiological Protection. "General Principles of Monitoring for Radiation Protection of Workers." ICRP Publ. No. 12, Pergamon, Oxford, 1969.
116. U.S.A.F. Nuclear Safety 65 (Part 2), Special ed. (1970).
117. Stevens, D. C., Churchill, W. L., Fox, D., and Large, N. R., A data processing system for radioactivity measurements on air samples. *Ann. Occup. Hyg. 13*, 177 (1970).
118. Sanders, M., *In* "Assessment of Airborne Radioactivity," p. 297. Proceedings of a Symposium. Int. Atomic Energy Agency, Vienna, 1967.
119. Lister, B. A. J., *In* "Assessment of Airborne Radioactivity," p. 37. Proceedings of a Symposium. Int. Atomic Energy Agency, Vienna, 1967. Brunskill, R. T., and Holt, F. B., *In* "Assessment of Airborne Radioactivity," p. 463. Proceedings of a Symposium. Int. Atomic Energy Agency, Vienna, 1967.
120. Holliday, B., Dolphin, G. W., and Dunster, H. J., Radiological protection of workers exposed to airborne plutonium particulate. *Health Phys. 18*, 529 (1970).
121. Sherwood, R. J., and Stevens, D. C., A phosphor-film technique to determine the activity of individual particles on air sample filters. *Ann. Occup. Hyg. 11*, 7 (1968).
122. Stevens, D. C., The particle size and mean concentration of radioactive aerosols measured by personal and static air samples. *Ann. Occup. Hyg. 12*, 33 (1969).
123. Langmead, W. A., and O'Connor, D. T., The personal centripeter – a particle size-selective personal air sampler. *Ann. Occup. Hyg. 12*, 185 (1969).
124. Tanaka, E., Iwadate, S., and Miwa, H., A high-sensitivity continuous plutonium air monitor. *Health Phys. 14*, 473 (1968).
125. Bradshaw, R. L., Empson, F. M., McClain, W. C., and Houser, B. L., Results of a demonstration and other studies on the disposal of high level solidified, radioactive wastes in a salt mine. *Health Phys. 18*, 63 (1970).
126. "Treatment of Airborne Radioactive Wastes." International Atomic Energy Agency, Vienna, 1968. National Agency for International Publishers, New York, 1968.
127. "Proceedings of the Eleventh AEC Air Cleaning Conference," CONF-70016, Clearinghouse for Federal Scientific and Technical Information, National Bureau of Standards, Springfield, Virginia, 1970.
128. Burchsted, C. A., and Fuller, A. B., "Design, Construction and Testing of High Efficiency Air Filtration Systems for Nuclear Application, ORNL-NSIC-65, 1970." Clearinghouse for Federal Scientific and Technical Information, National Bureau of Standards, Springfield, Virginia.

129. Craig, D. K., Adrian, H. W. W., and Bouwer, D. J. J. C., Effect of iodine concentration on the efficiency of activated charcoal adsorbers. *Health Phys. 19*, 223 (1970).
130. "Management of Radioactive Wastes at Nuclear Power Plants," Safety Series No. 28. International Atomic Energy Agency, Vienna, 1968.
131. Berke, H. L., The metabolism of rare earths I. *Health Phys. 15*, 301 (1968).
132. Morgan, A., Morgan, D. J., and Block, A., A study of the deposition, translocation and excretion of radioiodine inhaled as iodine vapor. *Health Phys. 15*, 313 (1968).
133. Furchner, J. E., Richmond, C. R., and Drake, G. A., Comparative metabolism of radionuclides in mammals. IV. Retention of silver 110 m in the mouse, rat, monkey and dog. *Health Phys. 15*, 505 (1968).
134. Thomas, R. G., Thomas, R. L., and Wright, S. R., Retention of cesium-137 and strontium-90 administered in lethal doses to rats. *Amer. Ind. Hyg. Ass. J. 29*, 593 (1968).
135. Fletcher, C. R., The radiological hazards of zirconium-95 and niobium-95. *Health Phys. 16*, 209 (1969).
136. Palmer, H. E., Nelson, I. C., and Crook, G. H., The uptake, distribution and excretion of promethium in humans and the effect of DTPA on these parameters. *Health Phys. 18*, 53 (1970).
137. Morgan, B. N., Thomas, R. G., and McClellan, R. O., Influence of chemical state of cerium-144 on its metabolism following inhalation by mice. *Amer. Ind. Hyg. J. 31*, 479 (1970).
138. Wenzel, W. J., Thomas, R. G., and McClellan, R. O., Effect of stable yttrium concentration on the distribution and excretion of inhaled radioyttrium in the rat. *Amer. Ind. Hyg. Ass. J. 30*, 630 (1969).
139. Phillips, R., and Cember, H., The influence of body burden of radiomercury on radiation dose. *J. Occup. Med. 11*, 170 (1969).
140. Quastel, M. R., Taniguichi, H., Overton, T. R., and Abbatt, J. D., Excretion and retention by humans of chronically inhaled uranium dioxide. *Health Phys. 18*, 233 (1970).
141. Leach, L. J., Maynard, E. A., Hodge, H. C., Scott, J. K., Yuile, C. L., Sylvester, G. E., and Wilson, H. B., A five-year inhalation study with natural uranium dioxide (UO_2) dust. I. Retention and biological effect in the monkey, dog and rat. *Health Phys. 18*, 599 (1970).
142. Hursh, J. B., Neuman, W. R., Toribara, T., Wilson, H., and Waterhouse, C., Oral ingestion of uranium by man. *Health Phys. 17*, 619 (1969).
143. Thomas, R. G., Transport of relatively insoluble materials from lung to lymph nodes. *Health Phys. 14*, 11 (1968).
144. Johnson, L. J., Bull, E. H., Label, J. L., and Watters, R. L., Kinetics of lymph node activity accumulation from subcutaneous Pu-O_2 implants. *Health Phys, 18*, 416 (1970).
145. Sanders, C. L., Maintenance of phagocytic functions following $^{239}PuO_2$ particle administration. *Health Phys. 18*, 82 (1970).
146. Sanders, C. L., The distribution of inhaled plutonium-239 dioxide within pulmonary macrophages. *Arch. Environ. Health 18*, 904 (1969).
147. Richmond, C. R., Langham, J., and Stone, R. S., Biological response to small discrete highly radioactive sources. *Health Phys. 18*, 401 (1970).
148. Dean, P. N., and Langham, W. H., Tumorigenicity of small highly radioactive particles. *Health Phys. 16*, 79 (1969).
149. Letavet, A. A., and Kurlyandskaya, E. B., eds., "The Toxicology of Radioactive Substances," Vol. 4. Thorium-232 and Uranium-238. Pergamon, Oxford, 1970.

150. Bender, M. A., Somatic chromosomal aberrations: Use in the evaluation of human radiation exposures. *Arch. Environ. Health 16*, 556 (1968).
151. Franco, S. C., A medical program for nuclear power stations. *J. Occup. Med. 11*, 16 (1969).
152. Tabershaw, I. R., Control of radiation hazards; role of the physician. *J. Occup. Med. 11*, 26 (1969).
153. "Medical Supervision of Radiation Workers," Joint Publication of the World Health Organization, the International Labor Office and the International Atomic Energy Agency, IAEA Safety Ser. No. 25, Vienna, 1968.
154. Trombropoulos, E. G., Bair, W. J., and Park, J. F., Removal of inhaled ^{144}Ce–^{144}Pr oxide by diethylenetriaminepentaacetic acid (DTPA) treatment I. *Health Phys. 16*, 333 (1969).
155. Sanders, C. L., and Bair, W. J., The effect of DTPA and calcium on the translocation of interperitoneally administered $^{239}PuO_2$ particles. *Health Phys. 18*, 169 (1970).
156. Johnson, L. J., Watters, R. L., Lebel, J. L., Lagerquist, C. R., and Hammond, S. E., The distribution of plutonium and americium: Subcutaneous administration of PuO_2 and the effect of chelation therapy. *In* "Radiobiology of Plutonium" (B. J. Stover and W. S. S. Jee, eds.). Univ. Utah Press, Salt Lake City, Utah, 1972.
157. Pfleger, R. C., Wilson, A. J., and McCellan, R. O., Pulmonary lavage as a therapeutic measure for removing inhaled "insoluble" materials from the lung. *Health Phys. 16*, 758 (1969).
158. Sagan, L. A., Radiobiological problems associated with adjudication of workmen's compensation claims. *J. Occup. Med. 11*, 335 (1969).
159. Jones, A. R., A portable area monitor for gamma rays. *Health Phys. 18*, 333 (1970).
160. Blanc, D., Mathieu, J., Bouet, J., and Prigent, R., A portable electronic system equipped with an ionization chamber filled with a liquid equivalent to the biological tissue. *Health Phys. 18*, 432 (1970).
161. O'Riordan, M. C., and Cott, B. R., Low energy x-ray shielding with common materials. *Health Phys. 17*, 516 (1969).
162. Jenkins, T. M., and Nelson, W. R., The effect of target scattering on the shielding of high energy electron beams. *Health Phys. 17*, 305 (1969).
163. Birchall, I., Gamma scatter from open-top cells. *Health Phys. 16*, 47 (1969).
164. Caldwell, R. D., and Cooley, R. C., Ventilation for control of tritium air contamination during reactor vessel repair. *Health Phys. 18*, 167 (1970).
165. "Manual on Safety Aspects of the Design and Equipment of Hot Laboratories." International Atomic Energy Agency, Vienna, 1969.
166. Morgan, K. Z., Ionizing radiation: Benefits versus risks. *Health Phys. 17*, 539 (1969).
167. "Basic Radiation Protection Criteria." National Council on Radiation Protection and Measurements, Washington, D.C., 1971.
168. U. S. Public Health Service Publication No. 494. Control of radon daughters in uranium mines and calculations on biologic effects. Washington, D.C., 1957.
169. Schulte, H. F., Ionizing radiation. *In* "Industrial Hygiene Highlights" (L. V. Cralley, L. J. Cralley, and G. D. Clayton, eds.), p. 118. Industrial Hygiene Foundation of America, Pittsburgh, Pennsylvania, 1968.

Engineering Approach to Analysis and Control of Heat Exposures

HARWOOD S. BELDING

In "Industrial Hygiene Highlights" physiologic criteria of heat strain were reviewed, certain heat indices and standards for occupational exposures were briefly described, and approaches to control were noted. Reference was made to forty-two key sources of more detailed information.[1]

The present article is intended as a procedural guide for industrial hygiene engineers. It is a statement of our own approach to the evaluation of specific hot jobs when the work is regularly scheduled and differences in daily heat strain are mostly attributable to changes in the outside weather. This situation applies to many hundreds of jobs in the metal and glass industries. It is characteristic that these require manual handling for transfer of materials at a pace and in a location fixed by the arrangement of the production machinery.

In directing this attention we are avoiding discussion of jobs which involve irregular heat exposures in undertaking repairs or maintaining plant operations. This is not because such exposures are seen less frequently, but simply because a different approach seems required. For unpaced, irregular exposures to work in heat it is not yet possible to assess the stress or "heat dose" in terms which predict the health consequences for the workers with any confidence. Knowledge of effects of such exposures must probably depend on measurements, directly on the individuals at risk, of related physiologic responses, including heart rate, body temperature, and sweating.

Here, we confine ourselves to those values determined from quantitative estimates of heat stress which do not require measurements on the workers; in other words, to consideration of the environmental factors and energy expenditure of the work as they affect heat stress. The purpose is to provide a rational basis for deciding priorities for adoption of control measures and what these measures should be. Our approach is best described by an example. Let us take a common type of job which requires material-handling at a regular pace.

Needless to say, the conditions of such jobs have been the target of labor disputes. Demands for addition of an extra worker during hot weather are not infrequent. While one may think of the problem represented by this example in the context of such dispute, hopefully industrial hygiene engineers will recognize potential problem situations of this sort before trouble develops.

More specifically, our example is a job which involves manual transfer of "billets" discharged every 40 sec from the portal of a furnace. (For another example, see ref. 2.) First, the worker eases the material out of the furnace; then he strips away the two halves of a 33-lb mold; and finally, he transfers the product, which weighs about 27 lb, across a 5-ft aisle to a conveyor. The pause between billets is only a few seconds. Handling is accomplished at waist level, with heavy tongs held in asbestos gloves. Each transfer position is manned by two operators who alternate at the task at 10-min intervals. Thus, each operator has 10 min of rest after each 10 min of work. An important part of the heat load on these workers is their own production of metabolic heat (*M*). The industrial hygienist should observe the nature and pace of the task, then consult a table such as Table I. (See also refs. 3, 4, 6.) In this case he estimated *M* while performing the task as about 1300 Btu/hr.

He then made the usual thermal measurements of the work environment[3] on the afternoon of a warm day, using a black Vernon globe (t_g), dry and wet bulb thermometers (t_a and t_{wb}), and a hot wire anemometer (V). Average readings were as follows:

Data	*Derivations*
t_g = 113° F t_a = 91° F t_{wb} = 70° F V = 300 ft/min	t_w = 163° F P_a = 13 mm Hg (P_a is water vapor pressure, from readings of t_a and t_{wb})

Appropriate formulas (see Table II, formulas for clothed man) were used to obtain estimates of heat loads due to radiation (*R*), and to convection (*C*) and these were added to *M* to obtain an estimate of the evaporation of sweat required for the worker to maintain heat balance (E_{req}). An estimate was also made of the maximum evaporative cooling (E_{max}) achievable under these conditions. These are given as hourly rates but apply for only one-half of each hour.

$$M = 1300 \text{ Btu/hr}$$
$$R = 1020 \text{ Btu/hr}$$
$$C = -80 \text{ Btu/hr}$$
$$E_{req} = 2240 \text{ Btu/hr}$$
$$E_{max} = 2130 \text{ Btu/hr}$$

The men rest in an aisle at the edge of the plant floor, near a water cooler. The thermal conditions of the rest area are of equal concern because each worker spends 50% of his time there. Here, the measurements revealed:

Data	*Derivations*
$t_g = 88°$ F	$t_w = 92°$ F
$t_a = 84°$ F	$P_a = 13$ mm Hg
$t_{wb} = 70°$ F	
V = 60 ft/min	

Note that this was not a particularly warm day (t_a was 82°F outside the plant when it was 84°F in the rest area). Also note that the humidity was moderate. M was assumed to be about 400 Btu/hr, including the short walk to the rest area. E_{req} and E_{max} in the rest area were calculated to be:

$$M = 400 \text{ Btu/hr}$$
$$R = -50 \text{ Btu/hr}$$
$$C = -80 \text{ Btu/hr}$$
$$E_{req} = 270 \text{ Btu/hr}$$
$$E_{max} = 810 \text{ Btu/hr}$$

The actual net hourly E_{req} and E_{max} were obtained by averaging the values for work and rest:

$$E_{req} = 1260 \text{ Btu/hr} \qquad E_{max} = 1470 \text{ Btu/hr}$$

On this day, heat balance should be maintainable by a fit, acclimatized "standard man" (154 pounds) without undue strain. The cost in sweat production of maintaining heat balance by evaporation is about ½ liter/hour (½ liter = 1200 Btu), whereas a reasonable upper limit for sweating rate over an 8-hr period has been given as 1 liter/hr.[5] Under these circumstances, demands on the circulatory system should not prove excessive and body temperature should not be elevated above 100.0–100.5°F, which is a normal level for this grade of activity even in cooler weather. Overall, these data suggest a job which requires intermittent heavy work with a moderate level of associated heat strain. Nevertheless, if a combination of an underestimate of E_{req} and/or overestimate of E_{max} amounted to 300 Btu/hr the situation would represent a threat to heat balance. The possibilities for such errors of estimate are real. The industrial hygiene engineer cannot be sure on the basis of these data alone that the job is "safe."

As has been noted, the weather was only seasonably warm (82°F) rather than hot on the afternoon when the assessment was made. With the heat stress estimates on this day indicating only a marginal safety allowance, it is desirable

TABLE I
Average Energy Cost while Performing Selected Activities [a]

Body position and activity	Energy cost typical	Btu/hr range
Sitting		
At ease	360	–
Light hand work (writing, typing)	410	380–430
Moderate hand and arm work (drafting, light drill press, light assembly, tailoring)	600	500–700
Heavy hand and arm work (nailing, shaping stones, filing)	840	720–960
Light arm and leg work (driving car on open road, machine sewing)	670	600–770
Moderate arm and leg work (local driving of truck or bus)	860	720–960
Standing		
At ease	460	–
Moderate arm and trunk work (nailing, filing, ironing)	890	720–960
Heavy arm and trunk work (hand sawing, chiselling)	1440	960–1920
Walking		
Casual (foreman, lecturing)	720	600–840
Moderate arm work (sweeping, stockroom work)	1080	960–1200
Carrying heavy loads or with heavy arm movements (carrying suitcases, scything, hand-mowing lawn)	1680	1440–1920
Transferring 35-lb. sheet materials 2 yds. at trunk level, 3 times per min.	890	–
Pushing wheelbarrow on level with 220-lb. load	1320	1200–1440
Level, 2 mph	770	–
3 mph	960	–
4 mph	1420	–
Up, 5° grade at 3 mph	2040	–
mailman climbing stairs	2880	–
Down, 5° grade at 3 mph	820	–
Jogging, level, 4.5 mph	1800	–
Running, level, 7.5 mph	3050	–
Lifting, 44 lbs., 10 cycles/min,		
Floor to waist	1970	–
Floor to shoulder	2590	–
Shoveling, 18-lb. load 1 yd with 1 yd lift, 10 times/min	1920	–
Heavy activity at fast to maximum pace	–	2400–4800

[a]Values apply for 70-kg (154-lb) man. For most activities adjustment for cost is proportional to body weight.

to procure readings for representative combinations of weather conditions over the course of a summer. Even if this is not possible, the probable situation in the worst hot weather can be modeled. This may be done on the basis of three assumptions: (*1*) that the process heat is not appreciably affected by the weather, which means that R would be unaffected by the weather; (*2*) that air temperature in the plant would be higher by the same amount that out-of-door air temperature is higher; and (*3*) that (since water is not used in the processing, and intake of outside air is large) the water vapor pressure inside the plant would be the same as outside.

In this case, a scan of local Weather Bureau records revealed that air temperature seldom exceeds 89°F, a level 7°F higher than the 82°F which prevailed on the afternoon that the data were collected. It was also noted that sometimes on very sultry days P_a reaches 20 mm Hg, which is 7 mm Hg higher than prevailed on the test afternoon.

The heat stress for such "hottest weather" was modeled accordingly using the higher values for calculating C and E_{max} at the work and rest sites. The results (Btu/hr) were compared with those for the test day:

	On warm (82° F) day	*Modeled for hottest weather*
At work site		
C	−80	60
E_{req}	2240	2380
E_{max}	2130	1610
At rest site		
C	−80	−30
E_{req}	270	320
E_{max}	810	620
Combined work and rest		
C	−80	20
E_{req}	1260	1350
E_{max}	1470	1120

Considering the figures for combined work and rest, the model indicates relatively small net effect of the higher air temperature on C and on E_{req}; however, a 24% reduction in E_{max} is predicted, attributable to the higher humidity.

In consequence, E_{max} is 230 Btu/hr less than the evaporation required for body heat balance. In such a case: (*a*) the skin temperature is forced upward (above the nominal 95°F on which the calculations of heat flow are based); (*b*) the sweating mechanism overresponds markedly, with corresponding increase in need for water and salt intake; (*c*) the blood circulatory system is put under substantial additional strain in its attempt to prevent accumulation of heat in the body by increasing flow of blood through the skin; (*d*) to the extent that this last physiologic adjustment is unsuccessful, internal body temperature is forced upward; the result is cumulative fatigue.[5] On the other hand, the rise in skin temperature on such occasions will increase the gradient for evaporative cooling by about 1 mm/°F. (In the model this is equal to a 60–70 Btu/hr increase in E_{max} per °F.) It will also decrease heat gain by R and C (totalling 30 Btu/hr per

TABLE II
Simplified Equations for Estimating Heat Load and Limit of Evaporative Cooling[a]

Metric units	British units (mixed)
Seminude (shorts, footwear)	
$R = 11(t_w-35)$	$R = 25(t_w-95)$
$t_w = t_g + 1.8\ V^{0.5}(t_g-t_a)$	$t_w = t_g + 0.13\ V^{0.5}(t_g-t_a)$
$C = 12\ V^{0.6}(t_a-35)$	$C = 1.08\ V^{0.6}(t_a-95)$
$E_{max} = 23\ V^{0.6}(42-P_a)$	$E_{max} = 4.0\ V^{0.6}(42-P_a)$
Clothed (cotton shirt and trousers added) effective heat transfer by R, C, E_{max} taken at 60% of seminude[b]	
$R = 6.6(t_w-35)$	$R = 15(t_w-95)$
$C = 7.0\ V^{0.6}(t_a-35)$	$C = 0.65\ V^{0.6}(t_a-95)$
$E_{max} = 14\ V^{0.6}(42-P_a)$	$E_{max} = 2.4\ V^{0.6}(42-P_a)$

kcal/hr	R, C, and E_{max}	Btu/hr
°C	t_w = temperature of solid surround (walls)	°F
°C	t_g = temperature of 6-inch black globe	°F
°C	t_a = air temperature	°F
m/sec	V = air speed	ft/min
35°C	Assumed skin temperature	95°F
mm Hg	42 = H_2O vapor pressure of wet skin	mm Hg
mm Hg	P_a = H_2O vapor pressure of air	mm Hg

[a] Values apply for a "standard man" weighing 70 kg (154 lbs).
[b] Note: recent data suggest a 40% reduction of heat transfer most commonly results from wearing ordinary clothing. McKarns and Brief have provided nomograms for estimating transfer based on 30% effect of clothing[6]; these are adequate for obtaining quick estimates.

°F). Thus, in this model, E_{req} and E_{max} might theoretically be brought into balance by a rise in average skin temperature of about 2°F; however, we know that the overall physiologic cost of such adjustments would be high and risk of heat injury would be real.

GENERAL INTERPRETATION OF STRESS

Based on the foregoing assessments this job should be regarded as stressful during quite a few days of the summer, and probably critically so in regard to elevation of body temperature for parts of a few days. Whether it is dangerously so will probably depend on the physical fitness of the individual workers. This in turn depends primarily on cardiovascular capacity and state of acclimatization to heat; and secondarily on age, state of hydration, nutrition, and general health.[5] If the hottest weather were sudden in onset in late spring, or protracted for several days, risk of heat injury would increase.

Certainly the industrial hygiene engineer cannot afford a laissez faire attitude about the heat exposures of this job. In addition to mutual sharing of available environmental and health records among the plant physician, nurse, safety engineer, and industrial (methods) engineer, the industrial hygienist should take steps on the basis of available knowledge to suggest heat control measures.

CONTROL MEASURES

1. An obvious and frequently practical solution for jobs like this during hot weather periods is to assign casual personnel (such as vacationing students) to give some extra relief time. This can be a successful device for reducing the heat load on the regular force. However, new workers must be broken in carefully in hot weather, particularly when the task is relatively heavy, demands some skill, and is paced.

2. The engineer who has analyzed the components of heat stress as in this example will consider other measures for reducing the heat stress. Principal options in this case are: (*a*) reduce the physical effort (*M*); (*b*) reduce exposure to radiation from the furnace (*R*); (*c*) reduce air temperature (*C*); and (*d*) modify air speed in order to increase E_{max}.

Reduction of M. In this case he concluded that the substantial physical effort could not be reduced without major rearrangement of the operation. Complete mechanization of the transfer would be costly and did not represent an immediately feasible solution. (However, we believe in the long run that repetitive, stressful tasks such as this would best be performed by machines.)

Reduction of R. The designer of the plant had included provision for aluminum shielding against furnace heat in various places but not where access to the portal of the furnace was required. The portal area was identifiable as the principal source of *R* at this task. The industrial hygienist might well consider the possibility of superimposing a mechanically activated screen between the portal and the worker, to be in place except during actual discharge of the billets. Such a device could protect during at least two-thirds of each cycle. Prediction of net reduction in heat stress from such an arrangement on a theoretical basis is difficult, but an estimate of the value could be obtained by manually interposing and removing a trial sheet of polished metal at appropriate intervals, with the blackened globe in the position where the transfer man works.

In this particular case, however, the engineer resorted to simulation of the situation for the purpose of such prediction. A globe thermometer was mounted opposite a battery of radiant heaters, at a distance equivalent to that of the worker from the furnace. The room air was controlled at 91°F and a fan was set to deliver air across the work area at 300 ft/min. Voltage to the radiant source was raised until the globe reached 113°F. This reproduced the environment of the job on the warm day used in the example; thus t_W was calculated to be 163°F and R during performance of the task to be 1020 Btu/hr.

A polished aluminum sheet was then interposed at a distance of about 6 inches from the heaters, far enough away to permit free convection between heaters and sheet. This reduced t_g to 93°F, t_W to 98°F and R to 40 Btu/hr. By time-weighting these values it was concluded that if this sheet were in place two-thirds of the time and were kept clean, the average value of R at the work site would be less than 400 Btu/hr, about 40% of the value without the shield. This would represent a net protective effect for an hour of work and rest combined of 300 Btu/hr.

The possibilities of aluminized reflective clothing may be considered as an alternative. The idea of wearing coat and trousers of this material must be rejected on the basis that such clothing would probably reduce E_{max} (which is critically low even in regular work clothing) even more than it reduced R.[2]

Another possibility would be an aluminized apron, since the worker faces the primary source of R. We have shown that a fabric apron with aluminized finish can reduce radiant heat load under such conditions by about 20%[2] without appreciably interfering with evaporation of sweat. Theoretically an apron could reduce R from 1020 Btu/hr at the work site to about 800 Btu/hr. Considering that each worker only spends half his time at the work site, the net reduction in hourly heat load by this means should be 100 Btu/hr. This means appears to be only about one-third as effective as a moveable heat shield but should prove better than nothing.

Reduction of air temperature. In the case of operations of this sort it seldom is feasible to reduce air temperature at the workplace. Open bays and ceiling vents are frequently provided to permit natural convection as provided by the chimney effect of the furnace and by the weather. Or, as in this case, air flow from outside may be forced by fans and carried by ducts to the site of the work. Characteristically such an arrangement involves ducts across the ceiling and, considering the very low specific heat of air, its temperature is raised close to that which prevails near the roof. In our example, the velocity is relatively high and the pickup of air temperature plus the effect of the furnace is such as to raise the air temperature above that outside the plant by 9°F. The rest site is another story; there is no active ventilation in that area but the air can move through a grill from a lower floor and its temperature is only 2°F higher than outdoors. The engineer can consider enclosing and air conditioning this rest area. But this would prove a real advantage to the worker only in the hottest weather, because if operated at 75–78°F (about as low as would be comfortable) the increase in convective and radiative losses would be of little consequence. Reduction of vapor pressure and increase of air movement inside the cubicle would substantially increase E_{max} but so also would simply increasing V alone.

In hot environments it is always important for the engineer to examine modification of air speed as a means for lowering the heat stress. C and E_{max} are both functions of $V^{0.6}$, which means that both can be halved or doubled by a threefold change in air speed. In our example, the existing air speed of 300 ft/min at the work site is reasonably high. However, at the rest site V is little greater than that of still air. Use of large, slow moving fans to raise V from the existing 60 ft/min measured for the rest site to a moderate 180 ft/min would double the E_{max} for half of the 8-hr shift. In the most humid weather it would bring a net increase in hourly E_{max} of about 300 Btu. (However, in hotter climates the engineer should bear in mind that, with plant temperatures of 95°F and above, raising V above the level which assures adequate E_{max} will unnecessarily increase the heat gain of the body by C. In our case, low E_{max} is the principal problem on the worst days, while C is not important.)

It seems logical that the engineer should give priority to immediate installation of fans in the rest area as a stop-gap measure. In the longer run, it would seem worthwhile to devise and install mechanically activated shields at the furnace portals.

The probable net effects of the more feasible protective measures that have been discussed are modeled below for the hottest weather. Values represent the average of work and rest conditions (Btu/hr).

	Before	*Fan only*	*Fan + apron*	*Fan + portal shield*
E_{req}	1350	1330	1230	1030
E_{max}	1120	1430	1430	1430

Third place significance is not claimed for these figures. Nevertheless the values can be used in considering trade-offs. It is obvious that the job would still be a hot job, even with adoption of these measures, but the risk of oversweating or build-up of body temperature due to excess of heat load over evaporative cooling capacity would be markedly reduced.

In conclusion, we must emphasize that this example was not chosen to show a general solution for materials transfer operations under hot conditions. The purpose has been to show how the results of an engineering analysis can not only point up the factors, M, R, E_{max}, etc., responsible for heat stress, but also indicate relative need for correction and directions to be taken in devising control measures.

Unfortunately, this analytic approach does not provide sure information as to the physiologic strain for individual workers. Variability of safe tolerance for heat and physical work is large; it even applies within each individual from day to day, depending in part on life style outside the plant. It is possible to learn much about this kind of variability from simple measurements on the workers,[5] but that must be the subject of another review at another time.

REFERENCES

1. Belding, H. S., Work in hot environments. "Industrial Hygiene Highlights," (L. V. Cralley, L. J. Cralley, and G. D. Clayton, eds.), p. 214. Industrial Hygiene Foundation of America, Pittsburgh, Pennsylvania, 1968.
2. Belding, H. S., Hertig, B. A., and Riedesel, M. L., Simulation of a hot industrial job to find effective heat stress and resulting physiologic strain. *Amer. Ind. Hyg. Ass. J. 21*, 25 (1960).
3. Hertig, B. A., and Belding, H. S., Evaluation and control of heat hazards. *In* "Temperature: Its Measurement and Control in Science and Industry" (J. D. Hardy, ed.), Vol. 3, Part 3, Biology and Medicine, Chapt. 32, p. 347. Reinhold, New York, 1963.
4. Leithead, C. S., and Lind, A. R., "Heat Stress and Heat Disorders." F. A. Davis, Philadelphia, Pennsylvania, 1964.
5. Health Factors Involved in Working Under Conditions of Heat Stress. WHO Tech. Report Series No. 412, Geneva, 1969.
6. McKarns, J. S., and Brief, R. S., Nomographs give refined estimate of heat stress. *Heat. Piping Air Cond. 38,* 113 (1966).

Evaluation of Chemical Hazards in the Environment

ROBERT G. KEENAN

This review is intended as a survey of the more significant accomplishments during the 1968–1970 period in the field of sampling and analysis of hazardous chemical substances in workroom atmospheres, in biologic fluids and tissues, and in industrial process materials. Such measurements are an integral part of the industrial hygienist's assessment of the degree of risk to a worker's health or well-being in the daily pursuit of the latter's occupation. The validity of the data resulting from the combined efforts of the industrial hygienist and the analytic chemist depends upon the overall efficiency, specificity, sensitivity, and accuracy of the sampling and analytic methods used in attacking each problem. The sophistication of our methodology continues to increase; it lies with the individual investigators to select the most appropriate devices and methods for the solution of each problem.

This review does not duplicate the extensive treatments given in *Analytical Reviews* published annually by the American Chemical Society in a special supplementary issue of *Analytical Chemistry*. The fundamental techniques and applications reviews published in alternate years by the ACS have provided excellent overviews in the entire field of analytic chemistry. These are extremely helpful to the practicing analyst faced with daily, new methodology requirements as in industrial hygiene.

STANDARDIZATION

It is most encouraging to note the progress of the Intersociety Committee on Methods for Ambient Air Sampling and Analysis in the publication of sixty tentative methods of analysis for substances of concern in air pollution studies. Many of these methods have been adapted from procedures devised previously for industrial hygiene studies. Hence, this collection of methods,[1] selected by expert subcommittees of the Intersociety Committee, should prove valuable to industrial hygienists and analytic chemists in their evaluation of hazardous substances in the workroom environment as well as for air pollution investigations. These tentative methods may be purchased from the American Public Health Association, Inc., 1015 Eighteenth Street, N.W., Washington, D.C.

During this period, the Committee on Recommended Analytical Methods, American Conference of Governmental Industrial Hygienists has added four approved additional methods to its "Manual of Analytical Methods." These methods for the sampling and analysis of iron, mercaptans, benzene, and toluene have been evaluated by the Committee's referee testing program, as have been all the methods in this manual. The manual may be purchased from the ACGIH by writing to the Secretary-Treasurer, whose office is located at 1014 Broadway, Cincinnati, Ohio 45202.

An important recent development in the standardization of methods of sampling and analysis is the establishment of Project Threshold by the American Society for Testing and Materials. This project is designed for the critical, statistically designed testing of the "35 Tentative Methods for the Sampling and Analysis of Atmospheric Contaminants" developed to date by ASTM's D-22 Committee. These methods will be tested under field sampling conditions in the presence of "real world" interferences. This type of evaluation will be extremely valuable in carrying these methods forward to a true standard status.

SAMPLING METHODS

It will suffice to note in this review that there has been a number of reports on the calibration and evaluation of samplers for airborne particulates, especially on a size-selective basis (see Engineering). The present review attempts to relate more closely with the analytic aspects of the evaluation of the environment and to limit its discussion of sampling largely to that which must be mentioned for a clear presentation of the individual methods for chemical substances. Thus, in the section on Methods for Organic Substances, certain improved techniques for the collection of organic solvent vapors for gas chromatographic and other physical methods of analysis are discussed.

METHODS FOR INORGANIC SUBSTANCES

Mercury

There have been numerous publications of analytic procedures for determining mercury in air and in biologic materials since 1967. The majority of these procedures preceded the March 1970 report of Norvald Nimreite, a zoology graduate student at the University of Western Ontario, of "total" (organic plus inorganic) mercury levels exceeding 0.5 ppm in fish from the St. Clair River–Lake Erie water system. Concentrations in excess of the 0.5 ppm level are considered unacceptable by the U. S. Food and Drug Administration and by the Canadian Food and Drug Directorate. Since the report of Nimreite's findings, there has been an escalation of mercury analytic activities in North America. This analytic effort has been directed to the determination of "total" mercury in fish tissue, human body tissues (in my laboratory and elsewhere), in river and lake waters, and in sediments in streams and lake bottoms. The story is of concern in view of our interest with man's total body burden of toxic elements. The source of the mercury in the inland waters and the seas has been accumulating for many years from the discharge of mercury-containing waste-water from the principal users of this element, notably the pulp and paper industry, chloralkali plants, paint and electronics factories, agricultural pursuits, and laboratories. The mercury deposited in the sediments of the lakes and streams is attacked by microorganisms which apparently ingest the metal.[2] Bottom-eating fish consume these organisms and the biologic food chain to man is thus established. Preliminary data published in June 1970 showed 0.08–0.28 ppm concentrations of mercury in the tissue of carp and other bottom-feeding fish from the Lake Erie–St. Clair River system, but significantly higher values were reported for the larger game fish which feed on the smaller bottom fish. For example, the mercury found in Coho salmon ranged from 0.24–0.96 ppm, in channel catfish 0.32–1.8 ppm and in walleye pike 1.40 to 3.57 ppm. As mercury is a cumulative poison, the degree of risk to man ingesting these food sources is being evaluated critically while fishing is banned from certain affected waters. Meanwhile, corrective controls are being instituted to reduce the levels of mercury discharged into the lakes and streams by users of this metal. In some cases it may be necessary to dredge certain sediments where extremely high concentrations of mercury have been deposited. The unanswered question remains: what will we do with the dredged material to preclude subsequent contamination of the ground waters and streams where the sedimental material containing 1 ppm (and possibly higher) concentrations of mercury is deposited or buried?

This mercury problem is not confined to the United States and Canada. Around 1955, ornithologists in different regions in Sweden noted a decrease in the population of seed-eating birds. In 1958, Borg[3] reported that the livers and kidneys of birds found dead and then analyzed at the State Veterinary Medical Institute contained 4–200 mg/kg (ppm) of mercury, remarkably high levels. Birds that were shot or trapped showed concentrations of 1–53 mg/kg of the same body tissues. Birds of prey, especially, showed high levels of mercury and the conclusion was reached that there was a widespread poisoning of birds resulting from the use of mercury compounds as seed-dressing agents in agriculture.[4,5] In 1966, Berg *et al.*[6] showed that the levels of mercury in birds of prey during the 1964–1965 period were ten times that in the years 1860–1870 in central Sweden.

The Swedish scientists addressed themselves to the problem of mercury in the aquatic environment since 1964 when high concentrations of mercury in fish from lakes, rivers, and along the coast of Sweden were reported to contain mercury ranging from 0.02–10 mg/kg (ppm) in Southern Sweden, 0.03–0.2 in Northern Sweden, and 0.01–0.15 in the Atlantic Ocean. An excellent treatment of the subject of mercury contamination of the environment in Sweden is available in "Chemical Fallout."[7] Through the joint efforts of Swedish scientists, a large quantity of data on mercury levels in many animal and fish species has been developed and published.[7] Of great significance is the finding by Westöö that the mercury in fish muscle is largely in the form of a methylmercury compound,[8,9] one of the more toxic forms of this element. By means of gas and thin layer chromatographic procedures developed for the isolation of the methylmercury, Westöö has found that the methylmercury content of numerous species of fish from Swedish freshwaters averaged about 85% of the total mercury.[8,10,11]

Westöö and associates[12] have also applied these chromatographic methods to other foodstuffs and found elevated concentrations of total and of methylmercury compounds (50–96% of total mercury) in meat, liver, and eggs. Other investigators, notably Smart and Lloyd[13] in 1963 and Tejning and Vesterberg[14] in 1964, reported 10 ppm levels of total mercury in eggs from hens fed seed containing 6 or 14 ppm of methylmercury dicyandiamide. This compound had been used as a seed disinfectant in Sweden from 1940 through January 1966 and was shown to be mainly responsible for the levels of methylmercury found in eggs.

Effective February 1, 1966 Sweden prohibited the treatment of seed with methylmercury compounds but permitted the substitution of methoxyethylmercury which had been used safely in Denmark as the main seed disinfectant. The effect of this action was a reduction between 1965 and 1967 of the total mercury content of Swedish eggs from 0.029 to 0.009 mg/kg (ppm), the same level as that found in eggs from Denmark and certain other European countries where methylmercury compounds were not used to treat seed. Furthermore, special feeding experiments showed total mercury levels in eggs of 0.010 and

0.012 mg/kg from feeding grains grown from methoxyethylmercury-treated seed and from untreated seed, respectively. This residual mercury is believed to have been derived from the remainder of the diet which contained a few percent fish meal along with other foodstuffs.

The substitution in Sweden of methoxyethylmercury for methylmercury compounds as the permitted mercurial seed disinfectant resulted in a significant decrease in the total mercury content of meats, viz., from 0.014–0.183 mg Hg/kg of pig liver in November 1965 (26 samples) to 0.011–0.049 in January 1968 (10 samples). Other meats showed similarly marked decreases in their total mercury contents during this period.[15] However, Westöö hastens to point out that a high percentage of this mercury was still in the methylated form (45–86% in pig liver and 80–100% in pork chops) whereas only one-third of the mercury in the 1968 egg yolks was present as a methylmercury compound.

In addition to the substitution of methoxyethylmercury for methylmercury dicyandiamide as a seed disinfectant on February 1, 1966, the Swedish authorities banned the use of phenylmercury compounds as a fungicide in the pulp and paper industry effective January 1, 1966.

Westöö's methods for separating methylmercury compounds from the sample is based upon Gage's benzene extraction of organic forms of mercury from a strong hydrochloric acid suspension of the food.[15,16] After purifying the methylmercury by means of an aqueous cysteine acetate extraction from the initial benzene extract, the acetate solution is strongly acidified with hydrochloric acid and the methylmercury compounds are reextracted with benzene. This final extract is dried with anhydrous sodium sulfate and then analyzed by gas chromatography using a polyethylene glycol (Carbowax 20M) or phenyl diethanolamine succinate on Chromosorb W (acid-washed DMCS, 60–80 mesh) column. Westöö states that by using the cysteine acetate separation procedure, no organic or inorganic forms of mercury interfere with the results. The limit of detection of methylmercury is 0.001 mg Hg/kg food. Recoveries of mercury added as methylmercury dicyandiamide to ten samples are reportedly 98 ±3%.[15]

Other methods for the determination of methylated forms of mercury include that of Kitamura *et al.* who used a similar gas chromatographic procedure in studying methylmercury poisoning of human beings from ingesting highly contaminated fish and shellfish in Japan (Minamata disease).[17]

In North America, efforts have been directed to improved analytic procedures for total mercury in biologic materials and air samples. Several procedures are based upon the liberation of metallic mercury vapor with a reducing agent, such as stannous chloride, following digestion of the sample with nitric acid or other oxidative reagent in the cold. The mercury vapor may be determined with an ultraviolet spectrophotometer, with a mercury vapor detector, or with an atomic absorption spectrophotometer. The principle of these procedures goes back to the work of Ballard and Thornton[18] who used cadmium sulfide-impregnated asbestos to isolate mercury from organic solvent samples, and to the work of Monkman *et al.*[19] who used a G. E. Instantaneous Vapor Detector equipped with a recorder to measure the mercury vapor released from a cadmium sulfide-impregnated glass filter pad through which the solution of a biologic sample was passed following cold digestion with acid permanganate and subsequent reduction of the permanganate with hydroxylamine hydrochloride.

Hatch and Ott[20] designed a "cold vapor" procedure for the determination of mercury with a reported sensitivity of 1 ppb in solution. Their method for the analysis of mercury in metallic cobalt or nickel consists of digestion with 7 *N* nitric acid in the cold, followed by the addition of dilute sulfuric acid, water, hydroxylamine sulfate, and stannous sulfate as reductant. The mercury is then aerated through a quartz cell mounted in the light beam of an atomic absorption spectrophotometer. The absorbance values at 2537 Å attains a maximum within a period of 3 min at an air-circulation rate of 2 liters/min. A modified procedure for the analysis of rocks involves the preliminary treatment of the finely ground sample with 50% hydrogen peroxide in sulfuric acid medium. Analysis of U. S. Geological Survey rocks G-1 and W-1 provided values in good agreement with those obtained with a dithizone method.

Rathje[21] devised a similar procedure for mercury in urine, using a Perkin-Elmer Model 303 atomic absorption spectrophotometer equipped with a digital readout accessory and a cell with quartz end windows. A 3-min digestion of 2 ml urine with 5 ml nitric acid at room temperature followed by dilution to 50 ml with water and subsequent reduction of the mercury with stannous chloride, was reportedly adequate for the recovery of all of the mercury contained in the sample. Magnesium perchlorate was chosen as the optional drying agent of those readily available commercially for use in the aeration stream carrying the mercury vapor to the absorption cell. The sensitivity of the method is stated to be 0.003 mg Hg/liter using a 2-ml urine sample.

Moffitt and Kupel[22] have reported on a rapid method using activated impregnated charcoal (Barneby-Cheney Co., Columbus, Ohio, Type 580-13 or 580-22) contained as two, 1-in. sections which are separated by a glass fiber plug in a straight tube used to sample atmospheric mercury. A fibrous glass plug at the inlet end of the tube is analyzed for total particulate-bound mercury and the charcoal sections are analyzed separately for volatile forms of metallic mercury and its compounds. The charcoal is removed from the tube and is analyzed for the collected mercury using a sampling boat in an atomic absorption spectrophotometer. The method is reported to provide an absolute limit of detection of 0.02 μg of mercury with a total analysis time of less than 3 min. Hence, the working range of the method, as applied to a 10-liter air sample, extends down to 4% of the 1970 Threshold Limit Value of 0.050 mg/m^3 for inorganic mercury and organic mercurials and to 20% of the 0.010 mg/m^3 TLV for alkyl mercury compounds. The authors also report good agreement with the method of Hatch and Ott for the analysis of mercury in duplicate series of samples of urine, body tissues, and water.

Kothny has designed a rapid spectrophotometric method for mercury in air, vegetation, and urine samples using crystal violet in ethyleneglycol monomethyl ether as the color forming reagent.[23] Mercury in air is sampled with 0.0075 *M* IBr_3 absorbing solution in a midget impinger preceded by a glass fiber filter to remove dust and other particulate. The excess oxidant is then destroyed with sulfite and the mercuric tetraiodide ion is converted to disulfitomercurate diiodide which is resistant to oxidizing and reducing substances but which reacts with crystal violet to form a blue-colored complex that is extracted with sulfur-free toluene. The absorbance is measured at the absorption maximum in the 590–610 nm range. EDTA may be added to complex interfering elements, prior to extraction. The author reports a limit of detection of 0.1 μg as Hg, using a 1-cm spectrophotometric cell.

Fluoride

Mu-Wan Sun has reported favorably on the application of the fluoride ion activity electrode for the determination of urinary fluoride.[24] Using a sodium citrate buffer to maintain the pH of unprocessed urine samples, this investigator obtained results in good agreement with those provided by the microdiffusion colorimetric method.

Neefus *et al.* have defined optimal conditions for the determination of fluoride in urine using a fluoride-specific ion electrode.[25] These investigators added EDTA to complex calcium and magnesium and thus prevent the precipitation of insoluble compounds of these elements formed with phosphate and fluoride present in urine. In addition, they employed a Total Ionic Strength Urinary Buffer (TISUB), mixed in a 1:1 ratio with urine samples, to provide test solutions which were close to 0.75 *M* in the region where the fluoride activity coefficient may be held constant rather than varying with the ionic strength of individual samples of urine. Working at a pH of 5.25, to ensure a 99% dissociation of HF and to eliminate hydroxyl ion interference, and standardizing against a simulated urine, Neefus *et al.* obtained excellent recoveries of fluoride from spiked samples of urine and good agreement between the results obtained on split urine samples (from industrial workers) analyzed by the described procedure and by a distillation-colorimetric method over the 0.30–13.70 mg/liter range.

Beryllium

Butler has designed two new methods for concentrating beryllium from biologic materials.[26] One is an exchange to the liquid cationic exchanger di-2-ethylhexylphosphoric acid (HDEHP) used as a 20% solution in toluene. At a pH 2–3, ^{7}Be in solutions of urine ash exchanged quantitatively to HDEHP during 30 min of mixing. The average recovery of ^{7}Be, introduced as 10^6 dpm into each of five 250-ml samples of urine, was 93%. Less than 3% of the solids (22 mg) accompanied the beryllium. Iron is one of the more common elements that exchange to HDEHP at a pH of 2–3. If it is found necessary to remove iron, as in the case of certain high iron-containing materials such as liver or blood, a triisooctylamine separation[27] of this element may be made from a hydrochloric acid solution of the sample prior to the beryllium exchange to the HDEHP reagent.

Butler's second concentration method for beryllium is based upon the exchange of beryllium oxalate to the liquid anionic exchanger triisooctylamine (TIOA) used as a 50% solution in xylene. The exchange takes place in 10 sec and does not require a critical adjustment of the pH. As applied to urine samples, the ash from a 250-ml specimen is dissolved in 25 ml of 0.25 N HCl, and 15 ml of 1.5 M oxalic acid is then added to the solution which is then mixed vigorously with 50 ml of the TIOA reagent for 10 sec. The organic layer is washed twice with 0.05 N HCl and the beryllium is then stripped from the TIOA with 4 N HNO_3. The nitric acid strip solution, containing the beryllium, is evaporated to sublime the oxalic acid and may be ashed, if required, prior to analysis. Recovery of ^{7}Be and stable beryllium from two, separate, replicate sets of urine samples was 98±1% and 90%, respectively, using the described concentration procedure. The ^{7}Be determinations were made using a γ well, scintillation counting method; the 1–12 μg quantities of stable beryllium in the other urine series were measured by atomic absorption spectrophotometry. The author suggests that gas chromatographic analysis of the beryllium–TIOA complex should provide "limitless" sensitivity.

Carbon Monoxide

Buchwald has adapted the spectrophotometric method of Commins and Lawther[28] for the determination of carboxyhemoglobin (COHb) in small samples of blood from workers exposed to carbon monoxide.[29] The method consists of the collection of specimens of blood by finger or ear lobe prick using 2-cm lengths of heparinized capillary tubes which contain a volume of 0.02–0.03 cm^3 when filled. The ends of the tube are then sealed with modeling clay in a manner that excludes air bubbles. The tubes are mailed to the laboratory for analysis. Upon removing the sealed ends, the analyst transfers the blood to an 18-cm^3 glass bottle filled with dilute aqueous ammonia reagent (1:800), seals the bottle with a polyethylene snap cap (again excluding air) and shakes thoroughly to obtain a homogeneous solution which is then divided into three parts, one of which (Solution I) serves as the sample to be analyzed, the second portion (Solution II) is saturated with oxygen, and the third (Solution

III) is saturated with carbon monoxide. Solutions II and III serve as the 0 and 100% standards in the measurement of the absorbance due to COHb in the solutions at 414, 421, and 428 nm. The percent saturation of Solution I is calculated simply from the absorbance measurements on the basis of a linear relationship between percent COHb and absorbance under the specified conditions. It is not necessary to measure the hemoglobin content of the blood separately or to know the exact volume of the specimen with this method of standardization of each individual sample. It is necessary to use boiled, deionized (or distilled) water to reduce errors resulting from the dissociation of COHb in the aqueous solution (Solution I), although the error is less than 1% at COHb saturations of less than 30%. The sealed samples may be kept without deterioration for several days preceding analysis. The author provides data showing an acceptable degree of accuracy which averages about 10% of the amount of COHb as determined by other methods, i.e., ±1.0% COHb at 2.5–10.0% COHb and ±1.5 at the 10-20% COHb level. The method has been applied successfully in field studies of garage and service station workers.[30] The data showed that cigarette smoke appeared to be a more significant source of COHb than that from motor vehicle exhaust gases.

McFee *et al.* have reported on the use of carbon monoxide tabs consisting of plastic squares containing a small circular depression, about the size of a dime, filled with a palladium chloride formulation.[31] These investigators found these tabs useful for detecting unsuspected sources of carbon monoxide in homes and automobiles. They are worn on the person and when freshly exposed for a few hours they produced a slight darkening at 30–70 ppm, a gray coloration at 80–120, and a black color at $>$ 130 nominal ppm of carbon monoxide. The tabs were made by Dansk Inpulsfysik in Denmark and were obtained through Intra-Port Limited in Allenhurst, New Jersey.

Breysse and Bovee have reported on the analysis of expired air samples with the MSA Appliances CO Poisoning Kit for the estimation of COHb levels in the blood of fork lift truck operators working in the holds of ships.[32] They found that the results obtained with this device correlated reasonably with those COHb determinations performed on blood specimens which were collected periodically as a check on the expired air sampling method. They too found that cigarette smoking provided a significant contribution to an individual's COHb level; some smokers reported for work with COHb percentages in the region of that expected for subjects who have been working at the TLV level of 50 ppm for carbon monoxide.

The Bureau of Occupational Safety and Health, Public Health Service, has published a report on the results of its evaluation of the carbon monoxide detector tube systems. This work was initiated in response to a request of the Joint Conference of Governmental Industrial Hygienists–American Industrial Hygiene Association Committee on Direct-Reading Gas-Detecting Tube Systems for a comprehensive study of the accuracy and precision of detector tubes used for the estimation of atmospheric concentrations of workroom contaminants. The study is extremely important because of the widespread use of detector tubes, many in the hands of inexperienced personnel who have no knowledge of the limitations of these devices. At the crux of this problem is the great variability in the results obtained with many of these tube systems. Hence, a critical evaluation of these devices is essential and this must include a valid assessment of the accuracy and precision of tubes for specific contaminants. As tubes for the estimation of carbon monoxide were the first to become available commercially in the United States during the 1940's and as tubes for this substance are generally regarded more highly than others by professional industrial hygienists, the investigators chose the CO detector-tube systems for the first evaluation. Using a dynamic mixing system for the preparation of known concentrations of carbon monoxide from a standard mixture of this gas in nitrogen by dilution with purified air and testing at 80° ±3°F and at relative humidities of 20, 50, and 80% within ±10%, the investigators set up a set of nine test conditions for measuring carbon monoxide at 25, 50, and 100 ppm or at one-half, one, and two times the Threshold Limit Value recommended at the time of the study. In addition, tubes from a single batch were retested within the expiration period at the test condition of 50 ppm of CO and 50% relative humidity. Also, tubes from a separate batch were tested initially with those of the first batch at 50 ppm of CO and 50% relative humidity.

A panel of three tube readers who had passed the Ishihara or Keystone IVS test for color-blindness read each tube independently and estimated the concentration indicated by the color change or length of stain, following the manufacturer's instructions. The average of the three readings, thus obtained from the panel, was recorded as the concentration of CO indicated by the tube.

None of the brands tested met the Joint ACGIH–AIHA Committee's approval criterion of providing estimated concentration values within ±25% of the calculated concentration of CO in the 25–100 ppm range. Eight brands met an alternate ±50% approval at the 95% confidence level. The identification of these eight brands is given in the published report.[33]

Lead

Keppler *et al.* have reported on the results of two studies conducted in 1966 and 1967 on the interlaboratory evaluation of the reliability of blood lead analyses.[34] This evaluation consisted of two studies, sponsored by the American Industrial Hygiene Association, to determine the reliability of the determination of lead present in its "natural" form in mixed, pooled samples of blood obtained from workers exposed industrially to lead. In addition, the samples included one set of blood-bank pooled specimens and two sets of "spiked" bloods. More than sixty laboratories participated in Study I, conducted in 1966, and in Study II, in 1967. Each laboratory used its analytic method of choice; these included dithizone spectrophotometry, atomic absorption spectrophotometry, spectrochemistry (photography), spectrochemistry (direct reader), and polarography. Prior to distributing the specimens contained in lead-freed vacutainers to the participating laboratories in Study I, the referee selected, by statistic design, a significant number of vacutainers filled from each blood pool and determined the blood lead content using both the spectrochemical and the dithizone spectrophotometric methods of analysis. The referee found a maximum variation of $\pm 5\ \mu g$ in any one pool whose separate lead concentrations approximated 30, 50, 70, and 90 $\mu g/100$ ml of whole blood, thus establishing the uniformity of the lead concentrations in the separate pools. In Study II, the referee conducted sextuplicate determinations of the lead content of the seven separate specimen lots obtained from blood donors (employees of a storage battery plant) to establish the following mean concentrations with the specified standard deviations: 121 ±6.2, 97 ±4.4, 83 ±5.9, 56 ±5.1, 23 ±1.9, 55 ±2.8, and 116 ±6.1 μg Pb/100 gm of whole blood in pools A, B, C, D, E, F, and G, respectively.

The participating laboratories included 23 industrial, 23 governmental, 6 private clinical or chemical, 4 medical college, 1 hospital, and 4 unspecified. The ranges of results reported on all blood pools were much too great to be considered satisfactory. The spread of values did not relate to the amount of lead present or to the analytic method used. The authors concluded from a statistical study of the variances in the data that many of the participating laboratories "had not adequately developed the techniques required for accurate measurement of blood-lead concentration." These comprehensive studies certainly support this conclusion. Of the sixty-one laboratories participating in Study II, only eighteen reported acceptable values (median value $\pm 10\ \mu g$ Pb/100 gm of whole blood) for blood B and forty-one for blood E, the blood-bank pool. The numbers of acceptable results for the other five sets of pooled bloods in Study II ranged from 21–33.

In view of all the emphasis that the experts have placed over the years on the precautions necessary for accurate quantitative determinations of metallic constituents, and particularly lead, in biologic materials, it is strange indeed that many analysts do not take the efforts required to prove the validity of their results to their own satisfaction before applying their methods to samples submitted for such important determinations.

Sulfur Dioxide

Scaringelli *et al.* have shown an enhanced stability of sulfur dioxide in the 0.04 *M* tetrachloromercurate (TCM) absorbing solution of the West-Gaeke method by adding 0.066 gm of the disodium salt of EDTA.[35] Whereas the decay rate of SO_2 in the absence of EDTA was shown to be dependent on the concentration of the collected gas and ranged from 3.9–1.2%/day for levels of 2–0.2 μg SO_2/ml, respectively, a constant decay rate of 1%/day was observed in the presence of EDTA over a period of 30 days. This refinement should be helpful in correcting for the decays associated with delayed analyses.

METHODS FOR MINERAL SUBSTANCES

Asbestos

Increased interest during recent years in biologically active fibers, particularly asbestos, has spurred epidemiologic and toxicologic researchers into accelerated programs in attempts to solve the unknown aspects of physiologic responses to fibrous substances. Cralley *et al.* have shown the surprising variety of sources of mineral, vegetable, and animal fibers to which man is exposed in his urban and rural environment.[36] There are more than 100 minerals with some degree of fibrous structure; these include the amphibole group of chain silicates with such examples as amosite, anthophyllite, actinolite, crocidolite, hornblende, and tremolite; the sheet silicates – the mica group, talc, serpentine (chrysotile), and the clay minerals; ortho- and ring silicates – mullite and sillimanite; framework silicates – the zeolite group; and nonsilicates – oxides (cassiterite, rutile), hydroxides (brucite), sulfates (gypsum, anhydrite, alums), and phosphates (apatite). From this limited listing of some of the more common minerals, which end up in the diverse commercial articles and chemicals used by man in the course of his daily activities, it is apparent that all of the current analytic tools may well be applied in identifying and making quantitative estimates of the fibers present in workroom and ambient atmospheres.

Keenan and Lynch have reported recently on the techniques now being used for the detection, identification, and analysis of fibers.[37] Their review article[37] provides a description of the methods used to prepare for analysis samples of fibers contained in dusts or in biologic tissues; it discusses the details of the light and electron microscope, the electron microprobe, and the x-ray diffractometric methods for the identification of fibers per se; and it provides detailed information on the applications of atomic absorption spectrophotometry, emission spectroscopy, and neutron activation analysis in the analysis of chemical elements associated with asbestos and other mineral fibers.[38]

Lynch *et al.* have evaluated the interrelationships of the gravimetric, count, and chemical analytic (for Mg) methods for chrysotile asbestos in airborne dust samples in an effort to define the best index of exposure to asbestos fibers as biologically most appropriate.[39] Their conclusions, based upon the analysis of approximately 10,000 samples, are quoted from their report:

1. There is, at present, no practical absolute method for the routine measurement of exposure to airborne asbestos fibers. The technical difficulties of the electron microscope for dust counting make its routine use unfeasible; when it is used, the results are heavily biased in favor of very small fibers.
2. Each of the methods considered measures something different and, therefore, they do not correlate well. Ratios between results by different methods should be used for qualitative prediction only.
3. Each method is strongly influenced by one attribute of the dust cloud. The biologically optimum method cannot be selected unless it is known which attribute is biologically most appropriate.
4. The preferred index of asbestos exposure is fibers longer than 5 microns counted on membrane filters at 430 × phase contrast. The method of counting is convenient and practical, and fibers > 5 microns constitute a direct index of asbestos fiber exposure.

Enzymatic Digestion of Tissue

Nenadic and Crable have studied the enzymatic digestion of human lung tissue to provide a simple technique for the recovery of minerals from this biologic matrix without the alteration of the mineral structures, as risked in the acid digestion, incineration, and other procedures for the destruction of organic matter.[40] These investigators have evaluated the optimal conditions for three enzymes, ficin, bromelin, and pronase, which have become commercially available recently. As both ficin and bromelin are reported to be inhibited by heavy metals, 0.025 *M* Versene was added to the 50-mg quantity of dried lung to complex the metallic constituents. A 1.5 ml volume of 0.025 *M* cysteine hydrochloride per milligram of ficin or bromelin promoted maximal activity of the enzyme. The digestion of the buffered mixture proceeded at 55°C for a period of 24 hr with continuous or intermittent agitation. The digestions are pH dependent and the investigators present data on this factor for each of the enzymes. A pH of 7.5 is optimal for ficin which was recommended as the enzyme which performed best at low-enzyme concentrations. Application of the ficin digestion to aliquot portions of ground, dried bituminous coal miners' lungs provided total dust and coal values that were in reasonably good agreement with results obtained by the King and Gilchrist sodium hydroxide digestion method. The latter method, according to its developers, was a compromise between practicability and accuracy[41] but it had been the method of choice in the authors' laboratory in a continuing study of metal and mineral concentrations in the lungs of bituminous coal miners.[42]

Taylor *et al.* have reported on the use of infrared spectroscopy as a complement to x-ray diffractometry for the identification of minerals.[43] Although this application of infrared was first reported in 1950, Taylor *et al.* have focused attention on the advantage of the pressed pellet technique for the infrared analysis of fine dusts in the respirable size range where the sensitivity of the x-ray diffraction method is reduced. This dust size is ideal for the preparation of a uniformly dispersed particulate in KBr pellets. Also, infrared spectrograms can provide valuable information on the chemical bond structures in the sample components. Finally, the technique provides a method for the identification of minerals and other inorganic substances to those industrial hygiene laboratories that have no x-ray diffraction instrumentation. The authors present infrared spectrograms of twenty minerals, including all of the common asbestos minerals, along with a tabulation of the wavelengths of their absorption bands expressed in cm^{-1}.

METHODS FOR ORGANIC SUBSTANCES

Air Sampling

Several investigators have been directing their efforts to the study of the optimal conditions for the sampling of organic solvent vapors by adsorption on activated charcoal and the subsequent desorption of the collected sample components for analysis in the laboratory. Fraust and Hermann have developed valuable breakthrough data on aliphatic acetates in their evaluation of the effects of mass flow rates and four separate mesh sizes of Pittsburgh Activated Carbon Company Type BPL activated carbon.[44] Working with 4–6, 8–12, 16–20, and 30–40 mesh sizes and using 5-in., 5 mm sampling tubes containing 0.5 gm of activated carbon, these investigators have shown that, with the exception of the 4–6 mesh carbon, typical breakthrough curves consist of two segments: (*a*) a plateau region of almost constant efficiency and (*b*) an S-curve portion where the efficiency drops to zero with continued sampling. The 4–6 mesh carbon proved to be a relatively inefficient absorber as it showed no plateau except at the lowest mass flow rate of 0.002 m*M* ethyl acetate per minute. The other mesh sizes each showed a plateau of high-efficiency adsorption that was independent of mass flow until breakthrough time. The 8–12 mesh adsorbent provided a definite plateau at 96% efficiency for butyl acetate sampled at a mass rate of 0.0204 m*M*/min for a period of approximately 45 min. In the experimental system used in this study, the 0.0204 m*M*/min flow rate was maintained by sampling a 500 ppm concentration of the ester at 1.0 liter/min. The time periods shown for the plateau regions with butyl acetate varied approximately from several minutes with a flow rate of 0.0409 m*M*/min to about 550 min at the lowest of the eight flow rates evaluated, i.e., 0.0020 m*M*/min.

The 16–20 and 30–40 mesh carbon each provided a plateau at 98% efficiency with amyl and butyl acetates separately. Again, the duration of the plateau varied with the mass flow rate; however, the sharp distinction between the plateau and S-curve sections should facilitate the selection of the upper limit of sampling time for anticipated concentrations of the esters.

Fraust and Hermann found that the times required for breakthrough and for the collection efficiency to drop to 50% are nearly inversely proportional to the mass flow rate expressed on a molar basis but that the adsorption process is not affected significantly by differences in the molecular weight. Other investigators have reported on the optional conditions for sampling organic vapors in the atmosphere with activated charcoal followed by desorption with carbon disulfide and gas chromatographic analysis of the sample components.[45,46]. Their procedures are based upon the method of Otterson and Guy.[47] Reid and Halpin[45] obtained average recoveries of mostly 90–100% in sampling a test chamber containing known concentrations of halogenated and aromatic hydrocarbons using Darco 12–20 mesh activated charcoal in 0.6 gm quantities in a 5-in., 6 mm i.d. tube, followed by desorption with 3 ml carbon disulfide and gas chromatographic analysis of an aliquot portion of the resulting solution. The method was applied to concentrations of the separately vaporized compounds equivalent to one-half, one, and two times the respective threshold limit values. The sampling rate was 2 liters/min in tubes packed to a prescribed pressure drop.

White *et al.*[46] have shown that the first two of four 1-in. sections of 180 mg each of activated charcoal supported in a 6-in., 4 mm i.d. tube adsorbed vapor concentrations of fourteen organic solvents completely at two times the TLV levels and that the overall recoveries were equivalent to those obtained by desorption of the corresponding liquid state of each compound with 1 ml of carbon disulfide. These investigators conducted experimental desorption studies with the single compounds and with multiple combinations of halogenated hydrocarbons, aromatic hydrocarbons, pyridine, ethanol, isooctane, 2-butanone, and ethyl ether. They concluded that a mixture of seven or fourteen of the compounds evaluated did not affect significantly the desorption efficiency for any one of the substances. The sampling rates ranged 0.5–2 liters/min with no apparent effect on overall efficiency. The 1-in. sections of charcoal reportedly have a saturation limit of 28–30 mg of a single solvent and 28–45 mg of the total group of fourteen solvents. In the case of butyl acetate, for example, a saturation limit of 30 mg/10 liters of air sampled (3000 mg/m^3) is more than four times the 1970 threshold limit value of 710 mg/m^3 for this ester, thus providing a comfortable margin of active adsorption sites when sampling at the TLV level.

Van Houten and Lee have reported their experiences in the use of 4-oz French Square bottles for the collection of atmospheric samples of solvent vapors for gas chromatographic analysis.[48] These bottles are filled by purging with an aspirator bulb fitted with two one-way valves. The bottles are capped quickly using a screw cap containing a 1/16-in. hole and provided with a set of four gaskets, one rubber and three Type 18 Saran, to prevent leakage of the collected sample. Aliquot portions of the sample are withdrawn through the port, using a calibrated syringe, and injected into the gas chromatograph for analysis. The authors have tested the effects of repetitive temperature changes over the 0–175°F range for a period of 7 days. The samples showed no appreciable loss from this treatment which was calculated to produce a pressure change equivalent to an altitude of 10,000 feet, considered as the effective altitude of the cargo compartment of commercial airliners. They concluded that the containers were suitable for shipping samples by air.

Benzo[*a*]Pyrene

Duncan has developed a gas chromatogrpahic procedure for the separation of benzo[*a*]pyrene from benzo[*e*]pyrene, perylene, and benzo[*k*]fluoranthene in the benzene extracts of atmospheric particulates collected on glass fiber filters.[49] A 10-foot by 1/8-in. stainless-steel column packed with a mixture 40% by volume 48–65 mesh sodium chloride and 60% 60–80 mesh Chromosorb G (A.W., D.M.C.S. treated) coated with 2% SE-30 effects a separation of benzo[*a*]pyrene in 15 min with recoveries of better than 95%. Analyses of the eluates by the electron capture detector (using peak height) and by ultraviolet fluorescence showed good agreement.

Ethyl Benzene and Styrene

Yamamoto and Cook have described an ultraviolet spectrophotometric method for ethyl benzene and styrene present as a mixture in air samples.[50] Isooctane serves as the collecting medium in a midget bubbler operated at a flow rate of 1 liter/min. The styrene absorbance is read at 291 nm; ethyl benzene does not absorb at this wavelength. The combined absorbance values of styrene and ethyl benzene are obtained at a wavelength of 268 nm. After correcting for the styrene contribution, using the standard curves for styrene alone at 291 and 268 nm, the net contribution of ethyl benzene to the 268 nm absorbance is calculated. Recovery data showed efficiencies exceeding 90% for styrene at concentrations of 25–200 ppm. Recovery of ethyl benzene was reduced to the 64–69% range when the atmospheric concentration of this compound was one-half that of styrene but it increased to 80% when the two vaporized substances were present at equal concentrations. The authors suggest that cooling of the collecting system may improve the efficiency.

Formaldehyde

Cares has investigated the interference of nitrite and nitrate with the chromotropic acid method for formaldehyde.[51] More than 0.3 μmole of nitrite and all quantities of nitrate decrease the recovery of formaldehyde in proportion to the concentration of the oxides of nitrogen in the sample. Collection of formaldehyde in a 10% solution of $NaHSO_3$ or $Na_2S_2O_5$ prevents the formation of nitrate; boiling the alkaline solution containing the bisulfite–formaldehyde complex, after adding 1.0 ml of additional 10% bisulfite, reduces nitrites to NO which is volatilized during the 1-min heating period. Cares has shown a 97.1 ±8.6% average recovery of 2.3 ppm concentrations of formaldehyde in four air samples containing 53 ppm of nitrogen dioxide when the samples were neutralized before boiling, as prescribed in the proposed modification of the chromotropic acid method. Conversely, untreated and non-neutralized samples gave recoveries ranging from 3.4–51.1%.

Halogenated Hydrocarbons

Baretta *et al.* conducted a continuous multipoint air sampling and analysis survey of vinyl chloride concentrations in a chemical plant and a concurrent breath sampling of workers exposed to this compound.[52] A job classification study, conducted initially, provided the information needed for the placement of five sampling probes at strategic points in each work area. Analysis of the sampled air fed continuously through Saran tubing to a centrally located infrared spectrophotometer-digitizer-computer assembly provided the mean and standard deviation of the concentrations of vinyl chloride for each 8-hr work shift. Calculations of the time-weighted average concentrations were made from the average shift concentration data and the time-location information obtained from the job surveys.

Each worker collected three breath samples daily, the first on his arrival at home from work, the second 5–10 hr later, and the third prior to returning to work on the following day. These samples were collected in glass pipets, 50 ml capacity, fitted with Saran-gasketed screw caps – one with a predrilled hole for subsequent sampling for gas chromatographic analysis of the vinyl chloride content of each breath sample. Breath decay curves derived from plotting these analytic data against the time-weighted average on-the-job exposure data showed remarkably good agreement with breath decay curves obtained from the experimental exposure of three human subjects to 50, 250, and 500 ppm concentrations of vinyl chloride in a chamber for 7.5 hr/day. This agreement demonstrates the validity of using the breath decay curves as an index of exposure to vinyl chloride vapor and the justification for extending this technique to evaluate exposures to other volatile organic chemicals.

Polycyclic Hydrocarbons

Smith *et al.* have described a procedure for the separation and spectrophotometric determination of polycylic aromatic hydrocarbons in rubber dust, furnace black, and lung tissue.[53] The furnace black and channel black recovered from the lung tissue digests are extracted with refluxed benzene. The other materials are extracted with redistilled benzene in a Soxhlet extractor for 24 hr. The rubber dust extracts are evaporated carefully to dryness and the residues dissolved in isooctane. After concentrating, extraction by *N,N*-dimethylformamide and back-extraction into isooctane is repeated three times. Microliter volumes of the final solutions are then subjected to a thin layer chromatographic separation of the polycyclic hydrocarbons. Standards are spotted on the same thin layer plate. The developer is a 1:1 mixture of *N,N*-dimethylformamide and water. After development, the zones are identified by comparison with the chromatograms of the polycyclic standards. The zones are then removed, dissolved in warm methanol, and their ultraviolet absorption spectra are prepared over the 240–340 nm range and compared with the spectra of the known compounds for additional confirmation.

Toluene Diisocyanate

Belisle has devised a portable field kit for toluene diisocyanate (TDI) in air.[54] The reagent is glutaconic aldehyde which provides a limit of detection of 0.005 ppm of TDI in a 1-ft^3 air sample. As the reagent is unstable, it is generated by adding a 4 *N* sodium hydroxide solution to a prescribed quantity of an aqueous solution of 1-(4-pyridyl)pyridinium chloride hydrochloride contained in a midget impinger. The reagent is then acidified but must be used for sampling within 8 hr. The impinger contains a spoonful of Dowex 50W-X1, 50–100 mesh cationic exchange resin on whose surface the color is produced by TDI with the reagent. The method provides results in good agreement with those obtained with the Marcali method but the analysis can be completed in 5 min at the sampling site instead of in the laboratory.

REFERENCES

1. *Health Lab. Sci. Suppl. 6,* 55–129 (April, 1969); *7,* 1–111 (January, 1970); and *7,* 128–193 (July 1970).
2. Mercury Stirs More Pollution Concern. *Chem. Eng. News 48,* 36 (June 22, 1970).
3. Borg, K., *Proc. VIII Nord. Veterinarmotet, Helsingfors, 1958,* p. 394. Cited in Ref. 7.
4. Borg, K., Wanntorp, H., Erne, K., and Hanko, E., Mercury poisoning in Swedish wildlife. *J. Appl. Ecol. Suppl. 3,* 171 (1966).
5. Borg, K., Wanntorp, H., Erne, K., and Hanko, E., *Viltrevy 5(5),* 1 (1968). Cited in Ref. 7.
6. Berg, W., Johnels, A. G., Sjöstrand, B., and Westermark, T., Mercury content in feathers of Swedish birds from the past 100 years. *Oikos 17,* 71 (1966).
7. Johnels, A. G., and Westermark, T., Mercury contamination of the environment in Sweden. *In* "Chemical Fallout" (M. W. Miller and G. G. Berg, eds.) Chapt. 10, pp.

221–241. Thomas, Springfield, Illinois, 1969.
8. Westöö, G., Determination of methylmercury compounds in foodstuffs. I. Methylmercury compounds in fish, identification and determination. *Acta Chem. Scand. 20,* 2131 (1966).
9. Norén, K., and Westöö, G., Methylmercury in fish. *Var Foda 19(2),* 13 (1967).
10. Westöö, G., Determination of methylmercury compounds in foodstuffs. II. Determination of methylmercury in fish, egg, meat and liver. *Acta Chem. Scand. 21,* 1790 (1967).
11. Westöö, G., Determination of methylmercury salts in various kinds of biological material. *Acta Chem. Scand. 22,* 2277 (1968).
12. Westöö, G., and Norén, K., *Var Foda 19,* 135 (1967). Cited in Ref. 15.
13. Smart, N. A., and Lloyd, M. K., Mercury residues in eggs, flesh and livers of hens fed on wheat treated with methylmercury dicyandiamide. *J. Sci. Food Agr. 14,* 734 (1963).
14. Tejning, S., and Vesterberg, R., Mercury in tissues and eggs from hens fed with grain containing methylmercury dicyandiamide. *Poultry Sci. 43,* 6 (1964).
15. Westöö, G., Methylmercury compounds in animal foods. *In* "Chemical Fallout" (M. W. Miller and G. G. Berg, eds.), Chapt. 5, pp. 75-93. Thomas, Springfield, Illinois, 1969.
16. Gage, J. C., The trace determination of phenyl- and methylmercury salts in biological material. *Analyst 86,* 457 (1961).
17. Kitamura, S., Tsukamoto, T., Hamakawa, K., Sumino, K., and Shibata, T., *Med. Biol. (Tokyo) 72,* 274 (1966). Cited in Ref. 15.
18. Ballard, A. E., and Thornton, C. W. D., Photometric method for estimation of minute amounts of mercury. *Ind. Eng. Chem. Anal. Ed. 13,* 893 (1941).
19. Monkman, J. L., Maffett, P. A., and Doherty, T. F., The determination of mercury in air samples and biological materials. *Amer. Ind. Hyg. Ass. Quart. 17,* 418 (1956).
20. Hatch, W. R., and Ott, W. L., Determination of submicrogram quantites of mercury by atomic absorption spectrophotometry. *Anal. Chem. 40,* 2085 (1968).
21. Ratje, A. O., A rapid ultraviolet absorption method for the determination of mercury in urine. *Amer. Ind. Hyg. Ass. J. 30,* 126 (1969).
22. Moffitt, A. E., Jr., and Kupel, R. E., A rapid method employing impregnated charcoal and atomic absorption spectrophotometry for the determination of mercury in atmospheric, biological and aquatic samples. *At. Absorption Newslett.* December, 1970.
23. Kothny, E. L., A micromethod for mercury. *Amer. Ind. Hyg. Ass. J. 31,* 466 (1970).
24. Sun, M-W., Fluoride ion activity electrode for determination of urinary fluoride. *Amer. Ind. Hyg. Ass. J. 30,* 133 (1969).
25. Neefus, J. D., Cholak, J., and Saltzman, B. E., The determination of fluoride in urine using a fluoride-specific ion electrode. *Amer. Ind. Hyg. Ass. J. 31,* 96 (1970).
26. Butler, F. E., Concentration of beryllium from biological samples. *Amer. Ind. Hyg. Ass. J. 30,* 559 (1969).
27. Sanders, S. M., Jr., and Leidt, S. C., A new procedure for plutonium urinalysis. *Health Phys. 6,* 189 (1961).
28. Commins, B. T., and Lawther, P. J., A sensitive method for the determination of carboxyhaemoglobin in a finger prick sample of blood. *Brit. J. Ind. Med. 22,* 139 (1965).
29. Buchwald, H., A rapid and sensitive method for estimating carbon monoxide in blood and its application in problem areas. *Amer. Ind. Hyg. Ass. J. 30,* 564 (1969).
30. Buchwald, H., Exposure of garage and service station operatives to carbon monoxide: A survey based on carboxyhemoglobin levels. *Amer. Ind. Hyg. Ass. J. 30,* 570 (1969).
31. McFee, D. R., Lavine, R. E., and Sullivan, R. J., Carbon monoxide, a prevalent hazard indicated by detector tabs. *Amer. Ind. Hyg. Ass. J. 31,* 749 (1970).

32. Breysse, P. A., and Bovee, H. H., Use of expired air–carbon monoxide for carboxy-hemoglobin determinations in evaluating carbon monoxide exposures resulting from the operation of gasoline fork lift trucks in holds of ships. *Amer. Ind. Hyg. Ass. J. 30,* 477 (1969).
33. Morgenstern, A. S., Ash, R. M., and Lynch, J. R., The evaluation of gas detector tube systems: I. Carbon monoxide. *Amer. Ind. Hyg. Ass. J. 31,* 630 (1970).
34. Keppler, J. F., Maxfield, M. E., Moss, W. D., Tietjen, G., and Linch, A. L., Interlaboratory evaluation of the reliability of blood lead analyses. *Amer. Ind. Hyg. Ass. J. 31,* 412 (1970).
35. Scaringelli, F. P., Elfers, L., Norris, D., and Hochheiser, S., Enhanced stability of sulfur dioxide in solution. *Anal. Chem. 42,* 1818 (1970).
36. Cralley, L. J., Keenan, R. G., Lynch, J. R., and Lainhart, W. S., Source and identification of respirable fibers. *Amer. Ind. Hyg. Ass. J. 29,* 129 (1968).
37. Keenan, R. G., and Lynch, J. R., Techniques for the detection, identification and analysis of fibers. *Amer. Ind. Hyg. Ass. J. 31,* 587 (1970).
38. Cralley, L. J., Keenan, R. G., Kupel, R. E., Kinser, R. E., and Lynch, J. R., Characterization and solubility of metals associated with asbestos fibers. *Amer. Ind. Hyg. Ass. J. 29,* 569 (1968).
39. Lynch, J. R., Ayer, H. E., and Johnson, D. L., The interrelationships of selected asbestos exposure indices. *Amer. Ind. Hyg. Ass. J. 31,* 598 (1970).
40. Nenadic, C. M., and Crable, J. V., Enzymatic digestion of human lung tissue. *Amer. Ind. Hyg. Ass. J. 31,* 81 (1970).
41. King, E. J., and Gilchrist, M., Chronic pulmonary disease in South Wales coalminers. III. Experimental studies c. The estimation of coal and aluminum in dried lung. *Med. Res. Counc. (G. Brit. Spec. Rep. Ser. 250,* 21 (1945).
42. Crable, J. V., Keenan, R. G., Kinser, R. E., Smallwood, A. W., and Mauer, P. A., Metal and mineral concentrations in lungs of bituminous coal miners. *Amer. Ind. Hyg. Ass. J. 29,* 106 (1968).
43. Taylor, D. G., Nenadic, C. M., and Crable, J. V., Infrared spectra for mineral identification. *Amer. Ind. Hyg. Ass. J. 31,* 100 (1970).
44. Fraust, C. L., and Hermann, E. R., The adsorption of aliphatic acetate vapors onto activated carbon. *Amer. Ind. Hyg. Ass. J. 30,* 494 (1969).
45. Reid, F. H., and Halpin, W. R., Determination of halogenated and aromatic hydrocarbons in air by charcoal tube and gas chromatography. *Amer. Ind. Hyg. Ass. J. 29,* 390 (1968).
46. White, L. D., Taylor, D. G., Mauer, P. A., and Kupel, R. E., A convenient optimized method for the analysis of selected solvent vapors in the industrial atmosphere. *Amer. Ind. Hyg. Ass. J. 31,* 225 (1970).
47. Otterson, E. J., and Guy, C. U., A method of atmospheric solvent vapor sampling on activated charcoal in connection with gas chromatography. *Trans. 26th Annu. Mtg. Amer. Conf. Govt. Ind. Hyg., Philadelphia, Pennsylvania,* p. 37. American Conference of Governmental Industrial Hygienists, Cincinnati, Ohio (1964).
48. Van Houten, R., and Lee, G., A method for the collection of air samples for analysis by gas chromatography. *Amer. Ind. Hyg. Ass. J. 30,* 465 (1969).
49. Duncan, R. M., A gas chromatographic determination of benzo(a)pyrene using electron capture. *Amer. Ind. Hyg. Ass. J. 30,* 624 (1969).
50. Yamamoto, R. K., and Cook, W. A., Determination of ethyl benzene and styrene in air by ultraviolet spectrophotometry. *Amer. Ind. Hyg. Ass. J. 29,* 238 (1968).
51. Cares, J. Walkley, Determination of formaldehyde by the chromotropic acid method in the presence of oxides of nitrogen. *Amer. Ind. Hyg. Ass. J. 29,* 405 (1968).

52. Baretta, E. D., Stewart, R. D., and Mutchler, J. E., Monitoring exposures to vinyl chloride vapor: Breath analysis and continuous air sampling. *Amer. Ind. Hyg. Ass. J. 30,* 537 (1969).
53. Smith, C. G., Nau, C. A., and Lawrence, C. H., Separation and identification of polycyclic hydrocarbons in rubber dust. *Amer. Ind. Hyg. Ass. J. 29,* 242 (1968).
54. Belisle, J., A portable field kit for the sampling and analysis of toluene diisocyanate in air. *Amer. Ind. Hyg. Ass. J. 30,* 41 (1969).

Hazard Evaluation and Control

BERNARD D. BLOOMFIELD* AND JAMES C. BARRETT

INTRODUCTION

Recently enacted Federal health and safety legislation will provide greater motivation for a broad-base approach to industrial hygiene control wherever the health of the worker might be endangered. The potential problem sources to be considered run the gamut from noise to plasma jet fumes, from construction sites to metal mines. Imparting the message that "the health of workers in industry is not to be compromised," these Federal Acts are: the Federal Coal Mine Safety Act of 1969 (PL 91-173); the Federal Metal and Nonmetallic Mine Safety Act (80 Stat. 772, 30 U.S.C. 721-740); the Federal Occupational Health and Safety Act of 1970 (PL 91-596); and the Federal Construction Safety Act.

The key to occupational disease prevention is the provision of health hazard control at existing operations and at the design of new or modified processes. The basic principles of industrial hygiene control will always be applicable. These include substitution of less hazardous materials, process change where possible or isolation, enclosure of the process (abrasive blasting and plasma jet, for example), local exhaust ventilation, and respiratory protection where needed and preferably as a temporary measure.

In reviewing the literature on the subject of engineering control it is apparent that techniques to control potentially hazardous operations tend to be anticipatory where capable industrial hygiene engineers are called upon and little control information is revealed, i.e., the job is done successfully and there appears to be little reason to describe the control approach in any detail. Nevertheless, new information does appear, the major portion in the *American Industrial Hygiene Association Journal,* and much of what is discussed below has

*Deceased

been culled from this source. There are many other sources of industrial hygiene information: abstracts, indices, technical journals, agency publications, university publications, books, Federal, state and local agency rules, regulations and guidelines, professional organization material, and specific company industrial hygiene studies. Experience shows that where the protection of health is the governing objective, a spirit of cooperation exists among all professionals.

Where a workplace problem is to be evaluated it is a good practice to first search the literature and then exchange information with other investigators. Few problems are so new as to be devoid of some type of health hazard evaluation and analysis. Still, such a situation does occur from time to time. The information presented below represents engineering approaches to problems and reveals a variety of concepts, ideas, and procedures which should be of interest to industrial hygiene investigators.

SAMPLING AND MONITORING

Monitoring for Carbon-14 in Breath

There is extensive use at the present time of two radioactive β emitters, carbon-14 and tritium, in the physical and biologic sciences, establishing the necessity of monitoring possible exposures among personnel working with these materials. Film badges and dosimeters are primarily γ detection sources and give little or no indication of contacted or absorbed β radiation. Inasmuch as carbon compounds may metabolize to form carbon dioxide in the lungs a technique was investigated and developed by Doremus *et al.*[1] for the detection of $^{14}CO_2$ in human breath. It is described as convenient and rapid and can be used semiquantitatively or quantitatively.

Plastic bags are used to obtain and store the breath samples. A measured 1-liter breath sample is absorbed by one to two amines. A liquid phosphor is added to the CO_2–amine complex and this solution is counted in a scintillation spectrometer.

The breath sample is obtained by having the patient exhale into a plastic bag of approximately 3-liter capacity. One liter of the breath sample is transferred to a gasometer, absorbed in 5 ml of amine over a 15–25 min period of time, and then transferred to a 20-ml counting vial containing 15 ml toluene phosphor. The sample is capped, mixed, labeled, and then counted in a scintillation spectrometer. Phenethylamine (PHEA) is used in the procedure. The method is specific for carbon-14 and it also fulfills an AEC requirement for monitoring the breath of those working with carbon-14 compounds.

The Halide Meter

The ability of the halide meter to respond to any halogen has been based on the theory of halogen interaction with a copper electrode and subsequent copper excitation in the spark source. However, in special studies undertaken by Nelson,[2] the evidence collected did not support this theory. Apparently it is a more complicated mechanism involving an incompletely understood halide interaction with the nitrogen molecule. In the presence of the halide the nitrogen spectrum becomes more intense and the photodetector responds to this enhancement.

Experiments have shown that with the platinum and stainless-steel electrode couple and an ultraviolet filter system, a significant increase in sensitivity can be realized. Some solvents such as methyl iodide could not be clearly measured with the platinum and copper electrode couple; however, with stainless steel as an electrode material, such materials are easily and accurately detected. Stainless-steel electrodes have been found to increase the longevity of calibration by 50–100 times because of their resistance to spark decomposition products. These electrode and filter changes have made it possible to measure all halogens more accurately and thus greatly increase the general reliability of the instrument.

Analysis of Selected Solvent Vapors in the Industrial Atmosphere

Before the advent of the gas chromatograph, sampling for industrial solvents in the workroom was usually done with the use of vapor detectors of the explosion meter type. The results were at best approximate since the instruments were not sensitive to low concentrations and a time consuming calibration procedure was necessary for each solvent mixed. Today most workroom solvent sampling is done either with the use of plastic bags or with selected absorption devices such as glass tubes containing activated charcoal.

White *et al.*[3] have shown that the absorption of organic vapors on activated charcoal with desorption by carbon disulfide is a most satisfactory technique for collecting and desorbing organic solvent vapors. Liquid desorption enables multiple analysis of the same sample and the collection efficiency of activated charcoal is not affected by variations in relative humidity since water vapor is easily displaced by organic vapors. The collection system consists of a pump capable of drawing a measurable amount of air through a glass tube containing activated charcoal. The tubes are easy to make, handle, and use.

Such a system has been developed for the sampling of solvent vapor mixtures; optimum operating conditions on the gas chromatograph for the separation and analysis of fourteen selected solvent vapors have been determined. The sampling tubes are made with fire-polished straight glass 6 in. long, 4 mm i.d. packed with two 1-in. sections (180 mg each) of 20/40 mesh activated charcoal. The two sections are separated with a plug of fiberglass and retained with plugs of fiberglass at each end. Both ends are flame-sealed to enable easy breaking in the field. Desorption of the organic materials is accomplished by using 1 ml carbon disulfide per 1-in. section of activated charcoal. Each section is analyzed separately. Experimentation has indicated that desorption is accomplished in 30 min if the sample is shaken repeatedly or otherwise agitated.

Table I indicates chamber desorption percentages at sampling concentrations ranging from one to two times the TLV value. It also indicates the fourteen solvents which were utilized in experimental procedure.

Several gas chromatograph columns were compared experimentally to determine which would be most satisfactory for a complete yet rapid separation of the fourteen compounds. These compounds represented alcohols, acetates, amines, aromatics, ketones, ethers, halogenated hydrocarbons, hydrocarbons, and halogenated olefins. A 20-foot by 1/8-in stainless-steel FFAP column was found to be most satisfactory. This column has 10% FFAP stationary phase on 80/100 mesh acid-washed DMCS Chromosorb W solid support.

It was found that matrices of three different random sets of several or all fourteen components apparently do not significantly affect the desorption efficiencies of any of the compounds used. It was further found that a single tube (two 1-in. sections of activated charcoal) can be used to sample and efficiently collect each of the fourteen solvent vapors up to the concentrations of at least twice the threshold limit value.

A small portable pump was used in the field for accurately measuring the amount of air and analysis of the tubes indicated that many of the fourteen solvents could be determined with the flame ionization detector in the parts-per-billion concentration range. However, the method was found to be inadequate for the analysis of acetone and methanol. A flow rate range of from 0.5–2 liters/min can be used.

Combustible Gas Indicator Response in Low-Oxygen Atmospheres

Brief and Confer[4] conducted tests to determine the response of combustible gas indicators at lower-than-ambient oxygen concentration. The concentration value, as indicated by the combustible gas indicator, is based upon the actual combustion of the sample and may not be truly indicative of the concentration if there is insufficient oxygen to support combustion of the vapor over the filaments of the instrument.

A series of tests were conducted to determine the critical concentration of

TABLE I
Chamber Desorption Percentages

Solvent	*TLV (ppm)*	*½ TLV (av. range) (7 components)*		*TLV (av. range) (7 components)*		*2 TLV (av. range) (7 components)*	
Benzene	25	99	97–100	96	94–100	93	91–94
2-Butanone	200	96	95–97	96	90–99	90	88–96
n-Butyl acetate	150	96	93–99	94	88–97	95	92–97
Carbon tetrachloride	10	98	93–101	101	96–113	98	96–99
Chloroform	50	97	93–105	100	98–101	96	87–103
p-Dioxane	1100	85	78–87	88	84–98	94	91–96
Ethanol	1000	55	51–71	67	64–72	69	67–71
Ethyl ether	400	91	88–95	90	86–96	87	86–88
Isooctane	400	94	85–97	95	92–97	98	96–104
Perchloroethylene	100	93	88–95	96	95–97	99	99–100
Pyridine	5	57	56–60	48	46–50	56	51–61
Toluene	200	96	93–98	96	93–99	99	98–100
Trichloroethylene	100	92	89–93	97	93–99	98	96–100
Xylene	100	97	91–105	97	97–100	94	92–96

oxygen, above which a combustible gas detector responds in an acceptable manner. Three of the most widely used field instruments were checked for several combustible vapors. Bag samples of known mixtures at approximately the LEL (lower exposure limit) value were made up and meter action observed at different levels of oxygen. Typically, when the oxygen level was below the critical value, the pointer might pass into the LEL zone but then would drift downward to some lower value on the scale. Oxygen levels were measured with oxygen detectors which were in turn calibrated.

The oxygen level was increased from about 4% until the three combustible gas indicators responded to give a reading of 100% LEL, which was the true concentration. It was decided that the critical oxygen level occurred between 6.8% and 8.2%. The nearest whole percent of the average of the Teledyne and J-W Oxygen Meters corresponding with the true concentration was 8%. The Mine Safety Appliance oxygen indicator gave high results. It was concluded that with a wide variety of hydrocarbons, the combustible gas indicators tested would accurately record the lower explosive limit if the oxygen content of the atmosphere was at least 8%. Lesser values of oxygen are insufficient to provide normal meter response.

When measuring combustibles in an enclosed space, oxygen measurements should also be made. Table II indicates the critical oxygen determination for methane at the LEL (5.3%).

Location of Personal Sampler Filter Heads

A detailed study with the personal sampler indicated that location of the sampler on the worker's body was the most important in terms of the concentration determined. It was found by Chatterjee *et al.*[5] in a lead storage battery plant study that the mean concentration obtained with the filter heads in the upper position (upper left chest) was 0.18 mg/m^3 while that obtained in the lower position (5 in. lower) was 0.23 mg/m^3. The difference was 22% of the overall mean and statistically significant. It appears reasonable that both in the design of a sampling study using personal samplers and the development of threshold limit values based on such studies that the location of the personal sampler in the worker's breathing zone (on his body) be specifically detailed so as to eliminate the error which could otherwise result.

Methyl- and Ethylmercury Compounds

Personnel monitoring with the use of personal sampling devices appears to be the best procedure for evaluating a workroom where alkylmercury derivatives, used extensively as fungicides and seed disinfectants, are manufactured. Linch *et*

TABLE II
Critical Oxygen Determination for Methane at the LEL

Oxygen meter response			*Oxygen value used (Average Teledyne and J-W)*	*Percent of LEL, combustible meter response*[a]			
Teledyne	*J-W*	*MSA*		*Davis*	*MSA*	*J-W(1)*	*J-W(2)*
2.0	2.0	3.5	2.0	21	21	22	26
4.0	3.5	5.0	3.8	36	36	45	49
5.5	5.5	6.5	5.5	76	64	85	80
6.8	6.8	7.5	6.8	82	71	89	85
8.0	8.3	9.0	8.2	100	98	>100	>100

[a]The meter pointer first passed into the LEL region (100%) and then settled at the percentile shown.

al.[6] found that samples collected over a long enough time to span a statistically significant range of air contamination peaks and valleys and from a large enough number of exposed personnel to detect individual differences in mercury absorption and excretion would provide a reliable basis for the assignment of a realistic threshold limit value. In the study, it was found that the absorber design, whether impinger or porous glass diffusion type, contributed only minor differences to the recovery of dimethyl- or diethylmercury vapor by absorption in the reagent usually recommended for collection of mercury from air. However, 0.1 *N* iodine monochloride in 0.5 *M* hydrochloric acid gave quantitative recoveries of dimethyl- and diethylmercury, monomethyl- and monoethylmercuric chlorides, and mercury vaporized into moving airstreams. This same reagent can also be used for the analysis of mercurial bearing dust. While the absorber design was not critical, impinger recoveries were rate-dependent. The scrubbing rate was found to be critical. At the recommended rate of 0.02 ft^3/min collection efficiency dropped off to 65%. The critical rate was not determined but increasing the air flow from 100–200 ml/min should not decrease the efficiency below 90%.

Three permeation tubes were prepared with diethylmercury and calibrated by the gravimetric determination of weight-loss rate. These gravimetric calibrations then were used to determine the dynamic air dilution volume for the airborne mercury concentration range required to calibrate the microimpingers. The average recovery efficiency for eight determinations in the range of 0.1–0.4 mg/m^3 was 102% when the air sampling rate was held to 100 ml/min.

Sampling and Analysis of Toluene Diisocyanate

Several laboratory methods have been developed for the analysis of toluene diisocyanate (TDI), the most important being those of Marçali and Ranta.[54] The

Marcali method utilizes the hydrolysis of the TDI to the corresponding amine by bubbling the air sample through a hydrochloric–acetic acid absorbing solution. The amine is deazotized and finally coupled to *N*-1-naphthylethylene-diamine. The concentration is determined by measuring the final reddish blue color spectrophotometrically. In the Ranta method, the air sample is bubbled through a reagent containing aqueous sodium nitrite and Cellosolve. The collecting solution develops a yellow-orange in the presence of TDI.

Both methods are unsatisfactory for field kit determinations since the threshold limit value for TDI has been lowered from 0.1 ppm to 0.02 ppm. A new field kit detection procedure has been developed by Belisle[7] based on the staining of a chemically treated filter paper. Analysis of the TDI is accomplished by drawing the air through an acidified absorber solution containing an ion-exchange resin and glutaconic aldehyde. After the air sample has been taken, the color of the resin is matched to a standard color. Any pump capable of drawing air at a rate of about 0.1 ft^3/min through the absorber is acceptable. A color calibration chart can be constructed relating various volumes of air to each color standard thus permitting a direct reading from the color to parts per million TDI in the air sample. The colors were found to be best matched by placing the impinger directly over the color and viewing at a 45° angle. Owing to the instability of the generated glutaconic aldehyde, the sample should be taken within 8 hr after impinger preparation.

This new sampling technique has been proven to be satisfactory for toluene diisocyanate but additional calibration data is needed for other isocyanates.

Sampling and Analysis of Halogenated Hydrocarbons in Air

A technique for the sampling of chlorinated, brominated, and iodinated hydrocarbons at air concentrations of 1.0, 0.1, and 0.5 ppm respectively, in a 10-liter air sample has been described in the literature. The method was evaluted by Sidor[8] because the low threshold limit values of halogenated hydrocarbons tax the sensitivity of conventional analytic techniques used.

The technique is dependent upon the combustion of the halogenated hydrocarbon in the presence of a platinum catalyst. The formation of the molecular halide is catalyzed by the platinum. The combustion gases, when drawn through an aqueous phenol red solution, react, probably to form the tetrahalo derivative. The air sample from the environment under study is drawn through the furnace and then through a midget impinger containing the phenol red absorbing solution.

Various factors which might affect the technique were investigated, including sampling rate, combustion tube temperature, stability of the halogenated phenol red, and sampling efficiency. All three halides were found to produce maximum optical absorbances at a sampling rate of 1 liter/min. The effects of variations in combustion tube temperatures investigated showed that a temperature of 1000°C should be recommended. Sampling efficiencies of 100, 99, and 96% respectively for the bromide, chloride, and iodide were determined. These efficiencies are based on normal operating conditions and sampling at 1 liter/min for 10 min, using 10 ml of absorbing solution in the impinger.

However, the method does not permit differentiation between different compounds of the same halide. Caution must be used in the interpretation of results when mixed halides are present since the optical absorption peaks are quite close for the three phenol red halide derivatives.

Airborne Fibrous Glass Particles

The question of whether fiberglass exposure causes identifiable roentgenographic abnormalities was evaluated by Wright[9] in a plant where workers were exposed to an environment containing airborne fibrous glass coated with phenolic resin. The fibers are formed either by blowing a jet of steam along thin streams of molten glass or by spinning the stream of glass off a centrifuge. Soon after the formation of the fibers a formaldehyde resin binder is applied, petroleum oil is added as a lubricant to improve handling characteristics of the product, and the glass with its binder is then "cured" at temperatures ranging from 450–550°F. The fiberglass is then formed into the desired products, inspected, and packaged. This subject is treated in depth in the review on epidemiology.

Exposure to Fibers in the Manufacture of Fibrous Glass

Five plants manufacturing fibrous glass insulation and one making textiles were studied.[10] Membrane filter sampling was used principally for the dust studies. The sampling medium was a Millipore Type AA membrane filter with a pore size of 0.8 μm. A 37-mm plastic filter holder was used for sampling at a flow rate of 1.7 liters/min. Gross airborne samples were collected with the holder open. Respirable samples were collected with a membrane filter arrangement utilizing a 10-ml nylon cyclone prior to and in series with the membrane filter. Both personal and general air samples were collected and high-volume respirable dust samples were also collected with samples operating at a rate of 35 ft^3/min. To limit the high-volume air sample material to the "respirable" portion, an Aerotec-2 size selector was used prior to the filter.

The samples were analyzed by two methods. A portion of each of the membrane samples was rendered transparent and counted at 430 $\times$ magnification. Fibers defined as elongated particles with a length-to-width ratio of 3 or greater were counted in three size ranges: total fibers, fibers longer than

5 μm and fibers longer than 10 μm. The other portion of the membrane filters and the high volume filters were analyzed for total silica. On the basis of a composition of 50% total silica for fibrous glass the dust concentrations were computed in terms of the amount of glass dust present in the atmosphere.

The concentrations determined were well below the threshold limit value of 50 mg/m^3. Fiber concentrations as total fibers, fibers longer than 5 μm, and fibers longer than 10 μm were found to be well below those found in the asbestos textile industry – about one-fortieth of the proposed asbestos threshold limit value. The fibers produced in glass plants tend to be too large to result in significant concentrations of respirable fibers.

Use of Expired Air-Carbon Monoxide in Evaluating Carbon Monoxide Exposures in Holds of Ships

In an effort to ascertain the carbon monoxide buildup in the holds of ships attributable to the use of gasoline engine powered lift trucks, Breysse and Bovee[11] utilized the expired air technique for estimating carboxyhemoglobin percentage (COHb%). Sampling was conducted for a 5-day work week once per month throughout an entire year to account for seasonal variation. Samples were collected before work, before lunch, after lunch, and after work.

Smoking in the holds of the ship was prohibited but there were work rules which enabled smoking during certain break periods. In addition, mechanical ventilation by means of a flexible duct handling 1600 ft^3/min of air was provided for the holds of those ships where gasoline fork lift trucks were used.

It is estimated that healthy people have a blood saturation level of approximately 10% carboxyhemoglobin following continuous exposure to carbon monoxide at the threshold limit value of 50 ppm. Therefore, in evaluating worker exposure one could consider a carboxyhemoglobin concentration greater than 10% at any time after work starts to be indicative of an excessive exposure to carbon monoxide. Continuous exposure to a 100 ppm concentration is believed to result in an approximate carboxyhemoglobin saturation value of 20%. It was found on measuring the expired air of smokers before lunch that the carboxyhemoglobin percentage ranged on the average from 1.2–3.5% with maximum values ranging from 6–17%. It was found in nonsmokers that the maximum carboxyhemoglobin percentage was 6% and the average value was 1.2%.

The investigators concluded that bull operators and stevedores were in all probability exposed to concentrations of carbon monoxide greater than the TLV of 50 ppm. They further concluded that if the net increase in carboxyhemoglobin is considered, the hazard potential of gasoline fork lift trucks in the holds could be considered of low order of magnitude so long as mechanical ventilation is provided in those holds where the lift trucks are used.

Disposable Hypodermic Syringes for Collection of Mine Atmosphere Samples

In an attempt to develop a substitute for 250 ml evacuated sampling bottles used by the U. S. Bureau of Mines, Lang and Freedman[12] evaluated a series of different makes of disposable 10-ml syringes. It was theorized that the 10-ml syringe sample would be sufficient for quantitative flushing and filling of 2-ml gas sampling loops in gas chromatographs.

It is known that gases can be lost in plastic sampling containers through leakage and permeation. Leakage losses from syringes are typically by mechanical escape through pinholes and around edges of retaining barriers. Permeation refers to passage through the interstices of retaining surfaces and represents a combination of molecular diffusion and absorption-desorption involving chemical solubility. Leakage can usually be detected by the escape of light gases such as hydrogen relative to high molecular weight gases. Permeation rates through rubber and high-polymer materials vary greatly and are dependent upon the chemical properties of the gases and the barrier materials used. Polar gases such as carbon dioxide and sulfur dioxide tend to permeate several plastic materials more rapidly than light gases such as hydrogen and helium.

It was found that plastic syringes and glass barrel syringes with Teflon O-rings suffered very rapid losses of hydrogen and carbon dioxide. Syringes having glass barrels and butyl rubber plungers or O-rings showed promise and natural rubber, while demonstrating excellent leakage resistance (as does butyl rubber), also demonstrated a high permeation rate for carbon dioxide. It was concluded on the basis of efficiency and cost that a 10-ml disposable butyl rubber plunger-type syringe was most suitable. It was also concluded that (*1*) the 10-ml disposable syringes selected have excellent retention properties for all gases present in normal mine air – except carbon dioxide – over a storage period of one week; (*2*) the carbon dioxide loss is not significant for small percentages. The 8.15% mean loss, after 7 days, is equivalent to an absolute loss of 0.021% for 0.26% CO_2 present; (*3*) although seldom required, a correction for carbon dioxide level can readily be made by using a regression line. For example, analysis made on the fourth day after taking the sample will have a loss of 5%; (*4*) suitable sampling syringes have an initial cost of about one-fourth as much as gas bottles, are convenient, and are relatively safe to use. Syringes require much less time to manipulate both in the mine and in the laboratory. Considerable analytic time can be saved.

Some Problems Associated with the Storage of Asbestos in Polyethylene Bags

A very interesting finding concerning the storage of asbestos in polyethylene bags has been made by Gibbs.[13] It has been demonstrated that cyclohexane extracts of some natural ores of crocidolite, amosite, and chrysotile types of asbestos from South Africa, Swaziland, and Rhodesia yielded oil or waxes which contained traces of benzo[*a*] pyrene in limited and discrete pockets of ore. This is not, however, considered an important source of contamination. Samples of Canadian chrysotile were examined and although traces of benzo[*a*] pyrene were not found it was determined through ultraviolet spectroscopy that certain other hydrocarbons were present though none were recognized to be carcinogenic. It is not known where the hydrocarbons come from – whether they resulted from contamination during mining, drying, and milling, or whether the oil is present in naturally occurring fiber.

It has been found that the yield of oil or wax obtained from samples of milled chrysotile asbestos collected and stored in polyethylene was higher than that obtained from asbestos stored in glass, the implication being that the polyethylene adds to the oil content of the asbestos. Polyethylene bags are commonly used for collecting and storing samples and it is suspected that a reaction takes place with some substance within the polyethylene bag in the presence of asbestos. The samples collected in bags contained a yellow compound with an absorption peak at 421 mμ. It was concluded from the study that four possibilities need to be considered. The absorbed compounds from polyethylene may (*1*) themselves act as carcinogens; (*2*) enhance the carcinogenic activity of other substances such as trace metals, asbestos itself, or associated oils; (*3*) inhibit the action of carcinogens present in the fiber; or (*4*) have no influence on the biologic action whatsoever. To overcome any possible difficulty attributable to collection and storage in polyethylene bags, glass jars and aluminum foil are said to be suitable for asbestos fibers for most experimental purposes; however, where long-term storage is intended glass is recommended.

A Method for the Collection of Air Samples for Analysis by Gas Chromatography

Van Houten and Lee[14] developed a portable miniaturized bottle sample easy to use for the collection, storage, and shipment of solvent vapor samples to be subsequently analyzed with the gas chromatograph. The bottles are capable of holding a wide variety of solvent vapor samples for at least a week without the benefit of an auxiliary seal, and also the withdrawal of samples for analysis is possible without disturbing the remaining contents of the bottle.

Type 18 Saran was found to be suitable gasket material if it was made in the form of a seal containing three thicknesses of 2.0 mil Saran backed with one thickness of 8.0 mil latex rubber. A standard, narrow neck 4-oz French Square sample bottle with screw cap is used and a 1/16-in. hole is drilled in the cap before the gaskets are inserted.

A standard 50-mm aspirator bulb with two one-way valves is used for filling the bottles. The aspirator is inserted and the bulb squeezed twenty times. This purges the bottle approximately eight times, thus providing greater than a 99% air change. After sampling, the cap with gaskets is screwed on tightly and the sample placed in a carrying case upsidedown. In practice the investigators utilized mercury seals in the bottles both to serve as an auxiliary seal and to enable resealing once the bottle has been punctured for the withdrawal of aliquot sample for chromatographic analysis. After puncture the hole can be taped and the bottle stored upsidedown for further analytic work. In order to withdraw a sample for analysis the needle on a calibrated syringe is inserted through the hole in the cap to pierce the gasket to reach the atmosphere in the bottle. A 0.5-ml aliquot is withdrawn and injected into the chromatograph. Up to ten such aliquots can be withdrawn without changing the concentration in the bottle more than 5%.

The Adsorption of Aliphatic Acetate Vapors onto Activated Carbon

Fraust and Hermann[15] found the gas chromatograph to be a most important industrial hygiene tool for the separation and identification of mixtures of chemicals at a high degree of sensitivity. Since activated carbon adsorption techniques have been used for sampling, a study was developed to determine the adsorption system characteristics when utilizing activated carbon for sampling ethyl acetate, propyl acetate, butyl acetate, and amyl acetate. The study objectives were to determine the effect of variation of mass rate and particle size on the adsorption efficiency of the sampling tube, the breakthrough time and two breakthrough curve parameters, the 50% time, and the breakthrough curve spread, for each chemical studied.

Activated carbon in four size groupings 4–6, 8–12, 16–20, and 30–40 mesh was used in 5 mm i.d. glass sampling tubes, 5 in. long. Each tube was filled with 0.5 gm activated carbon and plugged at both ends with glass wool. It was found that the 4–6 mesh activated carbon exhibited efficiencies generally less than 90%. Further conclusions were (*1*) except for 4–6 mesh carbon size, the curve consists of a plateau portion of constant efficiency and a breakthrough portion that has the shape of an S-curve. (*2*) Except for the 4–6 mesh carbon size, sampling efficiencies are independent of mass rate and in excess of 95% until the occurrence of breakthrough. (*3*) On a mol basis, the adsorption process is not significantly affected by molecular weight, the difference being less than 10% in most cases. (*4*) Breakthrough time and 50% time are nearly inversely proportional to the mass rate. (*5*) The S-curve spread is proportional to the mass rate, the rate of decrease with increasing mass rate being independent of carbon size.

Sampling Airborne Solids in Ducts

In an effort to determine the effectiveness of isokinetic sampling in a duct downstream from a 90 degree bend, Sansone[16] conducted a study utilizing 25 μm monodisperse glass spheres. Particle measurements were made at one, two, four, eight, and sixteen duct diameters downstream of a 90 degree bend with air velocities of 2400 and 3200 ft/min (centerline measurements) and elbow centerline radii of two and four times the duct diameter. Twelve sampling points at each cross section were utilized.

It was concluded that the centrifugal force generated by the elbow results in particles being forced to the periphery of the duct even sixteen diameters downstream of the elbow. It was found that not only were the average concentrations measured under laboratory conditions as much as 25% in error but the concentrations near the center of the duct tended to be lower than those at the periphery of the duct. Since test particles of 25 μm diameter glass beads were used (density of about 2.4 gm/cm^3) it can be predicted that greater errors could be determined experimentally with larger sized particles but there is no way to predict the percent error that would occur as the particle size varies.

A Rapid and Sensitive Method for Estimating Carbon Monoxide in Blood

A technique for collecting very small blood samples for the determination of carboxyhemoglobin utilizing a modification of Commins and Lawther method of spectrophotometric analysis has been described by Buchwald.[17] The blood samples are collected in 2-cm lengths of heparinized capillary tube which contain between 0.02–0.03 ml when filled.

RESPIRABLE MASS DUST SAMPLING

"Respirable" Dust Sampling

It has been recognized that the method of choice for the sampling of dust to evaluate a health hazard is one in which there is designed into the sampling instrument a size-selecting feature which in essence recognizes the characteristics of the human respiratory tract. In the Greenburg-Smith impinger method of dust sampling (1922–1925) and the midget impinger (1928) which have been used by industrial hygienists for many years as a dust sampling technique, size selection is effected in the counting procedure which utilizes a wet counting cell and approximately 100 X magnification under a microscope. In general, particles larger than 10 μm as observed during the microscopic count have been rejected as "nonrespirable." Lippman[18] has noted that another important factor in utilizing the dust count technique and the Greenburg-Smith impinger for dust evaluation is the necessity of a determination of the free-silica content of the suspended dust at the sampling location. This has been accomplished with the use of high-volume air samplers. It has been found that in general the free silica content of a suspended dust sample is less than the free-silica content of settled dust in the same location.

Attempts have been made by several investigators and technical committees to estimate the regional deposition of dust in the lung structure based on lung model calculations. In 1965 a special task group on lung dynamics, a committee of the International Committee of Radiological Protection (ICRP), proposed revised deposition and retention models for internal dosimetry of the human respiratory tract based on nose breathing. The task group defined a model respiratory tract as follows.

> 1. The nasopharynx, (N-P) – This begins with the anterior nares and extends through the anterior pharynx, back and down through the posterior pharynx (oral) to the level of the larynx or epiglottis. . . .
> 2. Continuing caudally, the next component, (T-B), consists of the trachea and the bronchial tree down to and including the terminal bronchioles. . . .
> 3. We recommend the third compartment be entitled pulmonary (P). This region consists of several structures, viz. respiratory bronchioles, alveolar ducts, atria, alveoli and alveolar sacs. . . . This region can be reagrded as the functional area (exchange space) of the lungs. Its surface consists of non-ciliated, moist epithelium with none of the secretory elements found in the tracheo-bronchial tree. . . .

The task group also concluded that the regional deposition within the respiratory tract can be estimated using the mass median diameter of the dust sampled.

The toxic response of inhaled dust is dependent upon regional deposition which in turn is dependent upon particle size. Accordingly, dose estimates for any toxic material which deposits in the lung can be derived from a knowledge of the mass concentrations within various size ranges. Mass concentration determinations can be made by separating the aerosol into size fractions corresponding to anticipated regional deposition. This can be done during the actual sampling process by making a size distribution analysis of the airborne aerosol with an instrument such as a conifuge, cascade impactor, or light-scattering aerosol spectrometer, or by making a size-distribution analysis of a collected sample.

In 1952 the British Medical Research Council adopted a definition of "respirable dust" applicable to pneumoconiosis-producing dust. In essence it defined respirable dust as that which reached the alveoli, and it selected the horizontal elutriator as a practical size selector with that portion passing the horizontal elutriator defined as respirable dust. The same standard was adopted by the Johannesburg International Conference on Pneumoconiosis in 1959. In order to implement the recommendations of the BMRC and the JICP the following was specified.

1. For purposes of estimating airborne dust in its relation to pneumoconiosis, samples for compositional analysis, or for assessment of concentration by a bulk measurement such as that of mass or surface area, should represent only the "respirable" fraction of the cloud.
2. The "respirable" sample should be separated from the cloud while the particles are airborne and in their original state of dispersion.
3. The "respirable fraction" is to be defined in terms of the free falling speed of the particles, by the equation $c/c_o = 1 - f/f_c$, where C and C_o are the concentrations of particles of falling speed f in the 'respirable' fraction and in the whole cloud, respectively, and f_c is a constant equal to twice the falling speed in air of a sphere of unit density 5 microns in diameter.

A sampling device which meets these requirements would generate a curve of sampling efficiency vs. size as suggested by Davies. It is illustrated in Fig. 1 and is defined as follows:

Deposition (%)	*Diameter (μm)*[a]
10	2.2
20	3.2
30	3.9
40	4.5
50	5.0
60	5.5
70	5.9
80	6.3
90	6.9
100	7.1

[a] For spheres of density

A second standard has been established by the AEC office of Health and Safety at a 1961 meeting. "Respirable dust" was defined as that portion of the inhaled dust which penetrates the nonciliated portions of the lung. Respirable dust was defined physically as follows:

Size[a] (μm):	10	5	3.5	2.5	2
Respirable (%):	0	25	50	75	100

[a] Sizes referred to are equivalent to an aerodynamic diameter having the properties of a unit density sphere.

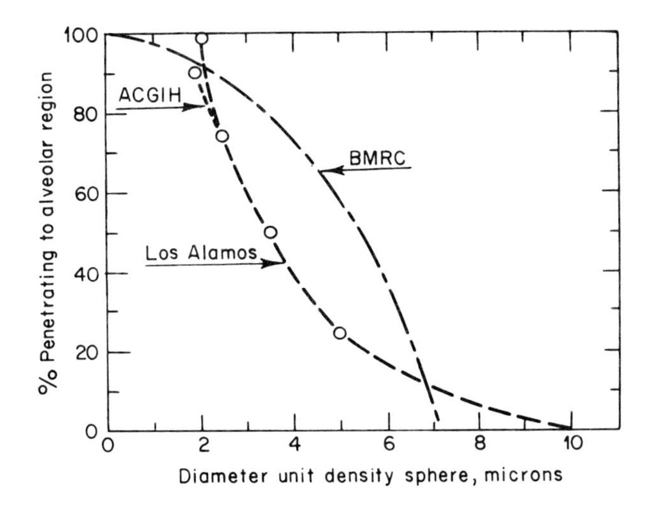

Fig. 1. BMRC, Los Alamos, and ACGIH sampler acceptance curves.

Lung Deposition of Aerosols

Generally the deposition of aerosols on any wall can take place by impaction, interception, sedimentation, or Brownian diffusion. The first three mechanisms tend to dominate the removal of larger particles of a 1-μm radius or larger. Hidy and Brock[19] note that in addition to the "classical" processes of deposition noted above, particles may reach surfaces by the influence of certain phoretic forces which produce drift velocities normal to the surface. In this phenomenon, small particles can develop a drift velocity in a non-uniform gas with gradients in concentration of one or more gaseous species. The diffusion transport of water vapor, oxygen, and carbon dioxide to and from air inhaled in the lungs must involve concentration gradients of these gases. Therefore, diffusiophoresis may have an important supplemental influence on the depositional processes of aerosols in the lungs. A computational study has indicated that the diffusiophoretic effect is most significant in retarding diffusional deposition of 0.1 μm and larger particles in the upper portions of the respiratory system. In this region diffusiophoresis involves the diffusion of water vapor into the inhaled air. With larger particles (above 0.5 μm) it is likely that the overall particle deposition from the nasal passages to the bronchi remain dominated by inertial effects and sedimentation since both Brownian diffusion and diffusiophoresis appear to be weaker mechanisms in this size range. In the lower parts of the lung diffusiophoresis depends on the diffusion of carbon dioxide and oxygen from and to the alveoli surfaces. Here the direction of the diffusiophoretic velocity is uncertain but its magnitude appears to become appreciable compared with Brownian diffusion only for 0.1 μm or larger particles inhaled during periods of heavy exercise. However, deposition of particles exceeding about 1 μm present in the alveoli will be controlled by gravitational settling which is much stronger than diffusiophoresis for such particles. It is likely that diffusiophoresis proceeds to the alveolar walls because of the strong influence of oxygen diffusion on particles.

Size-Selective Gravimetric Sampling in Dusty Industries

Ayer *et al.*[20] used size-selective gravimetric sampling equipment along with conventional impingers for sampling dust in twenty-two foundries. Personal samples were collected witha 10-mm cyclone sampler in series with a polyvinyl chloride membrane filter so as to enable separation of the nonrespirable dust. The respirable dust was collected on the membrane filter.

A total of 283 respirable mass–impinger count sample pairs were collected during the winter of 1966–1967. In addition, fifty-five gross air samples and fourteen respirable samples were analyzed for free silica. The percent of free silica in gross airborne dust samples ranged from 4–62 with a mean of 23%. The percent of free silica in the respirable fraction ranged from 1.3–13.7 with a mean of 7.2% as determined from samples taken with a horizontal elutriator. The correlation of the 10-min impinger samples with 150–180 min respirable mass samples for the entire 283 samples pairs was poor.

Samples were collected at five different types of operations including sand mixing, molding and pouring, hookout and shakeout, casting cleaning, and core room. Of ninety operations sampled utilizing both impinger and respirable mass sampling techniques, forty-two were judged equally hazardous by count and weight, thirty-two more hazardous by weight, and sixteen more hazardous by count. Only nine were judged safe by one method and hazardous by the other. There was a difference in the free silica content of air samples collected in different ways. For free silica determinations the Aerotec "2" cyclone sampling at a rate of 35 ft^3/min, the Unico cyclones sampling at a rate of 18 liters/min, a horizontal elutriator sampling at a rate of 10 liters/min, and a Del cyclone were used.

It was concluded that (*1*) size selected personal sampling would be easier and cheaper with the results more reproducible for most industrial hygiene units. (*2*) The silicosis hazard can be evaluated in the foundry utilizing respirable mass sampling with the use of the suggested threshold limit formula:

$$\text{Respirable dust (mg/m}^3\text{)} = \frac{10}{10(\%\text{RFS} + 2)}$$

where %RFS is the percent of free silica in the respirable fraction. (*3*) The 10-mm cyclone is a suitable size selector for personal sampling, but the horizontal elutriator should be used to determine free silica content.

Recalibration of Size-Selective Samplers

Initially the recommended flow rates for respirable mass sampling using the 10-mm nylon cyclone followed by the membrane filter or the ½-in. stainless-steel cyclone followed by a membrane filter were determined by Lippmann and Harris, who worked with polydispersed aerosols of uranium oxide (U_3O_8) having a specific gravity of 8.3. Since monodisperse aerosols provide a more accurate and reliable calibration, Knuth[21] undertook an investigation to determine the retention characteristics of the two cyclones using monodispersed latex, methylene blue, and iron oxide aerosols ranging in size from 1–8 μm in diameter.

The settling velocity of the monodisperse aerosol was measured by the horizontal tube method and converted to the equivalent size unit density sphere (the size of the unit density sphere having the same settling velocity of the test aerosol). The particulate settling velocities were converted to the size of the unit density sphere having the same settling velocity. The following Stokes-Cunningham equation where the slight density of air is neglected was used for this conversion:

$$p = \frac{18\mu Vg}{gd^2 \, (1 + 2A\lambda/d)}$$

where p = particle density; μ = viscosity of air; Vg = experimental settling velocity (cm/sec); g = acceleration of gravity; d = equivalent diameter unit density sphere (cm); A = 1.25; λ = mean free path.

It was determined that flow rates of 8 liters/min for the ½-in. cyclone and 1.4 liters/min for the 10-mm nylon cyclone will result in cyclone retention efficiencies that most closely approximate the "respirable dust standard." These flow rates are lower by a factor of about 2 than those previously determined by Lippman and Harris using a polydispersed aerosol.

Guide for Respirable Mass Sampling

The Aerosol Technology Committee of the American Industrial Hygiene Association has developed a guide for respirable mass sampling.[22] While dust can be quantified using the traditional microscopic count procedure, in recent years there has been a trend toward the use of respirable mass sampling procedures and instruments have been developed for this type sampling. In both Great Britain and the United States, industrial hygienists are relying more and more on respirable dust sampling techniques and there is a general belief that this procedure provides the best estimate of the health hazards attributable to the inhalation of particulates.

Size distribution, electrostatic charge, particle density, and the physical and chemical characteristics of the aerosols affect the measurement techniques, which could lead to difficulty where an attempt at correlation among the methods is made. Of particular interest is the fact that the recently published safety and health standards for Federal supply contracts of the U. S. Department of Labor specified the use of respirable mass sampling in the evaluation of workplace dust levels.[23] The limit as published was the same as that adopted by the American Conference of Governmental Industrial Hygienists as a tentative threshold limit value at the time of its 1968 annual meeting in St. Louis, Missouri. In its Notice of Intended Changes it listed alternate mass concentration threshold limit values for quartz, cristobalite, and tridymite (three forms of crystalline free silica) to supplement the threshold limit values based on particle count concentrations. For quartz, the alternative mass values proposed are:

1. For respirable dust, in mg/m^3:

$$\frac{10 \text{ mg/m}^3}{\% \text{ respirable quartz} + 2}$$

Note: Both concentration and % quartz for the application of this limit are to be determined from the fraction passing a size selector with the following characteristics:

Aerodynamic diameter in microns (unit density sphere)	2.0	2.5	3.5	5.0	10
% Passing selector	90	75	50	25	0

2. For "total dust" respirable and nonrespirable:

$$\frac{30 \text{ mg/m}^3}{\% \text{ quartz} + 2}$$

For both cristobalite and tridimite: use one-half the value calculated from the count or mass formulas for quartz.

It was noted in the "Guide for Respirable Mass Sampling" that two criteria have been specified for "respirable" dust measurement, the British Medical Research Council's "Johannesburg" criterion and that adopted at the 1961 Los Alamos conference sponsored by the U. S. Atomic Energy Commission, Office of Health and Safety. The British Medical Research Council's criterion is "measurements of dust in pneumoconiosis studies should relate the respirable action of the dust cloud, this fraction being defined by a sampling efficiency curve which depends on the falling velocity of the particles and which passes through the following points:

"Effectively 100% efficiency at 1 micron and below, 50% at 5 microns, and 0 efficiency for particles of 7 microns and upward; all the sizes refer to equivalent diameters (The equivalent diameter of a particle is the diameter of a spherical particle of unit density having the same falling velocity in air as the particle in question)."

The criterion can be met by some types of stacked parallel plate horizontal elutriator-type sampling devices. The rejection curve which is theoretical can be met by some existing commercial devices which are built with particular care. The Los Alamos criterion is based on dust deposition studies and is very much like that of the American Conference of Governmental Industrial Hygienists as adopted in its 1968 Notice of Intended Changes. The size selection curve for the ACGIH, the BMRC, and the Los Alamos criteria are shown in Fig. 1.

It has been shown mathematically that the weight concentration of particulate material passing the size selection portion of the dust sampling device and exactly meeting the Los Alamos criterion will be slightly less than that passing an instrument exactly meeting the BMRC criterion.

$$(C_{LA} = 0.8\ C_{BMRC} + 0.1\ C_{BMRC})$$

Experiments using dust clouds of coal, silica, pyrite, and mica covering a wide range of size distributions suggest that differences between weight concentrations passing commercial instruments of the two types (elutriators and inertial collectors) are somewhat less than predicted mathematically.

Instruments in use for respirable mass dust sampling include the Hexhlet sampler, which is a device using horizontally stacked plates for elutriation to meet the British Medical Research Council's criterion. The nonrespirable fraction of dust is collected on the elutriation plates and the respirable fraction which passes through the plates is collected on a Soxhlet thimble filter. Because of its hygroscopic nature, the thimble filter cannot be used to accurately measure weight changes. Filter disc holders are also available with the sampling device whereby glass fiber or membrane filters can be used for collecting the respirable mass fraction. The instrument operates at a flow rate of 50 liters/min (the original instrument with the flow rate of 100 liters/min tends to be unsatisfactory due to reentrainment of settled dust).

Another instrument in use is the ½-in. HASL cyclone that size selects the nonrespirable fraction in accordance with the Los Alamos criterion. The respirable fraction is collected on an external filter. The recommended flow rate is 9 liters/min. Another size selection device is the unmodified Dorr-Oliver hydroclone, a two-piece cyclone made of nylon with a 10 mm inside diameter. The cyclone size selects the nonrespirable fraction with the respirable fraction of dust collected on a filter device usually incorporated into a lapel sampler. The recommended flow rate is 1.7 liters/min. Still another sampler is the "Isleworth" unit which is a self-contained, battery powered, intrinsically safe dust sampler using a horizontal elutriator which meets the BMRC criterion. The respirable fraction which passes through the size collection device is collected on an internal glass fiber or 5 μm pore size membrane filter. The specified flow rate is 2.5 liters/min.

The dust passing a size selector is weighed rather than counted. Filters to be weighed should not be of the type which are affected by humidity. Glass fiber filter membranes and polyvinyl chloride membrane filters have been found to be satisfactory, whereas cellulose or cellulose ester filters can only be used with careful humidity control for relatively large weight differences and in general are unsatisfactory. Each filter must be weighed before and after sampling using a balance with a 0.01 mg sensitivity (semimicro or Cahn electrobalance).

Since the ACGIH threshold limit value is dependent upon the percent free silica, this analysis must be performed on the respirable fraction. Techniques have been developed for direct x-ray diffraction analysis of membrane filter samples for this purpose. A minimum sample weight of 2 mg is required for such a determination. Other analytic techniques require that the respirable fraction be removed from the filter. Care must be observed because of the silica background of some filter media. Glass fiber filters are obviously unsatisfactory and certain polyvinyl chloride membrane filters have relatively high and variable silica content.

There is always the possibility of reentrainment of particles collected in the size selection device (the first sampling stage). Consequently it is good practice to clean the sampling cyclone or elutriator at least daily. Also the cyclone and horizontal elutriator collection efficiencies are dependent on sampling flow rate and require frequent laboratory calibrations.

Comparison of Cyclone and Horizontal Elutriator Size Selectors

There is considerable interest in cyclones as size selection devices in respirable mass sampling because for a given flow rate they are considerably smaller than horizontal elutriators and do not require a fixed orientation. Knight and Lichti[24] carried out sample comparisons in a dust test chamber on thirty dust clouds of thirteen types utilizing four 10-mm nylon cyclones which were operated at airflow rates of 1.3, 1.65, 1.95, and 2.64 liters/min. These were compared with three (Medical Research Council) standard horizontal elutriators operated at 50, 2.83, and 2.5 liters/min air flow and with one "fine" horizontal elutriator (5.0 μm top cut) operated at 25 liters/min.

Most of the comparisons were made on the basis of a sampling period of

40 min while a few at higher concentrations were run for less time and some at lower concentrations were extended to 80 min. The mass of dust collected on the cyclone was between 0.3–1.5 mg. It was concluded that:

1. Dust samplers fitted with MRC standard horizontal elutriator or 10-mm nylon cyclone size selectors give the same mean estimate of the "respirable" dust concentration by mass when the cyclone is operated at 1.4 liters/min.
2. The estimate of "respirable" dust concentration by mass determined by a dust sampler fitted with a cyclone size selector is essentially the same when calibrated against either the United Kingdom MRC or the United States AEC specifications.
3. The size distribution may have a systematic effect on the comparison of the two size selectors. This amounts to less than a 10% change in the mass estimated in the range of size distributions likely to be found in mine dust clouds.

CALIBRATION AND REGULATION OF SAMPLING EQUIPMENT

A number of techniques have been suggested for calibration of the high-volume air sampler used in air pollution control work which generally handles a rate of about 50–60 ft^3/min. The Roots-Connersville positive displacement rotary type meter has been used as a primary standard. Other techniques include Venturi meters, rotating anemometers, the use of variable voltage transformer (Variac) for regulating the fan speed, and variable resistance calibration systems in the form of resistance plates or filter media.

Lynam *et al.*[25] have found that a variable voltage transformer (Variac) for regulating the fan speed produces a different calibration curve for the high-volume air sampler than the calibration system technique based on changes in air flow resistance. It was found that the calibration of the high-volume air sampler (Staplex) is dependent on the resistance at the sampler inlet whenever a Variac is used to vary the flow rate. The calibration curve obtained when the flow rates were varied by use of variable resistances differs markedly from those obtained by varying the flow rate by the use of a Variac. For equal reading on the sample flow meter, the Variac method produces higher actual flow rate and correspondingly lower particulate concentration than in the case when resistances are placed in the sampler inlet to vary the flow rate. The difference in calibration data was found to be due to the higher velocity pressures which existed at the orifice plate of the sampler when the Variac was used. Also, the variable resistance method of calibration produces somewhat higher temperatures in the motor housing. However, since the correction for temperature differences reaches a maximum of approximately 10% at the minimum flow rate of 20 ft^3/min, there appears to be little reason for temperature correction.

It was concluded that calibration of the high-volume air sampler should be done by use of resistances to simulate actual field operating conditions over the entire range of flow rate that might be used. This is recommended as the standard procedure.

Electronic Flow Regulation of High-Volume Air Samplers

The so-called high-volume air sampler which utilizes a single 8 by 10 in. glass fiber filter is widely used both in air pollution control work and in industrial hygiene air sampling activities where high volume samples are collected for specific analytic procedures, for example, for free silica determination. These units are not typically equipped with a means of maintaining a constant sampling rate and the rate therefore depends to a great extent upon the resistance offered by the filter. As the filter loads, the resistance increases, thereby decreasing the sampler rate.

Avera[26] has reported on the development of a technique whereby a constant rate of airflow can be maintained through automatic control of motor speed. The control circuitry is described in detail.

PARTICLE SIZING AND COUNTING

A New Instrument for the Evaluation of Environmental Aerocolloids

Goetz[27] has described a single-stage multiple-slit impactor designed to enable the sampling of aerocolloidal particles in the submicron range (greater than or equal to 0.2 μm). The sample is collected on the metallic mirror surface of the slide which moves perpendicular to two to four equal slit exits. Coagulation and mutual interference, inevitable in many other impactor processes, are avoided by control of the deposit density with the slide speed which is reproducible by synchronous drive. The deposit density is determined by measuring the light-scattering levels microphotometrically under reflected dark-field illumination. The multiple-slit arrangement enables comparison of deposits from different branches of the same air sample which could be exposed to high humidity, infrared, or ultraviolet radiation, thereby indicating the hygroscopic, thermal, and photochemical stability of the existing aerocolloids.

Aerosol Filtration by Means of Nuclepore Filters

STRUCTURAL AND FILTRATION PROPERTIES

Spurny *et al.*[28] have described a new type of porous analytic filter material, trade name Nuclepore, that resembles membrane filters in many of its properties. The filter material evolved from a technique used to study damages in solids. A thin (10-μm) sheet of polycarbonate placed in contact with uranium sheets in a nuclear reactor is bombarded with fission fragments as a result of the neutron flux which causes the fission of uranium-235, thereby leaving tracks of damage. Treatment in an etching bath removes the material and enlarges the pores to a degree determined by the bath reagents, the duration, and the temperature of treatment. Suitable geometry was devised to limit the effective fragments to those traveling nearly perpendicular to the film surface.

The new filter material is available with reasonably uniform graded pore diameters between 0.5–8.0 μm in a matrix with a density of 0.95 gm/cm^3. The upper surface of the filter is smooth, the pores are circular in cross section, and the matrix is mechanically very strong. Flow characteristics were determined experimentally and various aerosols were used to determine filtration efficiency. Liquid aerosols varying from pyrophosphoric acid droplets of mean radius of 0.02 μm and greater in size were generated as were solid aerosols with a mean radius of 0.04 μm for platinum oxide and 0.17 μm for selenium. A variety of sizes of particulate, as generated in a sodium chloride system, were also used.

The geometric standard deviation for uniformity of the pore sizes was less than 1.10 which is well within the desirable criteria range for uniformity.

As a sampling medium, Nuclepore is superior to other filters as a substrate for subsequent microscopy but is less retentive for some particle size ranges, depending upon pore size and filtration conditions. It could be concluded then that this material is complimentary rather than competitive to other filter types.

AEROSOL SAMPLING AND MEASUREMENT

Further work by Spurny *et al.*[29] with the Nuclepore filter has indicated that these filters are uniform in weight and virtually nonhygroscopic, since they weigh about 1 mg/cm. Samples as small as 10 μg can be weighed with acceptable accuracy on a Cahn electrobalance. Further, the high and uniform transparency of the Nuclepore material suggested that optical densitometry might be employed as an index of the particles collected. The primary advantage is the uniformity of the filter which makes it possible to attach meaning to very small optical densities. This method is definitely more sensitive than those using paper or membrane filter tapes.

The Nuclepore filter material is particularly advantageous in microscopy because it is durable enough to withstand handling, it is transparent, and large particles can be counted easily, sized, and examined morphologically under the optical microscope. It has the disadvantage of considerable birefringence, which limits its use in polarized light microscopy. It is excellent material for collecting particles for electron microscopy. Silicon monoxide replicas of the surface usually enclose and "extract" the collected particles for subsequent examination.

Aerosol Size Spectrometry with a Ring-Slit Conifuge

A modified conifuge for particle size determination which has an aerosol inlet in the form of a ring slit has been described by Stober and Flachsbart[30] The modification was developed to facilitate high aerosol sampling rates at reasonable size resolution in terms of aerodynamic diameters.

The determination of particle size distribution in aerosol work is somewhat ambiguous because the particle size is subject to special definition when the particles are not of homogeneous density and spherical shape. Typically, Stokes diameter of the particle is defined as the diameter of a sphere which would have the same density and, in the mechanical force field, attain the same settling velocity as the particle under consideration. This particle size definition is of significance and influence on a deposition pattern and thus on the potential health hazard of inhaled toxic or radioactive aerosol particles. Particle sampling devices for size separation consist of impactors such as the cascade impactor which produces deposits of defined size ranges which can be analyzed separately. Where continuous size spectrum in terms of aerodynamic diameters is desired it is more advantageous to use an experimental arrangement where laminar flow of air passes through a transversal mechanical force field which elutriates the particles in the air according to their settling velocity (kinetic diameter). In the case of dusts and other coarse aerosols where the kinetic diameter is not less than about 1 μm, the particle separation can easily be affected by the gravitational field acting upon a laminar flow of air passing through a horizontal duct. For smaller particles more powerful force fields which are technically available with centrifuges must be applied.

A ring-slit inlet was designed for the conifuge so as to facilitate higher rates than could be used with a conventional conifuge with a narrow aerosol inlet nozzle. The ring slit conifuge provides a much larger cross-sectional area through which the aerosol enters. Therefore at the same linear velocity, the ring-slit conifuge has a higher flow rate. Also, at the point of aerosol entrainment the cross-sectional area of the duct is greater for the ring-slit conifuge than for the original design. In the size range between 0.13–4.0 μm the modified conifuge permits sampling rates of 1 liter/min or more at a size resolution of better than 7%. For the size range between 0.06–0.13 μm the sampling rate has to be reduced to about 0.1 liter/min to give acceptable results. The rate is comparable with maximum sampling rates of other conifuge designs.

In checking the circular deposits of coarser particles in the 0.08–0.13 μm range, a microscope equipped with a Leitz Ultropak system verified the assumption that the less intense deposits consisted of aggregates of two and three spheres respectively. For the smaller latex sizes, micrographs obtained with a scanning electron microscope gave the same results.

Separation and Analysis of the Less than 10-μm Fractions of Industrial Dusts

Kupel *et al.*[31] have reported a technique which has been developed whereby an electroetched nickel 10-μm sieve can be used for removing the respirable or < 10-μm fraction of particulate matter from a dust sample. This has been done with bulk-milled samples of asbestos and with beryllium samples. The technique requires that a 1.5-gm sample be placed on top of the 10-μm sieve and subsequently placed in an ultrasonic transducer container. Fifty to 75 ml portions of distilled alcohol are placed on top of the sieve and the electrosonic generator is adjusted so that a slight ripple is noted in the liquid. The procedure is repeated until 1200 ml alcohol has been passed through the sieve. The alcohol containing the < 10-μm fraction is placed in evaporating dishes and the alcohol is evaporated. The residue is dried, scraped out, and weighed. The microsieving procedure appears to have value in that it enables the analysis of elusive respirable fractions of industrial dusts, both before and after the materials are placed in use.

Midget Impingers and Membrane Filters for Determining Particle Counts

Existing threshold limit values for respirable dust evaluated by counting techniques do not apply to samples of airborne dust collected by membrane filters. In order to compare dust counts made with membrane filters and midget impingers and thus to enable the possible use of membrane filters for determining compliance with threshold limits Renshaw *et al.*[32] conducted a study to establish a relationship between the two counting methods.

A polystyrene latex aerosol was generated and injected directly in an air sampling chamber. In order to eliminate variability of efficiencies of collection due to differences in particle size distribution, a system was set up to generate a monodisperse aerosol utilizing a polystyrene latex suspension with a particle diameter of 1.099 μm.

Sample pairs were collected simultaneously using two air sampling systems. Filter samples were collected on aerosol-type membrane filters with a pore size of 0.8 μm and diameter of 34 mm. Impinger sampling employed standard midget impingers containing 10 ml ethyl alcohol as the collection medium. Both samplers were operated at a flow rate of 0.1 ft^3/min.

For determining particle counts, impinger samples were prepared and analyzed according to the standard method of the American Conference of Governmental Industrial Hygienists. The samples obtained by membrane filter were prepared for dust counting by taking two sections of the membrane filter 8 mm in diameter (with a cork borer) from the center and from the periphery of each filter. Each section of filter was then mounted and fixed on a glass slide with a gummed reinforcement. The filter material was rendered transparent by inverting the mounted sections over a beaker of hot acetone. These were then counted by light-field microscopy in a manner similar to the impinger preparations.

The variations between filter and impinger methods included inherent variability in each sampling method, variation between the two individuals counting the dust, and the true variation between the two sampling methods. No attempt was made in this study to quantitatively separate the observed variation into its component parts. Dust count determinations from the membrane filter samples appeared to be the main cause of the observed variation. The transparent filter medium did not provide as smooth and clear a counting surface as did the Dunn cell preparations taken from midget impinger samples. Also, the dust did not appear to be uniformly deposited on the filter surface. For concentrations of aerosol of 20 million particles/ft^3 or greater, sampling times for filters had to be limited to 1 min at 0.1 ft^3/min in order not to overload the filter.

While statistical analysis, based on the data collected, would enable one to predict impinger dust counts from predetermined membrane filter counts or vice versa, the wide range of predicted dust counts for a measured value would be of very limited utility in most industrial hygiene situations.

HEPA Filter Efficiencies Using Thermal and Air-Jet Generated Dioctyl Phthalate

High-efficiency particulate aerosol (HEPA) filters are commonly used in those facilities where highly toxic materials are controlled with local exhaust ventilation. Typically, the quality of such filters is determined by the manufacturer, utilizing a 0.3 μm, monodisperse, thermally generated, dioctyl phthalate (DOP) test aerosol. The minimum acceptable filter efficiency is 99.97% as measured with a forward light-scattering photometer sampling upstream and downstream from the test filter. Since a monodisperse aerosol is used, the measurement represents filter efficiency either on a count or mass basis.

Since filters can be damaged during shipping, handling, and installation, testing the filter alone does not guarantee that the installation will be properly gasketed and sealed in place. It is a good procedure to install the filters and test them in place using a DOP aerosol which is polydispersed as generated. A forward light scattering photometer is used to measure the aerosol concentration upstream and downstream from the filter bank, but since the aerosol is polydispersed, it is not clear as to what efficiency (count or mass) is being measured. Because of this, Ettinger *et al.*[33] conducted a study at the Los Alamos Scientific Laboratory to compare filter efficiency using a 0.3 μm monodisperse, DOP aerosol with the efficiency measured using the polydispersed DOP aerosol which is routinely used for in-place testing.

Three commercial products, evaluated as potential materials for in-place testing of HEPA filters, included two types of DOP [di(2-ethylhexyl) phthalate] sold under the trade names of Octoil and Flexol Plasticizer DOP, and Octoil S [di(2-ethylhexyl) subacate] which, because of its similarity to Octoil, has been used for in-place testing. It was found that Octoil S has the highest aerosol count median diameter (cmd) 0.91–1.37 μm, while Octoil most closely approximates the 0.3 μm aerosol used during quality control testing. Flexol has a higher aerosol cmd (0.77–0.83 μm) and because of its considerably lower price than Octoil is used for in-place testing in the Los Alamos Scientific Laboratory. Size characteristics of the aerosol produced, using an air jet aerosol generator, were defined for the three commercial materials considered for in-place testing of HEPA filters. Filter penetration measured using a 0.3 μm monodisperse, thermally generated DOP aerosol was compared with filter efficiencies measured using a 0.8 μm polydisperse aerosol. Penetration of the 0.3 μm monodisperse aerosol was 10% higher than that indicated by the 0.8 μm polydisperse aerosol. The 10% variation between the two test methods is not in itself significant but it does point out that criteria set on the basis of acceptable efficiencies for quality-control tests may be adjusted to compensate for differences in the two test methods.

ENGINEERING EVALUATION AND CONTROL TECHNIQUES

Laboratory Hood Ventilation

Information obtained by Ettinger *et al.*[34] indicate that, while under ideal operating conditions lower airflow velocities at the face of the hood might be used, in practice a range of from 50–150 ft/min is applicable to laboratory fume hoods. The main consideration is that room air currents often exceed 50 ft/min that in turn can force or induce the flow of contaminated air out of the hood face. Also it is good engineering practice when working with hazardous materials to provide a margin of safety to insure that control will always be adequate.

Air supplied or auxiliary air hoods are sometimes suggested as a means of conserving conditioned air. This has been found not to be a generally satisfactory arrangement inasmuch as the control of contaminants released into the hood depends upon the velocity of air through the hood opening. Therefore the unconditioned air delivered inside the hood adds to the total volume of air exhausted and has no effect on the air velocity through the hood face and thus no effect on control. Another reason for the limited application of the auxiliary systems is that savings in refrigeration are overbalanced by the cost of a second supply ventilation system and by increased maintenance costs.

Relatively inert soapstone such as Alberene Stone is recommended for perchloric acid digestion service because of the potential explosive hazard of perchloric acid in contact with any oxidizable material. Where flammable vapors or explosive dusts may be encountered, an aluminum or bronze fan impeller is recommended to minimize chances of spark generation. Centrifugal fans are generally used for laboratory hood exhaust systems and a belt drive is preferred so that changes in the exhaust system can be made without replacing the entire fan assembly. To avoid recirculation of laboratory hood fumes the exhaust stack should be located carefully in relation to the building air intakes. It is preferable that the fan be placed at the extreme end of the exhaust system so that most of the ductwork is kept under suction rather than under positive pressure, which in turn could result in leakage into occupied areas. While individual exhaust fans are preferable, since they provide a flexibility of operation, it is usually satisfactory to connect several hoods into a single system if each hood has safety stops to prevent the door from being closed completely.

Generally, stacks extending at least 10 ft above roof level for one-story buildings and at least 15 ft above roof level for taller buildings should be employed. The discharge should be directly upward at a velocity of at least 3000 ft/min to minimize chances of recirculation.

Flame Cutting in a Granite Quarry

Flame cutting techniques have been developed for rough dimensioning or contouring of granite as well as for producing unique stone surface textures. Although limited in application, flame drilling can also be done. In flame cutting the burner cuts by a spalling action on the granite. The equipment is relatively simple, consisting of a lance with service connections for fuel and oxygen at one end and a water-cooled burner tip at the other. The burners are extremely noisy and the operations appear to be quite dusty, although dust sampling results are surprisingly low.

Burgess and Reist[35] found that the potential health hazards from flame cutting in a granite quarry include both noise and quartz dust exposure. Crystalline particles are generated in conventional cutting as the result of spalling action of the flame. A significant percentage of the particulates formed appear to be melted by the flame and subsequently form spherical particles. A second and much smaller-sized particle population is apparently formed by vaporization of the granite. The fume formed in this manner has a diameter of less than 0.1 μm and is probably in an amorphous state.

Samples collected from the various operations observed in this quarry and evaluated by conventional methods did not indicate hazardous exposures to the larger fractions. The fume exposure arising from flame cutting of granite requires additional study to identify the percentage of free silica, if any, in the fume and the concentration of the fume component.

Exposures in Repairing Microwave Ovens

To evaluate the occupational exposure of microwave oven repairmen, a clinical and environmental survey was conducted by Rose *et al.*[36]

The term microwave is applied to an arbitrary range of the electromagnetic spectrum generally agreed to encompass wavelengths of from 3000–0.3 cm or the equivalent frequencies from 10–100,000 MHz. Biologic responses to microwaves are primarily thermal but reports in the literature indicate that there may be other clinical responses (for example, the production of cataracts in the lenticular tissue of the eyes). Biologic responses in laboratory animals were found to be the most severe and to last longest at wavelengths of about 10 cm.

Exposure control for those working with microwave sources in the past utilized the concepts of distance and shielding, and more recently, the concept of time of exposure. The first exposure criterion proposed in the United States was a maximum exposure level of 10 milliwatts per square centimeter (mW/cm^2). This standard is based on energy density of 1 $mW\text{-}hr/cm^2$ for exposure times up to 0.1 hr and on power density of 10 mW/cm^2 for time periods of 0.1 hr or more. In addition, there is the latest U. S. Army and Air Force criteria of $T_p = (6000/W^2)$, where T_p is permissible exposure time in minutes during any 1-hr period and W is power density to which the worker is exposed, milliwatts per square centimeter. The army criteria is applicable between exposure levels of 1 – 100 mW/cm^2.

In modern microwave ovens the microwave source is a magnetron, the output of which is fed by a horn or waveguide into the cooking cavity. The door and cabinet interlocks are designed to insure that before the magnetron can be energized the unit is sufficiently shielded to protect the user.

In the survey one of eight workers, in a repair shop for microwave ovens, was noted to have a recurrent, blotchy, hemorrhagic eruption on the abdomen and thigh, which was histologically an anglitis (vasculitis). There was no known cause but its anatomic localization, pathologic picture, and temporal relation to the subject's work suggested the possible causal role of microwave energy.

Adequate protection can be afforded the worker by use of a simple, inexpensive device to intercept and absorb the microwave energy, i.e., a wooden frame covered with copper mesh screening which is placed on the oven. Such a device can reduce microwave energy to levels less than 5 mW/cm^2.

Ozone – A Cinema Production

Xenon bulbs, a new light source for motion picture projectors, produce a broad-spectrum light which includes ultraviolet radiation, below 2000 Å, that in turn ionizes the oxygen of the circulating (cooling) air. These disassociated ions combine and form ozone. The ozone is discharged into the room air by the fan in the projector.

As the result of complaints from machine operators, Felton[37] conducted an industrial hygiene study of a projection room. The equipment consisted of two Bell & Howell JAN 614CD 16-mm units modified and equipped with 450 W sealed xenon gas arc lamps. Air samples were collected in midget impingers containing 10 ml of 2% buffered potassium iodide as the absorbing solution. A flow rate of slightly more than 1 liter/min was maintained during sampling periods. The collected samples were read on a Beckman DU spectrophotometer within 3 hr after collection. The samples were taken and analyzed on random days over a period of 5 months. More than ninety samples were taken during the survey.

Atmospheric concentrations of ozone in the projection room varied from less than 0.01 ppm to more than 0.50 ppm. Average readings of ozone in air varied from nuisance or objectionable level of 0.04 ppm to levels above 0.1 ppm. The threshold limit value for ozone is 0.1 ppm.

It was readily shown that the ozone contamination could be controlled with local exhaust ventilation. A new line of ozone-free xenon lamps has recently been placed on the market.[38]

Biohazard Determination Aboard Two Naval Vessels

Recent modification of a navy destroyer with a new design concept for the surface vessel ventilation systems, with special toxicologic protection features, generated an interest in comparing the health conditions that might be imposed upon the crew during routine shipboard operations with that of a destroyer of normal fleet design. To assess the potential biohealth hazard created by possible buildup of airborne bacteria when activities were restricted to a closed environment, Wright *et al.*[39] measured the bacterial content of the air in selected compartments of both ships and compared findings within similar spaces.

The major difference between the ventilation systems is that in the new design concept, large volumes of recirculated air are subjected to high-efficiency cleaning by electrostatic precipitation with only a small quantity of highly filtered fresh makeup air provided. The second ship utilizing the conventional ventilation system provides a volume of coarsely filtered tempered fresh air drawn directly from the outside. Air sampling was conducted only while the ships were at sea in a normal operating status. Samples were collected by direct impaction samplers. One type used a four-stage cascading sieve-type device modified by the Naval Biological Laboratory. This provided aerodynamic classification in the ranges greater than 6 μm, 3–6 μm, 1.3–3 μm, and 0.7–1.3 μm. The size overlap between stages is minimal at stages 1 and 2 but increases to a level comparable to that reported for a five-stage sampler which was also used. The sampling rate was 20 liters/min for the four-stage sampler and 28 liters/min for the five-stage sampler.

Air samples taken on deck or from outdoor air supply ducts, both during daylight and at night, showed no bacteria and only occasional fungi on BHI agar. In addition to samples taken on BHI agar aboard the "new ventilation concept" vessel, forty-two samples were taken using EMB agar, fourteen on BAB agar and thirteen on MS agar.

Only a small number of bacteria grew on the selective media. No hemolytic streptococci were recovered. Many of the air samples contained fungi. About 90% of the organisms recovered on EMB agar appeared to be of the genus *Aerobacter* and the other 10% were unknown saprophytes. By far the greatest number of organisms recovered were nonhemolytic staphylococci.

It was found that personnel activity is a key factor in bacterial burden. In compartments with high occupancy and activity, maximum numbers of bacteria were found either in the early morning or afternoon, whereas the minimum counts tended to occur in the midafternoon and later evening. Peak counts were obtained at rising time in the morning and at bedtime. The counts noted in late evening when most personnel were sleeping were as low as those observed when there were relatively few persons in the compartment. In general the living spaces showed consistently higher levels of aerial contamination than did spaces used for operational purposes.

It was found that the airborne bacterial flora on the USS Thomas (newly modified system) was consistently less than that on the USS Southerland, although neither vessel provided an environment that could be considered contaminated under normal conditions. One could conclude that the shipboard ventilation systems can effectively reduce the normal flora generated by the crew but that the recirculation system in the USS Thomas was, for air sanitation purposes, as effective as or better than the air flow-through system of the USS Southerland.

Dust Control in Bituminous Coal Mines

There is much concern in the United States and Europe about the incidence of coal miners pneumoconiosis. Federal legislation[40] has been enacted in the United States which prescribes the techniques for evaluating work-area dust exposures on the basis of respirable-size fraction limits. Morse[41] has prepared a detailed review of dust control practices in the bituminous coal mining industry. He indicates that in general the methods used to achieve dust control in bituminous coal mining operations include the following. (*1*) Application of water to cutting bits, drills, to the coal face, to coal piles, and to coal on conveyors (cutting machine and hauling conveyors). Water is also applied to shuttle car and loading stations and on roadways. (*2*) Use of high pressure sprays on machines for suppressing dust. (*3*) Exhaust ventilation is provided with line brattice or tubing which is installed in such a way as to remove the dust from the breathing zone of the operator. Another approach is the use of a line curtain supplying air to dilute dust concentrations. (*4*) Ventilation of the continuous miner working space with a separate split of air which is coursed directly to the return without passing over other operators. (*5*) Respiratory protective equipment of either the filter mask type or a cryogenic air supply system to provide air to a face piece worn by operators.

It has been determined that face-cutting operations tend to be a major source of dust production in the mine and they in general impose a difficult and important dust control task. Roof bolting and drilling while being major dust producers can be successfully controlled through the use of wet drilling techniques. The dust produced at the face by a cutting machine is a function of the number of bits, the sharpness and speed of the bits, the depth of bit penetration, and very importantly, the experience of the machine operator. The operator should be instructed to minimize cutting into the roof or floor since this increases the dust concentrations. Blunt bits increase crushing action with the resulting generation of more dust.

In England the National Coal Board–Mining Research Establishment has given considerable attention to the theory of dust generation and has shown that the amount of splitting-off (coal chunks removed from the wall face) increased at a faster rate with deep bit penetration and with less dust produced. The NCB–MRE has also indicated that about 5% of the respirable dust created by a mining machine operation is dispersed into the ventilation air currents. The remainder aggregates with the larger particles soon after it is formed. Some of these fine dust particles are dispersed into the air by subsequent disturbance of the settled dust. This makes it important that there be wetting of the coal that falls to the floor by sprays which are mounted on the machine. This practice is widely used by coal operators in the United States.

Water-sprays on mining machines is one of the most widely used dust control methods. The sprays are located on each side usually behind the trim chain on boring machines, in the throat of the conveyor on all machines, on the cutting boom, and on the bar of ripping and milling machines. Sprays should also be applied at conveyor loading stations and along haulage belts to reduce dust dispersion. It has been found that dust suppression with water is most effective when the maximum tolerable amount of water is applied as early as possible to the coal. A nozzle pressure of 200–250 lb/in.^{2} is effective both for controlling dust and minimizing the plugging of nozzles. Droplet size has not been found to have any practical effect on dust removal where the dust is in the 2–20 μm size range. It has also been found that the use of wetting agents with the spray water does not appear to be better than water alone for the suppression of coal and rock dust in underground conveying. There are only marginal advantages when using wetting agents for the suppression of respirable dust at mining machines. However, there appear to be advantages to the use of wetting agents with hard-to-wet coals.

A relatively new mine ventilation technique involves the use of a two-split ventilation system. In this system the continuous miner working space is ventilated with a separate split of air. The air is coursed continuously over the miner and is returned directly to return airways without passing over other operating equipment. A minimum of 8000 ft^3/min is provided to the face and exhausted behind a line brattice. A minimum of 50,000 ft^3/min is provided in developing sections.

Three principal face ventilation techniques in use today include (*a*) line brattice on exhaust, (*b*) line brattice on supply, and (*c*) exhaust tubing with auxiliary fan. The effectiveness of any line brattice ventilation system depends upon good maintenance to reduce leakage to the minimum, maintaining the end of the brattice as close to the face as practical, and prevention of the mining machine operator from moving in by the end.

Where exhaust tubing and auxiliary fan ventilation techniques are used, a minimum exhaust air volume of 4000 ft^3/min is required. Care must be exercised in locating the discharge of the fan far enough into the return airway to prevent recirculation during a breakthrough and to add metered amounts of rock dust to the discharged air in the interest of preventing explosions of the coal dust in the return air courses.

It is sometimes necessary because of inadequate ventilation to use respiratory protective equipment. This should at best be considered a temporary measure and should not be considered a substitute for satisfactory ventilation control. Respiratory protective devices approved by the U. S. Bureau of Mines should be used.

Iron Oxide Fume at Arcair (Powder-Burning) Operations

Arcair, also known as jet-arc and powder-burning, is used in the foundry and steel fabrication industry for scarfing, trimming, and surface gouging of steel and cast iron parts. The process uses heat generated by an electric arc drawn between the work piece and a specially formulated copper-coated boron–graphite electrode. The intense heat creates an area of molten metal which is simultaneously removed by jets of compressed air.

Depending upon operational requirements, the compressed air pressure varies from 80–150 psi at a usual supply rate of 30 ft^3/min to remove molten metal. The power unit is dependent upon the size of the electrode being used. The rods vary from 1/8–1 in. in diameter and require from 50–220 A. In the study the arcair stations surveyed used 5/16, 3/8, and 1/2 in. rods at 400–800 A to remove an average of 0.8 pounds of metal per minute.

Powder-burning is a process in which an acetylene-burning torch which has been modified to accommodate a proportional mixture of acetylene gas, oxygen, and finely divided iron powder is used. The operation is commonly used for burning gates or protuberances from metal castings.

In the study by Sentz and Rakow[42] the environmental and preliminary medical data were obtained at five steel-manufacturing plants. The study was basically concerned with employee exposure to iron oxide fumes at arcair and powder-burning operation. Environmental samples were collected with membrane filters, 0.8 μm pore size, in plastic field monitors. The average sampling time was 10 min at 1 ft^3/min or less. Breathing zone samples were obtained by placing monitors both inside and outside the worker's helmet adjacent to the cheek. The threshold limit value (ACGIH) for iron oxide fume is 10 mg/m^3 based on an 8-hr working day.

Thirteen breathing zone samples, collected at six powder-burning operations, showed the operator to be exposed to an average iron oxide fume concentration of 31.1 mg/m^3 when no local exhaust ventilation system was utilized. With sidedraft exhaust ventilation, most concentrations were reduced to less than 7 mg/m^3.

About forty-five environmental samples collected at seventeen different arcair locations revealed that iron oxide fume concentrations were about 40% higher immediately outside the worker's helmet than inside. The operator was found to be exposed to an average concentration of 21.5 mg/m^3. When local exhaust ventilation is applied in the form of an effective sidedraft hood, exposures to iron oxide fume can be maintained well below the threshold limit value.

Ventilation Requirements for Gas-Metal-Arc Welding vs. Covered-Electrode Welding

A study was conducted by Alpaugh *et al.*[43] to compare the amounts of certain particulate and gaseous by-products generated by the shielded-metal arc and the covered-electrode welding processes. Air was sampled both inside and outside the helmet under similar circumstances to estimate the protection afforded by a "curved-chin" type welding helmet. Solenoid valves were set for inhalation and exhalation time observed in a normal adult male. A simulated breathing procedure was utilized together with a plaster dummy head mounted on an automatic welding head carriage that moved parallel to the welding direction so that its position remained fixed relative to the welding arc. Air samples, collected concurrently inside and outside the helmet during the welding process, were analyzed for iron oxide, nitrogen dioxide, ozone, fluorides, and carbon monoxide. Combinations of five wire electrodes, three of which were flux-cored, and five shielding gases were evaluated. One type of coated rod electrode was also used for comparison purposes.

It was found that the flux-cored wire electrode and the covered electrode generated concentrations of iron oxide of the same order of magnitude or greater than the shielded metal arc electrodes. Iron oxide concentrations ranged from 31.7–112.9 mg/m^3 determined outside the helmet and from 2.7–4.5 mg/m^3 inside.

Nitrogen dioxide concentrations inside the helmet were comparable for the different welding procedures. The average values inside of the helmet ranged from 0.54–1.49 ppm as compared to 1.62–9 ppm outside. The average fluoride values inside the helmet ranged from 0.43–1.04 mg/m^3 and those outside from 0.11–1.24 mg/m^3 with a peak of 2.73 mg/m^3. No measurable amount of ozone was detected when the covered electrode was being used, but significant concentrations were observed outside the helmet with the shielded gas welding process. Bare wire electrodes using argon in the shielding gas produced higher levels of ozone but it was found that the helmet significantly reduced the ozone concentrations inside the helmet. The ozone concentrations inside of the helmet ranged from nondetectable levels to 0.06 ppm and those outside from nondetectable levels to 0.49 ppm.

It was concluded from this study that contaminants generated by the shielded arc welding process were in general comparable to those generated by the covered-electrode welding process. On this basis it was further concluded that it would be unreasonable to specify a more stringent ventilation requirement for either process, but with the understanding that adequate ventilation is required for contaminant control. A secondary conclusion resulting from the study was that the welding helmet is an effective barrier against particulate and gaseous by-products of the welding process. It also should be noted that concentrations of contaminants which were determined were not time-weighted averages but represented a temporary elevated concentration of contaminants which build up shortly after the arc has been established, that is, conditions considerably more severe than encountered in routine production welding.

Evaluation of Tetraalkyllead Exposure by Personal Monitors

It was found by Linch *et al.*[44] in an extensive study of the airborne lead exposures in a large tetraalkyllead manufacturing plant that the air sampling results from seventy-seven fixed station monitors did not correlate well with the urinary excretion rates or routine medical examinations of personnel. It was concluded that fixed station monitors probably do not disclose the true inhaled air concentrations of lead in a highly variable ambient work atmosphere. In an attempt to validate this conclusion, an extensive study was conducted using personal monitoring sampling devices. The sampler used was made up of the following components: (*1*) bypass capillary; (*2*) activated carbon trap protection for the air sampler; (*3*) membrane filter, 0.8 μm pore size; (*4*) aluminum shield containing "spill proof" microimpinger; and (*5*) coated fabric carrying case which attaches top and bottom to the wearer's shirt by safety pins.

The inorganic (particulate) lead was collected on a membrane filter, ashed with nitric acid to dryness and analyzed by the standard dithizone method. The organic lead vapor was collected in 0.1 *M* iodine monochloride dissolved in aqueous 0.3 *N* hydrochloric acid and analyzed by a modified dithizone procedure. A Beckman Model DU spectrophotometer was used for the inorganic lead and a Beckman Model B spectrophotometer for the organic lead determinations. In the study, in an attempt to obtain correlation between biologic sampling, fixed station monitoring, and personal sampling, operators were monitored for 5 days on their day-shift rotations. Urine specimens were submitted 24 hr before start of the monitoring and at the end of each shift on which the monitor was worn. The fixed station monitors were operated on an 8-hr shift basis in the locations where the personal monitors were worn.

Linear and multiple regression analysis of the daily individual results disclosed no correlation between fixed station and personal monitors. Regression analysis also failed to indicate any correlation between individual paired urine and monitor (either personal or fixed station) results on a daily basis. It was found in the tetramethyllead operation that the average organic lead-in-air analysis was 0.179 mg/m^3, more than twice the current TLV (0.075 mg/m^3), but the average urinary lead concentration (0.071 mg/liter) was not significantly elevated above high normal (0.03 ± 0.03). Ten percent of 115 urine specimens gave results above 0.10 mg/liter but none exceeded 0.15 mg/liter.

Airborne organic lead levels for tetramethyllead operations were 0.21 mg/m^3 (1.5 times the threshold limit value). The average urinary lead content was found to be 0.06 mg/liter with only 3.5% above 0.10 mg/liter and none over 0.15 mg/liter.

It was concluded after a 3-month study that satisfactory health protection could be provided employees engaged in the production of tetraalkyl lead compound through a control program based on urine analysis and medical history. Fixed station air monitoring does not provide valid results. The use of the personal monitor type sampling device did enable the establishment of an approximate relationship between airborne alkyllead concentrations and urinary lead excretion. While such a program is costly to administer and the equipment is a source of annoyance to operation personnel, personal monitoring is the preferred procedure where air sampling is needed. It was also found that the revised threshold limit values for tetramethyllead (0.15 milligrams of lead per liter) and tetraethyllead (0.10 milligrams of lead per liter) can be exceeded by a factor of 2 before the urinary excretion control point of 0.10 ± 0.01 mg/liter urinary lead is reached.

Industrial Hygiene Engineering: The Process-Environment System

Mutchler[45] has suggested that the systems approach to occupational health can be used as a means of integrating the concepts applied by engineers, industrial hygienists, toxicologists, physicians, and others who make up the occupational health team. This approach would necessitate the use of automation, process control, and computer technology. A system, in terms of industrial health, is an organizational framework that defines the strategy for assuring the healthfulness of the worker population.

The author suggests that the design of a process should relate directly to the acceptable degree of environmental contamination. Since the operation of the process can vary in a manner that will cause changes in the level of contamination of the work environment, the acceptable exposure level serves as a guidepost toward which all future changes and improvements in the operation of the process should be directed. The industrial hygiene standard or the threshold limit value directly relates to the evaluation technique utilized and the threshold limit value is used as the standard against which the estimate of exposure is checked.

Continuous monitoring, which serves as an effective diagnostic tool for the identification of predominant sources of contamination in the work environment, and a digital computer coupled to an environmental monitor provide an effective means of reducing large volumes of data into a manageable form.

When a new process is to be placed on line, environmental surveillance should be repeated regularly or made a permanent feature of a process–environment system. A continuous monitor can be invaluable in characterizing the environment. The concentrations and working time at each of the work areas or sampling locations must be weighted to estimate the time-weighted 8-hr average concentration, the basis of most of the threshold limit values.

Other advantages of continuous monitors, if properly used, are that they can be used for diagnostic air sampling (including establishing time-dependent and location-dependent patterns) and they provide efficient documentation and reduced material loss. Where one considers the threshold limit value, acceptable ceiling concentration, acceptable maximum concentration for peaks above the acceptable ceiling, and the acceptable number of peak concentrations of five miniaturizations for an 8-hr period, a large number of samples is required for an appropriate evaluation of workroom exposure. Continuous monitoring coupled with computer programming and real-time data production could facilitate an evaluation based on the various limits noted above. From a systems engineering point of view there is really little difference between controlling the process and controlling the associated work environment.

Problems of Laboratory Glassblowers Working Fused Quartz

Fused quartz or silica glass has a very low thermal expansion coefficient and a high transmission of ultraviolet light that makes it useful in the fabrication of lasers. Skilled glassblowers are required to work with the material at a temperature range of from 1850°–1950°C. This necessitates an oxygen–hydrogen flame. Other types of glass are commonly worked at temperatures of 1000°–1200°C.

Because of questions concerning the health hazard of vapors released during the fabrication process, an environmental study was conducted by Schreibeis[46] to determine the concentrations of glass fumes, ozone, and oxides of nitrogen in the workroom environment. It had been reported that a metallic odor, and a dryness of the throat were produced during the working of fused quartz; references have been made to these operations having a silicosis potential. Both fused quartz and glass are amorphous.

Most glassblowing facilities consist of tables or lathes located under a side-draft type hood with an upper canopy type take-off. The factors evaluated in this study were the exposure to volatilized silica, ozone concentrations, and oxides of nitrogen. It was reported that no particulate could be determined in the breathing zone sample under most conditions but that a breathing zone concentration of the lathe operator was found to be 0.17 mg/m^3 when 4-in. diameter tubing was being worked. Ozone sampled through a period of 7 consecutive days showed a peak of 0.008 ppm with an average below 0.002 ppm. No oxides of nitrogen were detected in the worker's breathing zone but they were found in the plume inside the hood. It is concluded that when a properly designed local exhaust ventilation is provided with exhaust rates ranging between 300–500 ft^3/min, there is a negligible exposure to silica fume, ozone, and to oxides of nitrogen.

Air Supply Rates for Powered Air-Purifying Respirators

An investigation was carried out by Burgess and Reist[47] to determine the optimum supply air rates for half-mask respirator face pieces with powered air-purifying devices. The operational modes included an air supply rate which provided a positive pressure in the respiratory enclosure and a somewhat less desirable air supply rate which did not achieve positive pressure at all times but did result in reduction of breathing resistance and greater protection than conventional air-purifying respirators. A work rate baseline of 415 kg · m/min implies a volume of 30 liters/min (peak inspiratory flow would therefore be 90 liters/min). Higher work rates in the range of 622–830 kg · m/min are probably briefly experienced in industry and can be continued for about 10 min by a fit individual.

Pressures in the facepiece were measured at various supply rates under work conditions for the purpose of determining minimum air supply rates which will insure a positive pressure. Tests were conducted with the unit operated as a conventional air-purifying respirator (no air supply) and with air supply rates of 15, 44, 71, 98, and 125 liters/min. At the medium work rate of 415 kg · m/min, positive pressure is assured for all subjects at a supply rate of 110 liters/min. The parallel study at 830 kg · m/min demonstrated that a supply rate of 170 liters/min was required to achieve positive pressure.

The protection factor of the powered air-purifying respirators was tested by exposing an individual wearing the respirator to submicron sodium chloride aerosol in an exposure chamber. The concentration of sodium chloride was measured inside the mask and in the exposure chamber with a flame photometer. The percent of salt concentration noted inside the respirator was measured directly.

It was determined that an air supply rate of 116 liters/min is adequate to provide protection factors of 1000 on tight half-masked facepieces at a work rate of 415 kg · m/min. It was also concluded that a flow rate of about 113 liters/min (4 ft^3/min) appears to represent the optimum rate with minimum power pack size and weight while maximizing protection factors. This flow rate would match the maximum inspiratory flow at a work rate of 415 kg · m/min, would provide a positive pressure inside the face mask at all times at this work rate, and would provide a protection factor of at least 1000.

An Air Supply System for Operators of Motorized Street Sweepers

Schrag[48] found that the operators of motorized street sweepers are exposed to relatively high dust concentrations in the operating cabs. In an investigation, it was concluded that heated and filtered air supplied to the cab by means of a mechanical system would be advantageous and was installed. The system was superior to the use of U. S. Bureau of Mines approved respirators because of the discomfort and the operators reaction against them. The combination of hard hat and air-supplied helmet was considered to be clumsy and restrictive to the operator's vision.

The air supply system consisted of a vertically mounted filter box 24 by 24 by

10 in. high, a heating coil and plenum and a centrifugal fan capable of delivering 800 ft^3/min against a static pressure of a 2½ in. water gauge. The fan was driven by a belt-and-pulley arrangement from the engine crankshaft thereby assuring that air was available whenever the sweeper was used. A heating coil was connected to the engine cooling system to provide adequate heat for operation at outside temperatures as low as 10°F. Visual observation of the dust cloud raised while operating under various conditions of wind direction indicated that a location of the air intake on the vehicle central line just above the junction of the windshield and cab roof provided the least contaminated air. A 2-in. renewable type fiberglass filter followed by a 10-ply paper media type filter gave adequate filtration, satisfactory service life, and reasonable cost.

Motorized sweepers are used in Alberta, Canada where it was found that the road dust can contain up to 40% free crystalline silica in the respirable fraction. Based on the time-weighted exposure determinations, it was concluded that the threshold limit value should be about 14 million particles per cubic foot of air. The filter system was effective in reducing dust concentrations from over 60 million particles per cubic foot of air to approximately 7 million particles per cubic foot of air when operating on very dirty streets.

The Work Environment of Insulating Workers

Workers in the insulating trade handle asbestos and a variety of other materials. Evidence is being accumulated to verify that asbestos workers have a higher than average incidence of malignancies of the lungs, pleura and the peritoneum. Balzer and Clark conducted a preliminary environmental study[49] to determine the incidence of pneumoconiosis and malignancies in insulating workers in the San Francisco area. The preliminary study included measurements and determination of the materials used, methods of application, and dust and fiber concentrations.

There were approximately 500 members in the local union (International Association of Heat and Frost Insulators and Asbestos Workers), about 70% of whom were in the trade more than 10 years and approximately 42%, 20 years. Chest films had been taken on the voluntary basis for about 10 years with 80% of eligible workers participating. Review of the films had shown that about 25% of the members have roentgenographic changes strongly supporting the diagnosis of asbestosis with an equal percentage having suggestive changes.

> The major types of work which the men were engaged were classified as (1) commercial building, which involves the insulation of pipe and duct systems in office buildings, apartments, shopping centers, (2) heavy industrial building, which involves the insulation of turbines, boilers, pipes, duct systems, and processing equipment in power plants, factories, and (3) marine construction and repair, which involves insulating turbines, boilers, pipes and duct systems in ships in both naval and private vessels.[49]

A study of the insulating worker's environment and related work activity exposures indicates that he is predominantly working with calcium silicate and magnesium carbonate insulating material containing asbestos fibers, fibrous glass materials, plastics, foam, glass, cork, and adhesives. Chrysotile (hydrous magnesium silicate) is the most important commercial asbestos mineral accounting for about 90% of the world's asbestos production. Amphibole-type asbestos is found in five commercially useful forms: crocidolite, amosite, anthophyllite, tremolite, and actinolite. In the United States insulation industry, chrysotile (from Canada) and amosite (from South Africa) are most often used in manufacturing insulation materials. The investigators confirmed through x-ray diffraction analysis the presence of asbestos fibers in fourteen different asbestos insulating products, manufactured by seven major producers of insulating materials in the United States.

Fibrous glass consisting of a mixture of silicon dioxide, oxides of aluminum, calcium, magnesium, boron, and other additives was also a source of exposure. It generally was found that most of the commercial insulating products have a mean diameter of 4 microns or greater. Optical measurements indicated that the acoustical fibrous glass material has the largest diameter: 11–14 μm; building and duct insulation, 4–7 μm; pipe covering, 4–6 μm; and special high temperature material, 1–2 μm.[49]

In attempting to classify the time allocation of the asbestos worker the following was determined: (*1*) prefabrication: materials are precut and shaped using hand or power saws either on the job or at the contractors shop (10% of his time); (*2*) application: materials are fitted, hammered, or carved, and attached to the surface by wiring or gluing (40% of his time); (*3*) finishing: materials are coated with asbestos-containing cements, resins, asbestos or cotton cloth, or petroleum based sealers (30% of his time); (*4*) tearing out: removal of old or unusable materials in the process of insulating or reinsulating (10% of his time); (*5*) mixing: mineral wool, asbestos, fibrous glass, and cements or glues are mixed separately or in combination in buckets or troughs (5% of his time); (*6*) general: cleaning up old insulation, transporting materials (5% of his time).

Midget impinger samples were collected in order to enable comparison with the threshold limit value of 5 million particles of asbestos per cubic foot of air. It was found that the breathing zone concentrations exceeded the threshold limit value in three distinct areas – prefabrication, tearing out, and mixing. Personal and general samples were collected on membrane filters mounted in field monitor cases in order to help in the assessment of the role that the fibers play in the pathogenesis of asbestosis. Samples were collected for periods ranging from 30 min to 3 hr depending upon the estimated workplace concentration. Fibers were counted, by clearing a wedge-shaped segment of the filter using a modified technique of the United States Public Health Service, and sized by length and diameter at 430 X magnification using phase-contrast illumination. A fiber is defined as having an aspect ratio of 3:1 length to diameter. It was found that 98% ± 2% of the fibers have diameters smaller than 3.5 μm. These can be considered to be respirable fibers. Median fiber counts ranged from 0.8–8.4 mppcf with the highest concentration found in prefabrication, tearing out, and application.

The Plasma Jet: Industrial Hygiene Aspects and Current United States Practices for Employee Protection

Plasma may be defined as a mass consisting of free electrons and positively charged particles. Basically all plasma torch equipment utilizes an arc which is made to jump from one electrode through a confining space to another electrode, and a carrier gas which becomes plasmoid. According to Fannick and Corn[50] there are three types of plasma torches: the transferred (*1*) and the nontransferred (*2*) arc installations, which make up the majority of industrial units and (*3*) the radio-frequency or induction plasma torch. In the nontransferred plasma torch, the electron flow is impinged on the torch's hardware and only the plasma flame extends beyond the nozzle. The transferred arc is operated in the same manner; however, an incidental ionized path is formed beyond the end of the nozzle by inertia of the particles in the plasma stream produced. When the torch is brought close to the work piece connected to the opposite pole of the power supply, the main arc is established. Apparently little electrode heating occurs at the nozzle and it is possible to operate the unit at greater arc current densities (and therefore higher temperatures) than with the nontransferred unit. In the radio-frequency plasma torch, a stream of ionized gas contained in a quartz pipe is heated by induction. The plasma-forming gas is introduced tangentially, causing a swirling motion, and a fraction of the ionized gas is recirculated to stabilize the charge. After stabilization the system works as a transformer with the induction coil as a multiturn primary and the plasma as a one-turn secondary.

The heat developed in the plasma is great, thus when a developed plasma arc-flame comes in contact with the metal work piece a tremendous store of energy is released to the work surface. As an example, the heat transfer rate with a transferred arc plasma torch can be twenty times that of oxyacetylene flame (10 Btu/in./sec).

In an effort to develop information based on health evaluations of plasma jet installations in the United States, letters were sent to fifty-one state and territorial health agencies asking for survey information. About thirty-three replies were received and the following was indicated: measurements of ionizing radiation did not exceed 0.012 milliroentgen per hour, which is within accepted limits. It was found that the plasma jet produces intensive quantities of visible light. Personnel in the area should wear full face shields containing filters for eye protection. Minimum shade densities of H-5, H-6, H-9, and H-12 are recommended. Intense ultraviolet radiation is produced by the plasma arc torch, therefore all exposed parts of the body should be covered to protect against skin burns and possible harmful reaction to ultraviolet radiation and to protect the workmen being treated with certain drugs. Ozone, formed by ultraviolet radiation acting on oxygen in the air, ranged from 0.01–2.21 ppm. It was confirmed that while ozone concentrations can easily exceed the recommended threshold limit value of 0.1 ppm, control is readily achieved by local exhaust ventilation. Oxides of nitrogen are formed by the combination of oxygen and nitrogen atoms after molecular dissociation. The threshold limit value for oxides of nitrogen (as NO_2) is 5 ppm and air sampling data verified that concentrations determined at plasma jet operations could also exceed this value. However, such concentrations are readily controlled with local exhaust ventilation.

Sound-intensity levels produced in plasma torch operations are dependent on the rate of gas flow in the torch nozzle, which is dependent on the heat exchange properties of the gas and the requirements of the job. Measurements of sound intensities indicated that most of the plasma jet installations produce noise that exceeded the recommended criteria for prevention of hearing loss damage.

Chlorinated hydrocarbons should not be used in work areas where plasma arcs are utilized since the ultraviolet radiation from the arcs can readily decompose the solvent vapors to form irritating and/or toxic gases including phosgene. In addition, care must be taken to insure that residual solvent is not present on metal parts to be worked. It has also been found that carbon monoxide can be formed from the recombination of dissociated carbon dioxide. Concentrations were found that exceeded the threshold limit value of 50 ppm.

Local exhaust ventilation in form of side-draft hoods or booths is recommended for control of plasma jet operation. The face velocity should range from 200–1000 ft/min depending upon the character of the work being done and the size of the hood being used.

Copper Dust

A slight health debility among three workers, characterized by discomfort and distress, was associated with exposure to fine copper dust.[51] The operation consisted of lapping and polishing copper plates on a typical metallurgic polishing wheel while held in a fixture which rotated counter to the direction of the wheel. Four aluminum oxide abrasive wheels were in use on a fairly regular basis in a 15 by 25 ft room. A supply of filtered fresh air was introduced, producing a positive pressure within the room.

Several weeks after the operation was started, equipment operators "complained of discomfort similar to the onset of a common cold, with slight sensations of chills or warmth and stuffiness of the head" A detailed investigation was undertaken. Wipe tests of settled dust were taken by filter paper and analyzed spectrographically. These showed varying amounts of copper at all locations. Dust levels, determined with an electronic particle counter, indicated concentrations ranging from 50 thousand particles per cubic foot rising to peaks of 800 thousand particles per cubic foot as the operations were carried through the 25-min polishing cycle. Air samples were taken on membrane filters and analyzed spectrographically with the following results:

Number	*Location*	*Copper found (mg/m³)*
1	General air sample, middle of room, two wheels in operation	0.075
2	General air, middle of room, one wheel completing polishing cycle	0.030
3	Breathing zone at front of only wheel in operation	0.120
	ACGIH Threshold limit value	0.1

On the basis of the air sampling data it was decided to apply local exhaust ventilation to each wheel. Perforated copper tubing was placed around the inner side of the basin to provide a continual water flush to minimize the accumulation of the deposit. It had been determined (before exhaust ventilation was completed) by additional air tests that the "exposures arose from the drying out of the copper dust on the inside of the basin and from vibration of the basin and air currents induced by the rotation of the wheel."

Following installation of the local exhaust ventilation and water line, additional air samples were collected. They demonstrated levels of copper dust in the atmosphere less than 0.008 mg/m^3. Employee complaints ceased, and as far as could be determined there were no further feelings of discomfort, chills, and colds as had been previously registered. It is interesting to note that the concentrations of copper dust determined in the air were in the order of one-tenth the threshold limit value for copper dust. It may very well be that there were higher copper-in-air concentrations depending upon the amount of work being done at any time and that peak levels were not accurately determined.

Control of Airborne Dust Produced by Pneumatic Jackpicks

Scharf[52] was able to collect data on the usefulness of a dust control tool attachment at forty-eight excavation sites where jackpicks were in use. A cone-shaped continuous flow water attachment was utilized for dust suppression. It was found that dry picking was totally unacceptable because of the high dust concentration which developed and further that a down position of the water attachment (close to the work surface) was superior to the up position. No correlation could be determined between dust concentrations and trench depth under dry picking conditions. However, under wet picking conditions a correlation was shown. On the basis that the results obtained represented normal distribution, the threshold limit value of 6 million particles per cubic foot of air was exceeded 86% when dry picking, 24% with the cone in the up position, and 4.5% with the cone in the down position.

Monitoring Exposures to Vinyl Chloride Vapor: Breath Analysis and Continuous Air Sampling

In a chemical plant, where workers were exposed to vinyl chloride, an extensive environmental survey was conducted by Baretta *et al.*[53] to enable comparison of the time-weighted average exposure levels with breath analysis information. The plant atmosphere was monitored automatically to determine time-weighted average exposure levels for the job classifications to be studied. Five sampling probes were strategically placed in the work area and a sixth probe was placed outside the building. Charcoal and silica gel filters were placed in the sampling line to provide a clean-air reference. Sampling was executed sequentially with a set of solenoid valves controlled by a timer which advanced the sampling location every 5 min. The sample data could be viewed visually on the strip chart recorder. Data were recorded in digital form on paper tape by a tape punch and digitizer. These were programmed to record three equally spaced transmittances during the last half of each 5-min sampling period.

Breath decay curves constructed for several chemical solvents have been useful clinically as an index to chemical exposure. In this study breath decay curves were constructed from data obtained during experimental exposure to vinyl chloride both in an exposure chamber and in the chemical plant atmosphere. The validity of continuous monitoring data and the usefulness of breath analysis in assessing time-weighted average exposure was reflected to an appreciable extent by the close similarity between the two sets of breath decay curves.

The automated monitoring system enabled the development of time-weighted average concentrations calculated for each job classification. Three separate breath-sampling programs were conducted concurrently with the monitoring program. Each worker collected three breath samples daily, the first on his arrival home after work, the second 5–10 hr later, and the final sample before returning to work the following day. The samples were collected in pipets constructed from short lengths of 20-mm soft glass tubing to which had been welded at each end the threaded portion of an 8-mm screw cap glass vial. The plastic caps were lined with six layers of Saran film and then a 3/32-in. hole predrilled through one of the caps to provide an access for withdrawing samples. The Saran liners provided an effective gas barrier with vapor losses less than 10% for a holding period of 3 days. The worker was asked to remove the caps, place the pipet in his lips, and breathe normally in through his nose and exhale through the pipet three times. After expelling the fourth breath he quickly capped the tube, trapping a portion of the alveolar air.

Thirteen men participated in three experimental chamber exposures at nominal concentrations of 50, 250, and 500 ppm of vinyl chloride, producing a total of 160 breath data points. Ten workmen participated in the on-the-job breath sampling program producing a total of ninety-one sets of data. Absolute breath levels ranged from about 20 ppm in one sample (taken less than 1 hr after an 8-hr time-weighted average exposure of 250 ppm) to barely detectable levels (less than 0.05 ppm) in samples taken after exposures. Time-weighted averages were below 50 ppm.

It was concluded that a carefully conducted survey combining continuous analysis of the workroom atmosphere with a comprehensive job study could provide data valid for estimating time-weighted average exposure. However, breath decay curves are also very useful as a second method for assessing exposure to vinyl chloride vapor. The choice of whether one or both methods should be used is dependent upon the thoroughness of the study to be conducted and the type of information required. Breath analysis has the advantage of individualizing each worker's integrated daily exposure. It can be used to diagnose as well as measure an exposure which has already occurred. It is a relatively inexpensive and simple method that can be put into operation without extensive and costly preliminary preparations. However, postexposure breath analysis does not provide information on the daily fluctuations of exposure, and the peak exposure concentrations are not made evident by breath data. Also, breath analysis is not applicable to all solvents which may be encountered in the workroom. Continuous monitoring enables the determination of peak exposures, exposure trends, and concentration gradients but is extremely costly both in time and in the equipment required. Nevertheless, it does provide the data necessary for estimating time-weighted average exposure.

REFERENCES

1. Doremus, K. R., O'Brien, M. G., Gerber, A. M., and Reich, A., A rapid monitoring procedure for carbon-14 in breath. *Amer. Ind. Hyg. Ass. J. 30*, 161 (1969).
2. Nelson, G. O., The halide meter – the myth and the machine. *Amer. Ind. Hyg. Ass. J. 29*, 586 (1968).
3. White, L. D., Taylor, D. G., Mauer, P. A., and Kupel, R. E., A convenient optimized method for the analysis of selected solvent vapors in the industrial atmosphere. *Amer. Ind. Hyg. Ass. J. 31*, 225 (1970).

4. Brief, R. S., and Confer, R. G., Combustible gas indicator response in low oxygen atmospheres. *Amer. Ind. Hyg. Ass. J. 30*, 576 (1969).
5. Chatterjee, B. B., Williams, M. K., Walford, J., and King, E., The location of personal sampler filter heads. *Amer. Ind. Hyg. Ass. J. 30*, 643 (1969).
6. Linch, A. L., Stalzer, R. F., and Lefferts, D. T., Methyl and ethyl mercury compounds – recovery from air and analysis. *Amer. Ind. Hyg. Ass. J. 29*, 79 (1968).
7. Belisle, J., Portable field kit for the sampling and analysis of toluene diisocyanate in air. *Amer. Ind. Hyg. Ass. J. 30*, 41 (1969).
8. Sidor, R., A technique for the sampling and analysis of halogenated hydrocarbons in air. *Amer. Ind. Hyg. Ass. J. 30*, 188 (1969).
9. Wright, G. W., Airborne fibrous glass particles. *Arch. Environ. Health 16*, 175 (1968).
10. Johnson, D. L., Healey, J. J., Ayer, H. E., Lynch, J. R., Exposure to fibers in the manufacture of fibrous glass. *Amer. Ind. Hyg. Ass. J. 30*, 545 (1969).
11. Breysse, P. A., and Bovee, H. H., Use of expired air-carbon monoxide for carboxy-hemoglobin determinations in evaluating carbon monoxide exposures resulting from the operation of gasoline fork lift trucks in holds of ships. *Amer. Ind. Hyg. Ass. J. 30*, 477 (1969).
12. Lang, H. W., and Freedman, R. W., The use of disposable hypodermic syringes for collection of mine atmosphere samples. *Amer. Ind. Hyg. Ass. J. 30*, 523 (1969).
13. Gibbs, G. W., Some problems associated with the storage of asbestos in polyethylene bags. *Amer. Ind. Hyg. Ass. J. 30*, 458 (1969).
14. Van Houten, R., and Lee, G., A method for the collection of air samples for analysis by gas chromatography. *Amer. Ind. Hyg. Ass. J. 30*, 465 (1969).
15. Fraust, C. L., and Hermann, E. R., The absorption of aliphatic acetate vapors onto activated carbon. *Amer. Ind. Hyg. Ass. J. 30*, 494 (1969).
16. Sansone, E. B., Sampling airborne solids in ducts following a 90° bend, *Amer. Ind. Hyg. Ass. J. 30*, 487 (1969).
17. Buchwald, H., A rapid and sensitive method for estimating carbon monoxide in blood and its application in problem areas. *Amer. Ind. Hyg. Ass. J. 30*, 564 (1969).
18. Lippman, M., "Respirable" dust sampling. *Amer. Ind. Hyg. Ass. J. 31*, 138 (1970).
19. Hidy, G. M., and Brock, J. R., Lung deposition of aerosols: A footnote on the role of diffusiophoresis. *Environ. Sci. Technol. 3*, 563 (1969).
20. Ayer, H. E., Sutton, G. W., and Davis, I. H., Size-selective gravimetric sampling in dusty industries. *Amer. Ind. Hyg. Ass. J. 29*, 336 (1968).
21. Knuth, R. H., Recalibration of size-selective samplers. *Amer. Ind. Hyg. Ass. J. 30*, 379 (1969).
22. Aerosol Technology Committee, Guide for respirable mass sampling. *Amer. Ind. Hyg. Ass. J. 31*, 133 (1970).
23. *Fed. Regist. 34(96)* (1969). Department of Labor. Occupational Safety and Health Standards, National Consensus Standards, and established Federal Standards.
24. Knight, G., and Lichti, K., Comparison of cyclone and horizontal elutriator size selectors. *Amer. Ind. Hyg. Ass. J. 31*, 437 (1970).
25. Lynam, D. R., Pierce, J. O., and Cholak, J., Calibration of the high-volume air sampler. *Amer. Ind. Hyg. Ass. J. 30*, 83 (1969).
26. Avera, C. B., Jr., Electronic flow regulation of high-volume air samplers. *Amer. Ind. Hyg. J. 29*, 397 (1968).
27. Goetz, A., A new instrument for the evaluation of environmental aerocolloids. *Environ. Sci. Technol. 3*, 154 (1969).
28. Spurny, K. R., Lodge, J. P., Jr., Frank, E. R., and Sheesley, D. C., Aerosol filtration by means of nuclepore filters – aerosol sampling and measurement. *Environ. Sci.*

Technol. 3, 464 (1969).

29. Spurny, K. R., Lodge, J. P., Jr., Frank, E. R., and Sheesley, D. C., Aerosol filtration by means of nuclepore filters – structural and filtration properties. *Environ. Sci. Technol. 3*, 453 (1969).
30. Stober, W., and Flachsbart, H., Aerosol size spectrometry with a ring slit conifuge. *Environ. Sci. Technol. 3*, 641 (1969).
31. Kupel, R. E., Kinser, R. E., and Mauer, P. A., Separation and analysis of the less than 10-micron fractions of industrial dusts. *Amer. Ind. Hyg. Ass. J. 29*, 364 (1968).
32. Renshaw, F. M., Bachman, J. M., and Pierce, J. O., The use of midget impingers and membrane filters for determining particle counts. *Amer. Ind. Hyg. Ass. J. 30*, 113 (1969).
33. Ettinger, H. J., Defield, J. D., Bevis, D. A., and Mitchell, R. N., HEPA filter efficiencies using thermal and air-jet generated dioctyl phthalate. *Amer. Ind. Hyg. Ass. J. 30*, 20 (1969).
34. Ettinger, H. J., First, M. W., and Mitchell, R. N., Industrial hygiene practices guide. *Amer. Ind. Hyg. Ass. J. 29*, 611 (1968).
35. Burgess, W. A., and Reist, P. C., An industrial hygiene study of flame cutting in a granite quarry. *Amer. Ind. Hyg. Ass. J. 30*, 107 (1969).
36. Rose, V. E., Gellin, G. A., Powell, C. H., and Bourne, H. G., Evaluation and control of exposures in repairing microwave ovens. *Amer. Ind. Hyg. Ass. J. 30*, 137 (1969).
37. Felton, T. A., Ozone – a cinema production. *Amer. Ind. Hyg. Ass. J. 29*, 582 (1968).
38. American Industrial Hygiene Association, Ozone-free xenon lamps. *Amer. Ind. Hyg. Ass. J. 30*, 185 (1969).
39. Wright, D. N., Vaichulis, E. M. K., and Chatigny, M. A., Biohazard determination of crowded living–working spaces: Airborne bacteria aboard two naval vessels. *Amer. Ind. Hyg. Ass. J. 29*, 574 (1968).
40. Schlick, D. P., and Paluso, R. G., Respirable mass sampling requirements under the Federal Coal Mine Health and Safety Act of 1969. *U. S. Bur. Mines Inform. Circ. 8484* (1970).
41. Morse, K. M., Dust control practices in the bituminous coal mining industry. *Amer. Ind. Hyg. Ass. J. 31*, 160 (1970).
42. Sentz, F. C., Jr., and Rakow, A. B., Exposure to iron oxide fume at Arcair and powder-burning operations. *Amer. Ind. Hyg. Ass. J. 30*, 143 (1969).
43. Alpaugh, E. L., Phillippo, K. A., and Pulsifer, H. C., Ventilation requirements for gas-metal-arc welding versus covered-electrode welding. *Amer. Ind. Hyg. Ass. J. 29*, 551 (1968).
44. Linch, A. L., Wiest, E. G., and Carter, M. D., Evaluation of tetraalkyl lead exposure by personnel monitor surveys. *Amer. Ind. Hyg. Ass. J. 31*, 170 (1970).
45. Mutchler, J. E., Industrial hygiene engineering and the process–environment system. *Amer. Ind. Hyg. Ass. J. 31*, 233 (1970).
46. Schreibeis, W. J., Evaluation of laboratory problems of glassblowers working fused quartz. *Amer. Ind. Hyg. Ass. J. 30*, 89 (1969).
47. Burgess, W. A., and Reist, P. C., Supply rates for powered air-purifying respirators. *Amer. Ind. Hyg. Ass. J. 30*, 1 (1969).
48. Schrag, K. R., An air supply system for operators of motorized street sweepers. *Amer. Ind. Hyg. Ass. J. 31*, 74 (1970).
49. Balzer, J. L., and Clark, W. C, The work environment of insulating workers. *Amer. Ind. Hyg. Ass. J. 29*, 222 (1968).
50. Fannick, N. L., and Corn, M., The plasma jet: Industrial hygiene aspects and a survey of current United States practices for employee protection. *Amer. Ind. Hyg. Ass. J. 30*, 226 (1969).

51. Gleason, R. P., Exposure to copper dust. *Amer. Ind. Hyg. Ass. J. 29*, 461 (1968).
52. Scharf, A., Control of airborne dust produced by pneumatic jackpicks: report number II. *Amer. Ind. Hyg. Ass. J. 30,* **519**, (1969).
53. Baretta, E. D., Steward, R. D., and Mutchler, J. E., Monitoring exposures to vinyl chloride vapor: Breath analysis and continuous air sampling. *Amer. Ind. Hyg. Ass. J. 30*, 537 (1969).
54. Skonieczny, R., A field and laboratory evaluation of the Ranta and Marcoli method for TDI. *Amer. Ind. Hyg. Ass. J. 24*, 17 (1963).

Personal Protective Devices

WILLIAM A. BURGESS

INTRODUCTION

This review period witnessed a strong effort to identify the contribution of personal protective equipment to the control of environmental health hazards. Legislation in the United States including the Federal Coal Mine Health and Safety Act of 1969 and the Occupational Safety and Health Act of 1970 specify the use of personal protective equipment. Both regulations will recognize national consensus standards, and continued activity was noted by such organizations as the American National Standards Institute to provide such standards. In several cases, new standards represent a subdivision of a previous broad document thereby insuring more comprehensive coverage. Various informed user groups also voiced both their needs and possible solutions in a series of excellent symposia, conferences, and meetings. In short, the period 1968–1970 seemed a time to consider the proper application of available equipment and the direction of future research and development.

SYMPOSIA AND MEETINGS

Symposia play an increasing role in allowing interchange of significant new information and this review has, therefore, been devoted to developments in personal protective equipment at such meetings in the years 1968–1970.

The Ames Research Center Conference on Portable Life Support Systems held in 1969, for example, was a successful attempt to provide an exchange of information by various governmental agencies working on portable life support systems including respiratory support devices. In this symposium twenty-four papers covered subjects including thermal control techniques, oxygen supply systems, and contaminant control methods. In addition, designs of specific systems for astronauts, divers, aircrew personnel, tank crewmen, and ordnance technicians were described. The proceedings of this conference presents an excellent source of advanced information on portable life support systems and has specific application to respiratory protection.[1]

Similarly, a symposium devoted to survival equipment for aerospace personnel included several topics which have application in occupational health personal protection.[2] A flotation dummy was described for the evaluation of life preserver design. The dummy simulates an unconscious surviver and includes a telemetry system to provide test data. Two papers on fire-resistant and flameproof fabrics were presented at this meeting. In one paper, a history of fire-resistant materials used in the U. S. Navy protective flight clothing program was outlined, and case histories presented in which Nomex flight clothing prevented serious burns.

A symposium on the occupational health and safety hazards of the fire service included several papers on personal protective equipment.[3] It was the consensus of the symposium, which included over 300 fire fighters and invited experts, that additional effort was needed to provide better respiratory protection for the fire fighter and that modern advances in materials should be more quickly utilized to improve protective equipment.

Three papers on respiratory protection were presented at the Sympoisum on Respirable Coal Mine Dust in November 1969.[4] Schutz described the status of the Bureau of Mines approval schedules for dust and outlined some of the recent standards of the American National Standards Institute. Anticipated changes in approval schedules and design evolution should result in better respiratory protection for coal miners. While describing means of improving respirator acceptance by the wearer, Schutz cited the importance of a respirator program as outlined in ANSI Z88.2-1969. In the second paper, Burgess developed methods to protect both the mobile worker and the miner who works at a fixed work station. A number of techniques to supply a respirable atmosphere to various protective envelopes were described. The ultimate control was a supplied air system to an enclosed work station. Flanagan, in a third paper, described a respirator program underway for approximately a year in a coal mine in West Virginia. A program was implemented which increased respirator use from 40% of the miners to 85%. This increase in respirator use was felt due to the educational program and the improved respiratory equipment provided the miners.

In Britain control of asbestos in the working environment including the appropriate application of respiratory protective equipment[5] was considered by the British Occupational Hygiene Society and the Asbestos Research Council. The Asbestos Regulations of 1969 published in that country also specifies suitable types of respiratory protection for asbestos workers.

At another important British conference a series of brief papers discussing recent advances in personal protection against hostile environments were presented to the Royal Academy of Medicine in January 1970.[6] In the first paper Weiner proposed the establishment of a research center for the research and development of protective clothing. Such a center would carry out exploratory work for new materials, the evolution of new design techniques, and the evaluation of existing and new equipment. In supporting this proposal Weiner stated that the British Standards Institution had formed a panel for the research and development of protective clothing equipment and that governmental agencies working in this area demonstrated the advantage of this type of approach. The development of protective clothing for fishermen was described by Newhouse. The design goal is a marketable, warm, wet-proof, well-fitting garment that will provide positive buoyancy. Constable discussed the design evolution of a garment based on a closed cell foam neoprene rubber. Crockford discussed the evaluation of the clothing designs, now available for field trial, using physiologic measurements of body temperature, pulse rate, oxygen consumption, and weight loss. Physical measurements of the system include temperature of the microclimate, rate of air-change beneath the garment, and rain penetration.

The large air volumes required for suit cooling has led the British to extensive use of water-cooled suits. London described one system at the Royal Academy meeting which uses either a fixed refrigeration package for aircraft crewmen or a self-contained cooling unit with a consumable coolant in a portable system. The author cites advantages of a water-cooled system as high efficiency, small distribution duct size, and little ballooning. Such a system does have certain disadvantages; it cannot remove moisture, and clothing may become damp from condensed sweat. A portable cooling unit for industrial water-cooled suits which provides a work period of 1–2 hr in a basic cycle of 20 min work and 5 min rest was described by Hill. A water jacket surrounds a 7-lb Dry Ice source, and the gaseous carbon dioxide drives the cooling pump. A PVC garment for protection from radioactive contamination was described by Stott. This garment is provided with air through an air-distribution device in the head piece and air flows to the extremities. The air supply rate is 140 liters/min with a pressure differential of about 2.5 cm of water maintained between the inside and the outside of the suit.

Unfortunately there is little published information available on disposable protective garments. In this symposium Thomas described the manufacture of various types of garments and their applications in industry. Disposable underwear may be made from stretchable bonded rayon which is warm and can be worn under a complete survival suit. Thomas also indicated that work is under way on a bonded fiber fabric laminated with cellophane film which is not impervious and, therefore, should be more comfortable to wear. A bonded fiber fabric coated with emulsions of carbon black and other adsorbants has been developed by the Porton Chemical Defense Establishment. Disposable overalls for paint sprayers have been developed from a dry layed bonded fiber fabric which consists of Nylon, rayon, and polyester fibers bonded with acrylic resins. The author suggests that this type of garment will have wide application, which probably means that the solid-waste expert must face an additional task.

Harries described the respiratory protection devices used for protection against asbestos in shipyard operations. Workers are provided with a complete change in clothing, impervious overalls, and air-supplied respirators. Supervisors are equipped with overalls and a full face respirator with a plastic hood. Workers removing solid waste wear ordinary dust respirators in addition to air-supplied hoods. Techniques to improve the efficiency of air-purifying respirators were described by Schwabe. One unique technique uses the pressure pattern from breathing to operate a differential bellows forcing air at an elevated pressure into the void between an oral–nasal mask and a full face-piece, thereby providing a better facial seal. Schwabe also indicated that they were evaluating powered air-purifying systems using a unique oral–nasal mask.

RESPIRATORY PROTECTION

Application Information

The increased acceptance of airline respirators has resulted in their use in atmospheres immediately dangerous to life. The Joint Respirator Committee of the AIHA–ACGIH has prepared recommendations for the use of such devices under these conditions.[7] A listing of airline respirator types in order of protection ranging from pressure-demand airline respirators with full face-piece to demand flow airline respirators with half mask is presented. Wearers of airline respirators in atmospheres immediately hazardous to life or health must be equipped with safety harnesses and a worker must stand by with suitable self-contained breathing apparatus for emergency rescue. A detailed listing of cautions in the use of such equipment is provided by this committee.

A series of interviews with safety directors of gas utilities revealed a range of respirator programs involving supplied-atmosphere devices and self-contained equipment. The use of supplied-air respirators with air supply from truck-mounted air compressors seems to be a popular technique with many utilities.[8] Hyatt, meanwhile has carried out an evaluation of commercial respirators for use in radon daughter protection in uranium mines.[9] High efficiencies were obtained with densely packed fiber filters and resin-impregnated wool filters. Special problems have been encountered in the use of self-contained open circuit respirators in compressed-air tunnel work such as reduction in use time of the devices and the hazard of oxygen breathing at increased pressures.[10] The authors describe a Bureau of Mines training course on the use of self-contained closed-circuit apparatus available for rescue teams who must work under compressed air. Simple dust respirators should be worn by farmers to eliminate their exposure to airborne spores and resulting Farmer's Lung.[11]

One industry where single-use or disposable respirators have an application is the construction industry. Establishment of a respirator program in the sense that one can be established in a plant is difficult. Selikoff has stated that a single-use respirator for asbestos exposure would be a significant contribution to the protection of the health of the work force.[12] An International Labour Office report on dust control in mining, tunneling, and quarrying makes selective references to the status of respiratory protection in various countries.[13] In the United Kingdom respirators used in coal mines were only required to pass a Ministry of Power testing procedure. Now that British Standard Specification No. 2091 has been issued, all respirators must be supplied to this specification. A common statement from various countries was that dust respirators were available but used infrequently due to poor worker acceptance. Protection of an operator at a rock crushing station in Australia was provided by ducting filtered air to the control cab. This technique was also being considered in open-cut lead mines in Mt. Isa; however, the final solution was to provide demand-air line respirators from a manifold bottle supply.

Respiratory protective devices suitable for use against asbestos have been described by Luxon.[14] The new British Standard BS2091:1969 is cited, which states that the conventional filter respirator will have a protection factor of 20. This respirator will, therefore, provide protection against exposures to 40 fibers/ml of asbestos, which would reduce exposure to the TLV of 2 fibers/ml.

Grundarfer and Raber comment on a serious problem in respirator acceptance in their discussion of silicosis in granite crushing operations. Men with diagnosed silicosis stated they did not wear respirators except for very short periods, because the respirators decreased their work capacity to the extent that it affected their income.[15]

New Developments and Design Trends

The future of closed-circuit self-contained breathing apparatus suitable for extended work periods depends both on the development of solid state and chemical sources of oxygen and more efficient carbon dioxide absorbents. Respirator and aerospace interest has, fortunately, provoked a number of studies suggesting possible design directions.

As an example, active search has been underway in the aerospace industry for compounds containing oxidation states of oxygen which could be used for air revitalization.[16] These compounds would react with water vapor to produce breathing oxygen while the reaction products would absorb carbon dioxide. An apparatus has been described which evaluates chemicals for this purpose, including potassium ozonide, sodium peroxide, potassium peroxide, calcium superoxides, and lithium peroxide. This technique can be used to screen materials for closed-circuit breathing applications.

Alkaline hyperoxides are attractive chemical sources of oxygen for self-contained breathing apparatus. In a study of these compounds Steinert found that sodium dioxide had a low oxygen yield due to a surface crusting, but the yield can be improved by mixing it with potassium dioxide and a variety of catalysts.[17] Potassium dioxide, presently used by the Mine Safety Appliance Company in the United States in their Chemox apparatus, was also investigated by Steinert. A process has been developed to make porous grains which results in a higher oxygen conversion rate and reduces the resistance to airflow of the system. Another study of oxygen sources demonstrated that lithium peroxide systems would be significantly smaller and lighter than comparable lithium hydroxide–oxygen system for a 4 hr, 2000 Btu/hr extravehicular portable life support system. The major drawback of this system was that the lithium peroxide bed must operate under test conditions exceeding 600°F.[18]

One paper analyzed rebreathing concepts intended to conserve the oxygen supply of air crewmen.[19] The discussion contained an interesting review of electrochemical and diffusion methods of carbon dioxide removal which would eliminate the problems faced with chemical absorption methods which tend to form carbonates of metal oxides. Storage of oxygen as a cryogenic solid rather than in the gaseous or liquid state has been explored by an aerospace contractor who states that for extravehicular activities a 2-in. cube of the solid weighing 0.31 lb would provide about 3 man-hours of gas.[20] A study of thin-film solid regenerable adsorbants suitable for coadsorption of water vapor and carbon dioxide has been investigated.[21] The boundry gas film is the limiting factor in carbon dioxide adsorption. The equilibrium capacity for such a system was nearly 25% less than a conventional pellet bed system. A detailed study of the application of lithium peroxide for oxygen supply and carbon dioxide removal has been completed for underseas application.[22]

At the request of the Bureau of Mines, the National Academy of Engineering formed a committee on mine rescue and survival techniques in March 1969. In a final report the Committee defined a mine rescue and survival system which could be developed in 1 year from existing technology.[23] The system included an emergency breathing device which would be used if escape from the mine were possible. If an escape route were not available, a refuge chamber would provide isolation from the toxic environment and possibly from fire and explosion until the men could be rescued. Requirements for a more satisfactory emergency breathing device to replace the present self-rescuer used in coal mines are proposed in this document. A second part of this report considers research and development for long-range rescue and survival systems. Two concepts were proposed for an emergency breathing device. One was a plastic hood, of a material such as polyimide, using potassium superoxide as a source of oxygen. The second was a system using an oxygen candle for the oxygen source. In discussing self-contained breathing apparatus used by mine rescue teams, the committee stated that present equipment are merely updated versions of designs that are 25 years old. These systems have limitations of weight, bulk, and time of duration of use. Research is, therefore, recommended to define the life support requirements, new approval schedules, better methods of delivery of the supply to the wearer, and the evaluation of acceptance of the system.

For several years the U. S. Steel Corporation has been testing a self-contained cryogenic respiratory system (Fig. 1) to be used by continuous-mining machine operators.[24] The system consists of a 10 liter vacuum flask with an integral heat exchanger. A demand regulator is used to supply an oral–nasal mask worn by the

Fig. 1. Cryogenic system for continuous mining machine operators, Courtesy of United States Steel Corporation.

miner. This system provides a 4-hr operating time, which is adequate use-time for the continuous miner. An on-site liquifier is used to provide the liquid air used in the system. A number of these units are in routine use at a single coal mine. As a part of a contract with the U. S. Bureau of Mines, the Westinghouse Electric Corporation is developing a closed circuit self-contained breathing apparatus for 1-hr service for coal mines.[25]

A study of carbon dioxide build-up in space suits indicates the importance of reviewing this problem in any kind of total enclosure garment such as an air-supplied vest or suit.[26] In one study, a protective garment was provided a total flow of 11.5 ft^3/min. Approximately one-half the flow was directed to the head to provide oxygen for breathing and to clear the carbon dioxide exhaled by the wearer. This particular garment met an acceptable carbon dioxide partial pressure level of 7.6 mm Hg under nominal conditions. A breathing vest positioned inside a space suit permits significant reduction in the amount of oxygen required for a helmet supply in an open-loop life support system.[27] An air-supplied disposable plastic hood for protection against low-level radioactive contamination has been developed at a Navy facility.[28] Recent respirator developments in England include a half-mask face-piece with an inflatable pneumatic facial seal and a powered air-purifying blower.[29] An air-supply hood is in use in an asbestos textile operation.[30] No data are provided on airflow rate or protection factors obtained with this equipment. An air-supplied hood which provides a protection factor of 10,000 with an air supply rate of 3 ft^3/min has also been described.[31]

Costs of cleaning and maintaining a simple oral–nasal filter respirator is in the range of 50–75 cents/day. This is a significant fraction of the price of the respirator and suggests the need for a disposable or single-use respirator for dusts and mists. One such device with the Bureau of Mine's approval under Schedule 21B was introduced in 1970.[32] A number of other manufacturers are considering introducing such disposable masks.

It is convention on air-supplied respirators to require a minimum of 4 ft^3/min for tight-fitting respiratory enclosures such as half-masks, and full-face masks, and 6 ft^3/min for hoods and helmets. Data has not been available on the minimum airflow required for powered air-purifying respirators. A recent study has provided information on the flow rates required for a half-mask face-piece with such systems.[33] The supply rate will determine the success of this concept, since it influences filter and battery life which have a direct impact on size and weight of the unit. In this study, the airflow requirements were evaluated by determining the flow needed to match the maximum inspiratory flow, maintain a positive pressure in the mask at all times, and reduce leakage to 0.1%. At a medium work rate of 415 kg · m/min, the supply rate needed to achieve all three test conditions was a flow rate of 113 liters/min. This substantiates the flow rates previously recommended in the Bureau of Mines Approval Schedules for the minimum supply rates for air-supplied equipment.

A unique 10 min emergency escape apparatus has been developed for the U. S. Navy[34] that consists of a coiled tubing pressure reservoir, a regulator, and a hood. The device is self-contained in a cylindrical metal can. A polyvinyl chloride mask is pulled over the head and an elastic neckband holds the hood in place. The canister, permanently attached to the hood, fits at the nape of the neck. The operating pressure of this system is 3500 psi. This light weight system of 3.4 lb can be donned in less than 20 sec and contains an adequate oxygen supply for 10 min.

Ergonomic Studies

A number of interesting articles on respirator ergonomics were published in *Atemschutz-informationen* (E. Germany) during 1969 and 1970. An annual report on the activities of a Working Group on Respiratory Protection included studies of tolerability limits for the respirator wearer.[35] The weight of the respiratory protection device, breathing resistance, dead volume, and thermal burden all were classed as important parameters to be considered in evaluating the acceptance of respiratory protective devices. The group reported that the oxygen consumption of a man wearing a closed-circuit self-contained breathing apparatus weighing 20 kg is approximately 20% higher than if he were not wearing the protective equipment. An increase of approximately 7% in oxygen consumption was noted with a respiratory protective device weighing 10 kg. Techniques have been developed for measuring breathing resistance so that quantitative comparisons of various devices could be made, and methods were described to define the dead volume (dead volume is equal to the change in the physiologic dead volume due to the wearing of the device).

The impact of the psychrometric state point of closed-circuit breathing apparatus was investigated.[36] Both carbon dioxide adsorption and condensation of water in a closed-circuit system result in an increase of temperature and thermal burden on the wearer. A study was therefore carried out on conventional equipment in two thermal environments – one with a dry bulb temperature of 24–25°C and a relative humidity of 40%, the other with a dry bulb temperature of 30–32°C and a relative humidity of 95%. Detailed data including temperature of inhaled air, rectal temperature, respiration rate, pulse rate, and carbon dioxide concentration in the inhaled and exhaled breath were accumulated. The pulse rate and body temperatures achieved a steady-state condition in the first climatic situation under work conditions; however, in the second environment the steady-state condition was not achieved and the wearer suffered severe thermal stress.

A breathing simulator has been designed and constructed under contract from the National Aeronautics and Space Administration for use by the Bureau of Mines.[37] This unit, which simulates human breathing at various metabolic rates and reproduces air temperature, carbon dioxide concentrations, and absolute humidity, should be a valuable tool in respirator studies.

Few quantitative evaluations have been carried out on actual respirator acceptance by workers. One recent study compares the performance and acceptance of three full face masks with different peripheral seal designs.[38] To test for the differences in leakage at five seating forces, a uranine aerosol test was used to challenge the integrity of the facial seal and permit quantitative measurement of percent leakage. Differences in apparent seating force (and, therefore, comfort) between the three facial seal designs were studied utilizing a psychophysical technique. A suggested physiologic correlate of comfort, galvanic skin potential, was also measured during the second half of the session. Significant differences were noted in the performance of one of the three masks as demonstrated by the leakage tests. The psychophysical evaluation did not reveal significant differences between the masks; however, significant differences were revealed by the galvanic skin potential measurements.

Testing and Evaluation

Several new techniques for mask leakage have been described during the past 3 years. In one technique the wearer is placed in a plastic hood flushed with argon.[39] The subject inhales through a breathing tube from an oxygen supply and exhales through a second breathing tube to the atmosphere. The leakage of argon into the mask is determined by analysis of the exhaled gas using a mass spectrometer; a test carried out with this technique demonstrated that a pneumatic seal full face mask provides a better seal than conventional masks.

Another respirator leakage man test has been described which uses a sodium chloride aerosol as the contamination simulant.[33] This test differs from previous tests described by Hounam[40,41] and White[42] in two ways. The sodium chloride aerosol is generated from a 1% aqueous solution using an ultrasonic generator rather than the conventional pneumatic nebulizer and a peristaltic tubing pump is used to sample the air both from the exposure chamber and the mask under test. The authors state that this sampling method eliminates the difficulties faced with previous systems which sample exhaled breath. The test aerosol has a mean count diameter of 0.2 μm with a standard geometric deviation of approximately 2.0. The wearer is exposed to an aerosol concentration of approximately 20 mg/m^3. Leakages as low as 0.01% can be measured directly with this test method.

The above test has also been modified for testing respirator filter penetration suitable for Bureau of Mines approval and certification procedure.[43] The aerosol is generated by an ultrasonic generator and has the same characteristics as described in the man test.[33] The filter under test is placed in a pressurized stream and a portion of the downstream effluent is conveyed under pressure to a flame photometer for measurement of the sodium concentration. The authors state that the method has a sensitivity of 0.02% with an upper limit of 10%. Respirator filter penetration data obtained with this technique were compared to information obtained by another laboratory using a dioctyl phthalate aerosol. The authors state the correlation between the data showed the validity of the sodium chloride method.

The Safety in Mine Research Establishment (SIMRE) in England has carried out a study of outward leakage of oxygen from full face masks used with self-contained breathing apparatus.[44] Four full face masks with a flat rubber seal, reverted flap seal, air-filled cushion seal, and a water-filled cushion seal were evaluated with a closed-circuit breathing apparatus. The investigators found that leakage from the cushion seal mask was substantially less than from the flat seal and the reverted flap seal. On one particular mask, venting the bag by operating the relief valve and pressing the breathing bag caused high mask pressures and exfiltration from the maks. Dummy tests showed that leakage rates of 0.25 liter/min from the mask will result in ignition of human hair at the seal area of the face mask if a source of ignition is available.

An amendment to the Bureau of Mines Approval Schedule 21B includes a coal dust tightness test for respirators having a TLV of not less than 0.1 mg/m^3 or 2.4 million particles/ft^3 for radon daughters and attached particulates.[45] The test involves exposing a subject wearing the respirator to a concentration of 75±25 mg/m^3 of coal dust which has passed a 200 mesh screen. The respirator face-piece is modified to permit sampling directly from the face-piece to a membrane filter with a pore size of 5 μm at a sampling rate of 2 liters/min. During a 30-min test program, the wearer is required to perform certain facial and head movements, and samples are taken of the dust concentration inside the respirator and in the exposure chamber. The test requirements specify that the leakage shall not exceed 10% for a dust and mist having a TLV of less than 0.1 mg/m^3 or 2.4 million particles/ft^3, and 5% for radon daughters and attached particulates.

A comprehensive study of the performance of four self-contained breathing apparatus of the compressed air demand type was carried out at the Bureau of Mines at −25°F.[46] A number of serious functional changes were noted at the low temperature including malfunction of pressure regulators, high-pressure leaks, sealed exhalation valves, and fogged eyepieces due to frozen condensed moisture. Of the four demand-type devices tested, only two performed satisfactorily at this temperature when the nose cups and low-temperature regulator diaphragms were employed. A number of recommendations were proposed for equipment used at low temperatures. These involved a face-piece fitted with a nose cup, storage of the apparatus at room temperature, special materials for critical parts, and a series of special operating instructions.

Little data are available on the protection factors afforded by self-contained compressed-air breathing apparatus of the type used for high level protection by many industrial, mining, and fire services. One such study which does contribute valuable information has been carried out by Adley using a Freon leakage test technique.[47] The study was conducted on four Bureau of Mine approved devices. Two units were available in the demand mode and two could be used in both pressure and pressure-demand modes. These tests revealed significant differences in leakage when the systems were evaluated under the demand and pressure-demand mode. Operating in the pressure-demand mode normally increased the protection factor by at least 10 over the demand setting. As could be expected, a wide range of protection factors was found; a protection factor as law as 7 was obtained with a subject wearing glasses in the demand mode. Under the pressure-demand mode, median values of 300,000 were found even when eye protection was worn.

A study of the protection afforded by a powered air-purifying respirator against radon daughters demonstrated that the device could reduce the exposure to the radon daughters by factors of greater than 1000.[48] Since a particulate filter was used which did not remove the radon, the actual level of protection afforded by the device in a radon–radon daughter atmosphere was being evaluated. Based on dose calculations it is suggested that a conservative protection factor of 20 be used for such devices until the critical regions of the lung affected by exposure to radioactivity are identified.

Twenty masks from European countries and the United States were tested by the East German Committee for Respiratory Protection Devices for Toxic Mineral Dust.[49] The tests were carried out on manikins with exposure to quartz dust containing 12,000 particles/cm^3. A breathing machine was used at 20 respirations/min with a minute volume of 30 liters. The resistance to flow was checked at flow rates of 30 and 90 liters/min. These tests demonstrated that only a few of the masks met efficiency requirements although most of the masks did meet resistance to flow standards. A test bench has been developed which can be used to check the performance of respiratory protective devices.[50] The tests include mask leakage, efficiency of filter cartridges, and breathing resistance.

A detailed study of six Bureau of Mines approved and six unapproved particulate respirators was carried out with a variety of aerosols. As one would anticipate, the approved respirators provided higher efficiency; however, some of the approved respirators did not meet the Bureau of Mines Approval Schedules. In general, the respirators bearing Bureau of Mines approval performed uniformly. The nonapproved respirators varied widely in the protection they afforded. This paper should be read by anyone who is considering the use of unapproved respirators for industrial applications.[51] In tests of facial fit, the Bureau of Mines has determined that respirators with facelets have overall penetrations equal to or less than the same respirator without facelets when tested with the standard coal dust method.[52] Carbon tetrachloride is presently used by the Bureau of Mines in breakthrough testing of chemical cartridges. The use of Freon 113 has been evaluated as a substitute for this material.[53] A nondispersive infrared analyzer is used to monitor the breakthrough times of the cartridges.

A bioassay program to evaluate the effectiveness of half-mask face-pieces used for protection against a uranium dioxide dust exposure revealed an effective protection factor of approximately 2.[54] Apparently personnel wearing half-masks expose themselves to higher dust concentrations in the work area than when not wearing a respirator. This facility has converted to full face air-purifying respirators and air line respirators.

Aurich states that the facial seal in a respiratory protective device is the major contributor to the efficiency of the respirators.[55] The fit and sealing of the face mask depend on the ratio between the dimensions and shape of the mask and the wearer's face, the shape and material of construction of the seal design, the suspension adequacy, and the weight attached to the face-piece. The importance of quantitative methods of facial seal testing is stressed and a survey of applicable methods presented includes Freon, Uranine, *B. globigii,* sodium chloride, and nitrogen leakage techniques.

Training and Education

A training school for wearers of respiratory protection devices in the military and civil fire brigades in Norway has been described.[56] The training school consists of a classroom and a training room with various obstacles. The room can be heated to 95°F and filled with smoke. The obstacles, which are designed to train the students in moving through spaces and smoke, consist of stair-like structures and different timber structures. The obstacle training room can be viewed directly from an instructor's observation chamber. The training program, of 5 days duration, includes both theoretical and practical respirator training.

Revoir presented an excellent review of respirator use at the 1969 National Safety Congress.[57] Special attention was given the use of respirators in dangerous atmospheres such as those immediately hazardous to life, confined spaces, and under conditions of low and high temperatures.

To broaden the scope of their respirator program, the Oak Ridge National Laboratory has developed a slide presentation for their respirators training program covering all facets of a respirator use.[58] Hyatt outlined a respirator program for teaching and research facilities in a paper presented at a Campus Safety Association meeting.[59]

As a result of the Chalk River NRX reactor accident in 1952, an extensive protective equipment laboratory has been established at that facility.[60] The program includes a maintenance program involving washing and decontamination of face-pieces and components, and leakage and resistance testing. Both amyl acetate and quantitative sodium chloride aerosol man tests are routinely used in the training and education program.

Standards, Approval Schedules, and Health and Safety Regulations

One of the most active groups concerned with respiratory protection in the United States is the Joint Respirator Committee of the American Industrial Hygiene Association and the American Conference of Governmental Industrial Hygienists. This Committee, which consists of representatives from respirator users, governmental agencies, universities, research laboratories, and manufacturers of respirators, also includes members from Canada, and the United Kingdom. In a discussion of the activities of the Committee during the period 1967–1969, the Committee Chairman cites the variety of work which has been carried out.[61] A proposed reduction in the Bureau of Mines respirator testing and approval program prompted a Committee statement commenting on the need of increased activity in the respirator program which was forwarded to the Secretary of the U. S. Department of Interior and the Bureau of Mines by the parent associations. In 1967 a meeting was held at the Bureau of Mines to which manufacturers of respirators and users were invited to discuss activities in the field with the Bureau of Mines. As a result of this meeting, it was recommended that approval schedules be developed for chemical cartridge respirators for protection for acidic, alkaline, and certain other gases, chin-style canister gas masks, and air-purifying mouthpiece respirators. These recommendations were accepted and these devices have been or will be included in schedules being prepared by the Bureau of Mines.

It now appears that a single document covering revisions of current Bureau of Mines respirator approval schedules and new schedules for certain respirator types will be proposed which will eliminate individual approval schedules for various types of respirators. This document may include significant changes, in particular an automatic expiration of a Bureau's approval and a quality control test. Functional changes will include lowering breathing resistance and the use of quantitative fitting tests in the schedules.[62]

An amendment to Bureau of Mines Schedule 21B for approval of respirators for protection against radon daughters was issued in 1969.[63] This amendment was based on work done by the Los Alamos Scientific Laboratory and the Bureau of Mines which showed an excellent correlation between filter efficiency tests on lead fume in the laboratory and radon daughters tests in mines. This correlation made it possible to specify the lead fume test in the approval schedule. The amendment also included a face-piece leakage man test using a coal dust test atmosphere.

The Federal Coal Mine Health and Safety Act of 1969 was published in the *Federal Register* on April 3, 1970 (Volume 35, No. 65). Under this act miners exposed to concentrations of coal dust in excess of permissible levels must be provided with respirators approved jointly by the Department of Interior and the Department of HEW.[64] The act does not permit the use of respirators in lieu of engineering controls to achieve the required levels of dustiness. The Department of HEW presently accepts Bureau of Mines approved devices as approved by their Department.

Two important standards have been published by the American National Standard Institute during 1969. ANSI Z88.2, "Standards for Respiratory Protection" describes safe practices and requirements for respiratory protection against particulates, gases, vapors, and oxygen deficiency.[65] It is a revision of the respiratory protection portion of the American National Standard Safety Code for Head, Eye, and Respiratory Protection, Z2.1-1959. The Standard provides guidance on selection, use of respirators, maintenance and care, and the evaluation of respirator program effectiveness. This document should be a part of any environmental health library and available to all persons recommending the use of respiratory protection. The International Standards Organization is now considering ANSI Z88.2 as a basis for an international standard.

The American National Standard Z88.1, Safety Guide for Respiratory Protection Against Radon Daughters meets the needs of the uranium mining industry.[66] Other Z88 Subcommittees of the American National Standards Institute are presently working on standards for respiratory protection against asbestos-containing dust, performance requirements of particulate filter respirators, respiratory protective devices for fire fighters, and respiratory protection for coal dust.

The British Standard "Specification for Respirators for Protection Against Harmful Dust and Gases" was published during 1969.[67] Two filter masks are described: a type A low-resistance respirator for use against dusts of low toxicity and a more efficient type B high-resistance respirator. Cartridge and canister-type gas respirators are also covered in the standard.

A specification has also been proposed by the British Standards Institution for positive pressure powered respirators.[68] The Standard covers units equipped with high efficiency filters and standard particulate filters. It includes systems equipped with oral–nasal masks and those with full face-pieces. The powered respirator shall provide total efficiencies of greater than 99%. The "power off" leakage shall not exceed a mean value of 5% for ten test subjects when tested with the Standard BS2091 technique. The system shall be capable of providing 120 liters/min for a period of 4 hr without recharging. The inhalation resistance of the complete system with the blower off shall not exceed 50 mm of water at

85 liters/min and the exhalation resistance in this mode shall not exceed 12.5 mm of water at 85 liters/min.

Another British Specification, BS4455:1970, covers high efficiency dust respirators.[69] The overall leakage of the respirator system when worn by adults shall not exceed a mean value of 0.1% for the test subjects. The filter itself shall have a penetration of less than 0.01% at a flow rate of 30 liters/min when tested with the British Standard BS4400 test (sodium chloride particulate test). A specific requirement that penetration be less than 0.005% is cited for resin wool and resin felt filters. This rigorous test is stated to take into account the degradation of the electrostatic charge in such filters during storage. The standard also states that asbestos shall not be used in the filtering media.

A Polish standard for closed-circuit breathing apparatus provides an interesting comparison with United States standards.[70] It covers closed-circuit breathing apparatus for rescue work of ½, 1, and 2 hr duration, and closed-circuit breathing apparatus for escape purposes. This specification makes specific comments on acceptable interference with normal operations. For example, the mask must not reduce the wearer's field of view by more than 25%, nor should the face mask reduce the ability of the wearer to communicate by more than 15%. Specific comments on the requirements of the apparatus are of interest. The temperature of the inspired air at a moderate work rate where ambient temperature is below 40° shall not exceed 45°C. The carbon dioxide content during a rated service time should not exceed 0.5% and the maximum at the end of the reserve period of the apparatus should not be higher than 1.5% for apparatus used over an hour, 2% if used up to an hour, and 2.5% in emergency apparatus under conditions of very heavy work.

PROTECTIVE CLOTHING

An ergonomics approach has been used in the design of clothing for various occupations in England for the past decade. The clothing design is based on a study of the working environment and the worker's functional demands. Advanced textile materials are used to provide functional and attractive garments.[71-73]

The British Iron and Steel Federation have compiled a listing of protective clothing needs for various jobs indicating the type of hazard and the parts of the body exposed to injury.[74] The job classifications which were reviewed included those in the blast furnace, coke oven, and maintenance departments. Appendices to the reports list standards relating to protective clothing and statutory requirements under the Factories Act. A review of the work carried out by the All-Union Scientific Researchers Institute of the Sewing Industry in Russia includes testing and field evaluation of work clothing and physiologic testing of the clothing in test chambers to determine their suitability.[75]

A biologic isolation garment used to insure that the astronauts returning from a lunar surface would not contaminate earth was designed which contains 98% of the submicron particles shed by the wearer and his clothing. The cotton fabric garment does not require an external cooling supply.[76] A study of protective clothing design for glass handlers resulted in a scientific method for evaluating fabrics.[77] Considerable insight into the design of cold weather clothing is provided by Woodbury in a discussion of Eskimo clothing practice.[78]

Recent advances in fabric research hold promise for better protection clothing systems for industrial workers. In a review of fire retardant materials and coatings,[79] Radnofsky indicated that several modified aromatic polyamide fibers including Durette[80] and Fypro[81] have been used to make fabrics which have excellent qualities and are being considered for astronaut crew clothing. Kynol,[82] a phenolic-type fiber, has also been studied by NASA and used for protective garments for fire fighters and race drivers. NASA has manufactured garments for trial by the Houston Fire Department based on advanced fabrics which can withstand temperatures of 3400°F for short periods of time.

Heat-protecting Nylon (Nomex) has been used in several garment systems.[83] One system is a turnout coat for fire fighters, which includes five layers of material. Other protective systems have been developed for smoke jumpers in the U. S. Forest Service and for workers in the steel and aluminum industry. This material can replace flame retardant materials which require periodic treatment. The efficiency of Nomex in protecting against flash fires of workers employed in munitions manufacturing plants is due to the charring of the outer layer of the material and the trapping of air between the outer and inner layers of the cloth.[84]

A system for thermal protection rating of fabric based on pain and blister effects of human skin has been proposed.[85] The method is based on a tissue injury concept which correlates skin temperature–time history with transepidermal necrosis. The author outlines a simple method of rating fabrics which can be used by textile laboratories. The use of color for safety clothing for persons working near vehicular traffic has unfortunately been an art form to date. A recent study compared a number of fluorescent and nonfluorescent colors in order that an optimum color could be recommended. Under the varied circumstances studied, fluorescent orange was found to be the best color and white was the poorest color. Geometry did not seem to have any impact on visibility. The authors suggest that a 30-cm wide band around the body is probably the best pattern.[86]

A listing of "dos and don'ts" in the use of safety clothing has been presented which is a useful supervisory training aid.[87] A hazard which has been suspected but not evaluated to date is the radioactivity release from contaminated protective clothing. A study conducted by the Atomic Energy Authority of England demonstrated that air concentrations while handling protective clothing could be in excess of the MPC.[88] A comparison of cotton and Terylene overalls worn in a battery plant did not reveal a significant difference in the concentration of lead in air due to shedding of dust.[89] Tests have been conducted in England on the exposure of the wearer of protective clothing to asbestos. Workers wearing asbestos aprons and gloves can be exposed to concentrations of asbestos exceeding the British Threshold Limit Value of 2 fibers/cm^3.[90] An author recommends that dresses not be allowed where catch points exist, that trouser suits be recommended, and that special headwear and footwear be used by all women workers. Synthetic fiber which can cause static charge electricity should not be worn nor jewelry where the women are exposed to moving parts of machinery.[91]

Total protection of the body with impervious plastics and textile suits can cause severe heat stress to the wearer. A number of papers have proposed solutions to this problem. A microclimate-controlled clothing system (Thermalibrium suit) has been devised by the Clothing Personal Life Support Equipment Laboratory at the U. S. Army Natick Laboratories.[92] The system, shown in Fig. 2 consists of a protective helmet, body clothing, shoes and gloves, and a light weight, self-powered heat-regulating device. The design requirements for this system include an operating range of –40 to 100°F. Unique features of the system include a fabric layer system which permits a uniform air flow throughout the suit and a thermoelectric heating and ventilating system. Based on this design, prototype systems have been developed for space chamber work, air crewmen, tank crewmen, and explosive ordnance disposal personnel.

The U. S. Air Force has developed a water-conditioned suit using a Dry Ice cooler which permits resting subjects to maintain thermal balance at temperatures of 105–115°F for a period of 4 hr.[93] Kanz found that conductive cooling of the head with water tubes in a hood maintained a stable body temperature with reduced sweating in hot-humid environments.[94] An air supply protective suit has been developed at the Douglas Point Nuclear Power Station in Canada for maintenance work in a carbon dioxide–tritium exposure with high ambient temperatures.[95] The suit, a double garment consisting of an inner coverall and an outer plastic suit with a Vortex tube to supply cool air, permits working at temperatures up to 120°F. Crockford has discussed design parameters for supplied air hoods and suits including airflow requirements for thermal balance.[96] Flows of 8–10 ft^3/min are recommended for hoods. Suits can be ventilated based on different physiologic goals resulting in flow rates from 3–5 to 36–50 ft^3/min.

The Vortex tube has additionally been used in conjunction with respiratory protection and protective clothing in steel, aluminum, and glass manufacturing. In a recent study the device was evaluated as an air-cooling device.[97] A 15

Fig. 2. Thermalibrium suit. Courtesy of U. S. Army Natick Laboratories.

ft^3/min Vortex tube was used with an inlet air pressure of 100 psi which provided approximately 1300 Btu/hr of cooling. The discharge from the Vortex tube was connected to an Air Force ventilating garment. A series of eight unacclimatized volunteers participated in the experiment. In the first test they were seated at rest without Vortex tube in a test environment with a dry bulb temperature of 54°C and a wet bulb temperature of 40°C. Subjective comments were used to determine the limit of tolerance. In a second session, the cooling garment was supplied by a Vortex tube at 13°C, and the subjects were exposed to the same environment and evaluated for a period of time equal to the tolerance time in the control session. In the third test, the same study was carried out for a period equal to three times the tolerance time. The Vortex system reduced sweat loss, maintained acceptable rectal temperatures and pulse rates, and extended the subject's tolerance of the heat environment.

HEAD, EYE, FACE, FOOT, AND HAND PROTECTION

Customizing personal protective equipment has been considered by many specialists in the field as a means of increasing acceptance of the equipment by the wearer. A U. S. Air Force establishment has used this approach to provide a foamed in-place polyurethane helmet liner for crash or flying helmets.[98] It involves a specific foam formulation which can be used with safety and with a minimum of equipment. Since this material does contain a toluene diisocyanate catalyst, one must give close attention to the exposure of the wearer.

In a paper from Holland a history of head injuries in port accidents is used as an introduction to a discussion of requirements for safety helmets. The features included in the discussion are resistance to impact and perforation, dielectric strength, weight, and material.[99] The American National Standards Institute is presently preparing standards for head protection to replace the current Z2.1 Standard. Two standards will be adopted. ANSI Z89.1 will constitute the standard for all head protection other than electrical workmen's helmets and Z89.2 will deal specifically with electrical workers insulated headgear and will contain specific requirements on electrical resistance.[100]

Evaluating head protection for military aircraft, the authors state that the test method should include not only American National Standards Institute methods, but in addition a means to determine the impact protection capability of helmets and the effects of different types of impacts. A swingaway test method is proposed for defining protection capabilities using a statistical sampling method.[101] Other authors suggest that a helmet alone cannot provide effective protection against concussion; however, a high order of protection against skull fracture can be achieved by load spreading. A plastic foam and pneumatic liner could combine buffer and crash protection in one helmet.[102]

Health and safety personnel are routinely faced by difficulties in motivating persons to wear protective equipment. A review of the motivation of workers to use eye protection concludes that poor acceptance is due to inadequate eye protection, uncomfortable or cumbersome devices and, in many cases, glasses that are not aesthetically pleasing. The factors that motivate the individual to wear eye protection include the need for visual correction, fear of injury, and the effects of training.[103]

A series of articles on eye and face, foot, hand, and head protection covering application data and case histories is available in the *National Safety Council News.*[104-106] A thermal protective visor has been proposed for aerospace applications which consists of a sandwich of glass plate, gold coated polyester plastic, and a dead air space. This system combines thermal protection, impact resistance, noncombustibility, and visibility.[107] In a review of the energy density necessary to cause thermal damage to the retina, Clark found support for the present threshold value of 1.8 cal/m^{-2}/min. The author feels that the present infrared transmission limits for welding filters are adequate.[108] A study of mechanisms for a fast-acting filter for protection against high-intensity flash suggests a workable system might combine a gas bubble-electrochromic cell.[109] A dipole shutter offers a possible means of protecting the eye against high intensity flashes.[110] The range of dynamic change in optical density observed in tests did not approach that predicted by theory.

Dermatitis problems resulting from poor maintenance of cutting oils and coolants is a frequent problem in industry. One author has specified the hygienic care necessary to minimize this problem and recommends the use of equipment such as gloves for protection from harmful solvents and cutting oils.[111] Arndt has described the composition of cutting fluids and their contribution to industrial dermatitis.[112] Although several types of dermatoses are caused by cutting fluids, contact dermatitis is the most common problem. This author also recommends protective creams and protective clothing to control occupational dermatitis.

A series of thirty fishermen with a contact dermatitis from wearing rubber boots were followed for 2 years.[113] The specific etiologic agent in the rubber could not be identified. The authors suggest that an alternate to rubber be considered for waterproof long boots. The use of steel innersoles developed for military infantrymen is proposed as protection against foot puncture injuries in industry.[114]

REFERENCES

1. Portable life support systems, *NASA Spec. Publ. 234* (1970). Available from Clearinghouse for Federal Scientific and Technical Information, Springfield, Virginia.
2. "Proceedings of National Flight Safety, Survival and Personal Equipment Symposium." Survival and Flight Equipment Association, Las Vegas, Nevada (1969).
3. A symposium on occupational health hazards of fire service. *Int. Fire Fighter 54,* 7 (1971).
4. Proceedings of the symposium on respirable coal mine dust, Washington, D.C., Nov. 3–4, 1969, *U. S. Bur. Mines Inform. Circ.* 8458 (1970).
5. The control of asbestos in the working environment. *Ann. Occup. Hyg. 13,* 153 (1970).
6. Recent developments in personal protective clothing and equipment. *Proc. Roy. Soc. Med. 63,* 1003 (1970).

7. Joint AIHI–ACGIH Respirator Committee, Conditions for use of airline respirators in atmospheres immediately hazardous to life or health. *Amer. Ind. Hyg. Assoc. J. 30*, 305 (1968).
8. Respiratory protection for gas industry crews. *Safety Eng. 137*, 15 (1969).
9. Respiratory protection for uranium miners. *The Atom 5,* 12 (1968).
10. Bovee, H. H., and Breysse, P. A., Study of air quality and contaminant analysis for work under compressed air. *Amer. Ind. Hyg. Assoc. J. 29,* 432 (1968).
11. Gilson, J. C., Respiratory Diseases in Farming. *Ann. Occup. Hyg. 12*, 121 (1969).
12. The disposable respirator. *Insulation Hyg. Progr. 2,* 3 (1970).
13. "Fourth International Report on the Prevention and Suppression of Dust in Mining, Tunneling and Quarrying." Occupational Safety and Health Series, No. 24, International Labour Office, Geneva (1970).
14. Luxon, S., Respirators for protection against asbestos. *Ann. Occup. Hyg. 13,* 14 (1970).
15. Grundorfer, W., and Raber, A., Progressive silicosis in granite workers. *Brit. J. Ind. Med. 27,* 110 (1970).
16. Petrocelli, A. W., and Capotosto, A., Apparatus for kinetic studies of the reaction of air revitalization chemicals with water vapor and gaseous carbon monoxide. *Aerosp. Med. 41,* 1204 (1970).
17. Steinert, H., Alkalihyperoxide als Austauschmassen für Atemschutz geräte. *Atemschutz-informationen 9,* 6 (1970).
18. Pierce, R. N., and Dresser, K. J., Evaluation of lithium peroxide for oxygen supply and carbon dioxide control. Presented at the American Institute of Aeronautics and Astronautics Thermophysics Conference, San Francisco, 1969.
19. Kiraly, R. J., Babinsky, A. D., and Wynveen, R. A., Electrochemical aircrew oxygen systems. *Aerosp. Med. 41,* 1400 (1970).
20. Solid oxygen: Best breathing reserve for space? *Mach. Des. 12,* 14 (1968).
21. Spece, L. C., Rudek, F. P., Green, T. F., and Miller, R. A., Regenerable adsorbent study *NASA Contract Rep. 66534* (1970). Available from Clearinghouse for Federal Scientific and Technical Information, Springfield, Virgina.
22. Capotosto, A., and Petrocelli, A. W., "Use of Lithium Peroxide For Atmosphere Regeneration." Available from Clearinghouse for Federal Scientific and Technical Information, Report AD 678 076, 1968.
23. "Mine Rescue and Survival, A Final Report of the Committee on Mine Rescue and Survival Techniques." National Academy of Engineering, Washington, D.C., 1970.
24. Press release by the U. S. Steel Corp., New type of breathing apparatus (1970).
25. Contract H 010 1262 between Westinghouse Defense and Space Center, Special Defense Division, Baltimore and Department of Interior, U. S. Bureau of Mines (1970).
26. Michel, E. L., Sharma, H. S., and Heyer, R. E., Carbon dioxide build-up characteristics in spacesuits. *Aerosp. Med. 40,* 827 (1969).
27. Curtis, D. L., Open-loop life support system lightens crew load. *Space/Aeronaut. 52,* 68 (1969).
28. Bessmer, D. J., "Puget Sound Naval Shipyard, Bremerton, Washington." Personal communication. Available from Defense Apparel, Hartford, Connecticut, 1970.
29. Luxon, S. G., Recent developments of dust respirators in the United Kingdom. *Amer. Ind. Hyg. Ass. J. 29,* 333 (1968).
30. Briess, K., Vorbildlichen Atemschutz. *Atemschutz-informationen 9,* 11 (1969).
31. White, J. M., Beal, R. J., and Courneya, W. J., A supplied air hood for protection against very toxic air contaminants. *Amer. Ind. Hyg. Ass. J. 29,* 165 (1968).
32. Dust Demon, U. S. Bureau of Mines Approval No. BM-21B-107, American Optical Corp., 1970.

33. Burgess, W. A., and Reist, P. C., Supply rates for powered air-purifying respirators. *Amer. Ind. Hyg. Ass. J. 30,* 1 (1969).
34. Survival Support Device, Lear Siegler Electronic Instrumentation Division, Anaheim, California.
35. Ohl, R., Bericht Uber die Taigheit der Arbeitsgruppe Atemschutz. *Atemschutz-informationen 8,* 3 (1969).
36. Bucholz, C.H., Schulz, G., Leers, R., and Ritter, H., Der Einfluss von Temperatur und Feuchte der Einatemluft auf Trägen von Regenerations – Generäten bei Arbeit in Normalen und Feuchtwarmem Umgebungsklima. *Atemschutz-informationen 8,* 26 (1969).
37. Contract NAS W-2032 between National Aeronautics and Space Administration and International Business Machine, Gaithersburg, Maryland, 1970.
38. Burgess, W. A., Hinds, W. C., and Snook, S. H., Performance and acceptance of respirator facial seals. *Ergonomics 13,* 455 (1970).
39. Griffin, O. G., and Longson, D. J., The hazard due to inward leakage of gas into a full face mask. *Ann. Occup. Hyg. 13,* 147 (1970).
40. Manual Respirator Tester, Evans Electroselenium Limited, London.
41. Hounam, R. F., "A Method for Evaluating the Protection Afforded When Wearing a Respirator," Research Group Report AERE-R-4125, Health Physics Div., AERE Harwell, Berkshire, 1962.
42. White, J. M., and Beal, R. J., The measurement of leakage of respirators. *Amer. Ind. Hyg. Ass. J. 27,* 239 (1966).
43. Ferber, B. I., Brenenbrog, F. J., and Rhode, A., Respirator filter penetration using sodium chloride aerosol. *U. S. Bur. Mines Rep. Invest. 7403* (1970).
44. Griffin, O. G., and Longson, D. J., The hazard due to outward leakage of oxygen from a full face mask. *Ann. Occup. Hyg. 12,* 147 (1969).
45. Schutz, R., U. S. Dept. of Interior, Bureau of Mines, Personal communication (1969).
46. Kloos, E. J., Raymond, L. D., and Spinetti, L., Performance of open-circuit, self-contained breathing apparatus at –25°F. *U. S. Bur. Mines Rep. Invest. 7077* (1968).
47. Adley, F., and Uhle, R., Protection factors of self-contained compressed air breathing apparatus. *Amer. Ind. Hyg. J. 30,* 355 (1969).
48. Burgess, W. A., and Shapiro, J., Protection from the daughter products of radon through the use of a powered air-purifying respirator. *Health Phys. 15,* 115 (1968).
49. Regenverg, F., Comparative tests of fine-dust masks, *Atemschutz-informationen 8,* 66 (1969). Translated by Los Alamos Scientific Laboratory, Report, LA 4459-TR, July, 1970.
50. Respirator Tester, Prosec Association, Paris.
51. Revoir, W., and Yurgilas, V. A., Performance characteristics of dust respirators, Bureau of Mines approved and non-approved types. *Amer. Ind. Hyg. Ass. J. 29,* 322 (1968).
52. Schutz, R., U. S. Bureau of Mines, Personal communication (December, 1968).
53. Swab, C. F., and Ferber, B. I., "Freon 113 as a Test Material For Chemical Cartridge Respirators," Available from Clearinghouse for Federal Scientific and Technical Information, Report PB 192032, 1970.
54. Caldwell, R., and Schnell, E., "Respirator Effectiveness in an Enriched Uranium Plant." Presented at the 1968 American Industrial Hygiene Conference, St. Louis, Missouri.
55. Aurich, G., Der Dichtsitz von Atemschutzmasken. *Atemschutz-informationen 8,* 32 (1969).
56. Reier, B., The respirator training school of the Norwegian air force. *Drager Rev. 20,* 19 (1969).
57. Revoir, W., Use of respirators. Proceedings, 1969 National Safety Congress. Chicago, Illinois.

58. Bolton, N. E., and Whitson, T. C., Respiratory protective equipment, air-purifying devices. Presented at the American Industrial Hygiene Conference, Detroit, Michigan, 1969.
59. Hyatt, E. C., A respirator program for teaching and research facilities. Presented at the National Conference on Campus Safety, Burlington, Vermont, 1968.
60. White, J. M., and Merrett, K. W., The Chalk river nuclear laboratories respirator program. *Amer. Ind. Hyg. Ass. J. 29,* 601 (1968).
61. Revor, W., Activities of the AIHA-ACGIH respirator committee during the past three years. *Amer. Ind. Hyg. Ass. J. 31,* 221 (1970).
62. Revoir, W., Trends in respirators. Presented at the Annual Meeting of the New England Section of the American Industrial Hygiene Association, Marlboro, Massachusetts, 1970.
63. Amendment to Bureau of Mines Schedule 21B, *Fed. Regist. 34* (117) (1969).
64. Federal Coal Mine Health and Safety Act of 1969, *Fed. Regist. 35,* 65 (1970).
65. American National Standard Z88.2 - 1969, Practices for Respiratory Protection, American National Standards Institute, Inc., New York.
66. American National Standard Z88.1 - 1969, Safety Guide for Respiratory Protection Against Radon Daughters, American National Standards Institute, Inc., New York.
67. Specification for respirators for protection against harmful dust and gases, BS 2091:1969, British Standards Institution, London, 1969.
68. Specification for positive pressure powered dust respirators, BS 4558:1970, British Standards Institution, London, 1970.
69. Specification for high efficiency dust respirators, BS 4555:1970, British Standards Institution, London, 1970.
70. Closed-circuit oxygen breathing apparatus, requirements and technical tests, Polish Standard PN-65/Z-04077, Polish Standardization Committee, 1969.
71. Functional clothing moves off the drawing board. *Style Int. Mag. 24*, 16 (1968).
72. Ergonomic clothes make their transport debut. *Commer. Motor* (1968).
73. The clothes for the job. *Financial World,* 4 (1968).
74. "Report by the Working Party on Protective Clothing Requirements in Blast Furnaces, Coke Ovens, and Ancillary Departments," British Iron and Steel Federation, 1969.
75. Voskresenskaya, N., Work clothing. *Auchno-Tekhn. Obschchestra SSR 10,* 33 (1968). Translation available from Clearinghouse for Federal Scientific and Technical Information, Report AD 703 143.
76. "Biological Isolation Garment, National Aeronautics and Space Administration," Technical Utilization Div., Tech. Brief 68-10500 (1969).
77. Destafano, J. T., Material design tests produce protective sleeve for glass handlers. *J. Amer. Soc. Safety Eng. 14,* 14 (1969).
78. Woodbury, R. L., Clothing, its evolution and development by the inhabitants of the Arctic. *Arch. Environ. Health 17,* 586 (1968).
79. Radnofsky, M., Developments in fire-retardant materials and coatings. Presented at the 16th Annual Ohio Fire Prevention Seminar, Columbus, Ohio, 1970.
80. Durette, (X-400), Monsanto Corp., St. Louis, Missouri.
81. Fypro, Travis Fabrics Co., New York.
82. Kynol, Carborundum Corp., New York.
83. Johnson, R. H., Breakthrough comes with heat protecting fabric. *Safety Eng. 136,* 18 (1968).
84. Attaway, C. D., "Testing Flash Protective Clothing," p. 6 Dupont Textile Fibers Dept., New Products Division, Wilmington, Delaware, 1969.
85. Stoll, A. M., and Chianti, M. A., Method and rating system for evaluation of thermal

protection. *Aerosp. Med. 40,* 1232 (1969).
86. Michon, J., Ernst, J., and Koutstall, G., Safety clothing for human traffic obstacles. *Ergonomics 12,* 61 (1969).
87. The basics of safety clothing. *Environ. Control Management 136,* 64. 1969.
88. Butterworth, R., and Donoghue, J. K., Contribution of activity released from protective clothing to air contamination measured by personal air samplers. *Health Phys. 18,* 319 (1970).
89. Williams, M. K., A trial of terylene overalls for lead-acid electric accumulation posters. *Brit. J. Ind. Med. 25,* 144 (1968).
90. Bamber, H. A., and Butterworth, R., Letters to the Editor, Asbestos hazard from protective clothing. *Ann. Occup. Hyg. 13,* 77 (1970).
91. Himanen, M., Enkaise Tapatturmia Naisten Pukeutuminen Ja Tyoturvallisius. *Forebygg Olycksfall Helsinki 6,* 4 (1969).
92. Spano, L., "Microclimate Controlled (Thermalibrium) Protective Clothing System for Military Applications," U. S. Army Natick Laboratories, Clothing and Personal Life Support Equipment Laboratory, CSPLSEL-60, Technical Report 69-58-CE (1968).
93. Esposito, J. J., "Description and Evaluation of a Portable Dry Ice-Water Conditioned Suit System for Air Crew Members." Available from Clearinghouse for Federal Scientific and Technical Information, Report AD 700 915, 1969.
94. Konz, S., and Nentwich, H. F., "A Cooling Hood in Hot-Humid Environments." Available from Clearinghouse for Federal Scientific and Technical Information, Report AD 684 582, 1969.
95. Simmons, R. B. V., The Douglas point air-supplied vault suit. *Amer. Ind. Hyg. Ass. J. 29,* 605 (1968).
96. Crockford, G. W., Industrial pressurized suits. *Ann. Occup. Hyg. 11,* 357 (1968).
97. Van Patten, R. E., and Gaudio, R., Vortex tube as a thermal protective device. *Aerosp. Med. 40,* 289 (1969).
98. Allinikov, S., Ziegenhagen, J., and Morton, W., "Foam in Place Form Fitting Helmet Liners." Air Force Materials Laboratory, Tech. Rep. AFML-TR-70-21, Wright Patterson Air Force Base, Ohio. Available from Clearinghouse for Federal Scientific and Technical Information, Report AD 706 402 (1970).
99. Vroege, D., "En Studie Voor en over het Gebruik van de Veiligheidshelm in de Haven van Rotterdam." Bedrijfsgenesskundige dienst voor de Haven van Rotterdam (1969).
100. New standards for head protection proposed. *Safety Eng. 139,* 25 (1969).
101. Ewing, C. L., and Irving, A. M., Evaluation of head protection in air craft. *Aerosp. Med. 40,* 596 (1969).
102. Rayme, J. M., and Masten, K. R., Factors in the design of protective helmets. *Aerosp. Med. 40,* 631 (1969).
103. Wigglesworth, E. C., Motivation in eye protection programs. *Amer. J. Optomet. 47,* 9 (1970).
104. Index of 1968 Articles, *Nat. Safety Counc. News 100,* 108 (1968).
105. Index of 1969 Articles, *Nat. Safety Counc. News 101,* (1969).
106. Index of 1970 Articles, *Nat. Safety Counc. News 102,* (1970).
107. AEC-NASA Tech. Brief 68-10277. "Thermal Protective Visor for Entering High Temperature Areas," 1968.
108. Clark, B. A. J., Welding Filters and Thermal Damage To The Retina. *Aust. J. of Optomet. 42,* 91 (1968).
109. Ban, S. C., Ordway, F. D., and Swendells, F. E., "Research on Sub-Micron Metal-Fiber Solutions." Available from Clearinghouse for Federal Scientific and Technical Information, Report AD 672 004, 1967.

110. Carpenter, J. A., and Peters, W. R., "Dipole Shutter: A Transparency For Eye Protection." Available from Clearinghouse for Federal Scientific and Technical Information, Report AD 686-732, 1968.
111. Dunmire, R. W., Hand protection against cutting oil dermatitis. *Safety Eng. 136,* 27 (1968)
112. Arndt, K. A., Cutting fluids and the skin. *Cutis 5*, 143 (1969).
113. Ross, J. B., Rubber boot dermatites in New Foundland: A survey of 30 patients. *Can. Med. Ass. J. 100,* 13 (1969).
114. Errico, M. E., Steel insoles block foot puncture injuries. *Safety Eng. 135,* 19 (1968).

New and Recurring Health Hazards in Industrial Processes

J. A. HOUGHTON

In his Cummings Memorial Lecture, Herbert E. Stokinger[1] evaluated groups of air pollutants that had been widely discussed and stated: "These pronouncements (on threats to health and comfort), generally dire by implication for long term effects on health, appear before us as spectres of morbidity, mortality, and decreased longevity. More often than not, the pronouncements are not dispassionate factual statements, particularly if passed through the editorial mill of the news media." Studies reported in technical journals have been used as a basis for emotional articles without adequate attempt to put the problems in perspective.

No industrial hygienist will deny that chemicals and wastes can have adverse effects on man and his environment if the handling and use are not controlled, but publicity given to some investigations has resulted in demands for complete banning of a material or for zero tolerances that are completely unrealistic.

It has been pointed out[2] that the gaseous effluent from each human contains concentrations of carbon dioxide (4–4½%) that are well above the TLV for industrial exposure, and he exhales about 3×10^{13} atoms of radioactive carbon-14 per day. He also exhales traces of radon and of carbon monoxide. Thus, we would have to conclude that man under special conditions such as in confined spaces is a contributory source of air pollution.

We have difficulty in keeping up with the problems of our constantly changing technology, and in our ignorance we have been wrong about problems in the past and will be wrong about some in the future, but perhaps we can keep a perspective by constant review of problems and by reevaluations such as the discussion by Dr. Stokinger.

REDUCTION AND ELIMINATION OF EXPOSURES

The curtailment of the use of mercury and the discharge of mercury waste has been a main item of news,[3] but mercury consumption had decreased in many applications before the great discussions of its effect on the ecology. Manometers and other process control equipment requiring mercury have been almost completely replaced by other and more reliable types of measuring and control equipment; the use of mercury in antifouling paints on ships has been replaced by copper oxide and tributyl tin compounds[4]; the FDA banned the use of several organic mercurial compounds for seed treatment.

The finding of mercury in fish and in waterways near chlorine plants using mercury cells spurred the greatest activity in the control of mercury exposures. The amounts discharged as waste have been greatly reduced by the control measures set up by the plants, and further reductions will be made when additional control equipment is installed. Also, some mercury cells will be phased out and replaced by diaphragm cells for the production of chlorine. The concentrations of mercury in fish exceeded the maximum concentration of 0.5 ppm, set by the FDA, in many instances both in fish from inland waterways and from deep-sea fish. It was surprising to many to learn that inorganic mercury sediment in waterways could be changed to dimethyl mercury[5] and thus become concentrated in the aquatic food chain. With this information, it was not surprising to find mercury in fish from waters near outlets of mercury wastes, but it was surprising to find the heavy concentrations in deep-sea fish such as tuna and swordfish and the succulent New England lobster, which are remote from industrial sources. Concentrations of mercury in seawater range up to about 2 parts per billion[6] and it would appear that there has been a build-up in deep-sea fish and it is possible that similar concentrations of mercury have always been present in fish. This is covered in more detail in the review on Water Pollution.

Another target for the elimination of mercury is the mercurial fungicides used in paints, particularly in some types of water-base latex paints, and paint manufacturers are testing out the newer type of fungicidal additives to replace the mercurials. The types of mercury compounds used in paints, and the concentrations used, probably have not presented a serious exposure, but mercury now has a bad name with the public and this use will be restricted. Letters to the Editor[7] raised the old problem of the exposure to mercury from the amalgam used for dental work. Although it may be a significant source of the mercury found in individuals who have no industrial exposure to mercury,[8] the hazard is considered to be minimal.

Many studies have been made of the amounts of lead in the air of major cities since the time lead alkyls were first used as antiknock agents in gasoline, and these studies are still continuing. The results obtained have not exceeded the limit suggested by the AIHA[9] of 10 $\mu m/M^3$ averaged over a 30-day period. A recent report,[10] indicating a rising trend in the amounts of lead in air in San Diego, concluded from a determination of the ratios of lead isotopes in the samples that most of the lead in the air came from automobile exhausts. From such studies, and from publicized discussions on the environment, campaigns and pressures caused the reduction of tetraethyl lead in many brands of gasoline and elimination in a few. In Japan, the concentrations of lead alkyls in gasoline were cut in half in 1970 and TEL sales will be banned by 1974. The main valid reason for cutting down or eliminating the amounts of lead is to permit the use of available catalytic converters which will oxidize carbon monoxide and unburned hydrocarbons and hopefully reduce smog formation,[11] but it will also reduce potential hazards from the handling of the lead alkyls.

The easiest way to maintain the octane rating in a gasoline is to increase the content of the aromatic hydrocarbons as a replacement for the lead,[11] but this could introduce additional problems from the toxicity of gasoline vapors and also a greater potential for irritating exhaust gases if good catalytic mufflers are not used and maintained properly. The response of individuals to various solvent vapors and oxidant, in smog chamber experiments,[12] indicated that the aromatic compounds except for benzene are fairly reactive in the formation of eye-irritating compounds.

The use of lead in paint has been reduced, and has been eliminated in paints for children's toys. Cases of lead poisoning from ingestion of dried flakes of lead paint still continue and the National Commission on Product Safety has called the condition serious. Paint products are available for use in homes which contain less than 0.5% of lead.

Lead exposures from the casting of stereotype plates in printing has not been a serious problem, although some cases of occupational disease have occurred from the careless heating and handling of the lead alloy at remelt operations. Plastic printing plates have been introduced and will probably replace the casting of lead alloy plates to a large extent.[13]

For many years, industrial hygienists have emphasized the need for controlling the use of carbon tetrachloride in industry because of its serious effects on man and its use has been curtailed or eliminated in industry particularly for benchwork cleaning of parts. It was not surprising when the FDA took action in 1968 to ban the use of carbon tetrachloride in fire extinguishers and in interstate transportation, except for use in chemical manufacture. The effect of this ban was mainly on consumer products.

Newer types of resins for setting the shape of permanent-press clothing have reduced or eliminated the formation of formaldehyde during the cure and the resulting complaints of irritation.

POLLUTION – ITS BROAD ASPECTS

To the general public, the spread of a material throughout a neighborhood is pollution whether it is a waste material, an essential material used in a process, or a valuable product – it is unwanted. The emphasis of Federal and state legislation has been on the control of waste materials, but insurance costs from pollution claims are mainly for damage caused by the accidental release of materials that are ordinarily under control.

Only 25% of the pollution claims are from the scheduled discharge of wastes that have not been controlled or have not been controlled satisfactorily, and the number of claims from waste handling can be expected to drop when better control methods are installed as a result of pressure from regulatory bodies. The balance of the claims for damage from pollution have been the result of accidental release of materials. This type of claim is on the increase in number as well as in cost.

The failure of waste-control equipment accounted for approximately 15% of the claims. Bag dust collectors wore out and spewed dusts of all sorts over neighborhoods; scrubbers have corroded and failed so that contaminants were not removed from an exhaust; collection equipment has been taken apart for repair without shutting down the exhaust fan and the operations in the plant continued to discharge contaminated air through the collection equipment; settling basins and oxidation ponds have filled up with sludge and the liquid effluent has passed through essentially untreated. Many of these pollution claims could have been prevented by better maintenance of waste-control equipment and better planning.

Rupture of process equipment with escape of material was the cause in about 45% of the claims. Breaks or separation of flexible hoses while pumping fluids permitted ammonia to escape which spread to nearby homes; fill lines were broken by rolling freight cars on spur tracks and chlorine escaped; reaction mixtures have been vented to the atmosphere through relief valves or rupture disks; and leaks at flanges and pump connections have spread gases and liquids through neighborhoods. This type of accident is more difficult to control, but better maintenance of equipment, particularly of process controls, and better supervision of visitors on plant property can help control these potential exposures. An appreciable number of claims were from employees of contractors and from maintenance servicemen who were not aware of the potentials in the hazardous areas in which they were working.

Leaks from storage tanks comprised 10% of the claims. A frequent cause was an open drain valve in the dike around a tank, so that when the tank leaked or overflowed during filling, the liquid escaped through the open drain valve, generally into a waterway. Small leaks have been left unrepaired to attend to more pressing problems and liquids have accumulated in the surrounding ground. Seepage of gasoline into the ground has endangered men constructing sewers and digging trenches; seepage of fuel oil into a snow covered area created a mess when the snow melted and flowed into a nearby stream; seepage of a nitro-aromatic compound caused a serious exposure to men digging a trench in the contaminated ground; and seepages of chemicals have contaminated water tables and wells.

The balance of the claims were the result of fire and explosion in which there does not seem to be any common cause. It is not expected that incidents from the accidental escape of materials can be stopped completely. Insurance coverage will still be needed, but if the costs are to be kept in line, then additional controls must be set up to include all types of potential pollution and not just waste pollution.

UTILIZATION OF WASTES

The junkman who made door-to-door collection of paper, cloth, and metal scrap is gone, and scrap for reclaim now comes mainly from industry. But the need to recycle scrap is increasing because cities and towns are running out of dumping space for solid wastes. We can expect to see the junkman return, probably under a fancier name, because an excellent way of reducing pollution is to find use for the large volumes of solid wastes produced in urban and suburban areas.

There is no lack of ideas as to what can and should be done with waste materials. Research by the Bureau of Mines[14a,b,15] and others has developed ways to produce useful products with the emphasis on the development of profitable ventures. In industries where large amounts of wastes accumulate, there are chances for profitable recovery and reuse. Where wastes are scattered, the costs for collection, transportation, and sorting of wastes may exceed the value of the reclaimed products in most cases and subsidies will be required to encourage the development of waste-reclaiming industries. New York City made a start by opening a contract for the collection of waste paper along with a decision to purchase office stationery made from recycled scrap paper.[3]

A long list could be made of the many methods suggested or tried on a small scale in the utilization of cellulosic scrap, fly ash, and other wastes. If these do develop into viable industries it will be a slow process. A very promising procedure is the pyrolysis of organic wastes in preference to incineration. Destructive distillation[16a,b] of scrap tires, garbage, and other municipal refuse can produce useful gaseous and liquid by-products, and may become a new and important industry. By-products are reported to include paraffinic hydrocarbons including fuel gases and fuel oils, organic acids, phenolic compounds and other oxygenated aliphatics and aromatics, char for activated carbon or fuel, and ash. The composition of the distillates and residues vary widely, uniformity of product can not be expected, and it probably will be impractical to refine the products. Use as fuels will be one ready solution because rough cuts can be handled in furnaces, but there may be some solvent use.

These new waste utilization industries will be more than just an enlargement or extension of the present scrap reclaim business. However, the problems of the present will still be found in the new along with a host of additional ones. Since these will be marginal or subsidized industries for the most part, short cuts will be taken to keep down costs, and if good controls are not set up, then the industry could become the worst source of pollution.

Although the pressure is on to get reclaim programs started promptly, these must be carefully planned for safety and for control of pollution. Severe problems could develop from corrosion from organic acids formed in destructive distillation, from sulfur dioxide reclaim units, from hydrogen chloride formed from PVC plastics, and from fluxes used in remelting of metals; corrosion can permit the escape of contaminants. Exhaust ventilation will have to be carefully designed to control dust from sorting of wastes, from fly ash handling, and other sources. The planning of reclaim units should not be rushed. The experience of industrial hygienists is needed in the planning stage.

DETERGENT FORMULATIONS

Formulators of household detergents have been subjected to a long series of complaints about the ingredients used.[3] The sulfonates of dodecylbenzene (DDB) and tridecylbenzene (TDB) were the active detergents used in the first formulations, and these resisted biological breakdown. Their persistence in water and soil caused foaming in sewage plants, streams, and rivers, and in some areas the wellwater foamed as it came out of the tap. It was found that the polymerization step used to form DDB and TDB produced highly branched molecules which did not degrade readily in sewage treatment plants.[17] The analogous straight-chain molecule degraded with greater ease and a switch to the straight-chain forms of DDB and TDB solved the worst of the difficulties.

Then, the extensive use of the polyphosphate sequestrants was claimed to accelerate the growth of algae and other aquatic plants; concerted campaigns by conservationists forced the reduction in the phosphate content. The prime contender as a replacement for the polyphosphates, sodium nitrilotriacetic acid (NTA) was effectively banned by Federal action in December 1970. Tests made in Federal laboratories indicated that the chelating action of NTA increased the rate of transmission of heavy metals across the placenta into the fetus and "the administration of cadmium and mercury simultaneously with NTA to two species of animals, rats and mice, yielded a significant increase in embryo toxicity and congenital abnormalities in the animals studied." It was recommended that "NTA detergent products not be used in certain limited areas which have both well water supplies and septic tanks and in which these treatment systems are operating under completely anaerobic conditions and where short-circuiting of septic tank effluent directly into well supplies is occurring."[18] Further testing has indicated that this was an over-cautious opinion. There is no reason why NTA should not be used in industry because it has good sequestering properties, the toxicity is low, and it is biodegradeable. NTA will probably be used in household detergents as a partial replacement for the polyphosphates in selected areas of the country. The polyphosphates and NTA are safer for home use than some of the highly alkaline products which have come on the market.

Arsenic in detergents was found by analysis to be present[19] in concentrations ranging up to 70 ppm, and warnings were given about the possibility of detergent wastes raising the arsenic level, in sources used for drinking water, above the level of 10 μg/liter recommended by the Public Health Service.[20] There are greater potential sources of arsenic that could get into drinking water such as the traces of arsenic in superphosphate fertilizers which could drain from fields or sodium arsenite added to lakes to kill rooted aquatic plants.

None of these materials has a direct adverse effect on man, but the handling of concentrated proteolytic enzymes has caused difficulties, and there has been greater justification for the attacks on the use of enzymes in home laundry detergents. Enzymes, first used in household detergents in Europe, were widely incorporated in detergent products in the United States in 1968–1969. These enzymes were obtained from cultures of *Bacillus subtilis* and were a mixture of proteolytic enzymes which were useful in digesting protein stains on clothing. It has been known that some proteolytic plant enzymes, such as papain (from papayas) and ficin (from latex sap from fig trees), were irritating and that papain caused allergic reactions.[21,22] Similar reactions were reported in employees handling concentrated powdered enzymes obtained from the *Bacillus subtilis*[23–25] and pulmonary hemorrhage was reported in hamsters exposed to this enzyme.[22]

Severe asthmatic attacks were reported in plant employees lasting from several hours to several days in which the breathlessness was so severe in some that they had difficulty getting out of bed in the morning; they had to stop and rest while walking to the bus. There were symptoms of coughing, wheezing, nasal irritation, and some skin irritation. Immediate positive skin reactions were obtained when affected individuals were tested with solutions of the enzyme extract. These severe symptoms did not arise from exposure to other ingredients used in the detergents.

The investigators emphasized the extreme precautions needed to protect workers and prevent inhalation of the enzyme dust because extremely minute amounts could trigger the symptoms in sensitized individuals. The Soap and Detergent Association published a detailed list of precautions that should be taken.[26]

The concentrate as received by the formulator may contain 5–10% of the active enzyme along with other ingredients as inactive organic matter, sodium sulfate, and sodium chloride. Laundry products were reported to contain 0.1–1% of the concentrate. The Federal Trade Commission has taken action against formulators using the enzyme, mainly based on advertising claims, but also because of some cases of skin rash. The severe allergy has not been reported from use of the home product, probably because the enzymes become encapsulated in the blending and manufacturing processes so that only traces would be present in the respirable sizes of detergent powder.

GASES – HANDLING AND USE

The consumption of liquefied oxygen has increased markedly the past few years, mainly because of the availability of cheap, pure oxygen; a large use is in the basic oxygen furnace for production of steel. Newer uses give promise of greatly expanded consumption with wider exposures from handling and storage. Initial tests in the treatment of liquid wastes[27] show that the concentration of oxygen dissolved in water can be more than doubled by the use of pure oxygen, thereby increasing the rate of oxidation. The additional costs for oxygen can be offset by the smaller sizes of equipment needed.

A modernized version of the Deacon process for production of chlorine[28] can utilize waste hydrogen chloride from the production of chlorinated organic compounds; or from waste inorganic chlorides. Hydrogen chloride is oxidized with oxygen to form chlorine and water, using nitric oxides as a catalyst and sulfuric acid to strip off the water formed. The nitric oxides and sulfuric acid are recycled.

Tests have shown that oxygen can supplement chlorine and chlorine dioxide in the bleaching of pulp, aid in the nodulation of phosphate ores in the rotary kiln used in production of phosphorus, and aid in reducing odors from the manufacture of kraft paper by oxiding the sulfides in spent liquors.

Suppliers of liquefied oxygen have trained technical staffs to advise on the handling and use, and have trained men to deliver the product, but they cannot always dictate the location of storage tanks and vaporizers. Accidental spillage which can occur during transfer has been the cause of several serious fires. It is essential to follow the requirements set forth in the standard on the "Installation of bulk Oxygen Systems at Consumer Sites."[29]

The increase in the use of liquefied hydrogen has not been as dramatic but it did lead to the publication of a new standard in 1968 on liquefied hydrogen systems.[30]

Storage of tomatoes, apples, and other produce in an inert atmosphere appears to be feasible in reducing respiration rate, ripening of products, and extending storage life.[31] The inert gas is produced by propane burners and scrubbers to give an atmosphere containing less than 3% oxygen and less than 5% carbon dioxide – a potential asphyxiant mixture.

Liquefied petroleum gas (LPG) has been used as an automotive fuel in lift trucks, taxi fleets, and local delivery trucks and the latest proposal is to use compressed or liquefied natural gas. This has been fleet tested in California and Oregon, but will probably be slow in developing.

A proposal to improve mine safety[32] would take advantage of the techniques developed in space technology through use of an inert atmosphere in coal mines. The advantages claimed include the prevention of fire and explosion from coal dust and methane, dust protection for the miners who would be encased in life support systems, prevention of the oxidation of iron sulfides which causes acid mine drainage, and possibly the recovery of commercial quantities of methane. This sounds like a dream for the distant future, but who can predict what developments will come from this dream?

POLYMERS

Continued interest in the thermal degradation products of various types of polymeric materials has culminated in the publication of many papers and considerable speculation. Studies, made to evaluate mechanisms of breakdown in textiles and plastic products, have resulted in the development of excellent methods for fire retardant treatments of plastics and textiles, and for building fire resistance into the polymer molecules. There have been warnings about the hazards under fire conditions particularly from hydrogen chloride released from the decomposition of polyvinyl chloride, and cyanides from plastics containing nitrogen.

With a mixture of materials, such as would be found in the home, the decomposition products from plastics might add to the hazard under fire conditions, but this added hazard would be no greater than from many other materials in the home. The main consideration in this respect should be the fire resistance rather than composition. Exposure of rats to decomposition products of polystyrene, wood, wool, and leather[33] indicated that carbon monoxide was the killing agent and that the natural materials formed lethal concentrations of carbon monoxide at lower temperatures (ca. 300°C) than did foamed polystyrene (ca. 500°C).

All plastics will degrade to give off volatile products under moderately high temperatures; the products of decomposition mentioned most frequently are carbon dioxide, carbon monoxide, saturated and unsaturated hydrocarbons, and water, but irritating and other toxic products are also formed in various amounts. Cases of irritation have occurred from overheating the plastic in molding operations,[34] burning off polyurethane foam insulation from tank cars and other equipment, and from some coating operations requiring the use of heat.

Comparatively little has been done to evaluate the toxic and irritant properties of volatile degradation products at lower temperatures, such as would be found in industry. Eight excellent papers discussed the toxic hazards from the breakdown products of polytetrafluoroethylene,[35a-h] and the approximate temperatures at which various products are formed. These papers pointed out that there are marked differences in the products formed depending on the composition, temperature, time, oxygen available, and differences in formulations, so that precise values could not be given to cover all conditions. The same applies to all other polymers. Temperatures quoted in these papers indicate that a toxic particulate can be formed from the polytetrafluoroethylene at about 350°–400°C, the monomer about 450°C, the highly toxic perfluoroisobutylene at about 480°C and carbonyl fluoride about 500°C.

Studies on other polymers have not given as much consideration to potential toxic exposures, but do give some indication. A study on polystyrene[36] indicated that about 50% is depolymerized to form the monomer in the process of combustion. Although temperatures were not reported, it would be expected that appreciable amounts of the monomer would form under heating, particularly in the absence of oxygen. Soot formation, an outstanding characteristic of burning styrene, results essentially from the burning of the monomer formed.

Studies of decomposition products from a polyester (polyethylene terephthalate) indicate the formation of appreciable quantities of acetaldehyde and benzoic acid.[37] Pyrolysis of acrylic fibers formed hydrogen cyanide, acrylonitrile, and vinyl acetonitrile at 250°C.[38] Pyrolysis of various types of nylon formed large amounts of unidentified volatile products at 300°C,[38] with traces of ammonia. At higher temperatures, indefinite amounts of materials such as cyclopentanone, complex isocyanates, and acetamide derivatives were released.

The volatiles formed are invariably complex mixtures and the products mentioned are just a few of many mentioned in the literature.

There is need for a compilation and a toxicologic evaluation of degradation products mentioned in existing literature, and for further toxicologic studies on various plastic materials similar to the studies reported on the polytetrafluoroethylene plastics. Meanwhile, it would be advisable to control the heating of any plastic material in industry, not to apply any more heat than is necessary to form and fabricate it, and to heed reports of irritation from the application of heat to any polymer.

SOLVENT EXPOSURES

Many cases of rapid coma and death have been reported in youths from inhalation of massive amounts of vapors from model airplane glue, aerosol propellants and solvents. As a result, DuPont issued several strong warnings in 1968 about the dangers of self-intoxication from the deliberate inhalation of concentrated vapors of aerosols and other products. Also, there have been a few isolated cases in industry of massive exposures to methylene chloride and methyl chloroform. The sudden deaths have been difficult to explain because short, acute exposures to acetates, fluorinated hydrocarbons, stable chlorinated hydrocarbons, and petroleum hydrocarbons would be expected to produce narcosis and coma with recovery after removal from exposure.

Respiratory arrest was often given as the cause of death, but the combination of rapid death and negative autopsy findings suggested that cardiac arrhythmia might be a factor either from the solvent vapors alone or combined with the asphyxia. This postulate was tested[39] by exposing mice to high concentrations of vapors from glue solvents and from toluene; definite arrhythmias were found that were not found from asphyxia alone. The authors stated: "Regardless of whether ventricular fibrillation or arrest is present, external cardiac massage and vigorous mouth-to-mouth respiration, by eliminating these volatile hydrocarbons and alleviating asphyxia, may have special efficacy in acute deaths due to the inhalation of solvents. Defibrillation may then return the cardiac rhythm to normal. The reviver should not inhale the victim's expired air."

This should be considered when planning for emergencies where acute exposures to highly volatile solvents may occur so that the exposed individuals may have proper and immediate first aid.

A study of forty-five solvents used in paint manufacture was made by the Battelle Memorial Institute[12] to determine the relative tendency of each to promote the formation of photochemical smog. In general, the study confirmed the basis for Los Angeles' rule 66 and San Francisco's regulation 3, except for ketones which showed slightly lower irritation effects than originally suggested.

Mixtures of 4 ppm of solvent vapor and 2 ppm of nitric oxide were photooxidized under simulated sunlight atmosphere conditions in a smog chamber. Concentrations of nitrogen dioxide, oxidants (primarily ozone), and formaldehyde were determined at various time intervals, and subjective eye irritation was checked by a panel of seven members who reported in terms of time response. The evaluations were summarized in a series of figures and tables showing the relative ranking of the solvents in each of the four tests.

Solvents showing consistently high formation of irritants included mesitylene, mesityl oxide, methyl *tert*-butyl ketone, α-methyl styrene, styrene, tetrahydrofuran, *m*-xylene, and mineral spirits with 53% aromatics. Solvents which consistently rated low included acetone, benzene, cyclohexanone, isobutyl acetate, isopropyl alcohol, methyl *n*-propyl ketone, 2-nitropropane, *n*-octane, and triethyl amine. The report emphasized that there are other parameters that were not measured in these tests which could affect smog formation, such as mixtures of solvent vapors, effect of automobile exhaust, effect of sulfur dioxide, and humidity effects, and that there is a need for further studies.

TEXTILES

There will be a marked increase in the use of chlorinated solvents in the coming years for the scouring, dyeing, and finishing of textiles.[40] Equipment is now available but in limited use for solvent scouring of cloth, particularly for cotton and polyester–cotton blends, for solvent dyeing of the newer fabrics, and for the solvent application of special finishes such as stain repellant and flame resistant treatments. Organic solvents have advantages over the traditional water applications in that less heat is needed for evaporation of solvents and drying of the cloth, the scouring of the cloth is faster and more complete, there is better application of dyes, water-insoluble finishes can be applied that are resistant to washing, and there can be a great reduction in the amounts of liquid wastes because lint and other contaminants can be recovered as solids when the solvent is redistilled. With the present emphasis on water pollution, this is an important factor because scouring wastes and dye depleted wastes are discharged in large volumes when water is used as a solvent.

The industry has been slow in adopting the new solvent methods because these require the abandonment of proven methods, but the greatest factor is the large capital expenditures needed to replace equipment that is still serviceable. The industry prefers to apply several finishes from a single bath, or several baths in series, using the same solvent for all finishes instead of organic solvents in some and water in others.

Trichloroethylene, perchloroethylene, and 1,1,1-trichloroethane appear to be the ones that will be in greatest demand in the equipment now available, and exposures to these solvents can be anticipated from improper handling. The systems are enclosed and designed to contain the solvent vapors, but the residual solvent is removed from the cloth by hot water and steam (essentially steam distillation) which increases the chances of vapor escape. The possibilities of exposure should be similar to those from the operation of an automatic degreaser – carryout of residual solvent on the cloth, spills from transfer of the solvent to the equipment, handling of solvent contaminated sludge, and entering the equipment to clear material jamming the equipment.

STORAGE AND TRANSPORTATION OF HAZARDOUS CHEMICALS

Accidents occurring during the transportation of hazardous chemicals have been of continuing concern to the Department of Transportation (DOT), the Food and Drug Administration (FDA), the Manufacturing Chemists Association (MCA) and other groups.

FDA, responsible for the monitoring of foods and feedstuffs, has been concerned with the large number of incidents that have been occurring during transportation.[41] Pesticides have been mentioned most frequently as contaminating agents but the FDA has also cited cases of oatmeal contaminated with phenol, animal feed with arsenic trioxide, potatoes with a lead paint pigment, and many others.

Leaking drums or broken bags contaminated other goods in mixed shipments, and freight cars used to haul toxic materials were not cleaned out before reuse. The DOT issued regulations[42] (effective January 1968) which prohibited the transportation of Class A or B poisons in the same car or truck with "foodstuffs, feeds, or any other edible material intended for consumption by humans or animals." Also, inspection is required for every car which has been used to transport a poisonous material, and the car must be decontaminated before reuse if necessary. These regulations apply only to common carriers; chemical products are still stored near foodstuffs in warehouses and carried with foodstuffs in trucks carrying supplies to retail food stores.

The National Academy of Sciences–National Research Council, following studies on the transportation of hazardous materials for the Coast Guard and for the DOT, published a tentative guide[43] for evaluating the hazard of industrial chemicals for water transportation. This guide gives a good breakdown of the relative health, water pollution, and reactivity hazards of about 200 chemicals. It is also useful as a guide for potential exposures in industrial plants.

Amending the classifications of chemicals was backed by the MCA as a high-priority need because all other aspects of transportation are based on these classifications. A number of proposals were made to the DOT to reclassify chemicals and to simplify and strengthen warnings. The DOT published proposals for the classification of health hazards[44] which are similar to those in other federal laws and regulations, and also proposed the abandonment of the old familiar ICC labels with conversion to the U. N. system,[45] with minor exceptions. The new labels would include a pictorial warning in addition to the usual wording and color coding; this change makes sense not only for international shipments but also for the benefit of those in our country who cannot read English.

Railroad accidents involving large chemical shipments caused the evacuation of several communities during the past few years. The resulting headlines have caused alarm in many other communities. Details on some accidents were published by the MCA and the National Fire Protection Association, but the information is sparse at times and good statistics and information are not available. The DOT now requires reports to be made[46] on incidents involving hazardous chemicals where a person is killed or seriously injured, or damage costs exceed 50,000 dollars. It is hoped that the information compiled will be useful in pointing out changes that are needed.

The MCA has set up a transportation emergency information center, "Chemtrek," to provide immediate information on the hazards of materials involved in accidents and procedures to minimize possible danger.

The DOT has finally proposed a change in the required method for flash point determination from the open cup to the closed cup,[47] and the break point for the requirement of a red label from 80°F open cup to 73°F closed cup. This was done following a study and recommendations of the Bureau of Mines. One hundred proof bourbon (closed cup flash point 75°F) is still safe and will not require the use of a red label.

An NFPA code on the storage of oxidizing materials[48] has been prepared, but does not include ammonium nitrate (covered in NFPA 490), nor organic peroxides, which will be covered in a separate code. These code requirements are pointed toward the control of fire and explosion hazards. There is a need of standards for the storage of toxic materials, such as pesticides; information on controls and procedures is scattered and not readily available to warehouse men and others who need it. Manufacturers do supply information in their technical data sheets, but there are many good principles and practices which could be compiled as general guides. There are frequent arguments as to whether these should be written as "codes," "standards," "recommended good practices," or "guides," but there is a need.

HEALTH AND SAFETY LEGISLATION

New Federal legislation has strengthened the authority of Federal departments in regulating health exposures in industry, particularly the Coal Mine Safety Act of 1969 (PL 91-173) and the Occupational Safety and Health Act of 1970 (PL 91-596); the latter supersedes the Walsh-Healey Act, the Service Contract Act and the National Foundation on Arts and Humanities Act. Of greater interest to industrial hygienists are the regulations which have been and will be promulgated under these and other enabling Acts; anyone working in this field should be well acquainted with all of the provisions.

The Department of Labor has published new regulations[49] for the control of exposures in shipyards (effective July 1970). The definition of what constitutes a hazardous material was expanded and defined as "the term 'hazardous material' means a material which has one or more of the following characteristics: (1) has a flash point below 140°F, closed cup, or is subject to spontaneous heating; (2) has a threshold limit value below 500 ppm in the case of a gas or vapor, below 500 mg/m^3 for fumes, and below 25 mppcf in case of a dust; (3) has a single oral LD_{50} below 500 mg/kg; (4) is subject to polymerization with the release of large amounts of energy; (5) is a strong oxidizing or reducing agent; (6) causes first degree burns to skin in short time exposure, or is systemically toxic by skin contact; or (7) in the course of normal operations, may produce dusts, gases, fumes, vapors, mists, or smokes which have one or more of the above characteristics."

The concentration of 500 mg/m^3 for fumes given above is not a typographical error, unless it was an error as published in the Federal Register and reprinted in the regulations for shipyards[50]; this would be an abnormally large concentration for any fume. The entire definition is all inclusive. There are very few materials used in industry that would escape the classification as "hazardous."

These regulations also require the employer to obtain information on every material which could be classified as hazardous by the definition given. In effect, the shipyard will have to obtain information on everything used in the yard because the composition and properties must be known in order to classify the material as "hazardous" or "nonhazardous." This would include such things as solvent, paints and coatings, structural materials such as zinc or cadmium coated steel, plastics, and welding materials. The pertinent information must be recorded on the Department of Labor form LSB OOS-4 or an essentially similar form which has been approved by the Bureau of Labor Standards. It includes percentage compositions, data on physical properties, fire and explosion hazards, health hazards, reactivity, handling of leaks or spills, and special protection and precautions which may be required. The burden will fall on the suppliers of the materials to fill out the forms and supply the necessary information before they can sell the substances to the shipyards.

This type of information is desired when evaluating an industrial hygiene exposure, because it is very difficult to make an evaluation when the only obtainable information is the statement that the material is "nontoxic." It will be impossible to keep proprietary formulations secret with the widespread dissemination of information of this type. We can expect similar requirements to be included in all other Federal safety regulations.

The Bureau of Mines published safety regulations governing the exposures to airborne contaminants in metal and nonmetallic mines[51] (effective January 1971) and at surface work areas for coal mines.[52] The regulations for metal and nonmetallic mines establish the list of TLV values of the American Conference of Governmental Industrial Hygienists as official, stating that "C" values shall not be exceeded. Time-weighted values are used for all other values. The regulations require that "Employees shall be withdrawn from areas in which there is a concentration of an airborne contaminant given a 'C' designation, which exceeds the threshold limit value." These are more specific than the corresponding regulations under the Walsh-Healey Act.[53]

Regulations for surface work areas of coal mines incorporates the ACGIH publication on TLV values as part of the regulations, including any future revisions or amendments, but makes an exception for dusts. The concept of dust counting has been dropped; the requirement is for air sampling to be done for respirable dust (not total dust) determined on a weight basis. For respirable dust, the upper limit is set at 2.0 mg/m^3. Where the average concentrations of respirable dust contain more than 5% quartz then the amount of dust must be kept "at or below a level, expressed in milligrams per cubic meter of air, which shall be determined by dividing the percentum of quartz present in such concentrations into the number 10." This limit is similar to the TLV set by the ACGIH but permits slightly higher limits because the second factor in the divisor is dropped in this simplified form. The differences between the two methods of computing the TLV is probably not significant, but the simplified form is not valid for silica concentrations below 5%; the upper limit of 2 mg should not be exceeded.

REFERENCES

1. Stokinger, H. E., The spectre of today's environmental pollution – USA brand. Cummings Memorial Lecture. *Amer. Ind. Hyg. Ass. J. 30*, 195 (1969).
2. Williams, D. C., Letter to the editor. *Chem. Eng. News. 48*(44), 15 (1970).
3. News Items. *Chem. Eng. News 46, 47, 48* (1968–70); and *Chem. Week 103–108* (1968–71).
4. Hartley, R. A., Toxicity of new antifouling paints. *Proc. Marine Chem. Ass. Seminar, Nat. Fire Protection Ass., Boston, Massachusetts, July, 1970.*
5. Abelson, P. H., Methyl mercury (editorial). *Science 169*, 237 (1970).
6. Bache, C. A., Gutenmann, W. H., and Lisk, D. J., Residues of total mercury salts in lake trout as a function of age. *Science 172*, 951 (1971).
7. Letters to the Editor. *Chem. Eng. News 48*(49), 7 (1970); *49*(2), 8 (1971); *49*(4), 8 (1971); *49*(31), 34 (1971).
8. Stokinger, H. E., Mercury. *In* "Industrial Hygiene and Toxicology" (F. A. Patty, ed.), Vol. II, p. 1091. Interscience, New York, 1963.
9. Lead, Community Air Quality Guides. *Amer. Ind. Hyg. Ass. J. 30*, 95 (1969).
10. Chow, T. J., and Earl, J. L., Lead aerosols in the atmosphere; increasing concentrations. *Science 169*, 577 (1970).

11. Gasoline (Special Report). *Chem. Eng. News 48*(47), 52 (1970).
12. Levy, A., and Miller, S. E., Final technical report on the role of solvents in photochemical-smog formation. Tech. Div. Rept. 799 Nat. Paint and Varnish and Lacquer Ass., Washington, D.C. (1970).
13. Printing Shakeup Opens the Door to Plastics. *Mod. Plast. 46*(11), 68 (1969).
14a. Cservenyak, F. J., and Kenahan, C. B., Bureau of mines research and accomplishments in utilization of solid wastes. *U. S. Bur. Mines Inform. Circ.* IC 8460 (1970).
14b. Sullivan, P. M., and Stanczyk, M. H., Economics of recycling metals and minerals from urban refuse Bureau of Mines. *U. S. Bur. Mines Tech. Progr. Rep.* TPR33 (1971).
15. Ash Utilization. Proceedings: Second Ash Utilization Symposium, March, 1970. *U. S. Bur. Mines Inform. Cir.* IF 8488 (1970).
16a. Wolfson, D. E., Beckman, J. A., Walters, J. G., and Bennett, D. J., Destructive distillation of scrap tires. *U. S. Bur. Mines Rep. Invest.* RI7302 (1969).
16b. Sanner, W. S., Ortuglio, C., Walters, J. G., and Wolfson, D. E., Conversion of municipal and industrial refuse into useful materials by pyrolysis. *U. S. Bur. Mines Rep. Invest.* RI7428 (1970).
17. Silvis, S. J., Detergents, part II. *Chem. Week 105*(16), 80 (1969).
18. NTA Ban Creates Chemical Industry Void. *Chem. Eng. News 49*, 15 (1971).
19. Angino, E. E., Magnuson, L. M., Waugh, T. C., Galle, O. K., and Bredfeldt, J., Arsenic in detergents: Possible danger and pollution hazard. *Science 168*, 389 (1970); Pattison, E. S., Arsenic and water pollution hazard. Sollins, I.,V., Comment. Angino, E. E., Comment. *Science 170*, 870 (1970).
20. Drinking Water Standards. Arsenic. U. S. Public Health Service, Washington, D.C. (1962).
21. "The Merck Index," 7th Ed. Merck, Rahway, New Jersey, 1960.
22. Goldring, I. P., Ratner, I. M., and Greenburg, L., Pulmonary hemorrhage in hamsters after exposure to proteolytic enzymes of *Bacillus subtilis*. *Science 170*, 73 (1970).
23. Flindt, M. L. H., Pulmonary disease due to inhalation of derivatives of *Bacillus subtilis* containing proteolytic enzyme. *Lancet I*, 1177 (1969).
24. Pepys, J., Hargreave, F. E., Longbottom, J. L., and Faux, J., Allergic reactions of the lungs to enzymes of *Bacillus subtilis*. *Lancet I*, 1181 (1969).
25. McMurrain, K. D., Dermatological and pulmonary responses in the manufacture of detergent enzyme products. *J. Occup. Med. 12*, 416 (1970).
26. "Procedures for Preventive Plant Hygiene for Detergent Enzyme Products." Soap and Detergent Ass., New York (1969).
27. Oxygen Producers Zero in on Waste Treatment. *Chem. Week. 107*(21), 95 (1970).
28. Oblad, A. G., The Kel-Chlor process. *Ind. Eng. Chem. 61*(7), 23 (1969).
29. "Standard for Installation of Bulk Oxygen Systems at Consumer Sites." NFPA 566-1965. Nat. Fire Protection Ass., Boston, Massachusetts, 1965.
30. "Standard for Liquefied Hydrogen Systems at Consumer Sites." NFPA 50 B-1968. *Nat. Fire Protection Ass.*, Boston, Massachusetts, 1968.
31. Oxygen Starved Tomatoes Stay Fresher. *Chem. Eng. News 47*(53), 66 (1969).
32. Inert Atmospheres in Mines Could Abate Acid Drainage. *Chem. Eng. News 48*(21), 33 (1970).
33. Hofmann, H. T., and Oettel, H., Comparative toxicity of thermal decomposition products. *Mod. Plast. 46*(10), 94 (1969).
34. Cleary, W. M., Thermoplastic resins decomposition. *Mich. Occup. Health 14*(3), 5 (1969).
35a. MacFarland, H. N., Pyrolysis products of plastics – problems in defining their toxicity. *Amer. Ind. Hyg. Ass. J. 29*, 7 (1968).

35b. Waritz, R. S., and Kwon, B. W., Inhalation toxicity of pyrolysis products of polytetrafluoroethylene. *Amer. Ind. Hyg. Ass. J. 29*, 19 (1968).
35c. Kupel, R. E., and Scheel, L. D., Experimental method for evaluating the decomposition of fluorocarbon plastics by heat. *Amer. Ind. Hyg. Ass. J. 29*, 27 (1968).
35d. Coleman, W. E., Scheel, L. D., Kupel, R. E., and Larkin, R. L., Identification of toxic compounds in the pyrolysis products of polytetrafluoroethylene. *Amer. Ind. Hyg. Ass. J. 29*, 33 (1968).
35e. Scheel, L. D., Lane, W. C., Coleman, W. E., Toxicity of polytetrafluoroethylene pyrolysis products. *Amer. Ind. Hyg. Ass. J. 29*, 41 (1968).
35f. Scheel, L. D., McMillan, L., and Phipps, F. C., Biochemical changes associated with toxic exposures to pyrolysis products. *Amer. Ind. Hyg. Ass. J. 29*, 49 (1968).
35g. Coleman, W. E., Scheel, L. D., and Gorski, C. H., Particles resulting from polytetrafluoroethylene pyrolysis in air. *Amer. Ind. Hyg. Ass. J. 29*, 54 (1968).
35h. Birnbaum, H. A., Scheel, L. D., and Coleman, W. E., Toxicology of the pyrolysis products of polychlorotrifluoroethylene. *Amer. Ind. Hyg. Ass. J. 29*, 61 (1968).
36. Petrella, R. V., and Sellers, G. D., Fundamental combustion studies of styrene. "Notes on Flame Retardance of Fibers." Clemson University, Textile Department, Clemson, South Carolina (1970).
37. Roberts, C. W., Pyrolysis and combustion of polyesters. "Notes on Flame Retardance of Fibers." Clemson University, Textile Department, Clemson, South Carolina (1970).
38. Barker, R. H., Pyrolysis and combustion of nylon and acrylic fibers. "Notes on Flame Retardance of Fibers." Clemson University, Textile Department, Clemson, South Carolina (1970).
39. Taylor, G. J., and Harris, W. S., Glue sniffing causes heart block in mice. *Science 170*, 866 (1970). Copyright 1970, American Association for the Advancement of Science.
40. Hofstetter, H. H., Solvents in textile processing. *Text. Ind. 134*(6), 53 (1970).
41. Stringer, J. G., Pesticide contamination of food and drugs during shipment. *FDA Pap. 24*, 4 (1968).
42. Department of Transportation Regulation. Restrictions against loading and transporting poisons (class A or B) with foodstuffs. *Fed. Regist. 34*(224), (1969).
43. "Evaluation of the Hazard of Bulk Water Transportation of Industrial Chemicals (A Tentative Guide)." *Nat. Acad. Sci.*, Washington, D.C. (revised 1970).
44. Department of Transportation Regulation. Advance notice of rule making on classification of certain hazardous materials on the basis of their health hazards. *Fed. Regist. 35*(110), 8831 (1970).
45. Department of Transportation Regulation. Proposed amendments on the classification and labelling of hazardous materials. *Fed. Regist. 36*(101), 9449 (1971).
46. Department of Transportation Regulation. Amendments to requirements of reports of hazardous material incidents. *Fed. Regist. 35*(213), 16836 (1970).
47. Department of Transportation Regulation. Flash points of flammable liquids. *Fed. Regist. 35*(236), 18534 (1970).
48. "Tentative Code for the Storage of Liquid and Solid Oxidizing Materials, NFPA 43A-T." Nat. Fire Protection Ass., Boston, Massachusetts, 1971.
49. Bureau of Labor Standards. Hazardous materials. Amendments to safety and health regulations of 29 CFR, part 1501 for ship repairing, part 1502 for shipbuilding, part 1503 for shipbreaking. *Fed. Regist. 35*(1), 10 (1970).
50. Code of Federal Regulations, Title 29, Chap. XIII, Parts 1501, 1502 and 1503. U. S. Dept. Labor, Bur. Labor Stand., Washington, D.C., 1970.
51. Bureau of Mines, Health and Safety Standards. Part 55, metal and nonmetallic open pit mines, part 56, sand, gravel and crushed stone operations, part 57, metal and

nonmetallic underground mines. *Fed. Regist. 34*(145), 12503 (1969); and *35*(237), 18587 (1970).

52. Bureau of Mines, Coal Mine Health and Safety. Part 71, mandatory health standards – surface work areas of underground coal mines and surface coal mines. *Fed. Regist. 36*, 252 (1971).
53. Department of Labor, Safety and Health Standards. Part 50-204, safety and health standards for federal supply contracts. *Fed. Regist. 34*(96), 7946 (1969).

Contributions of Ergonomics in the Practice of Industrial Hygiene

BRUCE A. HERTIG

The Occupational Safety and Health Act of 1970 specifies that "Within two years of enactment of this Act, and annually thereafter the Secretary of Health, Education and Welfare shall conduct and publish industrywide studies of the effect of chronic or low-level exposure to *industrial materials, processes and stresses* on the potential for illness, disease, or loss of functional capacity in aging adults."[1] (*Italics added.*) This broad mandate carries an unescapable challenge to industrial hygienists: to the problems associated with the working environment, such as airborne contaminants (dusts, fumes, and gases) and physical stresses (heat, noise, radiation) must be added those problems caused by the stresses arising from the task itself. These are the new frontiers of the profession of industrial hygiene – a profession characterized over its history by ever-increasing responsibilities for maintenance of health and well-being of the industrial population.

Researches in ergonomics during the past 3 years have contributed new data and knowledge to help meet these challenges. In "Industrial Hygiene Highlights"[2] ergonomics concepts and techniques were described in general terms; because recent legislation has thrust *work-stress physiology* into the mainstream of industrial hygiene practice, that area of ergonomics will be emphasized here.

MANUAL MATERIALS HANDLING

How much should a worker be required to lift? In the early days of industrialization, lifting, carrying, pushing, and pulling heavy objects was expected by management and accepted as part of the job by employees.

Self-selection eliminated those whose physique or infirmities prevented them from meeting the demands of the job. Mechanization has substantially reduced those stresses so that strong muscles are no longer required for many tasks. Where manual materials handling is still routine, e.g., warehouses, shipping rooms, self-selection still plays an important part. It is the occasional demand for lifting, perhaps by an unfit or untrained individual, which provides the majority of the lifting overstrain and accidents. The National Safety Council reports that manual handling tasks represent the principal source of compensable work injuries.[3]

Several attempts have been made to define maximum permissible weights of lift. For example, the International Labour Office compromised on 88 lbs (40 kg) as the recommended maximum weight to be carried by one person.[4] But such a single figure ignores so many factors as to be largely invalid. It penalizes the trained load carrier. Porters in Alaska can carry immense loads over terrain impassable to vehicles or sleds; loads of 300 or more pounds are not unusual. On the other hand, the elderly, the untrained, or the unfit person will not be able to handle 88 lbs safely.

In recognition of the limitations of the single value for weight of lift, the American Industrial Hygiene Association (AIHA) has issued an *Ergonomics Guide to Manual Lifting** in which multiple factors are considered. Not only is the maximum weight of lift taken into account, but also the rate of lift, and whether the lift is from the floor, from a bench, or from a shelf. An additional feature of the Guide is that data are presented for the tenth, twenty-fifth, fiftieth, seventy-fifth, and ninetieth percentiles of the population. One can select for himself whether he will set the maximum weight to be lifted so that most employees will be able to handle the job, or whether he will select and train workers with special muscular attributes. The burden of determining standards within an organization remains with management; the Guide presents the data so that an informed decision may be made.

Figure 1 illustrates the use of the data for selecting maximum weight of lift. Note that the size of the object influences the weight which can be safely handled. This reflects the increased load on the spine which a given weight will exert as its center of gravity moves away from the body.[5]

These data were developed largely from laboratory studies conducted by Snook and Irvine.[6] Male industrial workers were allowed to choose the maximum weight which they could lift without excessive strain; psychophysical criteria were used.[7]

Example: Object width 20 in., assume fiftieth percentile worker. Read maximum object weight of about 62 lbs.

*Available from American Industrial Hygiene Association, 210 Haddon Avenue, Westmont, New Jersey; $0.50 per copy.

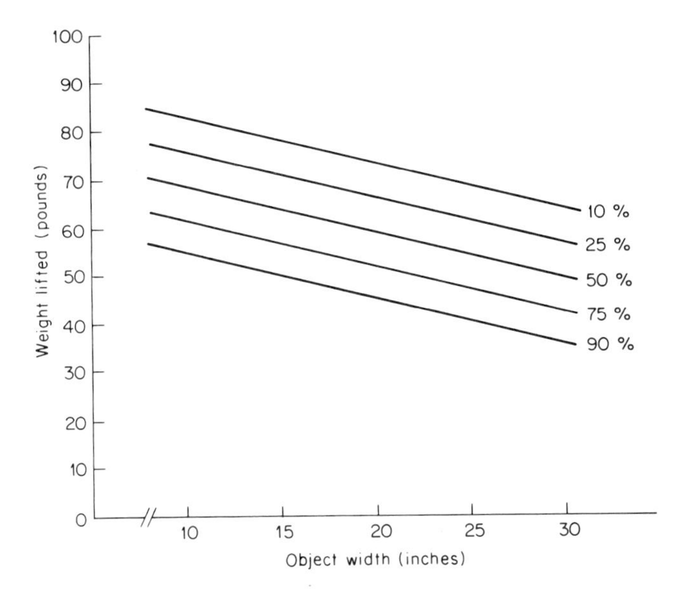

Fig. 1. The maximum weight that 10, 25, 50, 75, and 90% of the healthy male industrial population can be expected to lift between floor level and knuckle height while grasping the object midway along its width. For females, use 70% of suggested male values. (From *Ergonomics Guide to Manual Lifting,* American Industrial Hygiene Association, 210 Haddon Avenue, Westmont, New Jersey; used with permission.)

Note that less than 10% of the test population felt that they could handle safely the 88-lb limit recommended by the ILO.

Figure 2 presents data on maximum work load for the same lift (floor to knuckle height) as Fig. 1. First select the appropriate weight and size from Fig. 1 and then determine the frequency from Fig. 2.

Example: Object weight 40 pounds; assume fiftieth percentile worker. Read maximum work rate of 270 ft-lb/min. An average floor to knuckle lift is about 2.5 ft. so the frequency of lift calculates to be

$$\frac{270}{40 \times 2.5} = 2.7 \text{ lifts/min.}$$

Again, psychophysical criteria were used; the men selected rates which they felt they could maintain for an 8-hr day without suffering excessive fatigue. Note that a 50-lb weight seems to be the upper limit which the subjects felt they could handle repetitively.

Other graphs in the Guide provide data for bench and shelf lifts. In addition, characteristic injuries and industrial hygiene practices are outlined. Finally a

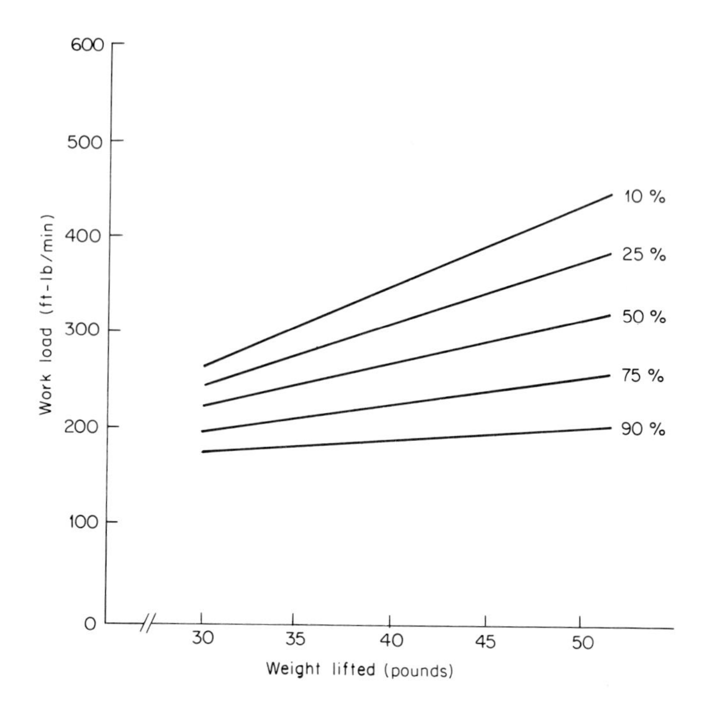

Fig. 2. The maximum work load that 10, 25, 50, 75 and 90% of the healthy male industrial population can be expected to maintain while lifting in a thermally comfortable environment, between floor level and knuckle height. For females, use 70% of suggested male values. (From *Ergonomics Guide to Manual Lifting,* American Industrial Hygiene Association, 210 Haddon Avenue, Westmont, New Jersey; used with permission.)

substantial bibliography is included for those who wish to pursue the subject more extensively.

The data discussed above apply to lifting. Snook *et al.* have extended their studies to cover pushing, pulling, carrying, and walking tasks.[8] The Ergonomics Committee of AIHA is preparing a new *Ergonomics Guide* to cover these other modes of manual materials handling.

WORK CAPACITY

How much physical effort should workers be required to expend? Human capacity to perform useful work involves many complex physiologic processes; however, the upper limits for physical work are ultimately determined by the capacity of the respiratory and cardiovascular systems to deliver oxygen to the working muscles. This capacity is termed maximal oxygen uptake (max$\dot{V}_{O_2}$). Units commonly used are liters per minute.

Tolerance times for maximal effort are measured in minutes. Some industrial tasks periodically may demand brief spurts of near maximal effort, e.g., changing anodes in an aluminum pot. However, over an 8-hr shift, the average work load will of necessity fall well below maximal.

Oxygen uptake provides a convenient measure of the energy demands of a task. One liter of oxygen consumed yields approximately 4.9 kcal (19.4 Btu). Trained athletes may reach maximal oxygen uptake capacities of 4.0 or 4.5 liters/min; some may even exceed 5 liters/min. Natural endowment largely determines max$\dot{V}_{O_2}$; training may increase this no more than 10–20%.[9] Industrial populations will have an average max$\dot{V}_{O_2}$ of 3–3.5 liters/min. Age and sex influence oxygen uptake; these effects will be discussed in a later section.

These figures translate into capacities for maximal aerobic energy production of 4600–5200 Btu/hr for the athletes, and 3500–4000 Btu/hr for the industrial population.[10] Demands for energy expenditure in excess of the max$\dot{V}_{O_2}$, such as all-out running upstairs, come from anaerobic sources. In the process, an oxygen debt is created which must be "paid off" during periods of recovery. It would be the unusual circumstance in industry, e.g., emergencies, for tasks to require mobilization of anaerobic processes.

Table I lists the energy costs of typical industrial activities, with resting levels included for reference. These values must be interpreted with caution. Adjustment for body weight is necessary if the individual is substantially heavier or lighter than the "average" man (154 lbs). Proportional adjustments may be made, as O_2 consumption is essentially linearly related to weight.

Few tasks require a steady rate of energy expenditure, thus it is necessary to calculate the time-weighted average to determine overall energy cost for a full work shift. Experience has shown that healthy individuals can maintain energy expenditure of the order of 40% of maximal aerobic capacity. For a max$\dot{V}_{O_2}$ of 3.0 liters/min (~3500 Btu/hr) the average work rate would be near 1400 Btu/hr. This is equivalent to walking at a brisk pace, nearly 3.5 mph; it is apparent that only physically fit individuals can tolerate prolonged work loads of this magnitude.

Where energy demands exceed this level, compensatory rest periods are required. The following expression allows calculation of percent rest time where an average energy expenditure in excess of 1400 Btu/hr is specified.

$$T_{rest}(\%) = \frac{1400 - M}{400 - M} \times 100$$

TABLE I
Metabolic Energy Costs (M) of Several Typical Activities[a]

Activity	*M (Btu/hr)*[b]
Resting, prone	330–360
Resting, seated	380–400
Standing, at ease	400–450
Drafting	430
Light assembly, bench work	430
Medium assembly	640
Driving automobile	670
Walking, casual	700–900
Sheet-metal work	715
Machining	740
Rock drilling	900–2250
Mixing cement	1100
Walking on job	1150–1600
Pushing wheelbarrow	1200–1650
Shoveling	1300–2100
Working with axe	1640–5700
Climbing stairs	1800–3100
Slag removal	2530–3000

[a] Values are for the "average" male worker. Data are based in part on Ref. 10.
[b] Divide by 3.97 to obtain M in kcal/hr.

where M is total metabolic cost of the task. The constant of 400 enters into the expression as an approximate value for resting metabolism.

Example: Assume M of 2000 Btu/hr

$$T_{rest} = \frac{-600}{-1600} \times 100 = 37.5\%$$

This means that for the assumed task, the individual would work five-eighths of the time and rest three-eighths. A somewhat different expression is presented in a new *Ergonomics Guide.*[11]

In setting work/rest cycles, one should keep in mind that the greater the energy demands the more frequent should be the rest pauses. Frequent short rests reduce cumulative fatigue.[11]

While Table I offers guide lines to approximate energy costs of industrial tasks, occasions may demand more accurate assessments of physiologic effort. One method involves sampling heart rate during the task, or in the first minute following the work. Counting the pulse at the wrist by palpation requires some practice, especially during active muscular movement. For more sophisticated studies cardiotachometers of various designs may be employed to count electrical impulses associated with the events of the cardiac cycle through electrodes fastened to the chest. Telemetry systems provide freedom of movement for the subject under study.

Energy expenditure may be approximated from heart-rate data. Table II outlines the ranges of values for several classifications of work. Note that orders of magnitude are given; extrapolation to an individual should be done with caution. Again, environmental factors, particularly heat stress, will add substantial increments to heart-rate values. Andrews[12] discusses the statistical relationship between heart rate and oxygen consumption in considerable detail. Consult also the Guide from which the data in Table II were taken.[11]

TABLE II
Energy Costs (M) Approximated from Heart-Rate Data[a]

Activity	*Heart rate (beats/min)*	*M (Btu/hr)*
Sitting quietly	60–70	380–400
Sitting, light to moderate arm and leg movements	65–75	380–600
Standing, moderate effort	75–100	600–1200
Walking about, heavier effort	100–125	1200–1800
Heavy work	125–150	1800–2400
Very heavy work	150–180	2400 to maximum

[a] Adapted from *Ergonomics Guide on Metabolic and Cardiac Costs of Physical Work.*[11] Values are for the "average" male worker.

Measurement of oxygen consumption directly provides, of course, much more precise estimates of energy production. Most methods involve the collection of samples of expired air for subsequent analysis of oxygen and carbon dioxide content. These well established techniques require some basic background in respiratory calorimetry. It is beyond the scope of this review to go into methodologic details. The reader is referred to Åstrand and Rodahl,[9] to Consolazio *et al.*,[13] and to the *Ergonomics Guide.*[11]

AGE AND SEX

What allowances are needed for women and for the older worker? Title VII of The Civil Rights Act of 1964[14] specifies that employment may not be denied on the basis of age or sex, among other characteristics of the applicant. The implications of this legislation for certain job classifications are disturbing, for there are anatomic and physiologic limitations imposed by the aging process and between the sexes which most assuredly should be considered in selecting applicants for industrial work. Particularly is this true for tasks which tax the body physically or impose substantial environmental stress.

Age

Extensive studies on physical work capacities have been conducted here and abroad, with subjects ranging in age from children to the elderly. Henschel, in his review of effects of age on work capacity,[15] summarized the findings in two statements: "(1) The capacity to perform light to moderate physical work is not grossly age dependent, at least up to 65 years of age; and (2) The capacity to perform hard exhausting work is strongly age dependent." These statements are valid for healthy individuals. With advancing age, of course, the likelihood that cardiovascular disease or other pathology will be present increases, further restricting work capacity.

Capacity for maximal oxygen uptake reaches a peak between the ages of 20–25 years. Thereafter it declines more or less linearly with advancing age. On the average, the 65-year-old man will have max$\dot{V}_{O_2}$ 70% of that at 25.[9] At any age, the variance is quite large, as one would expect. Thus it is not unusual to find older men with greater max$\dot{V}_{O_2}$ than many younger individuals.

The decline in aerobic capacity with age suggests that for tasks requiring substantial effort – approximating 40% of the max$\dot{V}_{O_2}$ – younger people should be selected. As a worker ages in a job, consideration should be given to increasing rest pauses; perhaps job redesign could reduce physical demands of the task. As noted above in Henschel's first conclusion, light to moderate work should be tolerated with no overstrain throughout the employment years, provided health is maintained.

Muscle strength follows somewhat the same pattern as does aerobic capacity; maximum strength is attained at age 25 declining to about 80% at 65. Decline in strength of leg and trunk muscles is greater than in the arm muscles.[9] As with max$\dot{V}_{O_2}$, the variance in a population is large, and, of course, training plays an important part in the abilities of individuals.

Sex

Much of the data on physiologic responses of human beings to environmental stresses have been gathered using young men as subjects. Qualitatively, women have always been branded "the weaker sex." Quantitative data on physiologic behavior of women to stress are becoming increasingly available. Maximal aerobic power of the female averages 70–75% of the male of equal age.[15] Muscle strength is about two-thirds that of men, though this primarily reflects difference in muscle mass; the strength per unit of muscle is really about the same in the two sexes.[9]

Besides their inherently lower capacity for physical work, recent studies show that women tolerate other stresses less well than men. For example, women exhibit greater strain than men in hot environments.[16] And their sensitivity to carbon monoxide appears to be substantially greater.[17] Does this mean that a separate set of TLV's are needed for women? The evidence increasingly points to that conclusion. The wise management will take this into consideration in assigning women to stressful jobs. For work and for heat stress, the "female correction factor" appears to be about 0.7. Whether this holds for other stresses remains to be determined.

SUMMARY

In the three years since "Industrial Hygiene Highlights" appeared, Federal legislation has accelerated the pace at which work-stress physiology has become incorporated into industrial hygiene practice. Concomitantly, researches in ergonomics have provided quantitative human data to aid in the solution of work-induced health problems. Publication by AIHA of a series of *Ergonomics Guides* on lifting, on metabolic costs of physical work, and on other topics makes conveniently available to industrial hygienists summaries of these new data.

Areas of concern for the immediate future include quantitative methods for assessing fatigue, establishment of work/rest cycles, advantages and disadvantages of the 12-hr day, 3-day work week. As these aspects of industrial hygiene practice increase, as they surely will, ergonomics research will continue to supply the basic information and techniques. In time, determination of oxygen consumption will be as much a part of analytic methods in industrial hygiene as atmospheric sampling and dust counting are now.

REFERENCES

1. Public Law 91-596, 91st Congress, s. 2193, December 29, 1970; Section 20. (a) (7).
2. Hertig, B. A., Ergonomics in the practice of industrial hygiene. *In* "Industrial Hygiene Highlights" (L. V. Cralley, L. J. Cralley, and G. D. Clayton, eds.), Vol. 1, pp. 229. Ind. Hyg. Found. Amer., Pittsburgh, Pennsylvania, 1968.
3. "Accident Facts." p. 31, Nat. Safety Counc. Chicago, Illinois, 1969.
4. International Labour Office, "Meeting of Experts on the Maximum Permissible Weight to be Carried by One Worker." M.P.W./1964/14, Geneva, 1964.
5. Tichauer, E. R., The biomechanics of the arm–back aggregate under industrial working conditions. Amer. Soc. of Mech. Eng. Publ. No. 65-WA/HUF-1 (1965).
6. Snook, S. H., and Irvine, C. H., Maximum acceptable weight of lift. *Amer. Ind. Hyg. Ass. J. 28,* 322 (1967).
7. Snook, S. H., and Irvine, C. H., The evaluation of physical tasks in industry. *Amer. Ind. Hyg. Ass. J. 27,* 228 (1966).
8. Snook, S. H., Irvine, C. H., and Bass, S. F., Maximum weights and work loads acceptable to male industrial workers. *Amer. Ind. Hyg. Ass. J. 31,* 579 (1970).

9. Åstrand, P.-O., and Rodahl, K., "Textbook of Work Physiology." McGraw-Hill, New York, 1970.
10. Durnin, J. V. G. A., and Passmore, R., "Energy, Work and Leisure." Heineman. London, 1967.
11. "Ergonomics Guide to Assessment of Metabolic and Cardiac Costs of Physical Work," *Amer. Ind. Hyg. Ass. 32*, 560-564 (1971).
12. Andrews, R. B., Indices of heart rate as substitute for respiratory calorimetry. *Amer. Ind. Hyg. Ass. J. 27,* 527 (1966).
13. Consolazio, C. F., Johnson, R. E., and Pecora, L. J., "Physiological Measurements of Metabolic Functions in Man." McGraw-Hill, New York, 1963.
14. Public Law 88-352. Section 2000E and following. July 2, 1964.
15. Henschel, A., Effects of age on work capacity. *Amer. Ind. Hyg. Ass. J. 31*, 430 (1970).
16. Hertig, B. A., Human physiological responses to heat stress: males and females compared. *J. Physiol. (Paris) 63,* 270 (1971).
17. Hertig, B. A., and Badger, D. W., Interaction of carbon monoxide and heat stress. Presented at American Industrial Hygiene Conference, Toronto, Canada, May 1971 (manuscript in preparation).

Air Pollution

GEORGE D. CLAYTON

Covering only the highlights of a subject as broad, popular, and active as air pollution presented a dilemma. Should this review concern itself with the details of instrumentation, analytic methods, control procedures, legislation, the effects on health, vegetation, economics? Obviously a book could be written and many have been written on any of the above-mentioned subjects. After serious consideration and much consultation it was agreed that for the audience this book was intended, the most useful facet of air pollution to be covered would relate to legislation at the Federal level, as it is the Federal Government which has taken the leadership and is directing the states on the methodology to be used for control. It is my considered opinion that the recommendations of the Federal Government will be followed by the states and eventually by cities and other smaller governmental segments. To make this review as useful as possible, related information such as present atmospheric contaminants in various cities are included along with comments relating to the future. The technical control of vehicular emissions is highly specialized and is of primary concern to the automobile industry and governmental agencies; therefore that area of air pollution regulations has been minimized in this review, because I am well aware of the auto industry's expansive knowledge of what is required in emission controls, as well as the Federal and state governments' roles in seeing that the controls are met. Thus, the main thrust of this review is related to emission controls from stationary sources and the broad concepts of air pollution control. During the course of reviewing this subject in "Industrial Hygiene Highlights," it was interesting to note the following (pp. 292-295).

> Air pollution in urban areas of the United States, if present trends continue, will become worse rather than better, irrespective of the millions of dollars being spent by Government and Industry, as air pollution control programs do not come to grips with the total problem. . . .
>
> The reason for the difficulty of the Los Angeles County engineering-type of control

program is that air pollution, in addition to being an engineering problem, is also a social problem. At the present time 70% of the 200,000,000 people living in the United States occupy 1% of the land area, and this trend is increasing on a continuing basis.

. . . Air pollution is a problem of activities associated with density – density of people and density of industry. Until this is clearly understood and acted upon, air pollution will continue. The control of air pollution will require broad, new, imaginative programs far beyond the scope being discussed today. Whether a community wishes to pay the price for clean air, however, is another matter. For a community to have clean air it will be necessary, in addition to having a stringent engineering program, to control the density of people and industry in the area.

. . . An individual should have a right to a clean, noise-free atmosphere in which to live. It is unreasonable to expect industry to maintain their atmosphere to the degree desired by a residential community living adjacent to the industrial property. Present zoning laws permit industrial and residential communities to live side by side. This is extremely unsound. One may observe heavy industry located within a few hundred feet of a tall school building in which the industry effluent entered directly into the school room. . . . The cities of the future should be planned to separate industry from the community; no defined industry should be permitted in the community area, and no residential buildings should be permitted in the industrial area. The two areas should be separated by a minimum of a one-mile wide green belt, which could be used for recreational purposes. The purpose of the "green belt" would be to isolate industry from the community, so that noise and normal air pollutants and a reasonable discharge of air pollutants from industry would be reduced before entering the residential area. The industrial complex should be located based upon meteorological considerations. The density of industry should be ascertained, and no larger number than had been predetermined be permitted to enter the industrial area.

. . . To achieve a dispersion of industries, favorable legislative action could be developed to give an incentive to industry to locate new plants in zones of low density; these zones, properly planned, would be free of air pollution and the multitude of ills currently found in large metropolitan areas.*

Although many people were skeptical of the author's projections back in 1968 it is indeed interesting to note the degree to which legislators have accepted these concepts and are moving in this direction. Industry, unfortunately, rather than being a leader in assisting the government to define air pollution controls and regulations, has allowed government to assume total leadership. This is indeed unfortunate. Industry should take a much more aggressive stand in relation to air pollution rules and regulations to assure a moderate balance

**Note:* An article appearing in the August 21, 1971 issue of the *Los Angeles Times* under the by-line of Harry Trimborn, *Times* Staff Writer in Moscow, indicates the Russian government has accepted this philosophy. Mr. Trimborn states: "MOSCOW – This 800-year-old city has embarked on a 20-year development program to transform itself into a star-studded metropolis adorned with a green necklace of parklands and forests. Some of the city's structures – including about 200 pollution producing industrial enterprises – will be demolished and not replaced.

"It's all part of an effort to prevent runaway urban sprawl and congestion."

between industry and government so that the public may have maximum protection; not only cleaner air but also being able to afford the environment we all desire.

LEGISLATION

Since the passage of the Air Quality Act of 1967, much legislation has been proposed and, in some cases, successfully passed. President Nixon has taken a firm stand as evidenced by the number of Executive Orders issued pertaining to air pollution.

On July 1, 1968, a Consumer Protection and Environmental Health Service was established. The most important aspect of this service pertaining to air pollution was the formulation of the National Air Pollution Control Administration. This agency has become an invaluable source of coordination of data and information and will be discussed further in a section of its own.

Then, on May 29, 1969, President Nixon issued an Executive Order which set up the Environmental Quality Council. This Council would be presided over by the President and would be composed of the Vice President and the Secretaries of Agriculture, Commerce, Health, Education, and Welfare, Housing and Urban Development, Interior, and Transportation, as well as other advisory personnel who may occasionally be called by the President to attend the meetings. The Executive Order states that "The Council shall advise and assist the President with respect to environmental quality matters and shall perform such other related duties as the President may from time to time prescribe."

The year of 1970 brought ever greater interest in the problem of environmental pollution. The widely publicized "Earth Day" focused the attention of citizens in every occupation in addition to reemphasizing it to the experts. As a result, increased pressure was brought to bear on the congressmen who could propose and pass more stringent regulations on pollution.

On January 1, 1970 President Nixon signed into law the National Environmental Policy Act. One of its major points was the formation of a three-member Council of Environmental Advisors. These advisors would be a full-time staff who would assist and advise the President in the completion of his Environmental Quality Reports which are required to be submitted annually to Congress. They will also be responsible for developing and recommending national pollution control policies and for the accumulation of data which would provide a continuous analysis of environmental trends.

A short time later, on January 19, 1970, Senator Gaylord Nelson of Wisconsin proposed an Amendment to the Constitution to the effect that "every person has the inalienable right to a decent environment." Although rather unfeasible in the sense that a "decent environment" may be defined in numerous ways by different people, it still conveys the idea that no person should be allowed to deliberately destroy the quality of the environment as set by local, state, or Federal restrictions. The proposal was not passed but is exemplary of the increasingly prominent attitude that a healthy environment is a right.

Also in the early part of 1970 (February 10), President Nixon gave an environmental message to Congress which contained the following:

> Air is our most vital resource, and its pollution is our most serious environmental problem. Existing technology for the control of air pollution is less advanced than that for controlling water pollution, but there is a great deal we can do within the limits of existing technology – and more we can do to spur technological advance.
>
> Most air pollution is produced by the burning of fuels. About half is produced by motor vehicles.

In this address to Congress, President Nixon proposed a 37 point pollution program. Of these, ten points were directed at air pollution.

Since much air pollution arises from automobiles, he warned that emissions standards would be much more stringent in the coming years and that emissions standards would include nitrogen oxides by 1973 and particulates by 1975 in addition to the standards already set for hydrocarbons and carbon monoxide.

Furthermore, since by present law, only manufacturers' prototypes are tested for compliance with emissions standards and that only voluntarily not mandatorily, he proposed that legislation be passed "requiring that representative samples of actual production vehicles be tested throughout the year." The President also suggested that there should be legislature to maintain lower emissions by controlling gasoline composition and that the Secretary of Health, Education, and Welfare should be authorized to regulate fuel composition and additives. Also, in order to encourage the development of other types of engines, President Nixon announced that he had ordered the Council on Environmental Quality to do research and conduct tests on unconventional vehicles, which are either developed by government or purchased from private industry.

With regard to stationary-source pollution, the President suggested that effective Federal standards should be established so that a company which voluntarily instituted more rigorous and more costly pollution controls would not be penalized by being at an economic disadvantage with competitors who refused to spend money on pollution control. In the same manner, a state or community with strict pollution control regulations should not be forced to lose industry to other states or communities which do not require as great an expenditure on control devices.

He further stated that over the past couple of years the administration discovered that Air Quality Control Regions and individual state legislation were too slow to be really effective and proposed "that the Federal Government establish nationwide air quality standards, with the states to prepare, within one year, abatement plans for meeting those standards;" "that designation of interstate Air Quality Control Regions continue at an accelerated rate, to provide a framework for establishing compatible abatement plans in interstate areas;" and "that the Federal Government establish national emissions standards for facilities that emit pollutants extremely hazardous to health, and for selected classes of new facilities which could be major contributors to air pollution."

In addition, the President felt that Federal authority to seek court action should "be extended to include both inter- and intrastate air pollution situations in which, because of local nonenforcement, air quality is below national standards, or in which emissions standards or implementation time tables are being violated," and "that failure to meet established Air Quality Standards or implementation schedules be made subject to court-imposed fines of up to $10,000.00 per day."

On March 5, 1970, the President continued his legislative support of pollution control by issuing an Executive Order that stated it was up to the Federal Government to initiate and to provide leadership for pollution control programs. He continued by changing the name of the Environmental Quality Council to the Cabinet Committee on the Environment.

On April 9, 1970, he issued another Executive Order creating the National Industrial Pollution Control Council, a body of members chosen to coordinate public and private pollution control efforts. It is hoped that such cooperation between government and industry will promote faster and more comprehensive pollution control.

At the July 9, 1970 meeting of Congress, President Nixon transmitted Reorganization Plan No. 3 of 1970 which establishes the Environmental Protection Agency. Since it was not vetoed by Congress within 60 days, the Agency was automatically established 120 days after President Nixon's proposal. In compliance with Section 901(a) of Titles of the United States Code, the Plan was proposed "to promote the better execution of the laws, the more effective management of the Executive Branch and of its agencies and functions, and the expeditious administration of the public business" as well as "to increase the efficiency of the operations of the government to the fullest extent practicable." With this in mind, the Environmental Protection Agency was established by transferring environmental quality units from existing departments.

The Bureaus of Radiological Health (in part), Solid Waste Management, and Water Hygiene, as well as the National Air Pollution Control Administration, were transferred from the Department of Health, Education, and Welfare. The Department of Agriculture gave up its authority over pesticide registration, and the Department of Health, Education, and Welfare and Department of the Interior no longer have control over pesticide research and standards setting of the Food and Drug Administration.

The Federal Radiation Council has been taken from the Executive Office of the President, and the Atomic Energy Commission no longer sets the Standards for Environmental Radiation Protection. The Agency also includes the Federal Water Quality Administration formerly with the Department of the Interior.

The result of this collection of agencies into one will hopefully be an avoidance of duplication of power and an increased ability to enforce the more stringent standards on environmental quality.

The President further took the initiative to help curb automobile pollution by issuing a notice on October 26, 1970 that all "Federally-owned vehicles use low-lead or unleaded gasoline whenever this is practical and feasible. The purpose of this regulation is twofold: to reduce air pollution and to increase the market for low-lead and unleaded gasoline, in order to make such fuels more generally available." In addition, he suggested to the governors of the fifty states tha they also institute a similar policy.

On November 18, 1969, President Nixon appointed a Task Force to study the effectiveness of present efforts to control air pollution and make recommendation on actions that are needed in the future.

Restoring and preserving the nation's air quality is a monumental task which requires the active cooperation of all levels of government and combined brainpower of our technical and scientific professions. Many problems overlap the artificial bounds of narrow provincial concerns. New governmental structures may have to be devised to meet the challenge of the future. Members of the Task Force on Air Pollution are: Arie Jan Haagen-Smit, Chairman, California Air Resources Board and Professor of Bio-organic Chemistry, California Institute of Technology; I. W. Abel, President, United Steelworkers of America; John W. Bodine, President, Academy of Natural Sciences; George D. Clayton, President, George D. Clayton & Associates, Inc.; Herbert J. Dunsmore, Director of Environmental Control, United States Steel Corporation; Martin L. Katke, Vice President-Technical Affairs, Ford Motor Company; Willard F. Libby, Professor of Chemistry and Director, Institute of Geophysics and Planetary Science, University of California; Norton Nelson, Director, Institute of Environmental Medicine, New York University Medical Center; Ernest S. Starkman, Professor of Mechanical Engineering, Thermal Systems Division, University of California; John W. Tukey, Chairman, Department of Statistics, Princeton University; and James L. Whittenberger, James Stevens Simmons Professor of Public Health, Harvard School of Public Health.

The Task Force met with experts on the many facets of air pollution including governmental organization, effects of pollutants, engineering controls of stationary and mobile sources, and research on measurements, meteorology and atmospheric chemistry.

As a result of their study, sixty-three recommendations were developed. They include both policy and technical matters and cover short-term as well as long-term needs.

In summary, the report presents the following:

General Considerations

Different levels of government – Federal, state, and local – and many separate agencies are involved in environmental control programs. Within the sphere of government, the variety of interests involved in environmental questions inevitably produces problems of coordination between levels and agencies. Some increase in effectiveness may be possible through reassignment of functions. However, organizational stability and functional integrity should be maintained. Stimulus to cooperation can often produce better results than attempts to change governmental structure.

FEDERAL GOVERNMENT

The Federal structure for the management of environmental affairs should be designed to ensure stability and coordination. Reconciliation of priorities, coordination, and program monitoring should be assigned to a single agency.

STATE/FEDERAL RELATIONSHIPS

Strong financial support should be provided by Federal government to cooperative programs that aim, on a region-wide basis, to assess air pollution problems and to enforce controls.

PUBLIC COOPERATION

Federal agencies must ensure that the environmental information they issue is accurate and so presented as to reflect adequately any necessary qualifications. Federal agencies should step up their information programs and do much more to encourage a broad range of channels of public education – such as schools, colleges, museums, citizen organizations – to disseminate information about what the ordinary citizen can do to reduce air pollution.

AIR POLLUTION – A FUNCTION OF POPULATION GROWTH AND DENSITY

Existing Federal programs aimed at controlling population growth should be vigorously strengthened. Consideration of this problem should be a high priority topic for the Federal Commission on Population Growth and the American future.

The Federal government, through its appropriate regulatory agencies, should encourage the development of regional plans for land use.

ECONOMIC INCENTIVES

Economic incentives should be introduced to help make further reduction in air pollution attractive and self-initiating. The uses of taxes and effluent fees in reducing air pollution need intensive examination. Even more rapid write-offs of the cost of pollution control equipment for older plants should be explored.

COMPLIANCE WITH STANDARDS

The feasibility of the principle of mandatory compliance with air pollution control standards in order to be eligible for Federal contracts requires intense study.

Mobile Source Emission Control

The President, in his February 10, 1970 environmental message, stressed the need to promote programs to minimize pollution from the automobile. The Task Force is in agreement with the emission goals announced by the Department of Health, Education, and Welfare for 1980. While there is basic confidence that compliance with intermediate standards can be achieved and maintained under the impetus of current initiatives in the public and private sectors, there is considerably less certainty for post-1975 requirements. The President's announced intent to intensify research and development and provide supplemental means to stimulate private activity will be a major insurance factor but other steps need to be taken.

FUEL ADDITIVES

Lead additives should be phased out of gasoline in an orderly manner, and beginning promptly. All motor fuel additives should be evaluated for their environmental effects. This evaluation should include a study of the effects of their decomposition products present in the exhaust. A fuel additive law is necessary to assure that these evaluations are carried out in a manner approved by a Federal agency prior to registration of any fuel additive for use in gasoline. The need for a Federal limitation on aromatic content of fuels should be studied carefully.

EMISSION TESTING AND INSPECTION

Present preproduction certification should be continued only until adequate assemblyline testing for certification has been implemented. Development of the short test cycle must be encouraged. Routine assembly line emission testing is essential. Better test methods for measuring automobile emissions must be developed and uniformly adopted. Maintenance of emission levels must be ensured by periodic, mandatory inspection. A rapid, accurate field emission test must be developed and applied.

DIESEL ENGINE EMISSIONS

Emission regulations for Diesels should be promptly promulgated.

AIR TRAFFIC

Pollution from aircraft should be brought under control without further delay.

MASS TRANSIT

Federal action is urged to promote development, installation, and trials of one or more advanced transit systems that will provide air pollution control benefits as well as a more efficient mode of transit for the public.

More intensive study is needed on (*1*) the effects of conventional choices in handling metropolitan traffic on total contribution to air pollution, and (*2*) the feasibility and effectiveness of really novel schemes of traffic management and operation.

CATALYSTS

A crash catalyst development program is required.

ADVANCED PROPULSION SYSTEMS

Federal support should be provided for unconventional propulsion system development, including better batteries, steam engine components, and hybrid systems.

Stationary Source Emission Control

Considerable research and technologic development has been done on control techniques for stationary sources. However, the increase in number and sizes of sources makes it imperative that research and development be accelerated to improve capability and provide adequate means of control where they do not now exist.

PLANT SITING

Regional plant siting authorities should be developed under Federal supervision to select and reserve sites long enough in advance to combine adequate electric energy and industrial development with maximum protection of the environment.

REGIONAL CAPACITY

The ventilation capacity of areas should be taken into account when emission standards are set.

LARGE FOSSIL-FUEL COMBUSTION SOURCES

The development and installation of stack-gas treatment processes is urgent.

An accelerated expansion of Federally sponsored research and development for air pollution controls is necessary to meet our needs for new and improved techniques for controlling even such common, major contaminants as sulfur oxides, nitrogen oxides, and particulates.

It is urgent that the utility regulating commissions consider such innovations as are necessary to enable the power industry to allow research costs related to environmental control to be part of the ordinary cost of doing business.

Physical-chemical combustion studies directed to the removal of oxides of nitrogen should have immediate attention.

Improved techniques able to remove fine particulates from stack gases should be developed on a priority basis.

The use of tall stacks should not replace stack gas treatment as a means of control.

FUEL CONVERSION – COAL GASIFICATION

A national fuels policy should assure that needs for low-sulfur fuel can be met. Development of commercially feasible processes of reducing sulfur and ash in fuels should be supported more actively.

NUCLEAR POWER

The resources of the Federal government and public and private utilities should be joined to provide the funds needed to sustain, for the next several years, the intensive development work needed to make breeder reactors competitive.

FUTURE POWER PRODUCTION AND TRANSMISSION

The Federal government should encourage and emphasize the development and testing of more efficient means for transmission of electrical power.

Efforts directed toward magnetohydrodynamics power generation and controlled fusion should be continued.

COMMERCIAL AND DOMESTIC HEATING

Federal support of an intensive program of development of heat pumping and techniques of heat-loss reduction is urgent.

The Department of Housing and Urban Development should report regularly on its program to reduce pollution from government-supported housing.

SOLID WASTES

Research and development on techniques for recycle and conversion of solid wastes should be intensified.

SMALL POWER UNITS

A broad study should be made of small sources of air pollution and of the ways to cope with them.

Biologic Effects

The present evidence related to the adverse effects of air pollutants on human beings and plant and animal life is meager and oftentimes subject to differences in interpretation. Research activity must be to a certain extent redirected and intensified to develop a solid foundation for decision-making.

EPIDEMIOLOGICAL STUDIES

A critical evaluation of present epidemiological evidence about the impact of air pollution on human health is needed. Assessment of health effects consequent to abatement procedure is needed.

CLINICAL INVESTIGATION

The effects of short-term safe exposures to air pollutants should be studied in human subjects.

PHYSIOLOGIC EXPERIMENTS

A broad range of animal studies at low-pollution levels is needed, including studies on mechanisms of action, factors which affect susceptibility to infection, interactions, and effects on animal models of human diseases.

LOW-LEVEL RADIATION HAZARDS

A complete review and, if necessary, revision of radiation standards is required. A national conference with effective public participation should be called to examine the implications of adequate radiation standards for a national policy on power production.

PLANT AND ANIMAL STUDIES

Federal research on air pollution effects on plants and animals must be continued and expanded.

Analytical Methods and Instrumentation Development

The need for better methods and less expensive and trouble-free equipment applies in all three categories of activities in which the measurement of air pollutants are involved: laboratory determinations, ambient air measurements, and source testing. Needs are especially pressing for determination of particulate matter, smoke and dust, oxidants, hydrogen sulfide, organic sulfur and mercaptans, and heavy metals.

AMBIENT AIR MONITORS

A vigorous program should be maintained to develop inexpensive, reliable, and for certain needs portable, automated instruments with adequate sensitivity for the ranges of concentration in ambient air.

The Federal government should take a strong lead in promoting the development, testing, and adoption of standard methods for calibration and measurement; it should also develop uniform data reporting and reduction systems to make the results conveniently accessible.

STACK SAMPLING AND ANALYSIS

Sampling and analysis of major stack sources should be made automatic and continuous. New methods and equipment are especially needed for sampling and analysis of the following substances.

1. *Particulate matter.* The sampling of particulate matter has been based on filtration methods which detect gross mass loadings of particulate matter in a size range depending on the porosity of the filter. However, the determinations of particle size distributions and chemical compositions are important, as well as mass loadings. There are laboratory methods for these determinations, but they usually involve complicated and expensive equipment which is not suitable for field use. A simple field method for size distribution determinations is needed. Such a method must eventually permit separation of particles by size and chemical analysis of separate size ranges as well as the total weight of particles sampled.

2. *Opacity measurements.* Smoke measurements are based on visual observations. The smoke opacities are compared with the Ringelmann Chart and expressed as Ringelmann numbers or the equivalent percent opacity. This method is subjective, and does not provide accurate measurement, particularly when the opacity is less than Ringelmann Number 1 (20% opacity). More objective techniques, effective over a range extending to lower opacities, are needed.

3. *Oxidants.* Oxidant concentration measurements are subject to interferences by both reducing and oxidizing agents. A measurement method which is specific for oxidants is needed.

4. *Hydrogen sulfide, organic sulfur and mercaptans.* There are no accurate and reliable techniques and instruments, which are also inexpensive, to measure these substances continuously and automatically. Such instruments are needed in areas where these odorous and noxious substances are emitted.

5. *Lead and other heavy metals.* Satisfactory laboratory methods exist for determinations of the heavy metals in air samples. The methods are not yet suitable for continuous monitoring or for automatic determinations.

Meteorologic Problems and Considerations

The concentration of pollutants in the atmosphere depends largely on meteorologic conditions which control the transport and dilution of material after it has been emitted to the atmosphere. In order to facilitate meaningful interpretation of atmospheric levels of pollution, detailed meteorologic data are required on a scale which is representative of the physical complexities of our major urban centers and their surrounding air sheds.

EXPANDED FORECASTS

Pollution-oriented meteorologic forecasts should be expanded to provide detailed forecasts for each large urban center.

Considering the importance of meteorologic data to the total air pollution control effort, the Environmental Science Services Administration should be authorized to expand its program of obtaining meteorologic data of concern to air pollution control programs.

Information on the air stability over cities and the variation in stability with location is needed. The effects of large buildings and large paved areas on the environment must be investigated.

Vertical atmospheric soundings must be made in all major urban centers, not just a few remote airport facilities.

Meteorologic data should be summarized specifically for air pollution purposes, and these summaries should be made readily available to state control agencies.

STRATOSPHERIC AIR POLLUTION

Study of the effects of man's activities on the earth's climate must be intensified.

CHEMICAL REACTIONS IN THE ATMOSPHERE

Knowledge of the fate of pollutants, the reactions they undergo, and their accumulative effects must be acquired.

CLEAN AIR AMENDMENTS OF 1970

One of the most important pieces of air pollution legislation, the Clean Air Act (Public Law 91-604) was enacted in December 1970. These amendments were proposed so that the Clean Air Act could be revised and expanded to provide a more effective program for the control of air pollution.

The major sections of the amendments are concerned with Federal Research Grants for State and Regional Air Quality Control Agencies; National Ambient Air Quality Control Regions, Criteria and Techniques; Standards of Performance for New Stationary Sources; Motor Vehicle Emission Standards; Aircraft Emission Standards; and Noise Pollution. Thus, by the categories listed, one can see that the amendments cover all phases of air pollution. The following discussion gives further details of the major topics.

The first section emphasized the need for an accelerated research program to improve knowledge "of the contribution of air pollution to the occurrence of adverse effects on health" and "of the short- and long-term effects of air pollutants on welfare." Later the research clause is expanded to include the "development of improved, low-cost techniques for the efficiency of fuels combustion so as to decrease atmospheric emissions, and producing synthetic or new fuels which, when used, result in decreased atmospheric emissions." The amendments also have a provision for partial funding of programs to develop low emission alternatives to the present internal combustion engine" and "to purchase vehicles and vehicle engines, or portions thereof, for research, development, and testing purposes."

Another section of the amendments reemphasizes that "each State shall have the primary responsibility for assuring air quality within the entire geographic area comprising such State by submitting an implementation plan for such State which will specify the manner in which national primary and secondary ambient Air Quality Standards will be achieved and maintained within each Air Quality Control Region in such State." It further designates Air Quality Control Regions assigned before and after the passage of these Amendments are to be left intact, but that any part of a state which is not already, or is not going to be, assigned as an Air Quality Control Region automatically becomes one which may be further subdivided by the state if it so chooses.

The amendments continue by redefining the duties of the administrator of the Environmental Protection Agency in the distribution of new air quality criteria, of national ambient air quality standards and of technologic advances in control techniques by which the criteria and standards may be met. In all cases, the additional data and information is to be announced in the *Federal Register* and made available to the general public. Once the air quality standards are established, it is the responsibility of each state to submit to the administrator within 9 months a plan for implementation, maintenance, and enforcement of their standards in each air quality control region in the state. Should the plan not be acceptable to the Federal administrator, the amendments detail the course of action which may be used to settle differences.

Another section deals with standards of performance for new stationary sources. It states that the administrator is responsible for listing categories (including classes, types and sizes) of stationary sources and for prescribing standards which may be adopted and implemented by the individual states. At the end of the section is the declaration that "after the effective date of Standards of Performance promulgated under this section, it shall be unlawful for any owner or operator of any new source to operate such source in violation of any Standard of Performance applicable to such source." Thus, this section catches air pollution even before the fact, whereas in other sections dealing with older sources, there is necessarily a "grace" period during which polluters may install control devices.

To insure that polluters comply with established standards, especially in cases in which the state involved has not taken appropriate action, the administrator may require the owner of any emission source to install, use, and maintain monitoring equipment and to establish and report records of data obtained from the equipment. Inspectors authorized by the administrator have the right of entry to premises in which an emission source is located and are to have access to monitoring equipment and records obtained thereby. If a company refuses to comply with the adopted regulations or an order issued by the administrator, legal action may be taken. Penalties may include permanent or temporary injunctions, fines of up to 25,000 dollars per day of violation for the first offense, up to 50,000 dollars per day of violation for the second offense and/or by imprisonment for not more than 1 year (not more than 2 years for the second offense). In all cases, Federal authority does not preclude or deny the right of any state to adopt and enforce standards of air pollutants. Before legal action can be taken by the Federal government, the state air pollution control agency responsible for the region in which the source of emission is located must be notified.

Appropriately, the amendments include a section pertaining to pollution from Federal operations. They all fall under the same restrictions as private industry and in addition, are supposed to set the example of having low pollution exhausts. In the case of national emergency or lack of funds appropriated by Congress for pollution control devices, the President may temporarily suspend compliance with the regulations. These exemptions and the reasons for them are to be reported to Congress each January.

The sections concerning motor vehicle emission standards are fairly similar to the Clean Air Act. One notable exception to this is the designation of carbon monoxide, hydrocarbon, and nitrogen oxide concentrations allowable in light duty motor vehicle exhaust. Hydrocarbon and carbon monoxide emission levels in engines manufactured during or after the 1975 model year are required to have at least a 90% reduction from emission concentration standards for 1970 models. Emissions of oxides of nitrogen from engines produced during or after the 1976 model year must have a 90% reduction from the average of emissions of nitrogen oxides actually measured on 1971 engines. Each year the administrator is required to report to Congress the developments made on systems necessary to implement these emission standards. There is also a provision for the Federal government to contract the National Academy of Sciences to do feasibility studies on meeting the emissions standards.

Violations of preceding section involving standards may result in fines up to 10,000 dollars. To determine whether the vehicles manufactured actually do conform with regulations, the administrator is authorized to test them and award certification for compliance. This certification may not be granted or may be revoked if the vehicles do not initially and continuously meet the standards. The manufacturer may appeal the decision, if he believes that his certification was denied or revoked unjustly. There are various other conditions which the manufacturer must meet, but any inspections carried out on a motor vehicle "after its sale to the ultimate purchaser, shall be made only if the owner of such vehicle or engine *voluntarily* permits such inspection to be made, except as may be provided by any State or local inspection program."

These amendments also regulate fuel additives. In addition to mandatory registration of the chemical composition of fuels and additives as specified in the Clean Air Act, the administrator may also require the fuel manufacturer to conduct tests to determine potential harmful health effects of additives and to furnish analytic techniques by which the additives may be detected. If the manufacturer does not comply, he will be fined 10,000 dollars for every day he continues the violation.

A new section added to this Act was one dealing with the development of low-emission vehicles. Such a vehicle would emit any air pollutant in amounts significantly below new motor vehicle standards but would also be considered for its safety, performance, reliability, serviceability, fuel availability, noise levels, and maintenance costs. If a vehicle is certified by a committee under this section, the Federal government will purchase or lease such vehicles up to 150% of the retail price or the least expensive model of certified substitutes.

A completely new section added to the Clean Air Act in 1970 was one concerning aircraft emission standards. The administrator is charged with investigating "the extent to which such emissions affect air quality in Air Quality Control Regions throughout the United States, and the technologic feasibility of controlling such emissions." After the conclusion of this study, the administrator may propose emissions standards but these may be accepted only after public hearings and consultation with the secretary of transportation. All standards set are to be of Federal nature only, forbidding states to adopt or to attempt to enforce standards. As a means to the control of aircraft emissions, the administrator may also carry out tests on and establish standards for aircraft fuels.

Finally, there are amendments dealing with general provisions of this act. In addition to the power of the administrator to take normal civil action against polluters who violate the standards, he is given emergency powers. The administrator may, in the event that a pollution source (including moving sources) is presenting "an imminent and substantial endangerment to the health of persons, and that appropriate State and local authorities have not acted to abate such sources, bring suit on behalf of the United States in the appropriate United States District Court to *immediately* restrain any person causing or contributing to the alleged pollution to stop the emission of air pollutants." In a similar, although nonemergency, manner, "Any person may commence a civil action on his own behalf" against any other person (including the United States and its agencies) who is in violation of air pollution standards or against the administrator himself where there is alleged a failure of the administrator to perform his duties under this act.

Another section regulates Federal procurement. No Federal agency may enter into any contract for goods, materials, or service with any person who is convicted of any offense regarding compliance with emission standards, inspections, or installation of control devices. This prohibition is in effect until the administrator certifies that the situation has been corrected. The only exemptions to the ruling may be made by the President where he determines that such an exemption is in the paramount interest of the United States.

Appropriations are divided into two parts. For the fiscal year ending June 30, 1971, 75 million dollars was appropriated to carry out the goals of the sections of this act pertaining to motor vehicle emission and fuel standards. This appropriation was raised to 125 million dollars for the fiscal year ending June 30, 1972 and 150 million dollars for the fiscal year ending June 30, 1973. Appropriations covering all other sections of this act came to 125 million dollars in 1970–1971, 225 million dollars in 1971–1972, and 300 million dollars in fiscal year 1972–1973.

Finally, contained under the Clean Air Amendments of 1970 is a relative newcomer to the field of air pollution, noise pollution. The relevant section was named the "Noise Pollution and Abatement Act of 1970." The provisions of this act call for the establishment of the Office of Noise Abatement and Control within the Environmental Protection Agency. Through this office, the administrator is to carry out full investigations of noise and its effect on public health and welfare. This includes psychologic and physiologic effects of both constant noise at various levels and of sporadic extreme noise. In addition, a study of the effects of noise and sonic booms on wildlife and property (including values) is to be conducted.

The administrator will hold public hearings and conduct research so that within a year of the passage of this act, he may report his results and make recommendations for legislature to the President and the Congress. "In any case, where any Federal Department or Agency is carrying out or sponsoring any activity resulting in noise which the Administrator determines amounts to a public nuisance or is otherwise objectionable, such Department or Agency shall consult with the Administrator to determine possible means of abating such noise." To achieve the purposes of this Act, it is authorized that up to 30 million dollars may be appropriated.

Thus, one is able to see that if the administrator of the Environmental Protection Agency actually completes all the duties conferred on him within the time guidelines set by this act, it will be a truly remarkable advance on air pollution.

The chronological schedule for clear air actions follows. These are maximum time limits. The appropriate actions must be taken on or before the noted date.

1. January 30, 1971. Administrator of EPA issued proposed national primary and secondary ambient air quality standards applicable to sulfur oxides, particulate matter, carbon monoxide, hydrocarbons, photochemical oxides, and nitrogen oxides. Criteria for all but nitrogen oxides had been issued previously.

2. March 16, 1971. Comments due on proposed air quality standards issued January 30.

3. April 1, 1971. Administrator to designate any interstate or intrastate area he deems necessary as an air quality control region.

4. April 1, 1971. Administrator to publish list of categories of stationary sources he determines may contribute significantly to air pollution.

5. April 1, 1971. Administrator to publish list of hazardous air pollutants for which he intends to publish emission standards.

6. May 1, 1971. Promulgate by regulation air quality standards proposed on January 30 (within 90 days after publication).

7. August 1, 1971. Administrator to propose regulations establishing Federal standards of performance for new stationary sources within each category (within 120 days of inclusion of a category on the list).

8. September 15, 1971. Comments due on regulations proposed by August 1.

9. October 1, 1971. Administrator to issue proposed regulations for emission standards for hazardous pollutants with notice of public hearings within 30 days (within 180 days after inclusion of a hazardous pollutant on a list).

10. October 31, 1971. Public hearing on emission standards for hazardous pollutants.

11. November 1, 1971. Promulgate by regulation the standards of performance for new stationary sources proposed on August 1 (within 90 days of initial proposal and after reasonable time for comment).

12. February 1, 1972. Each State to submit to administrator a plan for implementation and enforcement of standards (within 9 months after air quality control standards promulgated).

13. April 1, 1972. Promulgate by regulation the emission standards for hazardous pollutants proposed on October 1 (within 180 days of proposal).

TABLE I
Air Classification

Pollutant	*Concentration*		
	Priority I	*Priority II*	*Priority III*
Sulfur oxides			
Annual arithmetic mean	> 100	60–100	< 60
24-hr maximum	> 455	260–450	< 260
Particulate matter			
Annual geometric mean	> 95	60–95	< 60
24-hr maximum	> 325	150–325	< 150

14. June 1, 1972. Administrator to approve or disapprove state plans or portions thereof (within 4 months after state plan submitted).

15. July 1, 1972. Standards for hazardous pollutants from existing sources become effective. (Waiver for up to 2 years may be granted by EPA and President may exempt in 2-year periods indefinitely with periodic report to Congress.)

16. June 1, 1975. Primary air quality standards to be attained (no later than 3 years from date of approval of plan). (Administrator may extend up to 2 years upon application of a Governor for an air quality control region and up to 1 year for a source.)

The Environmental Protection Agency, by law, has until April 30, 1971 to promulgate national ambient air quality standards for sulfur oxides, particulate matter, carbon monoxide, photochemical oxidants, hydrocarbons and nitrogen oxides.

On Wednesday, April 7, 1971, the *Federal Register* (Vol. 36, No. 67) published information entitled "National Ambient Air Quality Standards Notice of Proposed Regulations for Preparation, Adoption and Submittal of Implementation Plans." This document spells out how the Federal government plans to enforce the ambient air quality standards. First, there will be a classification of regions. These classifications will be based upon measured ambient air quality or estimated air quality in an area of maximum pollutant concentration. For sulfur oxides and particulate matter, each region will be classified into one of three categories, defined as Priority I, Priority II, or Priority III. See Table I for ambient air concentration limits which define the classification system for sulfur oxides and particulate.

For carbon monoxide, nitrogen dioxide, and photochemical oxidants, each region will be classified into one of two categories, defined as Priority I or Priority III.

Ambient concentration limits which define the classification system are:

1. Carbon monoxide. Priority I: equal to or above 21 mg/m^3, 1-hr maximum, or 14 mg/m^3, 8-hr maximum; Priority III: below both of such values.

2. Nitrogen dioxide. Priority I: equal to or above either of the primary standards; Priority III: below both of the primary standards.

3. Photochemical oxidants. Priority I: equal to or above 170 $\mu g/m^3$, 1-hr maximum; Priority III: below such value.

In the absence of measured data to the contrary, classification with respect to carbon monoxide, photochemical oxidants and nitrogen dioxide will be based on the following estimate of the relationship between these pollutants and population: Any region containing a metropolitan area whose 1970 "urban place" population, as defined by the U. S. Bureau of Census, exceeds 200,000 will be classified as Priority I. All other regions will be classified as Priority III.

The report continues in establishing a plan of action which each of the states will be required to follow. Each state must submit a plan which shall include such "emission limitations and other measures as are necessary for attainment and maintenance of the national standards."

The plan requires adoption of emission standards and limitations. It also requires an establishment of statewide permit systems which will be required for construction and operation of new stationary sources of air pollution and the construction and operation of modifications to existing sources, including authority to prevent such construction, modification, or operation and any other necessary land use control authority as required.

The plan requires owners or operators of stationary sources to install, maintain, and use emission monitoring devices and to make periodic reports to the State on the nature and amounts of emissions from such stationary sources and to make such data available to the public as reported and as correlated with any applicable emission standards.

The plan requires a program of inspection and testing of motor vehicles be carried out to enforce compliance with applicable emission standards when necessary and practicable, and to impose other necessary controls on transportation provided for in the plan.

The plan shall show that the legal authorities specified in this section are available to the State agency or agencies directly responsible for implementing the plan at the time of submission of the plan.

The control strategy as proposed in the plan, for general emissions, is the following: in any region where measured or estimated ambient levels of a pollutant are above the levels specified by an applicable national standard, the plan shall include a control strategy which shall be shown to provide for the degree of emission reduction necessary for attainment and maintenance of such national standard, including the degree of emission reduction necessary to offset emission increases that can reasonably be expected to result from projected growth of population, industrial activity, motor vehicle traffic, or other factors that may cause or contribute to increased emissions.

The plan requires a detailed inventory of emissions from point sources and, as defined, a point source is any stationary source causing emissions in excess of 25 tons/year of any pollutant for which there is a national standard, or without regard to quantity of emissions, any of the following sources.

Major Pollutant Sources

Chemical Process Industries
Adipic acid
Ammonia
Ammonium nitrate
Carbon black*
Charcoal*
Chlorine
Detergent and soap
Explosives (TNT and nitrocellulose)*
Hydrofluoric acid*
Nitric acid
Paint and varnish manufacturing*
Phosphoric acid*
Phthalic anhydride
Plastics manufacturing*
Printing ink manufacturing*
Sodium carbonate*
Sulfuric acid*
Synthetic fibers
Synthetic rubber
Terephthalic acid

Food and Agricultural Industries
Alfalfa dehydrating*
Coffee roasting*
Cotton ginning*
Feed and grain*
Fermentation processes
Fertilizers*
Fish-meal processing
Meat smokehouses*
Starch manufacturing*
Sugar cane processing*

Metallurgic Industries
Primary metals industries:
  Aluminum ore reduction*
  Copper smelters*
  Ferroalloy production*
  Iron and steel mills*

*Major sources of sulfur oxides and/or particulate matter.

Lead smelters*
Metallurgic coke manufacturing*
Zinc*
Secondary metals industries:
Aluminum operations*
Brass and bronze smelting*
Ferro alloys*
Gray iron foundries*
Lead smelting*
Magnesium smelting*
Steel foundries*
Zinc processes*

Mineral Products Industries
Asphalt roofing*
Asphaltic concrete batching*
Bricks and related clay refractories*
Calcium carbide*
Castable refractories*
Cement*
Ceramic and clay processes*
Clay and fly ash sintering*
Coal cleaning*
Concrete batching*
Fiberglass manufacturing*
Frit manufacturing*
Glass manufacturing*
Gypsum manufacturing*
Lime manufacturing*
Mineral wool manufacturing*
Paperboard manufacturing*
Perlite manufacturing*
Phosphate rock preparation*
Rock, gravel, and sand quarrying and processing*

*Petroleum Refining and Petrochemical Operations**

*Wood Processing**

Petroleum Storage
(Storage banks and bulk terminals)

Miscellaneous
Steam electric power plants*
Municipal or equivalent incinerators*
Open burning dumps*

*Major sources of sulfur oxides and/or particulate matter.

Incineration

No person shall cause or permit to be emitted into the open air from any incinerator, particulate matter in the exhaust gases to exceed 0.10 lb/100 lb refuse burned.

Emission tests shall be conducted at maximum burning capacity of the incinerator.

The burning capacity of an incinerator shall be the manufacturer's or designer's guaranteed maximum rate or such other rate as may be determined by the director in accordance with good engineering practices. In case of conflict, the determination made by the director shall govern.

For the purposes of this regulation, the total of the capacities of all furnaces within one system shall be considered as the incinerator capacity.

Note: This mass emission rate is about equivalent to a grain loading of 0.1 grains/SCF at 12% carbon dioxide. More restrictive requirements are feasible for large incinerators.

Fuel Burning Equipment

No person shall cause or permit emission to the atmosphere from fuel burning equipment burning solid fuel of particulate matter in excess of 0.10 lb/million Btu/hr.

No person shall cause or permit emission to the atmosphere, from oil-fire fuel burning equipment rated greater than or equal to 250 million Btu/hr heat input, of particulate matter in excess of 0.025 lb/million Btu/hr.

For purposes of this regulation the heat input shall be the aggregate heat content of all fuels whose products of combustion pass through a stack or stacks. The heat input value used shall be the equipment manufacturer's or designer's guaranteed maximum input, whichever is greater. The total heat input of all fuel burning units on a plant or premises shall be used for determining the maximum allowable amount of particulate matter which may be emitted.

Process Industries – General[1]

No person shall cause, suffer, allow, or permit the emission of particulate matter in any 1 hr from any source in excess of the amount shown below for the process weight rate allocated to such source or in excess of 0.03 grains per standard cubic foot of exhaust gas (see Table II). Interpolation of this data for process weight rates up to 6000 lb/hr shall be accomplished by use of the equation $E = 4.10P^{0.67}$, $E =$ rate of emission in lb/hr and $P =$ process weight rate in tons/hr.

TABLE II
Particulate Emission Standards

Process weight rate (lb/hr)	*Rate of emission (lb/hr)*
100	0.551
200	0.877
400	1.40
600	1.85
800	2.22
1000	2.58
1500	3.38
2000	4.10
2500	4.76
3000	5.38
3500	5.96
4000	6.52
5000	7.58
6000	8.56
7000	9.49
8000	10.4
9000	11.2
12,000	13.6
16,000	16.5
18,000	17.9
20,000	19.2
30,000	25.2
40,000	30.5
50,000	35.4
60,000 or more	40.0

Process weight per hour is the total weight of all materials introduced into any specific process that may cause any discharge of particulate matter. Solid fuels charged will be considered as part of the process weight, but liquid and gaseous fuels and combustion air will not. For a cyclic or batch operation, the process weight per hour will be derived by dividing the total process weight by the number of hours in one complete operation from the beginning of any given process to the completion thereof, excluding any time during which the equipment is idle. For a continuous operation, the process weight per hour will be derived by dividing the process weight for a typical period of time.

Where the nature of any process or operation or the design of any equipment is such as to permit more than one interpretation of this regulation, the interpretation that results in the minimum value for allowable emission shall apply.

For purposes of the regulation, the total process weight from all similar process units at a plant or premises shall be used for determining the maximum allowable emission of particulate matter that passes through a stack or stacks.

Air Pollution Emergencies[1]

Notwithstanding any other provision of the air pollution control regulations, this episode regulation is designed to prevent the excessive buildup of air contaminants during air pollution episodes, thereby preventing the occurrence of an emergency due to the effects of these contaminants on the public health.

Episode Criteria[1]

Conditions justifying the proclamation of an air pollution alert, air pollution warning, or air pollution emergency shall be deemed to exist whenever the director determines that the accumulation of air contaminants in any place is attaining or has attained levels which could, if such levels are sustained or exceeded, lead to a threat to the health of the public. In making this determination, the director will be guided by the following criteria.

1. "Air Pollution Forecast." An internal watch by the Department of Air Pollution Control shall be actuated by a National Weather Service advisory that Atmospheric Stagnation Advisory is in effect or the equivalent local forecast of stagnant atmospheric conditions.
2. "Alert." The alert level is that concentration of pollutants at which first stage control action is to begin. An alert will be declared when any one of the following levels is reached:

SO_2 – 0.3 ppm 24-hr average.
Particulate – 3.0 COH's, 24-hr average.
SO_2 and particulate combined – produce of SO_2 ppm, 24-hr average and COH's equal to 0.2.
CO – 15 ppm, 8-hr average.
Ox – 0.1 ppm, 1-hr average.
NO_2 – 0.6 ppm, 1-hr average, 0.15 ppm, 24-hr average

and meteorologic conditions are such that this condition can be expected to continue for 12 or more hours.

3. "Warning." The warning level indicates that air quality is continuing to degrade and that additional abatement actions are necessary. A warning will be declared when any one of the following levels is reached:

SO_2 – 0.6 ppm, 24-hr average.
Particulate – 6.0 COH's, 24-hr average.
Combined SO_2 and COH's, 24-hr average, SO_2 and COH's equal to 1.0.

CO – 30 ppm, 8-hr average.
Ox – 0.4 ppm, 1-hr average.
NO_2 – 1.2 ppm, 1-hr average, 0.3 ppm, 24-hr average,

and meteorologic conditions are such that this condition can be expected to continue for 12 or more hours.

On or about April 30, 1971 pursuant to Section 109 of the Clean Air Act, the administrator will promulgate national air quality standards for sulfur oxide, particulate matter, carbon monoxide, photochemical oxidants, hydrocarbons and nitrogen oxides. The Environmental Protection Agency fulfilled the Law's requirement and has proposed ambient air quality standards as shown in Table III.*

To better understand the effects of these contaminants at various concentrations, Table IV was prepared.

The Clean Air Act required the Environmental Protection Agency to publish, no later than March 31, 1971, a list of air pollutants which, in the Administrator's judgment, may cause or contribute to an increase in mortality or an increase in serious irreversible, or incapacitating, reversible illness and to which no national ambient air quality standard is applicable.

The administrator, after evaluating available information, has concluded that asbestos, beryllium, and mercury are air pollutants which meet the above requirements.[2]

List of Categories of Stationary Sources[2]

The Clean Air Act *directs the administrator* of the Environmental Protection Agency to publish no later than March 31, 1971, a list of categories of stationary sources which he determines may contribute significantly to air pollution which causes or contributes to the endangerment of public health or welfare.

The administrator, after evaluating available information, has determined that the following are categories of stationary sources which meet the above requirements: contact sulfuric acid plants; fossil fuel-fired steam generators of more than 250 million Btu per hour heat input; municipal incinerators of more than 2000 lb per hour refuse charging rate; nitric acid plants; and portland cement plants.

Manpower and Training Needs for Air Pollution Control[3]

In the 1967 Amendments to the Clear Air Act, it was emphasized that air pollution control at its source is the primary responsibility of state and local

*Editors Note: Some changes were made in the standards promulgated – *Federal Register*, April 30, 1971.

TABLE III
Proposed Federal Standards

Pollutant	Standard	Concentration in Air: ppm	Concentration in Air: μg/m³
Sulfur dioxide			
3-hr concentration, not to be exceeded more than once per year	Secondary	0.50	1300
24-hr concentration, not to be exceeded more than once per year	Primary	0.14	365
24-hr concentration, not bo be exceeded more than once per year	Secondary	0.10	260
Annual arithmetic mean	Primary	0.030	80
Annual arithmetic mean	Secondary	0.023	60
Particulate matter			
24-hr concentration not to be exceeded more than once per year	Primary	–	260
24-hr concentration not to be exceeded more than once per year	Secondary	–	150
Annual geometric mean	Primary	–	75
Annual geometric mean	Secondary	–	60
Carbon monoxide			
Maximum 1-hr concentration not to be exceeded more than once per year	Primary and secondary	13	15
Maximum 8-hr concentration not to be exceeded more than once per year	Primary and secondary	8.7	10
Photochemical oxidants			
Maximum 1-hr concentration not to be exceeded more than once per year	Primary and secondary	0.064 as ozone	125
Hydrocarbons			
Maximum 3-hr concentration (6–9:00 A.M.) not to be exceeded more than once per year	Primary and secondary	0.190 as CH_4	125
NO_2			
24-hr concentration not to be exceeded more than once per year	Primary and secondary	0.13	250
Annual arithmetic mean	Primary and secondary	0.053	100

governments. They are expected to monitor ambient air, adopt acceptable control regulations, conduct inspections, answer complaints, and enforce regulations where there is noncompliance. At the time of the passage of the Clean Air Act in 1963, there were only eleven state and eighty-five local governments with basic air pollution legislation and programs. At the end of 1970, all of the states had basic statutes authorizing the state to control air pollution and over 200 state and local governments and regional areas had

TABLE IV
Effects of Various Air Contaminants

Contaminant concentration	*Expected effect*	*Affected species*
Sulfur dioxide		
10–50 ppm	Respiratory irritation	Man
1–10 ppm	Contradictory findings	Man
1–2 ppm	Damage	Plants
1 ppm	Pulmonary function change	Man
0.3–1.0 ppm	Perception threshold	Man
0.3 ppm	Maximum 1-hr concentration not to be exceeded more than 1 hr in any 3 months (recommended)	
0.16 ppm	Slight increase in pulmonary resistance	Animals
0.15 ppm	Tolerance limit	Plants
(~0.12 ppm) 365 $\mu g/m^3$	National standard (proposed primary) (sulfur oxides) 24-hr concentration, not to be exceeded more than once a year.	
0.1 ppm	24-hr concentration not to be exceeded more than one 24-hr period in any 3 consecutive months.	
(286 $\mu g/m^3$) 0.1 ppm	Wyoming standard, 24-hr average not to be exceeded over 1 day in any 3-month period.	
0.05–0.25 ppm	SO_2 may react synergistically with either O_3 or NO_2 in short-term exposures (e.g., 4 hrs) to produce moderate to severe injury to sensitive plants	Plants
260 $\mu g/m^3$ (~0.1 ppm)	National standard (proposed secondary) maximum 24-hr concentration not to be exceeded more than once per year.	
0.03–0.05 ppm	Yearly average concentrations below which acute symptoms will not occur (maximum 8-hr concentration of 0.3 ppm)	Plants
0.03 ppm	Recommended annual arithmetic mean for 24-hr samples	
(∿0.03 ppm) 80 $\mu g/m^3$	National standard (proposed primary) (sulfur oxides); annual arithmetic mean	
(∿80 $\mu g/m^3$ 0.03 ppm	Wyoming standard, maximum annual average	
(∿0.023 ppm) 60 $\mu g/m^3$	National standard (proposed secondary); annual arithmetic mean	
Oxidant (total)		
1.5–2 ppm	Pulmonary edema within 1 hr plus	Man
1.0 ppm	General irritation (acute)	Man
1.0 ppm	Early lung changes within 1 yr	Animals
0.2–0.5 ppm	3–6 hrs, several visual parameters change	Man
0.15 ppm	Wyoming standard, maximum allowable 1-hr value	
0.13 ppm	Aggravation of respiratory diseases, asthma	Man
0.1 ppm	Drying effect on mucous membranes and increased breathing difficulty	Man
0.1 ppm	5 hr, damage to pinto bean	Plants

TABLE IV (Cont.)

Contaminant concentration	Expected effect	Affected species
Oxidant (*cont.*)		
0.1 ppm	Recommended maximum for 1 hr/day average if human health is not to be significantly impaired	
0.053–0.6 ppm	4 hr, emergence tip burn of White pine needles	Plants
0.05 ppm	Normally exceeded in rural areas about 1% of the time	
0.05 ppm	Recommended maximum if complete freedom from irritation of all types is the objective	
0.03–0.3 ppm	1 hr, impaired performance of student athletes	Man
(∿0.06 ppm) 125 $\mu g/m^3$	National standard (proposed primary and secondary), maximum 1-hr concentration, not to be exceeded more than once a year.	
0.02–0.05 ppm	Odor threshold (ozone)	Man
0.01–0.03 ppm	Background atmospheric concentration of ozone in surface air at sea level	
Nitrogen oxides		
13 ppm	Acute eye and nasal irritation (NO_2)	Man
9 ppm	Lung pathology (NO_2)	Animals
1.0–3 ppm	Odor threshold (NO_2)	Man
0.5 ppm	Morphologic changes in rat lung mast cells (NO_2)	Animal
0.15 ppm	Wyoming standard, maximum allowable 1-hr value, not to be exceeded for 1% of the time during any 3-month period	
(∿0.15 ppm) 250 $\mu g/m^3$	National standard (proposed primary and secondary), annual arithmetic mean	
(∿0.05 ppm) 100 $\mu g/m^3$	National standard (proposed primary and secondary), maximum 24-hr concentration not to be exceeded more than once a year.	
Particulate		
260 $\mu g/m^3$	National standard (proposed primary)–maximum in 24 hr not to be exceeded more than once a year	
260 $\mu g/m^3$	National standard (proposed secondary) maximum 24-hr concentration not to be exceeded more than once per year.	
150 $\mu g/m^3$	(Size range 0.2–1.0 μm and relative humidity less than 70%)–visibility reduced to as low as 5 miles	Man
60–180 $\mu g/m^3$	(Annual geometric mean), in presence of SO_2 and water–corrosion of steel and zinc panels occurs at accelerated rate.	
80 $\mu g/m^3$	(Annual mean) adverse health effects noted	Man
75 $\mu g/m^3$	Wyoming standard–annual geometric mean	
75 $\mu g/m^3$	National standard (proposed primary) – annual geometric mean	
70 $\mu g/m^3$	(Annual geometric mean), in presence of other pollutants – public awareness and/or concern for air pollution may become evident.	Man

TABLE IV

Contaminant concentration	*Expected effect*	*Affected species*
Particulate (*cont.*)		
60 $\mu g/m^3$	National standard (proposed secondary) annual geometric mean.	
Carbon monoxide		
115 $\mu g/m^3$ (100 ppm)	1–3 weeks exposure – concentrations below which no detrimental effects on certain higher plants have been shown	Plants
115 $\mu g/m^3$ (100 ppm)	For 1 month – N_2 fixation by *rhizobium trifolii* innoculated into red clover plants has been reduced by about 20%	Plants
115 $\mu g/m^3$ (100 ppm)	Intermittent – impairment in performance of some psycho-motor tests at COhb level of 5%	Man
58 $\mu g/m^3$ (50 ppm)	For 90 min – impairment of time – interval discrimination in nonsmokers	Man
35 $\mu g/m^3$ (30 ppm)	For up to 12 hr – equilibrium value of 5% blood COhb is reached in 8–12 hr; 4% COhb is reached within 4 hr; evidence of physiologic stress in patients with heart disease	Man
12–17 $\mu g/m^3$ (10–15 ppm)	For 8 or more hours – produce blood COhb level of 2.0–2.5% in nonsmokers. Has been associated with adverse health effects as manifested by impaired time interval discrimination	Man
9–16 $\mu g/m^3$ (8–14 ppm)	Weekly average – some epidemiologic evidence suggests an association between increased fatality rates in hospitalized myocardial infarction patients	Man
15 $\mu g/m^3$ (~14 ppm)	National standard (proposed primary and secondary) – maximum 1-hr concentration not to be exceeded more than once a year	Man
10 $\mu g/m^3$ (~9 ppm)	National standard (proposed primary and secondary) –maximum 8-hr concentration	

established actual pollution control agencies.* The Federal government has the power to intervene if the state and local governments do not comply with the timetable of responsibilities set forth in the Act, but it also realizes that for compliance, there must necessarily be trained manpower. As a result, the Secretary of the Department of Health, Education, and Welfare submitted to Congress a report on the manpower and training needs for present and future air pollution control.

**Manpower and Training Needs for Air Pollution Control* p. 3, Report, Secretary of Health, Education, and Welfare, Document No. 91-98, U. S. Government Printing Office, Washington, D.C., 1970.

Even though there have been budget increases averaging 13% per year among local agencies and 32% per year among state agencies in the past 5 years, and even though many small agencies have combined to form larger, more efficient ones, half of the state agencies have fewer than ten full-time positions and half of the local agencies have fewer than seven. Using a model which assumes that the number of people to staff air pollution control agencies is directly related to economic, geographic demographic factors of the area served by the agencies and to the extent of the program of the agency, it appears that state and local staffs need to be increased from 2837 budgeted positions in 1969 to about 8000 in 1974 to do an effective job with conditions predicted by the model in 5 years. Unfortunately, even though there are enough engineers and scientists to fill the predicted positions, a 1967 study shows that the budgets of state and local pollution control agencies are so low that median salaries for these professional people are often from 20–50% lower than those in industry as a whole. This indicates that the Federal government will be called upon to supplement these agencies with financial aid.

The Federal Government itself will also be needing manpower for the upcoming expansion of its services to aid state and local agencies with specialized technical problems, to develop a more effective air quality monitoring system, and to establish new emissions standards. The projected model calls for an increase of manpower to 2900 in 1974, compared with the 1970 level of 1000.

Private industry, increasingly pressured by more stringent pollution legislation, is also in need of more manpower to devise, install, monitor, and analyze air pollution control devices. By 1974, it is estimated that about 40,000 man-years will have to be spent controlling emission, a figure which is double that spent in 1969. Even curbing air pollutants emitted from private automobiles is expected to take 500,000 competent auto mechanics in 1974, about 200,000 more than those employed in 1969. These auto mechanics will have to be trained to service emission control systems.

The manpower needs in the environmental field is summarized by Graber *et al.*[6] A study of Table V indicates the tremendous job that lies before us in the training of skilled professionals so that the environmental problems can be properly analyzed and controlled.

In regard to the cost of air pollution control programs, the Secretary of Health, Education, and Welfare submitted a projected budget to Congress in March, 1970.[4]

For governmental air pollution programs in the 1970 fiscal year, it was expected that 50 million dollars would be spent on research and development and 82 million dollars on abatement and control. For 1975, it was projected that research and development would cost 185.8 million dollars and abatement and control would be 304.6 million dollars. These estimates cover the research and

TABLE V
Environmental Manpower Needs Now Outstrip the Supply

	Supply	*Need*		
Occupation	*(1970)*	*1970*	*1975*	*1980*
Engineer	35,700	44,000	70,000	105,000
Scientist	11,300	22,000	28,000	32,000
Technologist	26,600	51.000	66,000	76,000
Technician	69,500	86,000	150,000	214,000
Aide	101,000	126,000	157,000	188,000

control programs of the Environmental Protection Agency, state and local programs, and the abatement of all pollution from Federal facilities (Ref. 14, p. 3).

The same study also made projected total cost estimates for controlling facilities in 100 selected metropolitan areas. For total area control, which includes waste disposal, fuel combustion, and industrial processes, for facilities operating in 1967, the figure projected for 1971 would be an initial investment of 116 million dollars and 96 million dollars annually in upkeep. For 1975 this cost will have risen to a 264 million dollar initial investment and 1.88 billion dollars upkeep (Ref. 14, p. 3). But without emission control enforced by the Clean Air Act as amended, the predicted levels of emission in 1975 would be: particulates, 4.656 million tons/year; sulfur oxides, 1.866 million tons/year; hydrocarbons, 1.498 million tons/year and carbon monoxide, 3.043 million tons/year (Ref. 14, p. 6).

In the forty-four states and the territories receiving Federal grants, total expenditures are at a present level of approximately 20.6 million dollars annually. The 142 local control agencies including Federal grants have total annual expenditures of about 31.4 million dollars (Ref. 14, p. 8).

These millions of dollars are necessary for air pollution control. In 1967 in the United States alone, industry and homes generated 329 million tons of solid waste which averages out to over 9 lb/day per person. Of this waste, 46% was disposed of by open burning, 16% by incineration and 38% by landfill, composting, and dumping at sea (Ref. 14, p. 17). Sanitary landfill and pollution controlled incinerators must be encouraged, although at an expected investment of 9.9 million dollars per year and an annual cost of 4.8 million dollars in 1971. Fuel combustion is another major source of pollution.

Steam-electric power plants alone consume 21% of the United States fuel, which includes 64% of the coal. The result of this is that these plants annually produce 20% of the particulates and 49% of the sulfur oxides which pollute the air (Ref. 14, p. 14). As a means to control this, such pollution control devices as electrostatic precipitators must be installed and maintained and low sulfur content fuels used or appropriate sulfur dioxide removal equipment installed. The resulting overall cost for the 100 selected metropolitan areas is expected to be an initial investment of 6.7 million dollars per year and an annual cost of 41.0 million dollars in the fiscal year 1971 (Ref. 14, p. 15). Industrial boilers, another major source of pollution in this category, annually produce 48% of the particulates and 32% of the sulfur oxides in the United States (Ref. 14, p. 15). It is estimated that to remedy this situation in 1971, a 22.0 million dollar per year investment and an annual cost of 19.0 million dollars per year would be required (Ref. 14, p. 17). Commercial, institutional, and residential heating plants are also significant sources of particulate and sulfur oxide pollution but pose the problem in that they are so small relative to the previously mentioned sources using fuel combustion that the only feasible means of control is the use of low sulfur-content fuels. The cost estimated for fiscal 1971 for substituting fuels and/or changing to electrical heat is an investment of 1.7 million dollars per year for commercial-institutional plants (Ref. 14, p. 19) and an investment of 35.1 million dollars per year and an annual cost of 9.8 million dollars per year (Ref. 14, p. 21).

A third major source of pollution is industrial processes. Sixteen industries contribute 21% of the particulates, 22% of the sulfur oxides, 7% of the hydrocarbons, and 7% of the carbon monoxide (Ref. 14, p. 22). The projected cost estimates for 1975 for all major industrial sources is 1.131 billion dollars initial investment and 1.335 billion dollars annually thereafter (Ref. 14, p. 6).

President Nixon, in his proposal for improving the environment, submitted to Congress on February 8, 1971, said in part:

> The course of events in 1970 has intensified awareness of and concern about environmental problems. The news of more widespread mercury pollution, late-summer smog alerts over much of the East Coast, repeated episodes of ocean dumping and oil spills, and unresolved controversy about important land-use questions have dramatized with disturbing regularity the reality and extent of these problems. No part of the United States has been free from them, and all levels of government – federal, State and local – have joined in the search for solutions. Indeed, there is a growing trend in other countries to view the severity and complexity of environmental problems much as we do.
>
> There can be no doubt about our growing national commitment to find solutions. Last November, voters approved several billion dollars in State and local bond issues for environmental purposes, and federal funds for these purposes are at an all-time high.
>
> The program I am proposing today will require some adjustments by governments at all levels, by our industrial and business community, and by the public in order to meet this national commitment. But as we strive to expand our national effort, we must also keep in mind the greater cost of *not* pressing ahead. The battle for a better environment can be won – and we are winning it. With the program I am outlining in this message, we can obtain new victories and prevent problems from reaching the crisis stage.

During 1970, two new organizations were established to provide federal leadership for the nation's campaign to improve the environment. The Council on Environmental Quality in the Executive Office of the President has provided essential policy analysis and advice on a broad range of environmental problems, developing many of our environmental initiatives and furnishing guidance in carrying out the National Environmental Policy Act, which requires all federal agencies to devote specific attention to the environmental impact of their actions and proposals. Federal pollution-control programs have been consolidated in the new Environmental Protection Agency. This new agency is already taking strong action to combat pollution in air and water and on land.

I have requested in my 1972 budget $2,450,000,000 for the programs of the Environmental Protection Agency – nearly double the funds appropriated for these programs in 1971. These funds will provide for the expansion of air and water pollution, solid waste, radiation and pesticide-control programs and for carrying out new programs.

In my special message on the environment last February, I set forth a comprehensive program to improve existing laws on air and water pollution, to encourage recycling of materials and to provide greater recreational opportunities for our people. We have been able to institute some of these measures by executive-branch action. While, unfortunately, there was no action on my water-quality proposals, we moved ahead to make effective use of existing authorities through the Refuse Act water-quality permit program announced in December. New air-pollution-control legislation, which I signed on the last day of 1970, embodies all of my recommendations and reflects strong bipartisan teamwork between the Administration and the Congress – teamwork which will be needed again this year to permit action on the urgent environmental problems discussed in this message.

We must have action to meet the needs of today if we would have the kind of environment the nation demands for tomorrow.

The Clean Air Amendments of 1970 have greatly strengthened the federal-State air-quality program. We shall vigorously administer the new program, but propose to supplement it with measures designed to provide a strong economic stimulus to achieve the pollution reduction sought by the program.

Sulfur oxides are among the most damaging air pollutants. High levels of sulfur oxides have been linked to increased incidence of diseases such as bronchitis and lung cancer. In terms of damage to human health, vegetation and property, sulfur oxide emissions cost society billions of dollars annually.

Last year, in my state-of-the-union message, I urged that the price of goods "should be made to include the cost of producing and disposing of them without damage to the environment." A charge on sulfur emitted into the atmosphere would be a major step in applying the principle that the costs of pollution should be included in the price of the product. A staff study under way indicates the feasibility of such a charge system.

Accordingly, I have asked the Chairman of the Council on Environmental Quality and the Secretary of the Treasury to develop a clean-air-emissions charge on emissions of sulfur oxides. Legislation will be submitted to the Congress upon completion of the studies currently under way.

The funds generated by this charge would enable the Federal Government to expand programs to improve the quality of the environment. Special emphasis would be given to developing and demonstrating technology to reduce sulfur oxides emissions and programs to develop adequate clean-energy supplies. My 1972 budget provides increased funds for these activities. They will continue to be emphasized in subsequent years. These two measures – the sulfur oxides emissions charge and expanded environmental programs – provide both the incentive for improving the quality of our environment and the means of doing so.

Leaded gasolines interfere with effective emission control. Moreover, the lead particles are themselves a source of potentially harmful lead concentrations in the environment. The new air-quality legislation provides authority, which I requested to regulate fuel additives, and I have recently initiated a policy of using unleaded or low-lead gasoline in federal vehicles whenever possible. But further incentives are needed. In 1970, I recommended a tax on lead used in gasoline to bring about a gradual transition to the use of unleaded gasoline. This transition is essential if the automobile emission control standards scheduled to come into effect for the 1975-model automobiles are to be met at reasonable cost.

I shall again propose a special tax to make the price of unleaded gasoline lower than the price of leaded gasoline. Legislation will be submitted to the Congress upon completion of studies currently under way.

The President proposed comprehensive noise-pollution control legislation that will authorize the Administrator of the Environmental Protection Agency to set noise standards on transportation, construction, and other equipment and require labeling of noise characteristics of certain products.

Before establishing standards, the Administrator would be required to publish a report on the effects of noise on man, the major sources, and the control techniques available. The legislation would provide a method for measurably reducing major noise sources, while preserving to state and local governments the authority to deal with their particular noise problem.

President Nixon proposed legislation to establish a national land-use policy, which will encourage the states, in cooperation with local government, to plan for and regulate major development affecting growth and the use of critical land areas. This should be done by establishing methods for protecting lands of critical environmental concern, methods for controlling large-scale development, and improving use of lands around key facilities and new communities.

The President announced that 100 million dollars in new funds would be authorized to assist the states in this effort – 20 million dollars in each of the next 5 years – with priority given to the states of the coastal zone. Accordingly, this proposal will replace and expand his proposal submitted to the last Congress for coastal-zone management, while still giving priority attention to this area of the country which is especially sensitive to development pressures. Steps will be taken to assure that Federally assisted programs are consistent with the approved state land-use programs.

The President also proposed a power-plant siting law to provide for establishment within each state or region of a single agency with responsibility for assuring that environmental concerns are properly considered in the certification of specific power-plant sites and transmission-line routes.

Under this law, utilities would be required to identify needed power-supply facilities 10 years prior to construction of the required facilities. They would be required to identify the power-plant sites and general-transmission routes under consideration 5 years before construction and apply for certification for specific sites, facilities and routes two years in advance of construction. Public hearings at which all interested parties could be heard without delaying construction time-tables would be required. Speaking of his plans to work toward a better world environment, President Nixon said:

> Environmental problems have a unique global dimension, for they afflict every nation, irrespective of its political institutions, economic system or state of development. The United States stands ready to work and cooperate with all nations – individually or through international institutions – in the great task of building a better environment for man. A number of the proposals which I am submitting to Congress today have important international aspects, as in the case of ocean dumping.
>
> I hope that other nations will see the merit of the environmental goals which we have set for ourselves and will choose to share them with us.
>
> At the same time, we need to develop more effective environmental efforts through appropriate regional and global organizations. The United States is participating closely in the initiatives of the Organization for Economic Cooperation and Development (OECD), with its emphasis on the complex economic aspects of environmental controls, and of the Economic Commission for Europe (ECE), a U. N. regional organization which is the major forum for East-West cooperation on environmental problems. . . .
>
> It is my intention that we will develop a firm and effective fabric of cooperation among the nations of the world on these environmental issues.

ATMOSPHERIC CONCENTRATIONS

The United States Public Health Service has been measuring air contaminants in selected cities for the past several years. A summarization of some of these data is presented herewith for the purpose of providing the reader with ready access to typical atmospheric concentrations in various locations of the United States.

Air pollution surveillance activities conducted by the National Center for Air Pollution Control include a nationwide network of stations equipped for periodic measurements of suspended particulate matter and, in some instances, such gaseous pollutants as sulfur dioxide and nitrogen dioxide; a six city network of stations equipped for continuous measurement of several gaseous pollutants; and a relatively new network of stations designed to provide a general indication of effects of air pollution on various types of materials in interstate regions.

The following tables summarize concentration data for the various pollutants for which detailed data are reported herein. Table VI presents the network-wide arithmetic average and maximum station average for suspended particulate matter and for the three gaseous pollutants sampled for at the 24-hr gas sampling

TABLE VI
National Summary of Air Quality Data[a]

Pollutant	Urban			Nonurban		
	Number of cities	*Arithmetic average*	*Maximum station average*	*No. of stations*	*Arithmetic average*	*Maximum station average*
Suspended particulate matter (1967)	279	97	299	34	35	61
Fractions						
Benzene-soluble organic (1967)	182	6.4	19.7	29	2.4	5.1
Benzo(a)pyrene (1967)	115	2.64	9.83	30	0.29	2.10
Ammonium (1966)	100	0.9	8.2	29	0.4	1.0
Ammonium (1967)	118	1.2	10.1	30	0.4	1.0
Nitrate (1966)	100	1.7	5.2	29	0.5	1.1
Nitrate (1967)	118	1.8	4.8	30	0.5	1.2
Sulfate (1966)	120	9.9	27.2	29	4.8	10.1
Sulfate (1967)	128	9.4	32.2	30	4.5	10.9
Gases						
Aldehydes (1967)	32	17	81			
Nitrogen dioxide (1967)	47	148	336			
Sulfur dioxide (1967)	47	61	364			

[a] All data are in $\mu g/m^3$.

TABLE VII
Estimated Nationwide Emissions, 1966 – 1968[a]

Source	*CO*		*Particulate*		SO_x[b]		*HC*		NO_x[c]	
	1966	*1968*	*1966*	*1968*	*1966*	*1968*	*1966*	*1968*	*1966*	*1968*
Transportation	71.2	63.8	1.2	1.2	0.4	0.8	13.8	16.6	8.0	8.1
Fuel combustion (stationary sources)	1.9	1.9	6.0	8.9	22.1	24.4	0.7	0.7	6.7	10.0
Industrial process losses	7.8	9.7	5.9	7.5	7.2	7.3	3.5	4.6	0.2	0.2
Solid waste disposal	4.5	7.8	1.2	1.1	0.1	0.1	1.4	1.6	0.7	0.6
Miscellaneous	8.6	16.9	7.2	9.6	0.6	0.6	6.5	8.5	1.4	1.7
Total	94.0	100.1	21.5	28.3	30.4	33.2	25.9	32.0	17.0	20.6

[a] Amounts are in $\times 10^6$ tons/year. Data on SO_x determined from samples collected during 1967 were available and are therefore included.
[b] SO_x expressed as SO_2.
[c] NO_x expressed as NO_2.

TABLE VIII
Summary of Average Concentration of Gaseous Pollutants at Camp Air Stations

	Arithmetic average (ppm)				
Pollutant	*1963*	*1964*	*1965*	*1966*	*1967*
Carbon monoxide					
Chicago	*a*	12.0	17.1	12.5	*a*
Cincinnati	*a*	*a*	4.0	4.9	5.6
Denver			7.2	7.9	7.6
New Orleans	*a*				
Philadelphia	*a*	7.1	8.1	6.8	6.3
San Francisco	5.4	5.2			
St. Louis		6.3	6.5	5.8	5.6
Washington, D.C.	6.9	5.7	3.7	3.3	4.9
Nitric oxide					
Chicago	0.097	0.100	0.096	0.101	0.072
Cincinnati	0.032	0.038	0.031	0.042	0.032
Denver			0.033	0.039	0.037
New Orleans	0.015				
Philadelphia	0.046	0.045	0.049	0.059	0.063
San Francisco	0.087	0.089			
St. Louis		0.036	0.026	0.032	0.035
Washington, D.C.	0.039	0.034	0.032	0.036	0.047
Nitrogen dioxide					
Chicago	0.041	0.046	0.043	0.057	0.050
Cincinnati	0.030	0.032	0.035	0.036	0.028
Denver			0.036	0.034	0.037
New Orleans	0.017				
Philadelphia	0.039	0.038	0.037	0.039	0.043
San Francisco	0.049	0.057			
St. Louis		0.033	0.026	0.034	0.024
Washington, D.C.	0.034	0.037	0.035	0.035	0.043
Sulfur dioxide					
Chicago	0.150	0.175	0.130	0.084	0.125
Cincinnati	0.026	0.038	0.030	0.031	*a*
Denver			0.019	0.011	0.005
New Orleans	0.008				
Philadelphia	0.070	0.081	0.084	0.091	0.098
San Francisco	0.009	0.017			
St. Louis		0.059	0.047	0.042	0.029
Washington, D.C.	0.047	0.047	0.046	0.044	*a*
Total hydrocarbons					
Chicago	3.2	3.0	2.6	2.8	3.0
Cincinnati	3.3	3.0	2.8	*a*	2.5
Denver			2.5	2.4	2.4
New Orleans	*a*				
Philadelphia	2.5	2.2	2.2	2.5	2.4
San Francisco	2.7	2.8			

TABLE VIII (Cont.)

Pollutant	Arithmetic average (ppm)				
	1963	*1964*	*1965*	*1966*	*1967*
St. Louis		3.1	2.9	3.0	*a*
Washington, D.C.	2.8	3.0	2.1	2.4	*a*
Total oxidants					
Chicago		0.030	0.026	0.021	0.029
Cincinnati	Oxidant	0.028	0.033	*a*	0.031
Denver	data for		0.033	0.030	*a*
New Orleans	1963 not				
Philadelphia	included	0.026	0.026	0.032	0.026
San Francisco	because	0.020			
St. Louis	of SO_2	0.031	0.032	0.037	0.035
Washington, D.C.	inter-ference.	0.030	0.028	0.027	0.025

[a] Average not calculated because of insufficient or poorly distributed data.

stations. National Air Surveillance Network (NASN) data shows the annual geometric mean concentrations of suspended particulate matter in urban areas range from 60 $\mu g/m^3$ to about 200 $\mu g/m^3$. For nonurban areas the annual geometric mean is typically 10 $\mu g/m^3$ to 60 $\mu g/m^3$. These data are higher than previous National Air Pollution Control Administration (NAPCA) estimates because of the inclusion of sources not previously considered. For example, emission estimates have been included for forest fires, burning coal refuse banks, and an increased number of industrial process sources. The increasing availability of new and revised emission factors and additional basic data sources are responsible for making these improved estimates possible. Table VII presents the main sources of the five major air pollutants estimated on a nationwide basis for 1966 and 1968.

When comparing emission rates of 1 year with another it is important to bear in mind that two basic differences can occur which will influence the data. The first, which is based on increases or decreases in fuel usage, industrial production, population, vehicular travel, or refuse disposal rates is referred to as an "actual" change. These "actual" changes, in combination with changes in emission factors or the inclusion of new sources, result in the second difference, which is called an "apparent" change. For this reason a source category may appear to show a decrease in emissions when in reality the emissions are the same and are only the result of a revision in the emission factor used to convert the basic data into emission data. In addition, the emission rate of a category can increase significantly because of the inclusion of a source that was not previously considered.

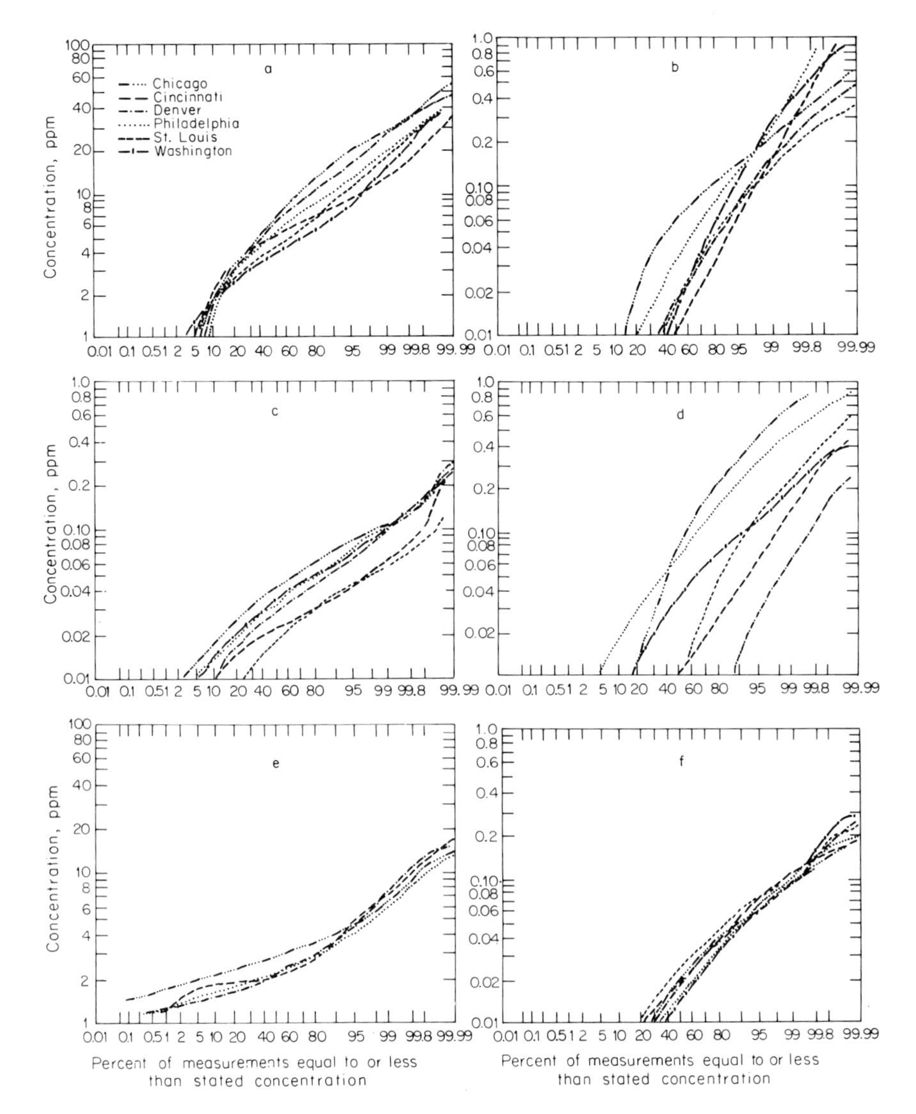

Fig. 1. Frequency distribution of gaseous pollutant data (CAMP), 1967. (a) Carbon monoxide; (b) nitric oxide; (c) nitrogen oxide; (d) sulfur dioxide; (e) total hydrocarbons; (f) total oxidants.

The Continuous Air Monitoring Program (CAMP) stations produce a wealth of data – a summary of which is given in Table VIII. The average concentrations for six gaseous pollutants (CO, NO_x, NO_2, SO_2, HC and O_x) are listed for each station. Figures are given for the years 1963 through 1967 so that comparisons

TABLE IX

Distribution of Cities by Population Class and Particulate Concentration

Population class	*Average particulate concentration (μg/m³)*										
	Less than 40	*40-59*	*60-79*	*80-99*	*100-119*	*120-139*	*140-159*	*160-179*	*180-199*	*More than 200*	*Total*
Over 3 million							1		1		2
1–3 million							2	1			3
0.7–1 million			1		2		4				7
400–700,000				4	5	6	1	1	1		18
100–400,000		3	7	30	24	17	12	3	2	1	99
50–100,000		2	20	28	16	12	6	5	1	3	93
25–50,000		5	24	12	12	10	2	1	2	3	71
10–25,000		7	18	19	9	5	2	3	1		64
Under 10,000	1	5	7	15	11	2	1	2			44
Total	1	22	77	108	79	52	31	16	8	7	401

may be made. The frequency distribution of these six pollutants is given in Fig. 1.

Further evaluations may be made by comparing a city's average particulate level with those of other cities of the same general size. Table IX presents the distribution of average particulate values for 400 cities by population. Thus, a city of 300,000 with a suspended particulate average of 75 $\mu g/m^3$ would have relatively low particulate levels for its size, while a city of similar size with an average of 175 $\mu g/m^3$ would have relatively high levels.

Keeping in mind that the accuracy of projected estimates will vary with the pollutant, some trends have been noted. Detailed studies have been completed for three of the five primary pollutants: CO, SO_x, and NO_x. In 1968 it was estimated that 102 million tons of carbon monoxide were emitted in the United States. By weight this amounts to 50% of all the major air pollution emissions that year. The principal source of carbon monoxide (58%) is fuel combustion in gasoline-powered motor vehicles. However, largely because of the increased use of controls on vehicle exhausts, estimates indicate that no further increases in vehicular emissions of carbon monoxide would be expected before the year 2000.

As indicated in Table VII, in 1968 approximately 33 million tons of sulfur oxides, primarily sulfur dioxide, were emitted into the nation's atmosphere. The principal share (about 74%) comes from the combustion of coal, mainly for the generation of electric power and for space heating. Optimistic forecasts indicate that approximately 37 million tons will be emitted in 1971, and unless adequate control measures are initiated, a 60% increase is expected by 1980 and a 400% increase by 2000. This will result from the nation's increasing demand for electric power which industry spokesmen are forecasting.

Approximately 21 million tons of nitrogen oxides were emitted into the atmosphere in 1968. This figure is believed to have increased in 1970 (basic data have not yet been completed) to about 23 million tons. Since fuel combustion is the major source of technologic nitrogen oxide air pollution, this change is primarily the result of a yearly increase of about 4% in motor vehicle exhaust emissions.

REFERENCES

1. *Fed. Regist. 36*(67), (1971).
2. *Fed. Regist. 36*(62), (1971).
3. Report of Secretary of Health, Education, and Welfare to Congress of U. S. in Compliance with Public Law 90-148 – The Clean Air Act as Amended (Report – June, 1969).
4. Second Report of Secretary of Health, Education, and Welfare to Congress of U. S. in Compliance with Public Law 90-148 – The Clean Air Act as Amended (March, 1970).

5. "Environmental News." Environmental Protection Agency, Washington, D.C., December 16, 1970.
6. Graber, R. C., Erickson,, F. K., and Parsons, W. B., Manpower for environmental protection. *Environ. Sci. Technol.* *5*, 314 (1971).
*7. Air Quality Data from the National Air Surveillance Networks, APTD 69-22, 1967 Edition. Division of Air Quality and Emission Data. U. S. Dept. of Health, Education and Welfare, Public Health Service. Consumer Protection and Environmental Health Service. National Air Pollution Control Administration, Raleigh, North Carolina, November, 1969.
*8. Nationwide Inventory of Air Pollutant Emissions – 1968. U. S. Department of Health, Education, and Welfare, Public Health Service. Environmental Health Service. National Air Pollution Control Administration, Raleigh, North Carolina, August, 1970.
*9. Middleton, J. T., Air pollution – where we are and what we can do. An address given before Southern Universities Student Government Association, Misenheimer, North Carolina, February 8, 1971.
*10. Air Quality Criteria for Carbon Monoxide, AP-62. U. S. Department of Health, Education and Welfare, Public Health Service. Environmental Health Service. National Air Pollution Control Administration, Washington, D.C., March, 1970.
*11. Air Quality Criteria for Nitrogen Oxides, AP-84. Environmental Protection Agency, Air Pollution Control Office, Washington, D.C., January, 1971.
*12. Air Quality Criteria for Particulate Matter, AP-49. U. S. Department of Health, Education, and Welfare, Public Health Service. Consumer Protection and Environmental Health Service. National Air Pollution Control Administration, Washington, D.C., January, 1969.
*13. Air Quality Criteria for Sulfur Oxides. U. S. Department of Health, Education, and Welfare, Public Health Service. Bureau of Disease Prevention and Environmental Control. National Center for Air Pollution Control, Washington, D.C., March, 1967.
14. "Cost of Clean Air." Report, Secretary of Health, Education, and Welfare, Document No. 91-65, U. S. Government Printing Office, Washington, D.C., 1970.

*References 7 through 13, although sometimes not specifically cited, were used as source material.

An Empirical Approach to the Selection of Chimney Height or Estimation of the Performance of a Chimney

R. C. WANTA

We start with the notion that a chimney can be an acceptable means to put practical limits on the occurrence at ground level of continuously emitted pollutants. Even if it be granted that with great enough distance from the chimney the ground-level concentration tends toward values which would have resulted from release of the same effluent close to the ground, realistic consideration takes into account the diluting effect of travel distance. Realistic consideration also requires that one strive to think in terms of the triad: pollutant concentration (C), the averaging period (T) which characterizes this concentration, and a frequency (f) which can indicate how often such an averaged concentration occurs. Because the generally turbulent character of atmospheric flow results in fluctuating ground-level concentrations such that (*1*) pollutant concentration decreases as averaging time increases, and (*2*) the higher pollutant concentrations are associated with lower frequencies of occurrence, it should be obvious that fallacies can be produced in abundance if one (or two!) components of the triad are neglected. Hence some standards designed for the control of pollutants have the form, concentration C of pollutant A as averaged over period of duration T is not to occur more often than once in N such periods.[1] This review describes an empirical approach toward meeting such standards. Although sulfur dioxide emission from a power plant of moderate size is used for illustration, other pollutants might be similarly treated.

DETERMINATION OF GOVERNING STANDARD

Suppose that we are given the following standards, all referring to ground-level concentrations of sulfur dioxide at a sampling site, and that we seek a chimney height which would meet these requirements in the absence of other sources of sulfur dioxide:

Sulfur dioxide concentration[a]	*Occurrence*
0.02 ppm	Maximum annual arithmetic mean
0.10 ppm	24-hr average not to be exceeded more often than one day in any 3 consecutive months
0.25 ppm	1-hr average not to be exceeded more often than 1 hr in any 24-hr period

[a]At ambient temperature of 25°C and pressure of 1013.2 mbar, these concentrations convert to 52.4, 262, and 655 $\mu g/m^3$.

Experience with the long-term averages attributable to power plant effluent from adequate chimneys permits us to assume that the annual standard will be readily met if the more stringent of these requirements on the 24-hr and 1-hr average concentrations are met. (See also the section below on the estimation of long-term averages.)

We shall now define $\overline{C}^{(T)}$ as the ground-level concentration in ppm of sulfur dioxide averaged over a period of T hours, and the *standard triad* $(C, T, f)_s$ as the statement that $\overline{C}^{(T)}$ may not be exceeded at any sampling site more often than once in T/f hours. Thus the second and third standards are expressible by the standard triads $(0.10\text{ ppm}, 24\text{ hr}, 1/91)_s$ and $(0.25\text{ ppm}, 1\text{ hr}, 1/24)_s$. The *observed triad* (C, T, f) and the *design triad* $(C, T, f)_d$ will represent the statements that $\overline{C}^{(T)}$ is observed or expected with frequency f.

To determine which standard is governing, we need to put them on some comparable basis. We do so by (*1*) reducing $\overline{C}^{(24)}$ to an equivalent $\overline{C}^{(1)}$, and (*2*) making the values of f in the different standard triads comparable. For the first reduction, we make use of the power-law relationship $\overline{C}^{(T_1)} = \overline{C}^{(T_2)} \cdot (T_2/T_1)^q$ and give exponent q the empirically supported value 1/2 for reduction from $T_2 = 24$ hr to $T_1 = 1$ hr.[2] Having observed that the frequency f with which a ground-level concentration $\overline{C}^{(T)}$ is exceeded in records of order of length one year and averaging times of 1/2 or 1 hr exhibits a relationship $f^p\overline{C}^{(T)} = \text{constant}$ for the higher concentrations in the most exposed zone, with the exponent p equal to about 1/3 (Fig. 1)[3], we next compare the differing values of the product $f^{1/3}\overline{C}^{(1)}$. The first reduction of the 24-hr standard leads to an equivalent concentration $\overline{C}^{(1)}$ of $0.10\ (24)^{1/2}$ or 0.490 ppm; the corresponding product is $(1/91)^{1/3}\ (0.490)$ or 0.109. The comparable product for the 1-hr standard is 0.0867. Because the 1-hr standard yields the lower value, it is regarded as the governing standard.

The steps here have been: (*1*) to accept applicable standards with their

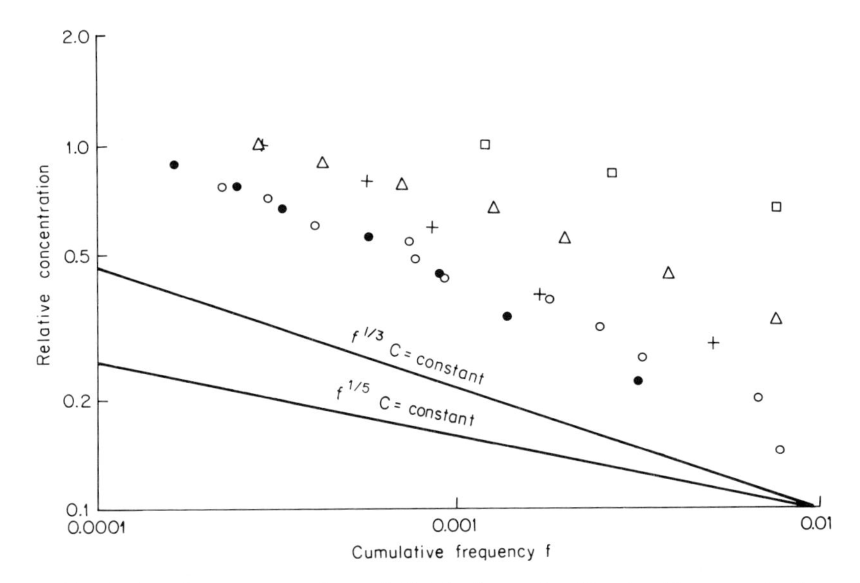

Fig. 1. Low-frequency portions of distributions of sulfur dioxide concentrations at fixed locations near major sources. The abscissa gives the frequency with which the ordinate was exceeded. Concentrations are relative to the highest in each distribution. Refer to 3 in References and Notes for further explanation.

associated averaging times and frequencies of occurrence, (*2*) to transform these standards to a common averaging time and comparable frequency of occurrence, and (*3*) to identify the governing standard.

SAMPLING FLUCTUATIONS AND THE MOST STRINGENT STANDARD

We might now ask what expected (average) frequency of occurrence of a relatively high concentration is required to make the violation of a standard triad $(C, T, f)_s$ a rare event. Clearly, if for example the average frequency of occurrence of $\bar{C}^{(1)}$ were once in 24 hr, we should expect sampling fluctuations to result in occasional 24-hr periods with 2 or more hours above $\bar{C}^{(1)}$. We seek an average frequency of occurrence such that the latter event happens rarely. As a rough first estimate we ignore the nonindependence of successive days' weather, and assume that the probability that $\bar{C}^{(1)}$ is exceeded 2 or more hours in 24 is analogous to the probability of 2 or more defectives in batches of 24 when the expectation of a defective in the sampled population is z, namely, that it is given by $1 - e^{-z}(1 + z)$.[4] Let us suppose that an acceptable value for this probability is 0.00274 or once in 365 days. Solving for z we obtain the expectation 0.0744 defectives in a sample of 24, i.e., a frequency of 0.00310 or 0.310%. We shall accept the latter frequency as a suitable design frequency.

It should be noted that with longer averaging times or smaller frequencies in the standard triad we have seemingly less drastic reductions to a design frequency. Thus attention generally must be given to both the governing standard and the standard which exhibits the sharpest sampling fluctuation. In our example, the 1-hr standard triad is both, and thus is clearly the more stringent standard.

The steps here have been: (*1*) to consider sampling fluctuations in their effect on the governing and other standard, (*2*) to identify the more stringent standard, (*3*) to accept a particular very low frequency with which the standard may not be met, and (*4*) to estimate the corresponding design frequency.

DEVELOPMENT OF THE DESIGN TRIAD $(C, T, f)_d$

We shall now anticipate that empirical performance data for chimneys, expressed by the triad (C, T, f), are abundant for values of T equal to 1/2 or 1 hr. We have already developed a suitable design frequency for 1-hr data, and if such data were abundant and reliable at a frequency of 0.3%, we might write directly the design triad $(0.25 \text{ ppm}, 1 \text{ hr}, 0.00310)_d$. This signifies that a 1-hr average concentration of 0.25 ppm at any one sampling site is expected (on the average) to be exceeded with a frequency of 0.310%, i.e., once in 323 hr.

If our empirical data were expressed in 1/2-hr averages, then to proceed further we reduce $\bar{C}^{(1)}$ to $\bar{C}^{(1/2)}$ by the power law as before, though now we take the exponent equal to 1/3, an empirically derived value for the change of T from 1 to 1/2 hour[5], and obtain $\bar{C}^{(1/2)} = 0.315$ ppm. Let us further suppose that our empirical data are reliably expressed at a frequency of 0.1%. As before we compute the product $f^{1/3}\,\bar{C}^{(1/2)} = (1/323)^{1/3}\,(0.315) = 0.0459$ and divide by $(1/1000)^{1/3}$ to obtain 0.459 and hence the design triad $(0.459 \text{ ppm}, 1/2 \text{ hr}, 0.001)_d$.

The steps here have been: (*1*) to choose an averaging time and a reliable (low) frequency characterizing a body of available data on chimney performance, and (*2*) if required, to transform the more stringent standard and its design frequency to obtain the design triad.

EMPLOYMENT OF A BASE CURVE TO SELECT CHIMNEY HEIGHT OR ESTIMATE CHIMNEY PERFORMANCE

Ground-level concentrations of a substance emitted continuously from a chimney are expected to be proportional to the emission rate Q, i.e., the quantity of the substance released in a given period of time. Hence if we have ground-level concentration data for several different emission rates, we might combine such data by converting concentrations $\bar{C}^{(T)}$ to specific concentrations $\bar{C}^{(T)}/Q$. We might then speak of specific observed triads (C/Q, T, f), specific design triads $(C/Q, T, f)_d$, and so on. In our example for sulfur dioxide, we shall use for specific concentration the units ppm per 100 tons of sulfur dioxide emitted per day. (At ambient temperature of 25°C and pressure of 1013.2 mbar, this unit is equivalent of 2.49×10^{-6} sec/m³.)

For present purposes let us assume the existence of a relationship between the specific observed triad $(C/Q, T, f_i)$ and chimney height h, where f_i is a fixed (low) frequency of occurrence. Figure 2 illustrates an estimate[6] of such a relationship (which we shall call a base curve) for T = 1/2 hr and f_i = 0.1%, with the following properties: (*1*) the base curve is developed for a particular regional climate, uncomplicated terrain, and chimney height adequate with respect to the height of nearby structures; (*2*) the base curve is developed from measurements near power plants with effluent gas temperature approximately 143°C or 289°F, and effluent gas velocity approximately 15 m/sec or 49 ft/sec; (*3*) frequency is adjusted to the condition that all wind directions are equally likely; (*4*) observed values of $\bar{C}^{(T)}$ have been reduced to a common heat emission rate; (*5*) the base

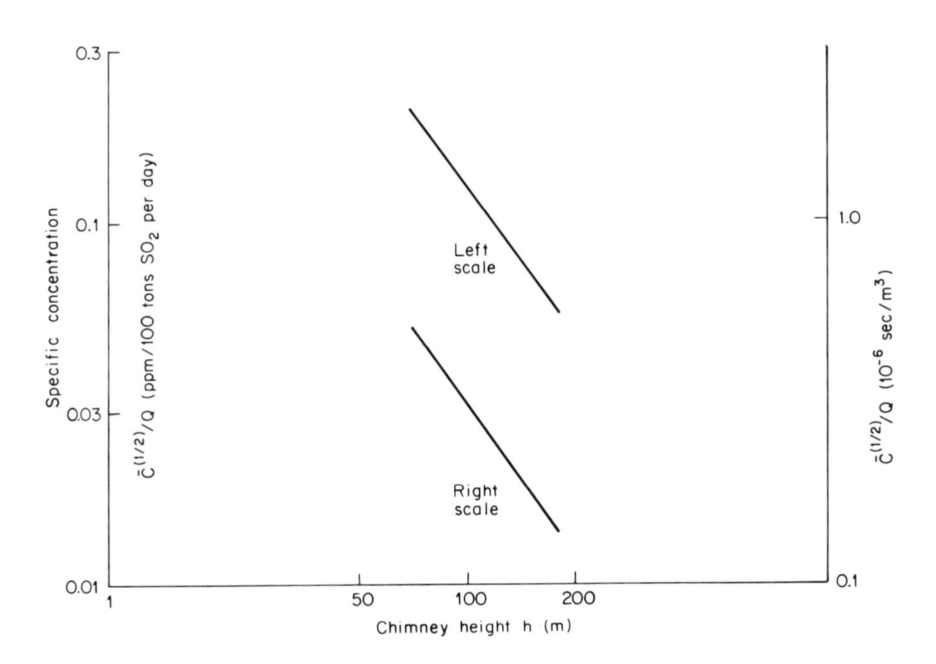

Fig. 2. Specific concentration (in two units) vs. chimney height for f = 0.1% (see text).

curve represents at any value of chimney height the highest specific observed triad ($C/Q, T, f_i$) in the field of the chimney; and (*6*) the base curve applies primarily to a zone of greatest ground-level exposure ranging in distance over roughly 15 to 75 chimney heights, with the proviso that the specific observed triad ($C/Q, T, f_i$) and certain of its T and f transformations are exceeded nowhere else either.

Assuming that Fig. 2 represents a base curve such as the one sketched above, the principal difficulties in its application to the estimation of required chimney height or performance of a given chimney would be meteorologic. Allowance must be made for differences (if any) in regional climate, for terrain complications, and for chimney heights which are inadequate relative to neighboring structures and topography. Allowance must also be made for departures from the assumption that all wind directions are equally likely. Sometimes the allowances are first expressed in their effect directly on chimney height (e.g., for neighboring structures), sometimes on specific concentration (e.g., for heat emission rate), and sometimes on frequency (e.g., for particular ranges of wind direction or speed). Such adjustments of the base curve for specific concentration and frequency can be reduced to an adjustment for frequency alone by having regard for the constancy of $f^{1/3} C/Q$. (Cf. earlier acceptance of constancy of $f^{1/3} C$.)

If the base curve applies to a heat emission rate equivalent to 19 (thermal) MW, then it might be overly conservative to ignore greater rates, and adjustments of C/Q can be made by using an empirically determined relationship between relative C/Q and jointly the number of chimneys and heat emission rate, but interpreting the result in terms of heat emission alone. The adjustment factor is 1.0 for unit heat emission (19 MW), 0.95 to 2 units, 0.84 for 4, and so on.

Suppose that the various allowances lead to an estimated 10-fold increase in frequency over the base curve. We can then transform the design triad $(0.459 \text{ ppm}, 1/2 \text{ hr}, 0.001)_d$ to $(0.459 / 10^{1/3} = 0.213 \text{ ppm}, 1/2 \text{ hr}, 0.001)_d$. If we further take 200 tons of sulfur dioxide per day as our emission rate, then our design triad might be expressed as $(0.107 \text{ ppm} / 100 \text{ tons } SO_2 \text{ per day}, 1/2 \text{ hr}, 0.001)_d$, and using the base curve we find a required chimney height of 112 m.

The steps for the estimation of performance of an existing chimney should now be evident.

ESTIMATION OF LONG-TERM AVERAGE GROUND CONCENTRATION IN THE MOST EXPOSED ZONE

The relationship $f^p C = \text{constant}$ readily permits the estimation of an upper limit for long-term average ground concentration in the most exposed zone.

Suppose the relationship holds as indicated by the solid line in Fig. 3 for an averaging time of 1/2 hr. In other words, it holds down to frequency zero, and up to cumulative frequency m where C drops abruptly to zero. A frequency of 0.05 is usually conservative for such an upper limit. Furthermore, the assumed relationship is conservative because measurements in areas with low background show curvature as suggested by the dashes. An elementary integration then yields the average concentration

$$\frac{(f_o^p \; C_o)}{1 - p} \times m^{1-p}$$

or, if $p = 1/3$, $m = 0.05$, and $\overline{C}_o^{(1/2)} = 0.46$ ppm at $f_o = 0.001$ (the example above), the average long-term concentration becomes 0.0094 ppm. The reader who wishes to pursue other implications of the f^pC = constant relationship should recognize that it is an example of the curious Pareto distribution, whose properties have already been given some study.[7]

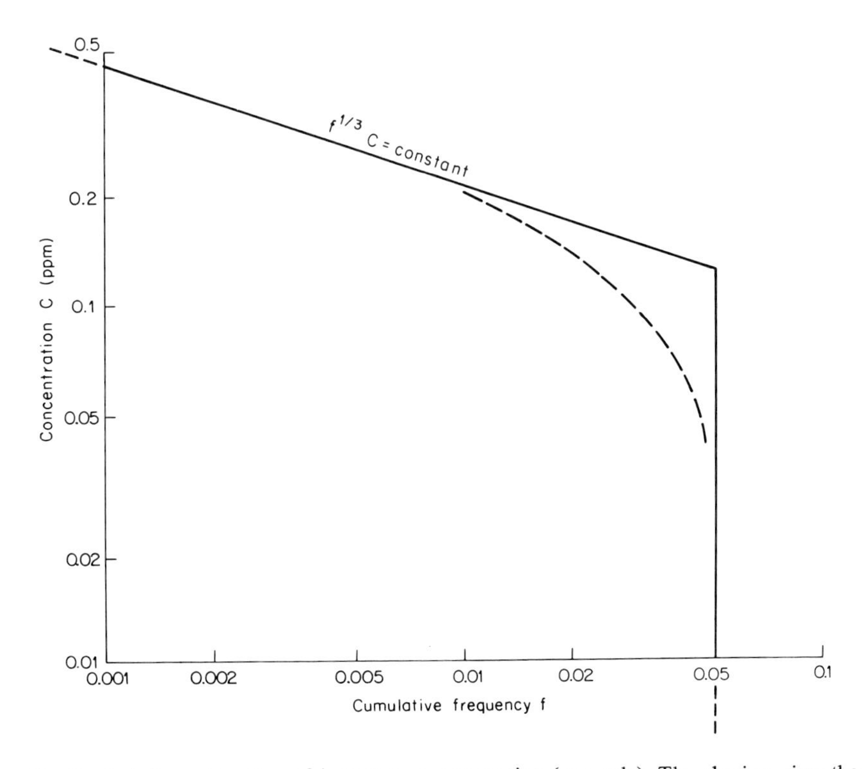

Fig. 3. Basis for estimation of long-term concentration (example). The abscissa gives the frequency with which the ordinate is exceeded.

DISCUSSION

Although this empirical approach to a practical aspect of atmospheric diffusion has thus far withstood the appearance of some data newly available, the user should exercise appropriate caution. The approach yields rough first approximations for the conditions described. It will be beneficial if this note gives impetus to trial applications where both climatic and air quality data are abundant and results in more published data on observed performance of chimneys in triad form. A larger data base would support efforts to determine accuracy of the crucial exponents and range of validity of the approach with respect to chimney heights, buoyancies, climates, etc. Modification of the approach to take into account pollutant reactivity (compared with sulfur dioxide) could be developed. The synthetic approach of the present note, to the extent that it is valid, should converge to valid estimates made by the customary analytic approach.[8] A union of the two approaches has already been studied by Högström.[9]

The method used here to allow for buoyancy of the chimney effluent is admittedly primitive and should be developed further. Observed triads near sources with effluents of small or negative buoyancy may be significantly different from those presented.. Fluctuations in source strength Q should be given explicit attention in future investigations. Whereas the pollutant background level and emissions from nearby sources was for the most part negligible in the data base used to develop Fig. 2,[6] applications of the present approach will frequently deal with such backgrounds.[10] Broadly, when elements of a buoyant elevated plume can mix down to the ground, diffusion of effluents emitted at much lesser heights above ground is quite good. Transitory and transitional changes in atmospheric stability may require special consideration. Other interesting investigations would include: (*a*) the effect of meteorologic control on observed point concentration distributions; (*b*) the relationship between ambient air quality and pollutant emissions from the synthetic and analytic viewpoints; and (*c*) implications of the observed f vs C relationship for the economics of air pollution control.

The range of chimney height given in Fig. 2 is limited. For very tall chimneys, i.e., greater than about 175 m in height, does the upper tail of the point concentration distribution become an imperfect illustration of the observed f vs C relationship, in the sense that the latter is an upper limit? The limitation to averaging times of 1/2 or 1 hr is probably overly conservative. See for example note 3e and Fig. 1. It is not yet clear what will happen to the f vs C relationship as the duration of record N increases, say for 1/2 and 1 hr averaging times, beyond the order of 1 year. If one assumes that it will persist, some light is thrown on the variation of exponent q with averaging time. It is clear that additional links between this empirical approach and observed changes of concentration with averaging time remain to be educed. What is observed at a fixed continuous recorder might well be described by a convolution,[11] one component of which is the distribution of wind speeds,[12] especially but not only for buoyant emissions. Some time ago it was suggested that the statistical tools which hydrologists apply in flood forecasting will find application in air quality analysis.[13] We may well see such terms as recurrence or exceedance interval and return period[14] employed increasingly in pollution studies generally. The example given in connection with Parzen's discussion of Bayes's theorem[15] may help explain why the observed occurrence of the higher concentrations of pollutant at a fixed location near an elevated continuous source seems often to fall short of expectation, when that expectation is guided by the fixed weather categories characteristic of the analytic approach. It is to be hoped that investigations into the effects of air pollutants on receptors in the environment will become more productive as the description of the insult at the receptor is made clearer.

ACKNOWLEDGMENTS

I gratefully acknowledge the experience gained during my assignment by the U. S. Weather Bureau to the air pollution studies group of the Tennessee Valley Authority in the 1950's and my associations there with F. E. Gartrell, S. B. Carpenter, and F. W. Thomas, among others; partial support of travel and study by the Aluminum Company of America; partial support of study and use of data by the Northern States Power Company; discussions with D. K. A. Gillies and his associates and use of data of Ontario Hydro; discussions with D. J. Moore, J. R. Mahoney, M. Benarie, U. Högström, and G. Manier; and comments by F. E. Gartrell, F. Pooler, and R. E. Welch. Responsibility for the contents of this note rests of course with the author.

REFERENCES AND NOTES

1. For example, see National Ambient Air Quality Standards, Environmental Protection Agency, Washington, D.C., April 30, 1971 (36 F.R. 8186), in which one of the primary standards for sulfur oxides (pollutant A) puts $C = 365\ \mu g/m^3$, $T = 24$ hrs, and N equal to 365-6.
2. Ross, F. F., Clarke, A. J., and Lucas, D. H., Tall stacks – how effective are they? Preprint ME-14A, Second International Clean Air Congress, Washington, D.C., December 6–11, 1970 (proceedings to be published). Data in their Table 1 (after Catchpole) lead to $q \doteq 1/2$ for observed mean and "normal maximum" concentrations.
3. For comparison, the two straight lines indicate p-values of 1/3 and 1/5 in f^pC = constant. Sources for data presented in Fig. 1 are:

   a. (○) Wanta, R. C., unpublished report on observed sulfur dioxide concentrations near TVA steam power plants, U. S. Weather Bureau, Cincinnati, Ohio, March 1956, revised June 1956.

b. (·) Djurfors, S. G., and Clark, C. H., Ontario Hydro, Toronto, personal communications, December 1970 and March 1971, respectively.
c. (Δ) Martin, A. and Barber, F. R., Investigations of sulphur dioxide pollution around a modern power station. *J. Inst. Fuel 39,* 294–307 (1966).
d. (+) Wanta, R. C., unpublished report, Northern States Power Company, Minneapolis, May 1969.
e. (□) Högström, U., A statistical approach to the air pollution problem of chimney emission. *Atmos. Environ. 2,* 251–271 (1968).

Location or name of source, height of chimney h, distance x of sulfur dioxide sensor from source, averaging time T, duration of record N expressed in units of T, and remarks on the concentration data from which relative concentrations were derived as follows:

Source location or name	*h (m)*	*x (km)*	*T (hr)*	*N*	*Remarks*
a. Kentucky	76.2	1.67	½	27269	C/Q
b. Ontario	150.3	8.0	1	12384	C
c. High Marnham	137	5.3	1	7070*	C (power station only)
d. Minnesota	239	9.7	1	3508	C
e. Reymersholms–verken	12 to 65	0.75	2	2602	C

*Estimated.

4. Moroney, M. J., "Facts From Figures," 3d revised ed. Penguin Books, Baltimore, Maryland, 1965.
5. For example, see Ross, F. F. *et al.,* Ref. 2 above. Their Table 1 yields q ≐ 1/3 in the present case by interpolation.
6. The data base from which the estimate in Fig. 2 was prepared includes: (a) Ref. 3a above; (b) personal communications from S. B. Carpenter; and, (c) Gartrell, F. E., Thomas, F. W., and Carpenter, S. B., Transport of SO_2 in the atmosphere from a single source, pp. 63–68 in Monogr. 3, Atmospheric Chemistry of Chlorine and Sulfur Compounds, American Geophysical Union (1959). See also Gartrell, F. E., Control of air pollution from large thermal power stations, Revue Mensuelle 1966 de la Société Royale Belge des Ingénieurs et des Industriels, Brussels. All measurements were made in the region of the Tennessee River valley. Reductions to a common heat emission rate of approximately 19 thermal MW was accomplished through use of the empirical relationship given in reference c immediately above. The relative flatness of the maximum of the ground concentration with change of distance from the source and the dynamism of the atmosphere tend to support the concept of a *zone* of greatest exposure. See a nondimensional plot of C vs. x, e.g., Fig. 82 on page 180 in "Air Pollution Manual," Part I, Evaluation, 1st ed., Amer. Ind. Hyg. Ass., Detroit, 1960; this figure appears also in the 2nd ed. now in press.

7. Cramér, H., "Mathematical Methods of Statistics," Princeton Univ. Press, Princeton, New Jersey, 1946. See also Feller, W., "An Introduction to Probability Theory and its Applications," Vol. 2, 2nd ed., Wiley, New York, 1971.
8. For recent examples of analytical approaches, one may consult:
   a. Mahoney, J. R., Models for the prediction of air pollution, Study Group on Models for the Prediction of Air Pollution, Organization for Economic Cooperation and Development, Paris, 1970.
   b. Manier, G., "Die Bestimmung von Schornsteinhöhen im Hinblick auf die Luftreinhaltung, Theoretische Grundlagen einer neuen VDI – Richtlinie," to be published in VDI Fortschritt Berichte.
9. Högström, U., see Ref. 3e above.
10. See for example Martin, A., and Barber, F. R., Ref. 3c above, or *Atmos. Environ. 1,* 655–677 (1967).
11. Feller, W., "An Introduction to Probability Theory and its Applications," Vol. 1, 3d ed., Wiley, New York, 1968.
12. See for example Benarie, M., Le calcul de la dose et de la nuisance du pollutant emis par une source ponctuelle. *Atmos. Environ. 3,* 467–473 (1969).
13. Wanta, R. C., Application of extreme-value theory to analysis of community air pollution. *Bull. Amer. Meteorol. Soc. 37,* 186 (abstract) (1956).
14. For example, see Linsley, R. K., Jr., Kohler, M. A., and Paulhus, J. L. H., "Applied Hydrology." McGraw-Hill, New York, 1949.
15. Parzen, E., "Modern Probability Theory and its Applications." Wiley, New York, 1960.

Water Pollution

S. CHARLES CARUSO, FRANCIS CLAY McMICHAEL,
AND WILLIAM R. SAMPLES

INTRODUCTION – THE HYDROLOGIC CYCLE

Water is considered polluted if it is not suitable for its intended use: domestic, industrial, or agricultural water supply; propagation of fish and wildlife; recreation and others. Natural water bodies are dynamic systems continually changing in response to numerous stimuli. The simplest description of natural waters centers on the concept of the hydrologic cycle. There is evaporation from the oceans, clouds form, and precipitation on the land feeds the groundwater basins, flows in the rivers, and returns to the oceans. Some water is stored and delayed in transit in lakes and glaciers. Brooks[1] emphasizes that an often neglected aspect of the hydrologic cycle is its transport characteristics. In passing from one segment of the cycle to another, the water carries dissolved chemicals and suspended particulate matter as well as heat. This fundamental transport capacity of the hydrologic cycle has enormous importance to nature and man's activities. For example, Brooks[1] points out the Colorado River Aqueduct of the Metropolitan Water District brings about 1 million acre-feet of water per year from the Colorado River to Southern California. This freshwater has about 700 ppm dissolved solids, or about 1 ton of solids per acre-foot of water. Thus into Southern California man brings a million tons of salt per year, about 200 lb per resident, equivalent to a seventy-car trainload per day. Particulate matter, called sediment by river engineers, is moved in large quantities also. Brooks[1] states that in the year preceding the completion of the Glen Canyon Dam, the Colorado River moved 63 million tons of sediment through the Grand Canyon. In a single day the largest amount transported exceeded 1.5 million tons.

Material transported by the rivers ultimately gets to the ocean. While nature can predominantly effect the transport characteristics of the hydrologic cycle, man has his influence too. Use of water by man is primarily through the addition of dissolved and suspended matter to the load created by the weathering of soils and rocks. Processes that result in water evaporation result in a loss of pure water and result in a more concentrated load in the remaining water. Figure 1 is Brooks[1] schematic description of man's interaction with the hydrologic cycle. The thin arrows with hollow points denote the transport of nearly pure water without salts. This is not wholly the case since precipitation is the principal mechanism for removal of aerosols and dissolved gases from the atmosphere. Solid arrows show inputs containing salts. The diagram emphasizes the need to ask where everything goes as well as what effect it has in getting there. Water is reused by man and nature again and again to effect material transport. When the transport capacity associated with a given beneficial use is exceeded, a significant environmental problem is created.

Water is abundant on the earth but 97% of it is in the ocean. Of the remaining 3%, 87% of that is in the ice caps. The fraction that remains, about 200,000,000 gal freshwater for every person on earth,[1] is not evenly distributed. Data on precipitation[2] show that it averages about 30 in./year on conterminous United States. Of this 30 in., about 21½ in. evaporates and the remaining 8½ in., or 1250 billion gal/day, is the manageable water supply. In 1960 in the United States, a little less than 2 in. was used of which about ½ in. evaporated in use. The remaining 1½ in. joined the unused by man portion to make a total of 8 in. (about 1165 billion gallons/day) that returned to the oceans. The figures are large and not easily comprehended.

Pecora[3] emphasizes the need to recognize fundamental earth processes in our ambition to maintain an acceptable environment. Without excusing man's

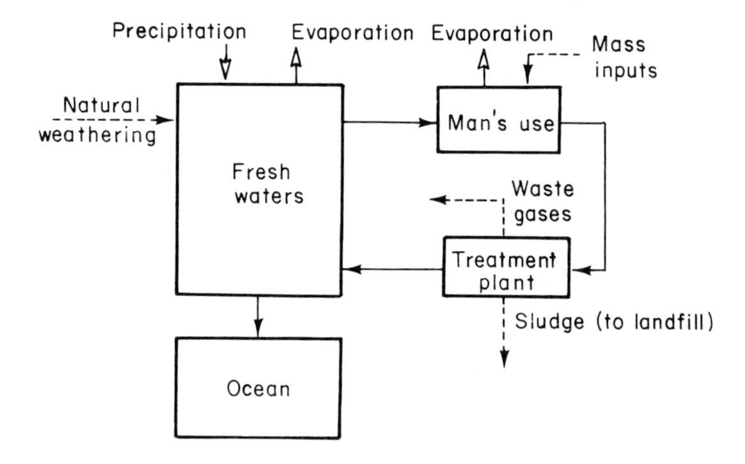

Fig. 1. Conventional treatment of waste water, inland disposal. After Brooks.[1]

activities, he cites numerous examples in which natural processes are the principal agents in modifying the environment. The Lemonade Springs in New Mexico carry 900 ppm sulfuric acid which is about ten times the acid concentration of most acid mine streams in the United States. Hot Springs in Yellowstone is found to be many times more acidic than a typical acid stream in a coal mining district. Lakes are dynamic throughout geologic history. They have life cycles and are not permanent. Great Salt Lake, once 20,000 square miles, is now only 950 square miles. It is in its later stages. Thousands of years ago it was a freshwater lake, while today it is ten times as salty as ocean water. In contrast, Pecora states that Lake Erie, the shallowest of the Great Lakes, is not dead in the same sense. It is very alive, accounting for about 50% of the Great Lake fish catch consistently over the past 100 years.

In total context, natural earth processes dwarf the effect of man; however, man is or can be a dominant geologic agent in a local context. Pecora unequivocally states that the most important issue of future years is the interreaction of man with nature. The people will pay the cost for the acceptable landscape changes in taking from the environment what mankind needs. Water supply and waste disposal play interacting parts in the hydrologic cycle. It is not possible to discuss rationally one problem without the other.

WATER QUALITY STANDARDS

Natural waters are not pure dihydrogen oxide, H_2O. They contain a variety of chemicals and biologicals in a wide range of concentrations and vary also in physical attributes. These differences represent variations in water quality. For the many uses which man has found for water, he has discovered that waters having certain qualities are better suited to serve a particular purpose than others. To judge the suitability of a water for a particular purpose, limitations on the chemical, biologic, and physical quality, based upon technological and economical parameters, have evolved. These parameters are called water quality criteria. Specific limits on water quality established by legislation or other governmental action are called standards. Only when realistic decisions are made concerning the purposes a body of water is to serve and standards established that are compatible with the criteria for these purposes can maximum benefits be derived from that water.

Historically, the term water standards was reserved for those limitations placed on drinking waters used on interstate carriers and administered by the United States Public Health Service. Today's usage of the term most often refers to the interstate stream standards established by the states as a result of the Federal Water Quality Act of 1965. The material in this review will use this latter definition.

The first Federal legislation specifically designated to prevent water pollution was not enacted until 1948. Legislation in 1956 authorized planning, technical assistance, grants for state programs, and municipal waste-treatment construction grants. Following a series of amendments, in 1965, more amendments moved the Federal authority for pollution control from the Public Health Service of the Department of Health, Education, and Welfare to the Federal Water Pollution Control Administration of the Department of the Interior. The most important provisions in the 1965 act called for the establishment of water quality standards and implementation plans for clean-up of all interstate and coastal waters.

The Water Quality Act of 1965 called upon the states to establish standards for their interstate waters to be approved by the Secretary of the Interior as Federal standards. To assist the states in establishing these standards and the Federal government in judging them, the Secretary of Interior appointed the first National Technical Advisory Committee on Water Quality Criteria to collect into one volume the basic information available on water quality criteria. This volume[4] is divided into five sections based upon the major beneficial uses of water and the differences in the criteria for each use. The major beneficial uses of waters as defined by this committee include recreation and aesthetics; public water supplies; fish, other aquatic life, and wildlife; agricultural uses; and industrial water supplies. For each of these beneficial uses, an annotated list of water quality parameters is provided. Each state with the aid of this and other volumes of a similar nature[5] and their experience, proposed a set of standards for interstate waters and submitted them to the Federal government for approval. To establish these standards, the States were required to decide which beneficial uses the particular stream was to serve, the quality needed for this purpose, and to formulate a plan for meeting this quality.

The selection of the specific beneficial uses that a stream or other body of water is expected to serve is the most critical decision in establishing standards. To some, water should serve all purposes at all times. This is impossible because some uses are incompatible with others. For example, a water cannot simultaneously be expected to meet the salinity criteria for both fresh- and saltwater fish. To consider the use of a river just after receipt of even well-treated municipal sewage for bathing purposes is almost equally absurd. The establishment of water-contact sport as a beneficial use in such an instance would subject the bathers to the unnecessary risks of water-borne diseases since no treatment system can be expected to have a realiability of 100%. The point is that beneficial uses should be selected only with the knowledge of the criteria for each and the associated problems involved in maintaining the aggregate quality demands of these uses.

Initially, over one-half of the state standards were not acceptable to the Federal government because, in general, they were not strict enough. Major stumbling blocks pertained to temperature and dissolved oxygen limits for the protection of aquatic life. Other problems developed because of the universal requirement for secondary treatment of waste waters and the insistence that stream standards should not allow any decrease in existing water quality. Presently, the standards of all states have been approved by the Federal government, some, however, with exceptions that still have not been resolved. In one case, where no agreement has been reached, the Federal government has taken steps to impose standards. In addition, most states have now established standards for their intrastate waters.

Even though standards of water quality have been formulated for essentially every sizable stream or lake, much discussion continues. For example, there are those who consider that the establishment of standards has become a game. Abel Wolman,[7] whose remarks were specifically about standards for drinking waters but are equally applicable to other beneficial uses, formulated the following argument:

> The change in water quality perspective is best illustrated by contrasting ancient criteria with those of 1970. At the turn of the century, a liquid that flowed was accepted. Later, the search was extended to include a commodity which was clear, tasteless, and odorless – although not always either available or served.
>
> Today the criteria already exceed several dozen, some required, some desired, some optional. They cover physical, chemical, and organoleptic ingredients. As refinements in analytical procedures emerged, levels of permissible limits usually were lowered. It has been said that, as testing procedures become more and more delicate, more ingredients will be added to the prohibited list.
>
> Little evidence is so far available in the literature that this trek to zero risk is being tested realistically against resulting benefit – or at what price. The tacit assumption prevails with many that water, having no external constituents, natural or man made, is a universal desideratum. This concept, which needs serious scientific debate, is already becoming embedded in law, regulation, and voluntary association policy.

On the other hand, the President's Council on Environmental Quality[6] has stated that

> Even when standards have been approved, there is need to improve knowledge of water quality characteristics so that the standards can be upgraded. For example, general criteria in toxic materials are written into all of the approved water standards. But in many of the standards, specific limits on such products as pesticides have not been spelled out. Nor have definitive measures for implementing standards been detailed. The standards generally require secondary treatment – or its equivalent – for industrial wastes, even though this is a crude measure for industrial wastes, and may not be adequate to forestall pollution in particular areas. In his February 10 (1970) *Message on the Environment*, the President said that he was requesting authority to require the States, with Federal approval, to set specific effluent discharge requirements as part of water quality standards. These would arm the States and the Federal Government with a tool to assure that water quality goals would be met. It would provide a standard against which the States and the Federal Government can enforce compliance.

These two views of current standards and their implications for water quality and their attainment were chosen because they represent the opinions of rational and competent individuals. Yet, some honest differences exist, and since certain consequences are inferred by each side in their discussions, it is imperative that these differences be understood and ultimately resolved. A few of the major points made in these two discussions that are particularly pertinent concern future changes in the standards, universal secondary treatment, costs in meeting standards, surveillance and enforcement procedures, and effluent standards.

Many criticisms of specific water quality standards could be enumerated. Some maintain that the standards were often set under the duress of public emotionalism, that others were set without an adequate technological basis, and others with no thought to possible economic implications. No doubt examples of these can be found. However, for the most part, the standards were established based upon the best available information at the time. The standards also were established under the condition that changes would be made as conditions warranted. The Federal government has apparently taken the attitude that changes in standards can only reflect an upgrading of quality. The assumption must be made, however, that changes in the standards will be made based on up-to-date criteria with prudent selection of beneficial uses to be served. Only in this way can the maximum benefits from our water resources be obtained.

One of the major problems involved in the discussion of water quality enhancement is finding a way to compare the benefits to be obtained with the benefits lost by maintaining the status quo. This comparison, as normally made in formulating technologic decisions, is almost always based on an economic analysis of dollars to be spent vs. dollars to be gained. This remains as the only basis on which many want to compare pollution abatement procedures. Many estimates of the cost for water pollution control have been made and all point out the high costs involved. One estimate by the Federal Water Quality Administration is given in Table I along with the monies expended over the past few years. These cost estimates were made for the 5 years 1970–1974 and indicate that municipal expenditures for waste treatment plants to meet current standards is about 10 billion dollars. Operating costs for these facilities are expected to escalate to almost 0.7 billion dollars by 1974. The cost to solve the problems associated with overflows from combined sewers has been estimated at between 15–48 billion dollars. Costs to abate industrial pollution are estimated at about 3.3 billion dollars in the 5 years; to include the costs for thermal pollution from power plants, add an additional 2 billion dollars. Industrial operating costs are expected to increase to about 0.7 billion dollars/year by 1974. Thus, water pollution abatement is expensive and is going to become more expensive. In addition to these costs for municipal and industrial pollution must be added the costs of controlling pollution from other sources such as

TABLE I
Annual Outlays for Water Pollution Control[a]

Year	Industrial Outlays		Municipal outlays			Total Annual outlays
			Investment			
	Invest-ment	Operating charges	Treatment works	Collec-tion	Operating charges	
1965	640	200	476	355	270	1940
1966	780	270	520	400	295	2215
1967	565	365	550	505	320	2306
1968	530	430	655	500	350	2465
1969[b]	740	515	880	450	410	2995
1970	655	555	2000	–[c]	470[d]	3680
1971	655	595	2000	–[c]	530	3780
1972	655	635	2000	–[c]	590	3880
1973	655	675	2000	–[c]	650	3980
1974	655	715	2000	–[c]	710	4080

[a] In millions of current dollars. From "Environmental Quality."[6]

[b] 1970–1974 expenditures represent those associated with an investment level that will achieve controls required by water standards within the period.

[c] No estimates available.

[d] No estimate of incremental collection operating and maintenance costs.

agricultural runoff, abandoned mines, highway construction, animal feedlots, etc. These items are not insignificant. For example, the wastes generated by cattle alone in this country are almost seven times that produced by humans. Fortunately, only a fraction of this amount is handled as a solid waste problem, but even a fraction of this amount will be costly to control. In total, the costs for clean water are immense.

Whereas the costs of pollution abatement can be evaluated with reasonable accuracy, the benefits to be gained through pollution abatement are not so well defined. Matters such as savings in municipal water treatment costs and industrial water supplies are easily evaluated but the recreational value to be gained from cleaner water is much more difficult. However, these evaluations need to be made. Rational decisions on changing quality standards and on treatment priorities must be based on this type of information. Dr. Lee DuBridge,[9] former Science Advisor to the President, has asked a question as to how pure our air and water must be. He partially answered his own question by stating, "Absolute purity will be obtained only at infinite cost, represented for example, by prohibiting the use of all automobiles and other smog-producing vehicles, by shutting down all power plants and industrial plants and, indeed, by bringing our whole economy to a grinding halt, leaving our entire population without the means to even survive. Absolute purity is, therefore, unobtainable."

This is obviously an exaggerated stand but points out that costs certainly increase with increasing demands for water cleanliness. A balance is needed between current conditions and DuBridge's pollution-free "utopia"; this balance can only be made by a careful evaluation of benefits gained through pollution abatement and the benefits lost through the expenditures of monies for that abatement.

One of the relatively new Federal concepts in pollution abatement is the establishment of effluent standards or requirements.[8] Effluent standards are specific limits placed upon the quality of the discharge to a body of water whereas the water standards previously discussed pertained to the conditions in the receiving body of water itself. Effluent quality has long been recognized as being important in pollution abatement. The current problem has evolved only because the Federal government is requesting the authority to establish standards, whereas, previously this was a state function. Currently, it appears that effluent standards are the method of choice for assuring that stream standards are reached and maintained. A concern among many is that effluent standards will be established without regard to the assimilative capacity of the receiving water. If for example, effluent standards are established that are equal to the existing stream standards, then essentially no wastes can be discharged. This may indeed be the goal of some individuals but is an untenable solution if any real use of the water is to be allowed. The fact that water has the capacity to cleanse itself is undeniable. Dr. Hoak[10] calls this "large capacity for inoffensive assimilation of wastes" a "valuable economic resource." Professor Fair[11] calls it "the healing power of nature." The use of waters to transport, disperse, and assimilate waste approaches the age of the waters themselves. The major task today is to utilize this resource prudently.

The computation of the amounts of materials that a stream may receive and the allocation of these amounts to existing and anticipated dischargers is not an easy task. A discussion of the details involved in this computation are far beyond the scope of this review. Suffice it to say that much care must be exercised in this computation as the economic and water quality ramifications are immense. If, for example, effluent standards are established which allow the discharge of more pollutants than the stream can assimilate, the stream is degraded and some established beneficial use no longer can be maintained. On the other hand, a severe economic penalty can be subjected on a discharger by establishing a limit that is too strict. The cost of waste treatment increases exponentially with increased degree of treatment. To force unneeded levels of treatment on a discharger is indeed a waste of our resources. The need for effluent requirements is apparent but much caution should be exercised in establishing and administering them.

The only reasonable method by which realistic effluent standards can be established is through a detailed understanding of the effects the receiving water has and will have on the area served. The only real way to understand these effects is to utilize basin planning. Other synonomous terms are watershed and drainage area planning. These terms infer studies that encompass a complete stream system with the necessary depth to understand the multiplicity of interrelationships involved in maximizing the benefits provided by the water course. This concept is basically sound and is accepted by most water users and water quality administrators. The implementation of this plan, as is so often the case, has problems. There are a few examples of river basins for which comprehensive plans have been developed. However, the extensive application of this valuable tool has not taken place. Much of this delay has resulted from the reluctance of state officials to relinquish any control to a basin authority. The legitimate questions of high costs involved and timing, however, must be resolved. Many states have already invested much money and time in planning. To begin a new round of planning, costing more money, may simply add more confusion just when implementation of previous plans is getting underway. However, in most instances, small sums spent in planning yield big dividends. Regardless of the arguments against the basin approach voiced by the states, it appears that the Federal government is going to insist that basin planning become an integral part of water quality management. For example, the FWQA (Federal Water Quality Administration, now the Water Quality Office in the Environmental Protection Agency) has announced that no construction funds for waste water treatment facilities would be approved unless they were tied to a comprehensive basin plan. This ruling, if not rescinded, will undoubtedly force the states to comply.

Another edict from the Federal government that concerns effluent standards is the guideline that required treatment of wastewaters to a level which would not degrade the quality of the receiving stream. In practice, this has been interpreted to mean secondary treatment. This essentially establishes secondary treatment as an effluent standard for every waste treatment plant discharging to an interstate stream or coastal waters. This then becomes a uniform effluent standard. On the other hand, the Federal government, at least to date, has recognized the need for many exceptions to mandatory secondary treatment. For example, many coastal cities have not been required to provide secondary treatment.

Great caution should be exercised in the formulation of effluent requirements. Technically, effluent requirements can only be established through basin planning and the determination of the effects of discharge on the stream in light of all the other dischargers to and users of the water. In addition, effluent standards should not be established in terms of pollutant concentrations as the secondary treatment policy essentially does. Rivers and lakes are not harmed by concentrations of pollutants in waste discharges. They are affected by the mass of pollutant discharged, which in turn is the product of the concentration and the volumetric flow rate. The effect on the receiving water, as a first approximation, may be assumed proportional to the mass discharged divided by the amount of water in which it is dispersed, the pollutant concentration in the stream. Thus, effluents should be limited on the mass of pollutants rather than concentration.

The possibility of uniform effluent standards, for the entire country and undoubtedly based on concentrations, is even more questionable. However, this concept is gaining supporters from among dischargers who fear that if standards are not uniform, a round of gamesmanship similar to what is happening in establishing air pollution codes will take place.

The establishment of water quality standards and effluent requirements have no meaning unless compliance with the standards and requirements is maintained. To judge compliance demands much monitoring of water quality. To accomplish this, the Water Quality Office, the U. S. Geological Survey, and state pollution control agencies are jointly developing a nationwide surveillance system. According to "Environmental Quality"[6] some 1320 stations now exist. This number, estimated to cover about 20% of the nation's streams, is expected ultimately to contain at least 10,000 stations. This may seem like a large number of stations until the length of streams, lakes, impoundments, and shoreline to be monitored is fully considered. These proposed stations, if equally spaced, would be 350 miles apart. In addition to these stations, many other sources of water quality data are available. One of the biggest sources is from municipal water treatment plants.

As stated in "Environmental Quality,"[6] "Vigorous State and Federal enforcement is critical if the Nation hopes to attain water quality standards." The first line of enforcement lies with the states. Most states have substantial means of enforcing pollution control regulations, such as through the discharge waste permit system. However, the Federal government also has certain powers and is seeking even more. On August 30, 1969 the Federal government took its first action under the Water Quality Act of 1965 against violations of the water quality standards by ordering six alleged violators to clean up within 180 days or face court action. Since then, several other similar actions have taken place. The government found a new source of enforcement in the Refuse Act of 1899 which provides that an Army Corps of Engineers permit is needed to dump wastes into navigable waters. Violation is punishable by both fines and criminal sanctions. Recent applications of the act indicate that its role in pollution abatement may be a major one. However, there are some limitations to current Federal power. For example, current acts generally cover only interstate waters. The President has requested that this power be extended to intrastate waters and ground waters. In addition, the Federal government now has jurisdiction only if the pollution of one state endangers the health and welfare of another state or the Governor of the State requests assistance. Also at present the only force the Government can wield against a polluter is a cease-and-desist order, and the only option available to the court is a contempt-of-court action. This does not, however, include the power under the Refuse Act of 1899. However, the President has asked for a streamlining of rules and the power to fine offenders. As William D. Ruckelshaus, Administrator of the Environmental Protection Agency, has stated "standard setting and firm enforcement are key pollution control approaches the Environmental Protection Agency will employ."

There are certain principal problem areas for water pollution. Several of these categories will be discussed in more detail in order to highlight the scope and complexities of the pollution problems.

EUTROPHICATION

Eutrophication is simply the process of adding nutrients to a body of water. Nauman[12] a Swedish biologist, was the first to designate a nutrient-poor water as "eutrophic." Nutrients are introduced naturally at a relatively slow rate principally through rain water carrying dissolved organic and inorganic matter obtained from land drainage. These nutrients are responsible for the biologic community present in surface waters. Thus, eutrophication is the natural "aging" process of a body of water such as a lake, i.e., the progression from a relatively deep, sterile environment to a shallower biologically productive system. The major concern is that through man's activities nutrients are entering many lakes at a greatly accelerating rate thereby rapidly increasing biological activity with subsequent deleterious effects. The primary effect is the excessive production of algae and other plants leading to a serious deterioration in water quality. Algal blooms discolor the water, produce turbidity, and can cause biologic catastrophies. The decomposition of the plants by bacterial action deplete the dissolved oxygen causing fish kills and the formation of anaerobic zones where unpleasant odorous compounds are produced. Consequently, the fish population changes from desirable to coarser species. In addition, large quantities of decomposing algae are washed up on beaches causing disagreeable odors which are a nuisance to swimmers, fishermen, boaters, and shore inhabitants.

The amount of mineral and organic matter present in a body of water is the limiting factor in the abundance of aquatic life. Aquatic systems are nutrient sensitive and respond rapidly to the introduction of fertilizing compounds. The main nutrients required are carbon, hydrogen, oxygen, sulfur, potassium, calcium, magnesium, nitrogen, and phosphorous. In addition, trace amounts of organic and inorganic nutrients are also necessary. According to Liebig's Law of the Minimum, growth is limited by the essential nutrient present in the lowest relative amount. The limiting nutrient for algae and other aquatic plants has been the subject of some controversy. Nitrogen, phosphorous, and carbon are considered to be important limiting nutrients. Since nitrogen is usually available in sufficient quantity or can be introduced by nitrogen-fixing organisms,[13] the current controversy concerns the roles of carbon and phosphorous in the eutrophication process. Until recently most aquatic scientists believed that phosphorous was the controlling nutrient for algal growth. Recently, Lange[14] and Kerr[15] have reported evidence that organic carbon may be the limiting nutrient. These authors point out that algae exist in a symbiotic relationship with bacteria so that the algae utilize carbon dioxide produced by the bacterial decomposition of organic matter while they provide oxygen required by the bacterial cells. Others believe that the major source of carbon is from the dissociation of bicarbonates present in many lakes. The arguments of both schools of thought are summarized in Table II.[16] Both groups may be correct, being dependent on the specific aquatic system under consideration. For example, in lakes very low in bicarbonates but high in nitrogen and phosphorous, the quantity of organic carbon would be limiting whereas for lakes high in dissolved bicarbonates and nitrogen but low in phosphorous, the amount of phosphorous would be the limiting nutrient.

The current emphasis by regulatory agencies is on limiting the amount of phosphorous entering the aquatic environment. In Canada detergent manufac-

TABLE II
Carbon vs. Phosphorus as Key Nutrient in Eutrophication[a]

Carbon-is-key school believes:	*Phosphorus-is-key school believes:*
Carbon controls algal growth.	Phosphorus controls algal growth.
Phosphorus is recycled again and again during and after each bloom.	Recycling is inefficient: some of the phosphorus is lost in bottom sediment.
Phosphorus in sediment is a vast reservoir always available to stimulate growth.	Sediments are sinks for phosphorus, not sources.
Massive blooms can occur even when dissolved phosphorus concentration is low.	Phosphorus concentrations are low during massive blooms because phosphorus is in algal cells, not water.
When large supplies of CO_2 and bicarbonate are present, very small amounts of phosphorus cause growth.	No matter how much CO_2 is present, a certain minimum amount of phosphorus is needed for growth.
CO_2 supplied by the bacterial decomposition of organic matter is the key source of carbon for algal growth.	CO_2 produced by bacteria may be used in algal growth, but main supply is from dissociation of bicarbonates.
By and large, severe reduction in phosphorus discharges will not result in reduced algal growth.	Reduction in phosphorus discharges will materially curtail algal growth.

[a]From Bower.[16]

turers have until August 1 to reduce the content of phosphate in their products to 20% (expressed as P_2O_5) with complete elimination by 1972. In the United States, Congress is considering a bill to limit phosphorous in detergents. However, if all phosphorous were removed from detergents, Ferguson[17] estimates that this would remove only 30% of this nutrient presently reaching surface waters in this country. The remaining sources would contribute many times the amount required to promote excessive algal growth. Another complicating factor is that nutrients already deposited in sediments[18] of many lakes can be recycled to the water to promote plant growth. For many systems the only input required to keep the cycle operating indefinitely is solar energy, since the normal addition of natural nutrients is greater than the discharge losses. Thus, eutrophication is largely an irreversible process for lakes, but rivers and streams can eventually recover by flushing nutrients downstream with eventual discharge into the oceans.

Other approaches for algal control include the use of chemical and biologic agents. For example, copper sulfate, chlorine, and 2,3-dichloronaphthoquinone have been employed as algicides. Recently, researchers[19] have succeeded in isolating a virus that is capable of selectively destroying blue-green algae. These workers report that additional studies should uncover other algicidal viruses.

Some other measures that have been proposed include[20] harvesting fish crops; removing algal growths and plants; aerating deep water in lakes; and removing nutrients released by biodegradation of organic substances from lakes by siphoning off the bottom layer of water. Although these expedient measures might prove helpful in restoring oligotrophic conditions in some cases, at present the most effective control of eutrophication seemingly depends upon the removal of nutrients from domestic and industrial waste waters. However, additional research on the basic chemical and biologic parameters of eutrophication, on adequate processes for removal of all essential nutrients from waste waters, and on developing an adequate substitute for phosphorous in detergents will be required to provide a lasting and economical answer to this problem.

WASTE WATER RECLAMATION

Reclaimed water may mean water which as a result of treatment of a waste is suitable for a direct beneficial use, or a controlled use that might not otherwise occur. Deaner[21] summarizes the operating details and features of forty-five water reclamation systems in California. The reclaimed water was used for crop irrigation at four locations, groundwater recharge at eight locations, industrial uses at four locations, golf course and landscape irrigation at twenty-nine locations, and for recreational lakes at four locations. Deaner[21] emphasizes that public health protection at reclamation operations is provided by treatment which is capable of producing a reclaimed water which meets established quality standards, by reasonable precautions in the use of reclaimed water, and by provisions for the reliability of the treatment.

The first large scale plant designed to produce a reclaimed water for groundwater recharge to supplement municipal water supplies has been in operation since 1962 at Whittier Narrows in Los Angeles. The plant produces about 15 million gallons/day of reclaimed municipal waste water. The chlorinated effluent from conventional primary and secondary activated sludge treatment is discharged to a floodwater spreading grounds where it percolates through the soil, receiving tertiary treatment, and recharges the underlying groundwater basin. There has been no evidence of marked or harmful effects on the groundwater at or below the recharge area.[22]

McGauhey[23] includes a chapter on water reclamation in his book "Engineering Management of Water Quality." He quotes the extensive work of the Los Angeles County Sanitation District (Parkhurst[24]) and their field experience with water reclamation systems. Parkhurst lists four criteria to justify the construction of separate reclamation facilities: (*1*) the chemical quality of the water must be suitable for reuse, (*2*) the quantity available must be sufficient to permit economical production costs, (*3*) reclaimable water must be near a project which can utilize it, and (*4*) a benefit must be derived from the project to provide interest in the purchase of water at a price to compensate for all or part of the cost of production. Salinity is often a controlling factor. Domestic use adds an increment of about 300 mg/liter (ppm) of total dissolved solids. Today a commonly accepted limit of dissolved solids is about 1000 ppm. Therefore, unless some form of demineralization is practiced, raw water supplies cannot contain more than 700 ppm dissolved solids for practical reclamation for domestic use. This suggests the merits of studies on the technology of blending waters and partial demineralization.

THERMAL POLLUTION

Thermal pollution has been defined[25] as a deleterious change in the natural temperature of a body of water through man's activity. Temperature changes occur through discharges of waste heat, road building, logging operations, the creation of artificial lakes, and use of water for irrigation purposes. Industrial cooling water is by far the major source of waste heat reaching surface waters. Approximately one-half of all water utilized in this country is used for cooling and condensing purposes. Of this quantity electric utilities account for about 80%. In 1964 American industry used about 50 trillion gal for cooling purposes.

Future demand for electrical power is expected to increase greatly. In the past, power generation has doubled each decade and similar increases are projected over shorter time spans. Some of this future demand will be met by nuclear plants, which are less efficient than the fossil fuel power plants. For example, a 1000 MW nuclear plant will require 750,000 gal of cooling water per min and produce an 11°C temperature increase. Some experts[26,27] have predicted that waste heat from this source will increase almost ninefold by the end of this century. Thus, it is important to find an adequate solution to the management of waste heat so that it will not seriously impair our water resources or impede industrial and economic progress.

Temperature has many chemical, physical, and biologic effects on a receiving waterway. Normally, temperatures in surface water vary due to climatic changes. Aquatic life is found to exist over a range from supercooled seawater at −3°C to thermal springs at 85°C. However, all organisms are not able to tolerate temperature changes over this full range. The tolerance range is an important factor governing the geographical distribution of a species.

Temperature is a very important environmental variable in its effect on aquatic organisms. Chemical reaction rates are determined primarily by temperature. In general, the reaction rate doubles for a 10° rise in temperature. Similarly, temperature controls the rate of metabolism and activity of living organisms. Excess heat can accelerate and distort the biochemistry of aquatic biota with a resultant decline in reproductive patterns. However, the relationship of an organism to its aquatic environment is very complex and the total effects of temperature changes are not fully understood at the present time. For example, a fish species may be able to tolerate a temperature increase, but a type of plankton or insect species which affect the fish population indirectly may be destroyed.[28]

Temperature changes can seriously decrease a stream's capacity to safely assimilate waste materials. Most organic matter entering a water system is assimilated by bacterial action. In this process oxygen is consumed by the bacteria and the organic substances are broken down to carbon dioxide and water. If the quantity of organic matter is not excessive, this process causes a partial depletion of oxygen (oxygen-sag) followed by recovery of oxygen concentration. An increase in temperature of the water promotes a more rapid assimilation with a corresponding greater depletion of oxygen.[29] If the organic matter present is sufficiently high, the oxygen may be completely depleted or depleted to the point where fish-kills and/or anaerobic conditions will result.

Physical properties of water are also affected by temperature changes. Density, viscosity, vapor pressure, and the solubility of oxygen in water are temperature-dependent. Density (above 4°C) and viscosity decrease with an increase in temperature promoting an increase in the settling rate of particles and at times thermal stratification, especially in lakes. Vapor pressure also increases with a rise in temperature, thereby causing an increase in the evaporation rate. This loss may be significant in water-poor regions.

The solubility of oxygen is inversely proportional to temperature. An adequate concentration of dissolved oxygen is necessary to maintain a healthy aquatic environment because most organisms require oxygen to maintain their life processes. Thus, the effect on the oxygen balance in water is perhaps the most harmful result of waste heat. In most cases dissolved oxygen concentration is below saturation[30] and a significant increase in temperature could bring this concentration below the danger point for the aquatic community.

Thermal pollution should be controlled because of its deleterious effects on the environment and to conserve expensive energy. Improving the efficiency of electrical generating plants[31] has been proposed as an economical control measure. Modern fossil-fuel plants are about 40% efficient and nuclear-fuel plants are about 33% efficient. Additional research effort is required to improve the utilization of energy for increased production of electrical energy. Current research is in progress on advanced heat converters for nuclear plants and on the development of new methods such as gas turbines, fuel cells, and magnetohydrodynamics to convert heat directly into electrical energy.[31,32]

Other proposals involve the use of waste heat for practical purposes.[31] These include aquaculture (the farming of plants or commercial fish and shellfish), prevention of shipping lanes from freezing in winter,[31,33] irrigation for growth of tropical and subtropical crops, warming recreational areas, heating seawater for desalination purposes, and furnishing steam for a nearby industrial complex.

The most conventional solution to thermal pollution is the use of cooling towers or cooling ponds. A cooling pond is the simplest method of cooling but is less efficient than a cooling tower. Fogging and drizzle are side effects of these methods so that selection of a suitable location is important to minimize hazards to vehicular traffic from this source.

TOXIC METALS

In the last few years much concern has been expressed about the presence of toxic metals in the environment. The elements of most concern to public health officials, as a potential danger to humans, include antimony, arsenic, barium, beryllium, cadmium, chromium, cobalt, lead, mercury, and nickel. Although these are generally found in surface waters in very low concentration, the danger exists that in some cases they are entering the algae–fish–animal food chain in increasingly higher concentrations. In addition, some of these metals have a tendency to accumulate in the human body. Paradoxically, many of these elements are nutritionally essential since they are necessary to human health in trace quantities. It is well known that copper, zinc, chromium, iron, magnesium and arsenic are essential in low concentrations. The problem, then, is to keep the intake of these metals below their toxic level. However, in some cases, the level of safe consumption is not accurately known nor is the mechanism of toxic action clear. Table III lists some toxic effects which have been attributed to some of the heavy metals.

The presence of mercury in surface waters and fish has been receiving a great deal of attention in recent months. The finding of mercury in fish caught in Lake St. Clair early in 1970[34] was the beginning of the widespread search for this element in the Nation's waters and food supplies. In this case, the source of the mercury was waste discharges from a chloralkali plant in Sarnia, Canada. Since this finding, the chlorine industry has reduced the mercury in its effluents by more than 95%.[35] Losses of mercury for the entire North American chloralkali industry have been reduced to less than 15 lb/day at a cost of 20 million dollars.

Mercury has been found in fish caught in many bodies of water throughout the world. For example, mercury has been reported in Swedish sportfish, Japanese fish, Great Lakes perch, canned tuna, and frozen swordfish. Although the mercury present in some of our lakes and rivers is derived from man's activities, that found in the oceans is largely attributed to natural sources.[36] A recent report[37] states that "the U. S. Coast and Geodetic Survey has concluded that at least 50 percent of the mercury found in freshwater fish probably comes from natural sources." Preserved fish caught 43 years ago and fish from landlocked lakes free of industrial wastes were also found to contain significant levels of mercury. Even though it is rarely found in the elemental form, mercury occurs widely in organic and inorganic combinations. The most common form is mercuric sulfide. Most rocks and soils contain low concentrations, but some shales, crude oils, and coals show relatively high concentrations.[36] The industrial sources of mercury are the chloralkali industry, the manufacture of electrical

TABLE III
Toxicity of Metals to Man

Element	*Toxic effect*
Barium	Muscular and cardiovascular damage, kidney disorders
Cadmium	High blood pressure, cardiovascular disorders, gastrointestinal disorders
Chromium	Liver damage, skin problems
Lead	Brain damage, colic
Mercury	Damage to central nervous system, liver and kidney damage, birth defects
Nickel	Dermatitis, cancer

equipment, antifouling paints, some types of pesticides, seed treatment agents, combustion of fossil fuels, some types of chemicals, and medical and scientific laboratories.

Mercury metal and its salts cause damage to the liver and kidney,[36] but generally are excreted before accumulation to dangerous levels. The ingestion of methyl mercury is far more dangerous since it damages the central nervous system and is retained in the human body with a half-life of 70 days. Thus, toxic amounts can build up by receiving low dose rates over a period of time. Methyl mercury is thought to be formed by the action of organisms on the metal and its salts present in the sediments of rivers and lakes.

It has been stated that "everyone has mercury in his hair, blood and urine."[37] Research is urgently needed to establish safe levels of mercury for humans, to find ways of decontaminating the existing deposits in sediments, to develop methods to remove it from waste discharges, to locate all significant sources of the metal, and to develop techniques for reversing its toxicity.

Arsenic has also received some attention recently. Arsenic was found in the Kansas River[38] in concentrations from 2–8 μg/liter and in ten Minnesota lakes[39] in concentrations of 7–224 μg/liter. The source of the metal in the river samples was attributed to detergent products. For the lake samples it was likely due to the addition of sodium arsenite for control of rooted aquatic plants. The Public Health Service's recommended limit for drinking water is 10 μg/liter. Preliminary tests[38] with two common water-treatment methods indicated an arsenic removal factor of only 70–85%. These authors conclude that a potential danger can exist for river systems in which the water is extensively reused and they recommend further study of this problem.

As analysts develop more sensitive techniques to measure the heavy metals in water, there appears to be a tendency towards a "zero-risk" world in setting standards. Recently, Dr. A. Wolman[40] editorialized against this negative approach in favor of the rational procedure used prior to the present clamor over the state of the environment. He stated that "the framing of standards was an orderly process by voluntary professional societies and governmental bodies in summing up scientific information, epidemiologic inquiry, empirical experience, and value judgments. These were carefully put together as guides for control and design purposes, subject to regular review and modification as new data justified." A return to such a wise and logical process would provide the most meaningful standards without penalizing progress.

PESTICIDES

Serious concern has been widely expressed over the effects of pesticides in our environment. The term pesticides includes fungicides, herbicides, insecticides, fumigants, and rodenticides. Of these most concern has been centered on the presence and effects of the insecticides or residues of these substances on biologic systems. Most of the insecticides in use today include the chlorinated hydrocarbons and the organic phosphates. These are synthetic organic compounds and the chlorinated hydrocarbons cause greater damage to water quality because they are refractory substances which have been widely dispersed in the environment. Although herbicides can cause plant damage, as a class they present the least concern to water quality, since generally they are rapidly inactivated by soil microflora.[41] Obviously pesticides are extremely toxic to biologic systems, since this is the purpose for which they were manufactured.

Pesticides contaminate surface waters in several ways: by direct application for pest control, by aerial spraying of adjacent land, by runoff from treated surfaces, by percolation through soil to the groundwater, by waste discharges and accidental spills by manufacturers, from food-processing waste waters, by the return of irrigation waters to a stream, and through misuse. Very sensitive analytic procedures have been developed to measure pesticide residues. An analyst can usually measure these substances in the parts per trillion range. However, the isolation and concentration from a water sample, prior to analysis, is a problem. It is difficult to determine the percent recovery from water because the pesticides can be present in several forms. They can be adsorbed on particulate matter, dissolved in fatty material, associated with organic substances as complexes, or dissolved in water. The most common techniques for isolating pesticides are solvent extraction and carbon adsorption.[42]

Pesticides have been found widely dispersed in surface waters. This is especially true of the chlorinated hydrocarbons since they have enjoyed widespread use and persist in the environment. For example, Dieldrin, Endrin, DDT, and DDE have been reported[43–45] in all the major river basins in the country with concentrations ranging from 4–118 parts per trillion. Aldrin, BHC, DDD, Heptachlor, and Heptachlor epoxide were also present in some of the major rivers. In a 7-year survey consisting of 6000 samples, Dieldrin was the dominant insecticide reported. The level found ranged from 1–22 parts per trillion.[43] Endrin has been measured in concentrations as high as 214 parts per trillion in the Mississippi River. DDT has been found worldwide and is probably present in all surface waters of this planet. It has been reported in local fishes 2000 miles from the nearest land and a search for DDT-free chlorella by University of California researchers has been unsuccessful.[46]

The primary effect of pesticide residues on water quality is that they endanger aquatic life.[47,48] Following aerial spraying with DDT of a forested area, the insect populations in trout streams were drastically reduced along with 90% of the bottom fauna. Many fish kills have also been caused by the chlorinated hydrocarbons.[49] Of these, Endrin is the most toxic and TDE and BHC the least toxic to fish. The organic phosphates are considered less dangerous to fish since they are rapidly biologically degraded in water.

A major concern is that pesticides may be consumed by man through drinking water and edible aquatic life. Although the levels generally found are low, the potential danger lies in the concentration of these substances in the food chain by "biologic magnification." For example, the insecticide DDD present in a body of water becomes selectively absorbed by plankton, which are eaten by small fish and they in turn are eaten by trout or other game fish. The insecticide concentration increases at each step and can reach dangerous levels. For example, the FDA has banned the sale of some fish from the Great Lakes because of high DDT concentrations.

The long-term effects to humans of consuming low concentrations of pesticides are not known, but man derives considerable benefits from the control of pests. These include a plentiful food supply, control of disease-bearing pests, and protection of our forests.[46] However, research efforts should be accelerated to provide an understanding of the long-term effects of pesticides to man and his environment; to develop or synthesize pesticides which will be degraded after fulfilling their function; and, to seek nontoxic methods to control pests.

TASTE AND ODOR PROBLEMS

In addition to the chemical, physical, and biologic properties of potable water, the physiologic properties of taste and odor are extremely important to water quality. Consumer complaints of water quality are most often due to a mal-flavor of their drinking water. Many compounds can cause tastes and odors in water and impart an off-flavor to fish and shellfish. Generally, trace quantities of some types of compounds are sufficient to produce a noticeable organoleptic effect. Psychologists claim there are only four basic taste sensations. These are salty, sour, sweet, and bitter. The wide flavor variations experienced by man are ascribed chiefly to the sense of smell.

The taste sensation is due to the excitation of the receptor cells by a foreign substance on the tongue. Inorganic salts of copper, iron, manganese, potassium, sodium, and zinc dissolved in water produce a taste response. Some compounds can impart a taste in very low concentrations. For example, threshold odor concentrations for metals vary from 0.04–256 mg/liter.[50] Most sensory problems in water quality are due to odorous substances and not to taste-producing compounds.

Organic compounds are the main source of odor problems in potable water and tainted fish. Some volatile inorganic substances, such as hydrogen sulfide or ammonia, can also cause a specific odor response. Odor-producing contaminants are contributed to waters by industrial discharges, municipal wastes, decay of vegetation, metabolic products of microorganisms, agricultural runoff, and flush-out of anaerobic bottom deposits. Many substances can impart an odor response in trace quantities. For example, chlorophenols have been reported[51] to taint fish at a 0.1 μg/liter level.

Organoleptic studies have resulted in significant developments during the last few years. The adaptation of modern sophisticated analytic instrumentation and techniques to the trace analysis of water contaminants has resulted in the identification of specific compounds affecting sensory quality and provides the methods for tracing pollutants to their source.[52,53] Of primary concern to water plant operators are the odor-producing compounds elaborated by some aquatic microorganisms. These are responsible for the earthy, musty, or alga-like odors frequently encountered in potable supplies. Some of the metabolites formed by odor-producing algae include alcohols, esters, acids, and ketones.[54-56] Several sulfur compounds were also identified in bacterially contaminated blue-green algal cultures.[57] The earthy and musty odors have been attributed to metabolites of *Actinomycetes.* Studies with many species of *Actinomycetes* indicate that the characteristic earthy odor produced is due mainly to a compound which has been named geosmin.[58] This compound has also been isolated from two species of blue-green algae[59,60] and its structure was later determined to be 1,10-dimethyl-9-decalol.[61] Another substance isolated from cultures of *Streptomyces* having a strong musty odor has been named mucidone.[62] This compound differs somewhat from geosmin in molecular formula. The infrared spectrum of mucidone ($C_{12}H_{18}O_2$) shows the presence of a carbonyl group, which is not present in geosmin ($C_{12}H_{22}O$). The complete molecular structure for mucidone has not been established as yet, but several groups are actively working on this problem. However, the discoverers of mucidone have developed a treatment procedure to effectively remove the musty odor from water.[63]

Although studies with many species of *Actinomycetes* have shown that geosmin is largely responsible for the characteristic earthy odor, the overall odor contributed by microorganisms is often the result of a combination of many compounds. In three species, the major odor-producing component has been identified as a camphor-smelling compound, 2-exo-hydroxy-2-methylbornane.[64] This compound was present in concentrations ten times higher than geosmin and has some structural similarities. Both compounds contain a tertiary, sterically hindered hydroxyl group and a saturated ring moiety with methyl substituents. These structural components may be necessary for the characteristic earthy-musty odor.

The elimination of tastes and odors in potable water is currently accomplished by several treatment practices. The most commonly used include adsorption on activated carbon; oxidation with chlorine, chlorine dioxide, or ozone and aeration. The application of one or more of these procedures is usually successful in controlling most problems. The main problems requiring continued study are development of improved sensory methods for the measurement and classification of odors; obtaining information on the chemistry of organoleptic compounds in water, i.e., identification of odorous compounds and correlation with specific problems; and development of odor-treatment procedures based on the identification of specific odorants.

SLUDGE DISPOSAL

The ultimate disposal of waste water concentrates poses a challenging problem. The residue of material which remains, after waste water has been treated sufficiently to allow its return to a river or lake, generally has little or no economic value.[65] Often such concentrates or sludges require further treatment before they can be safely disposed of to the environment. Since these concentrates can be adequately disposed of only on land, air, or the oceans, an important consideration for any disposal procedure is that it does not cause a pollution problem elsewhere. Obviously neither freshwater nor air are acceptable disposal media for noxious substances. However, it is accepted practice to use the atmosphere as a sink for carbon dioxide disposal and the oceans for some sludge disposal. Also, the controlled burial of waste organic matter and radioactive substances is widely accepted.

The oldest disposal method for sewage sludge is spreading it on the land surface.[66] Most soils have a considerable capacity for dewatering and oxidizing organic matter. Soil microorganisms oxidize organic substances to produce carbon dioxide and humus. However, care must be exercised to minimize odor problems and avoid pollution of surface waters through seepage or leaching. The organic matter of sewage sludge is useful for improving sandy soils and some farmers find it desirable as a soil nutrient. The sludge is usually digested or stabilized before it is applied to the land. It is also sterilized, if the land is to be used for root or leaf crops, to destroy pathogenic organisms.[67] It was recently proposed[68] to transport sludge some distance to improve poor quality land and to reclaim strip mined areas. Farmers near Pittsburgh are using wet sludge from two treatment plants and find it a highly desirable fertilizer.[69]

Combustion is another method for the disposal of organic sludges after they have been dewatered by a suitable method such as sedimentation and filtration. The fuel value of dry sludge is equivalent to that of a low grade coal[70] and can provide heat for dewatering additional sludge. Incinerator design is an important consideration for the efficient removal of water.[71] In addition, control of particulate matter from the stacks is required to avoid air pollution problems. Thermal disposal has excellent potential because it can handle most of the difficult disposal problems.

Combustion can also be carried out by the Zimmerman wet oxidation process.[72] This involves heating the sludge in the presence of dissolved oxygen under pressure. For example, one plant using the high-temperature version processes a liquid sludge of 3% solids content at 500°F and 1750 psi. This results in a sterile inert sludge, acids, aldehydes, carbon dioxide, and a liquid effluent containing an ash that settles rapidly. A low temperature version that is carried out at about 350°F and 300 psi has also been suggested for treating sludges without dewatering. This process results in a solid fraction that is easily filtered to a sterile sludge cake. A disadvantage of wet oxidation is that the liquid effluent still contains a relatively high BOD (biological oxygen demand) and must be recycled. The advantages are a high flexibility in the type of sludge treated and the extent of oxidation attainable.[73]

The oceans have long been utilized as a sink for many types of wastes. Many seacoast communities dispose of sewage into the ocean and some cities load heavy sludge on barges and deposit it far from shore where it generally settles to the bottom. The addition of biologic nutrients to the sea should be beneficial since they increase the world's food supply; however, in large quantities they promote objectionable problems such as excessive algal growths and a decrease in dissolved oxygen content. Proper disposal would require their discharge far from shore and adequate dilution with seawater. Ocean water has a high buffer capacity and is, thus, capable of assimilating large amounts of acids and bases. Also, most heavy metals precipitate from solution and settle out preventing widespread dispersion in the oceans.[74]

The most attractive disposal schemes are those which allow the recycling of components to the bioindustrial process of origin. Obviously, this planet contains a limited amount of resources and conservation of this supply, where possible, is highly desirable. Recovery of usable components is not generally profitable; however, since pollution control measures are necessary, a treatment process that can recover a reusable or saleable product can defray part of the overall costs. As an example, dry sludge from some sewage treatment plants is sold for agricultural purposes. Although this revenue does not pay for the drying costs, it provides a useful material rather than a sludge requiring disposal. A more favorable situation is the recovery and reuse of lime used to soften raw water at water treatment plants. The lime is recovered from the sludge by calcining at 1600°F; this converts the calcium carbonate to calcium oxide and carbon dioxide and has proved more economical than buying lime and disposing of the sludge.[75] Recovery processes can be expected to increase since the problems associated with sludge disposal are intensifying.

The disposal method of choice depends on local conditions and requires an evaluation of the economics and technology of the alternative methods available.

There exists a need to expand research on new methods of handling sludges and on the basic parameters involved in flocculation, sedimentation, and conditioning of sludges.

AREA FOR RESEARCH

From consideration of the hydrologic cycle one is aware of the natural recycle and reuse of water. This concept of recycle must be explored, exploited, and researched with regard to man's activities. Sections of our activities must be blocked out and attempts to delineate a mass balance must be made. Inputs and outputs must be quantitatively described as well as attempts to make sinks for waste products rejected from the recycle loops. The concept of making mass balances can be applied on many different scales: a single industrial process stream, an integrated industrial facility, a whole municipality, a watershed, a region of water basins, the entire country. The concept of complete reuse or recycle without special treatment is not acceptable or feasible. Water has only a finite transport capacity for all substances. When this capacity is exceeded, material is rejected and some type of sink must be provided. In most cases the ocean acts as the ultimate sink for water transported materials. An inland facility not having access to the ocean sink may face as an alternative some demineralization process. The blowdown from this operation, be it solid or highly saline liquid, must be accounted for in the mass balance. By alternatives it may be possible to reduce the mass inputs that man adds to the hydrologic cycle. This may be one of the hardest decision areas since it may require significant changes in our life style. On the other hand, natural rivers and bodies of water may not be the best vehicles for transporting man's wastes to the oceans. It may be worthwhile to construct specific pipelines for this purpose – more concentrated and direct lines from waste sources to the sinks. Again, Brooks[1] points out this concept in the 300 miles of drain to provide continuity of the solids balance for the San Joaquin Valley from Bakersfield to San Francisco. In southern California the Los Angeles County Sanitation District transports municipal and industrial wastes by sewer nearly 60 miles to a coastal treatment plant and finally disperses wastes into the ocean several miles more by pipeline and outfall.

The application of recycle ideas to a single industrial process stream may have some illuminating benefits. There are often two types of variables to be considered: process control parameters not generally considered pollutants and, vice versa, pollutants to the environment but not necessarily harmful or influential for the process. Going to water reuse may be generally motivated as a means of reducing the mass discharge of pollutants. The reused water picks up additional mass inputs with each cycle of reuse. Some variable will ultimately limit the degree of recycle and demand the loop be blown down and treatment provided. Each system must be considered for its own peculiarities. McMichael *et al.*[76] describe the limitations of a blast furnace gas washer water recycle system limited by calcium carbonate saturation. The purpose of the recycle system was to reduce the water discharge of cyanide, phenolics, and suspended particles. The concepts of limited blowdown, limited treatment, and provision of material sinks is fundamental to all reuse systems. It takes very careful scrutiny to account for a mass balance for large scale systems. This basic concept is most necessary to research the interaction of man and the hydrologic environment.

A comprehensive report[77] was prepared by the Subcommittee on Environmental Improvement, Committee on Chemistry and Public Affairs, American Chemical Society on "Cleaning Our Environment: The Chemical Basis for Action." Twenty-three recommendations are presented about the water environment designed to exploit and upgrade man's knowledge of his environment and of how to control it. It is pointed out that growth is needed in fundamental understanding in the broad areas: (*1*) the flow and dispersion of water pollutants and their degradation and conversion to other chemical and physical forms, (*2*) the means of abating water pollution where generation of the pollutants cannot be avoided, (*3*) the effects of pollutants on plant and animal life and on inanimate objects, and (*4*) the means of detecting and measuring water pollutants and their effects.

REFERENCES

1. Brooks, N. H., Man, water, and waste. *In* "The Next Ninety Years," Proceedings of a conference held at the California Institute of Technology, March 1967.
2. American Society of Testing and Materials Committee D-19 on Water, "Manual on Water," 3rd ed., Tech. Publ. No. 442, 1969.
3. Pecora, W. T., Science and public affairs. *Bull. At. Sci. 62*(8), 20 (1970).
4. Federal Water Pollution Control Administration, U. S. Department of the Interior. "Report of the Committee on Water Quality Criteria," April 1, 1968.
5. McKee, J. E., and Wolf, H. W., "Water Quality Criteria," 2nd ed., The Resources Agency of California, State Water Quality Control Board Publ. No. 3-A, 1968.
6. "Environmental Quality," The First Annual Report of the Council on Environmental Quality, Transmitted to the Congress August 1970, U. S. Gov. Print. Off.
7. Wolman, A., Water supply and environmental health. *J. Amer. Water Works Ass. 62*, 746 (1970).
8. Dominick, D. D., Projected direction and new developments in water quality standards. Talk presented to Industrial Waste Forum on Water Quality Standards, Water Pollution Control Federation, 42nd Ann. Mtg., Dallas, Texas, October 8, 1969.
9. DuBridge, L. A., Federal research problems. *Amer. Sci. 57*, 546 (1969).
10. Hoak, R. D., Who pays the clean water bill. 13th Nat. Watershed Congr., May 16, 1966. National Association of Manufacturers, New York, 1966.

11. Fair, G. M., and Geyer, J. C., "Water Supply and Waste-Water Disposal," Wiley, New York, 1954.
12. Nauman, E., Nagra synspunkter angaende plankton okalogie. *Sv. Bot. Tidskr. 13* (1919). Cited in Ref. 20.
13. Sources of Nitrogen and Phosphorous in Water Supplies. Task Group Report. *J. Amer. Water Works Ass. 59*, 344 (1967).
14. Lange, W., Effect of carbohydrates on the symbiotic growth of planktonic blue-green algae with bacteria. *Nature (London)* 215, 1277 (1967).
15. Kerr, P., Paris, D. F., and Brockway, D. L., Interrelationship of carbon and phosphorous in regulating heterotrophic and autotrophic populations in aquatic ecosystems. Presented at the 25th Purdue Industrial Waste Conference, Purdue University, May 5-7, 1970.
16. Bowen, D. H. M., The great phosphorous controversy. *Environ. Sci. Technol. 4*, 752 (1970).
17. Ferguson, F. A., A nonmyopic approach to the problem of excess algal growths. *Environ. Sci. Technol. 1*, 573 (1967).
18. Frink, C. R., Nutrient budget: Rational analysis of eutrophication in a Connecticut lake. *Environ. Sci. Technol. 1*, 425 (1967).
19. Safferman, R. S., and Morris, M., Control of algae with viruses. *J. Amer. Water Works Ass. 56*, 1217 (1964).
20. McGauhey, P. H., "Engineering Management of Water Quality," p. 105. McGraw-Hill, New York, 1968.
21. Deaner, D. G., "Public Health and Water Reclamation," Water and Sewage Works, 1970 Reference Number, p. R7-R13, November 28, 1970.
22. State of California. "Waste Water Reclamation at Whittier Narrows," California State Water Quality Control Board Publ. No. 33, 1966.
23. McGaughey, P. H., "Engineering Management of Water Quality." McGraw-Hill, New York, 1968.
24. Parkhurst, J. D., Progress in waste water reuse in southern California. *J. Irrigation Drainage Div. Amer. Soc. Civil Eng.* 4523, IR1, *91*, 79 (1965).
25. U. S. Department of the Interior, FWPCA, "Thermal Pollution – Its Effect on Water Quality," U. S. Dept. Interior, 1968.
26. Kolflat, T., Thermal discharges. *Ind. Water Eng. 5*, 26 (1968).
27. Remirez, R., Thermal pollution – hot issue for industry. *Chem. Eng. (New York) 75*, 43 (1968).
28. Hemens, J., Waste heat and the water environment, water – 1969. *Chem. Eng. Progr. 65*, 47 (1969).
29. Dysart, B. C., III, and Krenkel, P. A., "The Effects of Heat on Water Quality," p. 18, Proc. 20th Ind. Waste Conf., Purdue, 1965.
30. Phelps, E. B., "Stream Sanitation." Wiley, New York, 1960.
31. U. S. Department of the Interior, FWPCA, "Industrial Waste Guide on Thermal Pollution," 1968.
32. Hauser, L. G., "Advanced Methods of Electrical Power Generation," The Cooling Tower Institute Semi-Annual Meeting, June, 1968.
33. Thermal Wastes May be Aid in Melting Seaway's Ice Jams. *Chem. Eng. News 46*(26), 24 (1968).
34. Mercury Mars the Catch. *Chem. Week 106*(14), 16 (1970).
35. Chlorine Makers Clutch at Last Drops of Mercury. *Chem. Week 108*(8), 75 (1971).
36. Hammond, A. L., Mercury in the environment: Natural and human factors. *Science 171*, 788 (1971).

37. Mercury: More Than a Fish Tale. *Chem. Week 108*(4), 57 (1971).
38. Angino, E. E., Magnuson, L. M., Waugh, T. C., Galle, O. K., and Bredfelt, J., Arsenic in detergents: Possible danger and pollution hazard. *Science 168*, 389 (1970).
39. Shapiro, J., Arsenic and phosphate: Measured by various techniques. *Science 171*, 234 (1971).
40. Wolman, A., Someone put a number on it. *Chem. Eng. News. 48*(50), 5 (1970).
41. "Pesticides in Soil and Water, An Annotated Bibliography," U. S. Pub. Health Service Publ. 999-WP-17, July, 1964.
42. Faust, S. D., and Suffet, I. H., Recovery, separation, and identification of organic pesticides from natural and potable waters. *Residue Rev. 15*, 44 (1966).
43. Breidenbach, A. W., Gunnerson, C. G., Kawakara, F. K., Lichtenberg, J. J., and Green, R. S., Chlorinated hydrocarbon pesticides in major river basins, 1957–1965. *Pub. Health Rep. 82*, 139 (1967).
44. Breidenbach, A. W., and Lichtenberg, J. J., DDT and Dieldrin in rivers: A report of the national water quality network. *Science 141*, 899 (1963).
45. Weaver, L., Gunnerson, C. G., Breidenbach, A. W., and Lichtenberg, J. J., Chlorinated hydrocarbon pesticides in major U. S. river basins. *Pub. Health Rep. 80*, 481 (1965).
46. McGauhey, P. H., "Engineering Management of Water Quality," Chapt. 10. McGraw-Hill, New York, 1968.
47. Webb, F. E., "Aerial Forest Spraying in Canada in Relation to Effects on Aquatic Life," Biological Problems in Water Pollution, R. A. Taft San. Eng. Cen., Tech. Rep. W60-3 (1960).
48. Graham, J. R., "Effects of Forest Insect Spraying on Trout and Aquatic Insects in Some Montana Streams," Biological Problems in Water Pollution, R. A. Taft San. Eng. Cen., Tech. Rep. W60-3 (1966) 62.
49. Tarzwell, C. M., "Pollutional Effects of Organic Insecticides," Trans. North American Wildlife Conf., pp. 132–142, 1959.
50. Cohen, J. M., Kamphake, L. J., Harris, E. K., and Woodward, R. L., Taste threshold concentrations of metals in drinking water. *J. Amer. Water Works Ass. 52*, 660 (1960).
51. Boetius, J., "Foul Taste of Fish and Oysters Caused by Chlorophenol," *Medd. Denmacks. Fishlag. Havundersdg. [N. S.] 1*, 1 (1954).
52. Rosen, A. A., Skeel, R. T., and Ettinger, M. B., Relationship of river water odor to specific organic contaminants. *J. Water Pollut. Contr. Fed. 35*, 777 (1963).
53. Caruso, S. C., Bramer, H. C., and Hoak, R. D., The analysis of trace constituents in water by spectroscopic methods. *Develop. Appl. Spectrosc. 6*, 323 (1968).
54. Gaines, H. D., and Collins, R. P., Volatile substances produced by *Streptomyces odorifer. Lloydia 26*, 247 (1963).
55. Collins, R. P., and Kalnins, K., Volatile constituents of *Synura petersenii*. I. The carbonyl fraction. *Lloydia 28*, 48 (1965).
56. Collins, R. P., and Kalnins, K., Volatile constituents produced by the alga *Synura petersenii*. II. Alcohols, esters and acids. *Int. J. Air Water Pollut. 9*, 501 (1965).
57. Jenkins, D., Medsker, L. L., and Thomas, J. F., Odorous compounds in natural waters. Some sulfur compounds associated with blue-green algae. *Environ. Sci. Technol. 1*, 731 (1967).
58. Gerber, N. N., and Lechevalier, H. A., Geosmin, an earthy-smelling substance isolated from *Actinomycetes. Appl. Microbiol. 13*, 935 (1965).
59. Medsker, L. L., Jenkins, D., and Thomas, J. F., Odorous compounds in natural waters. An earthy-smelling compound associated with blue-green algae and *Actinomycetes. Environ. Sci. Technol. 2,* 461 (1968).
60. Safferman, R. S., Rosen, A. A., Mashni, C. I., and Morris, M. E., Earthy-smelling

substances from a blue-green alga. *Environ. Sci. Technol. 1*, 429 (1967).

61. Gerber, N. N., Geosmin, from microorganisms, is *trans*-1, 10-dimethyl-*trans*-9-decalol. *Tetrahedron Lett.* 2971 (1968).
62. Dougherty, J. D., Campbell, R. D., and Morris, R. L., *Actinomycete:* Isolation and identification of agent responsible for musty odors. *Science 152*, 1372 (1966).
63. Dougherty, J. D., and Morris, R. L., Studies on the removal of *Actinomycetes* musty tastes and odors in water supplies. *J. Amer. Water Works Ass. 59*, 1320 (1967).
64. Medsker, L. L., Jenkins, D., and Thomas, J. F., Odorous compounds in natural waters. *Environ. Sci. Technol. 3*, 477 (1969).
65. Weinberger, L. W., Stephan, D. G., and Middleton, F. M., Solving our water problems – water renovation and reuse. *Ann. N.Y. Acad. Sci. 135*, 131 (1966).
66. Kershaw, M. A., and Wood, R., Sludge treatment and disposal at maple lodge. *J. Proc. Inst. Sewage Purif. (London) 34*, 75 (1966).
67. Irgens, R. L., and Halvorson, H. O., Removal of plant nutrients by means of aerobic stabilization of sludge. *Appl. Microbiol. 13*, 373 (1965).
68. Dalton, F. E., Stein, J. E., and Lynom, B. T., Land reclamation – a complete solution of the sludge and solids waste disposal problem. *J. Water Pollut. Contr. Fed. 40*, 789 (1968).
69. Wolfel, R. M., Liquid digested sludge to land surfaces. Experiences at St. Mary's and other municipalities in Pennsylvania. Presented to 39th Annual Conference Water Pollution Control Association of Pennsylvania, August, 1967.
70. Owen, M. B., Sludge incineration. *J. Sanit. Eng. Div. Amer. Soc. Civil Eng. 83*, Paper 1172 (1957).
71. Sebastian, F. P., and Cardinal, P. J., Jr., Solid waste disposal. *Chem. Eng. (New York) 75*, 112 (1968).
72. Teletzke, G. H., Wet-air oxidation. *Chem. Eng. Progr. 60*, 33 (1964).
73. Teletzke, G. H., Gitchel, W. B., Didarns, D. G., and Hoffman, C. A., Components of sludge and its wet air oxidation products. *J. Water Pollut. Contr. Fed. 39*, 994 (1967).
74. Dean, R. B., Ultimate disposal of waste water concentrates to the environment. *Environ. Sci. Technol. 2*, 1079 (1968).
75. "Cleaning Our Environment – The Chemical Basis for Action," p. 121, American Chemical Society, Washington, 1969.
76. McMichael, F. C., Maruhnich, E. D., and Samples, W. R., Recycle water quality from a blast furnace. *J. Water Pollut. Contr. Fed. 43*, 595 (1971).
77. "Cleaning Our Environment – The Chemical Basis for Action," p. 249. A Report of the Subcommittee on Environmental Improvement, Committee on Chemistry and Public Affairs. American Chemical Society, Washington, 1969.

Agriculture Now

KEITH R. LONG and DAVID L. MICK

Agriculture in the 1970's is facing a dilemma created by the increasing pressure from a society that is demanding more abundant and higher quality agricultural products. Throughout most of man's existence he has been limited by his food supply. For the first million years or so he lived as a predator, hunting and gathering. There were no large centers of population and there were perhaps a total of five million people spread over the earth as a whole at the beginning of the agricultural revolution. With the beginning of the agricultural revolution the population multiplied 100 times in 8000 years.[1] Some 10,000 years ago man began to shape his environment to his own needs and, in his desire for more food, began to upset the delicate equilibrium of the ecosystem in which he lived. As man's primitive techniques for the production of food and animal crops gave way to more efficient methods, the world's food supply expanded permitting man to increase in numbers to his present level of nearly four billion with an expectation of reaching nearly six billion by the year 2000.

The continuing expansion of land under cultivation with the evolution of a mechanically and chemically oriented modern agriculture is producing a potential threat to man's health as well as ominous alterations and threats to world ecology.

According to the Malthusian philosophy, a population tends to grow exponentially. The checks on subsistence and the checks which suppress the superior power of a population to keep these effects on a level with that of its food supply are all resolvable into moral restraint, vice, and misery.[2] In the terms of practical problems posed by agriculture that must be faced within the next few generations with foreseeable technology, it is clear that human misery will increase greatly if we do not immediately assume that the world available to man is finite.[3]

The theme "Agriculture Now" is a phrase which is intended to focus attention upon the problems associated with the food-producing capacity of agriculture at a time when environmental concern has never been greater. The complexities and interactions with the environment and man's future health are heightened by an intensive agricultural program which is attempting to feed an ever more populous world.

One fact has to be kept in mind – food is one of man's basic needs. A starving man is easy prey to anyone who promises him relief from his plight. Population control may be an answer to providing food for future generations but improved agricultural practices have been considered to be of greater importance. We say we engage in these practices for the good of man even though poor food distribution has resulted in inadequate diets and malnutrition for many. One person may consider wilderness as "good" or ski lodges for thousands as "good," or estuaries for ducks, or factory or agricultural lands. We as a people want the maximum good with little understanding of the definition of the word.[4]

The production capacity of United States agriculture since World War II is unmatched in history. For example, from 1940 to 1964 the amount of corn raised for grain increased 61.4%, wheat production jumped 59.3%, and meat production rose 60.6%.[5] This was accomplished despite a decreased farm population and a stable farm acreage. Some of the factors responsible for this production are increased mechanization, use of modern fertilizers, pesticides and drugs, improved crop varieties, and improved practices such as water conservation, irrigation, cultivation, and optimum plant populations as well as improved breeding and feeding, etc.

The outlook of the nation's overall economy has direct implications for agriculture.[6] Expectations are for the gross national product to pass the trillion dollar mark in 1971 which is an estimated 6.5% increase over 1970. Any improvement in business activity obviously creates demands for meat, poultry, and other farm products. All of these demands are coming at a time when agriculture is undergoing dramatic changes heavily influenced by new technology and increases in farm size.

A look at the Iowa scene reveals that farming units declined 24% (about 4200 fewer farms per year) from 1960–1969.[7] The percent of land under cultivation remains essentially the same. These changes are a direct result of new technologic developments in crop production through increased mechanization and the use of pesticides, fertilizers, and other growth-promoting substances. One can only forecast that technology will continue to improve agricultural efficiency at present or even faster rates as population pressure increases. Land is a nondepreciating and a nonliquidating asset.[7] These factors have a direct influence in demanding increased efficiency through use of new technologic developments which have a direct potential in developing the health and survival of man. In Iowa it is postulated that most farming units will become a combination of owned and leased land with an increasing number of two-man livestock farms.

The position of agriculture in the microcosm created by man in his manipulation of the environment has been that of producing adequate food for an enlarging population without a great deal of concern about how the increasing production and its demands for efficiency affect man and his environment. Man's health is not without jeopardy as agricultural efficiency and technology develop. The farmer is caught up in an economic whirlwind of consumer demands and farm production. Coming to his aid are a battery of chewing, cutting, digging, and biting machinery supported by protoplasmic poisons to retard, defoliate, control, or destroy. These machines and poisons make no distinction between product and producer nor between farmer and pest. Injuries and fatalities resulting from accidents involving farm machinery take their toll in human misery each year. During 1968, 7400 farm residents died as a result of accidents and another 650,000 were injured.[8]

Epidemiologic investigations of man, a machine, and the environment with respect to tractor accidents showed that in spite of all the learned and involuntary responses of man, he was frequently incapable of acting correctly.[9] This may be partially attributed to an increasing average age of the farm population. Perhaps the "mechanical revolution" has produced machines both in quantity and quality that are not totally compatible with the reflexes of older operators. Safety design characteristics should always be under close scrutiny because exposure to injurious agents is the primary cause of farm accidents.

The environment should not be overlooked in the man–machine–environment complex.[10] The farmer often has no choice but to move his machinery on modern highways for access to his fields. The dangers of such slow-moving, and often cumbersome-sized equipment is evident.

Protection of the farm worker from the dangers of his machines in an often hostile environment takes many different forms but should never be substituted for carelessness. Protective shields are of value only when in their proper place, warning stickers are valuable only when heeded, slow-moving vehicles are better identified by the slow-moving-vehicle emblem, sophisticated machines should be operated only by alert and healthy individuals, etc.

An area of potential health involvement not as easily identified as farm accidents involves the health aspects of the farm population in relation to the ever increasing use of synthetic chemicals. Until recently, the need was unmet for studying not only the relationship between pesticides and illnesses in farmers but also in the general population. The Environmental Protection Agency through its Office of Pesticides is the only branch of the Federal government now studying the effects of pesticides on human health. These studies, initiated in 1965, include studying of the occupationally exposed farm workers, commercial pest control operators, and formulators of chemicals as well as the general population.[11]

A few years ago one could not determine the incidence of injury from economic poisons (pesticides) because of inadequate data. Presently there has been but little improvement. Facts are still meager and unavailable from a single source. There have been a variety of explanations expressed for this lack of reported illnesses from pesticides, particularly among farmers. Perhaps farmers are not made ill, or those who are ill do not see a physician or are not properly diagnosed, or if a diagnosis is made the results are not reported. The latter three reasons are the most probable.

Interviews conducted among Iowa farmers revealed that over half believed they had been adversely affected by farm chemicals.[12] There are a number of recent reports that indicate that pesticides adversely effect farmers. Manifestations of illnesses have included effects on the central nervous system resulting in impaired mental alertness, increased tension, anxiety, and nervousness which provokes sleep difficulties with excessive dreaming and nightmares, impairment of memory, and slowing of reactions.[13] Withdrawal, apathy, and depression have occurred with exposure to organophosphorus substances.[14,15] Two such cases were studied at University Hospitals, Iowa City. Both subjects were farmers, white men in their early 40's, that were judged by medical and clinical laboratory examinations to be of sound mind and normal responses. These two subjects complained of anorexia, anxiety, hyperirritability, and depression for which symptomatic treatment was prescribed. Complete epidemiologic follow-up including occupational histories and site visits to their work areas revealed no unusual activities. The only influencing factor appeared to be their rather high and sometimes careless use of organochlorine and organophosphorus pesticides.

Dyscrasias of the central nervous system are not the only effects. Dermatitis and other skin diseases have been reported among farmers employing various pesticides.[16,17] In a 1967 report from California, more than 80% reported occupational disease among farmers fell into one of four categories of skin conditions, eye conditions, systemic poisonings, or chemical burns. Reports of respiratory conditions accounted for 5% of the cases, about the same as in previous years, while the remaining reports described digestive disorders and other signs and symptoms due to toxic materials.[18]

The California report provides several examples of the kinds of poisonings farmers are incurring. One is that of a farm laborer who was cleaning a vat where potatoes were dipped. He felt nauseated and experienced a burning sensation in his lungs from breathing the chemical vapors. His physician kept him under observation for suspected chemical poisoning from mercury bichloride. Another example is that of a ranch foreman who was using an organophosphorus pesticide and experienced nausea and vomiting for about a month. He also reported having blurred vision and respiratory difficulty. His condition was diagnosed as organophosphorus poisoning.

In a similar example, a farm laborer exposed to agricultural chemicals, including sulfur sprays, suffered severe erythemateous, oozing, crusted, and edematous lesions of the exposed skin areas of his face, neck, and arms. His physician estimated that the worker would need 6 weeks treatment and would be disabled from work for 1–3 weeks. Another example is that of a swamper who was wearing a mask and goggles while loading a plane with sacks of a carbamate insecticide and sulfur. When he transferred the chemicals to a bucket to take to the plane, he felt sudden weakness, was dizzy, and could not get his breath. Two days later the same thing had occurred after which a diagnosis was made of chemical toxemia. The physician noted that the employer reported four other employees who had also been sick or had similar attacks. None had been to see a doctor.

In Iowa, a farmer had used Dichlorvos in a systemic hog-dewormer feed additive and had cultivated ground treated with an organophosphorus pesticide for the control of corn rootworms.[19] This man, suffering diarrhea and upper respiratory effects, was hospitalized for a time under intensive care. The signs and symptoms coincided exactly with the time sequence of employing the agricultural chemicals. Another Iowa subject accidentally spilled concentrated organophosphorus insecticide on his lower extremities while loading a spray tank. He subsequently suffered a complicated series of effects including marked paralysis of the lower part of his legs, numbness in his fingers, and marked depression and anxiety, all of which were treated symptomatically. The time sequence of these effects fitted into the farmer's use of agricultural chemicals. We have seen one case in which a farmer experienced irritation of the bronchii and lungs resulting in a secondary infection and pneumonia after having cultivated soil treated with an allylacetamide herbicide.

No distinction has ever been made between the extensive user of pesticides and one who uses them little or occasionally. It is becoming quite obvious that the health professions must express concern for occupational exposure of commercial operators, farm operators, or uninformed persons employed in agriculture as well as the general population. Exposures are as simple as inhaling the dust from a paper bag at the time of loading the hopper of an applicator or skin contact from cleaning a plugged nozzle of a spray machine. Contact exposure of migrant workers results in a build up of residues in their blood through working in fields where pesticides have been sprayed.[20] These residues are derived from close contact with the treated soil and crops being processed.

Not only are symptoms described, but other changes have been observed in the central nervous system as reflected by alterations in the electroencephalographic tracings,[15] in kidney and liver functions as revealed by altered amino acid values and phosphorus readsorption indicies in urine,[21] and in blood values where differences in hemoglobin and hematocrit levels have been observed in people occupationally exposed to pesticides.[22] Scientists have recognized for many years that organophosphorus pesticides inhibit cholinesterase activities with subsequent cholinergic effects. Whether these observed conditions are prologues to chronic disease or enhance existing disease is a matter for speculation. The fact remains that a variety of acute illnesses and physiologic changes have been observed in farmers pursuing their usual activities, including the handling of agricultural chemicals. These exposures may be subtle or great. The farmer's method of handling agricultural chemicals is rooted in attitudinally conditioned action. Farmers often accept the uncomfortable working conditions, the skin rashes, the headaches, and the upper respiratory effects suffered in their farming enterprise as their lot in life.

An understanding of exposure or the development of an exposure index is complicated by the attitudes of the farmer and the conditions under which economic poisons are used. The factors having a direct bearing on exposure to pesticides are the conditions of use: how much, when, in what form they are used and what protective measures are taken. The amounts and hours or days of pesticide use or pesticide residue levels in blood or body fat or a combination of these three factors have been employed in attempts to develop an "exposure index." A typical example is afforded by the work of Selby *et al.* who developed a pesticide exposure index based upon detailed information concerning mixing, application, and physical presence when a pesticide was being used.[23] They concluded that the toxic potential of pesticides upon persons in a general population cannot be examined by correlation with a pesticide to which the individual states that he has a known exposure. Similar attempts were made in which the type of chemical and the protective measures were given a rank, based on toxicity of the compound and degree of protection afforded by the protective clothing.[24] This index was then equated with pesticide residue levels in the blood and tissues of the subjects concerned. At present there has been no generally accepted pesticide exposure index developed.

In addition to occupational exposure, there is an unanswered question of the affects of residues in blood and fatty tissues of man and other animals. Reports from the Food and Drug Administration have shown pesticide residues to be present in most items of food used by man where he obtains about 70% of his body burden of chlorinated hydrocarbon pesticides[25] with the remainder coming from other sources. The affect of these residues on human health excite the layman but as late as 1969 the National Academy of Sciences' Committee on Persistent Pesticides[27] concluded: "Available evidence does not indicate that present levels of pesticide residues in man's food and environment produce an adverse effect on his health."

Pesticide residues have been observed in almost all tissues and body fluids of man including fat, liver, kidney gonad, and brain as well as blood, lymph, amniotic fluid, and mother's milk.[26-48] Some of these residues have been shown to pass the human placental barrier into the developing embryo. Our understanding of the role played by these residues in the health and welfare of man and his offspring is hypothetical. However, the limited information presently available on long-term exposure to varying levels of pesticides in the environment indicates the necessity for continued systematic longitudinal studies on man. There is evidence that there are apparent differences in hemoglobin and hematocrit levels between persons who have low use of pesticides and those who have high, that there is an apparent metabolic interaction between certain drugs and pesticide residues as shown by the example of Dilantin interacting with the metabolism in the lessening of the DDT and DDE body residue levels,[22,49] and that there are changes in pulmonary and renal physiology.[21] These observations are cause for concern about long-range relationships of pesticides to disease. Man desperately needs to know the effects of these long-term relationships with pesticides in the enhancement of existing disease or the development of other disease.

The practice of administering antibiotics to animals scheduled for human consumption has received increased attention in recent years. Residues in meat, eggs, and milk and the possibility of a build-up of drug resistant organisms are the two primary concerns to human health.[50] Research so far has shown that neither occurrence is a threat to human health.

Controlling disease and increasing food production are two advantages derived from using antibiotics. These advantages will be maintained only as long as producers use antibiotics properly. The Food and Drug Administration constantly monitors agricultural produce to insure its wholesomeness for the consumer. In fact, the FDA had reason to prohibit certain injectable antibiotics. The problems that have arisen in the United States are a direct result of improper use of pesticides, tranquilizers, antibiotics, and other chemicals.[51]

It is undoubtedly important to study the health ramifications of the xenobiotics to which man may be subjected. Another important aspect is how these compounds act when a living system is subjected to them simultaneously or in sequence. The classic example is the discovery and elucidation of the mechanism explaining the increased toxicity of Malathion when administered with *O*-ethyl *O*-*p*-nitrophenyl phenylphosphonothioate (EPN).[52,53] This potentiation was explained on the basis that EPN inhibits the enzymatic hydrolysis of the carboxyester linkages of Malathion. Continued research involving these interactions is imperative if we are to use the pesticides, drugs, and other chemicals safely. More recent studies have shown that many pesticides and drugs stimulate snythesis of hepatic microsomal enzymes with resulting alterations of the metabolic activities in the liver, causing alterations of normal metabolic pathways for many chemicals that are frequently encountered.[54-59]

Certain segments of our society are now asking for total bans of many of the chemicals used to increase food production and alleviate human health problems. Indictment of these substances for their accumulation in and damage to the environment is not without foundation. But how do we feed a growing population if the tools of production are banned? The National Academy of Sciences' Committee on Persistent Pesticides reported in May, 1969[26]:

> During the past quarter of a century, nations in all parts of the world have benefited from increasing use of synthetic organic pesticidal chemicals. Through the use of these chemicals, spectacular control of disease caused by insect-borne pathogens have been achieved, and agricultural productivity has been increased to an unprecedentable level. No adequate alternative for the use of pesticides for either of these purposes is expected in the forseeable future. Modern agricultural productivity depends on coordinated increase in the use of pesticides fertilizers, machinery and better crop varieties.

This does not mean that we have to continue using the same chemicals in the same way. Some will be restricted, better chemicals will be discovered, better application methods will reduce quantities needed, and other than synthetic organic chemicals will find use in agriculture. However, it does mean that for "now," chemicals will continue to remain in the farmer's arsenal for insuring his livelihood for profitable food and fiber production as well as attempting to feed the populace on decreased acreage.

If we accept the findings of the Commission on Persistent Pesticides, it seems logical that the next step is to select and use these chemicals more wisely than we have in the past.[26] Proper use ultimately rests with those individuals involved in the actual application process, but what products may be used is regulated primarily by Federal agencies. The recent creation of the Environmental Protection Agency should decrease some of the red tape currently involved with Federal pesticide regulations.

At this point it is appropriate to raise the question of a "cost vs. benefit" equation. It should be quite clear by now that any activities of man, particularly agricultural ones, produce alterations, threaten his environment, and raise very poignant questions about effects on his health. It is obvious that some cost-vs.-benefit ratio or equation must be established. The question is often asked, "Who establishes such an equation?" Perhaps not so obvious is the answer – society establishes the equation. How much cost to his health and environment is man willing to allow in receiving a given amount of benefit from increased productivity is a question that apparently has no ready answer. The concern of society about the cost equation is becoming manifested in some of the legislation relating to agriculture that is beginning to appear. Perhaps some of the most meaningful examples are found in the legislative actions revealed in the accomplishments of various state legislatures during their regular sessions in 1970. At least twenty states have enacted legislation creating commissions or review boards in relation to control of agricultural or other chemicals. In addition, twenty-eight state legislatures have enacted some sort of pesticide legislation during their regular sessions of 1970. From the standpoint of interpreting the future impact of these actions, it is perhaps worthwhile to examine some of the actions that were taken.

Generally speaking, the pesticide or chemical review boards or commissions were created to collect, analyze, and interpret information relating to agricultural chemicals and their uses. In most instances, these boards have coordinating capabilities with other state agencies created within them. The most important responsibility of these boards has been the authority to adopt rules and regulations relating to the sale, possession, use, and misuse of agricultural chemicals. In general, they have the responsibility for considering the toxicity, hazards, effectiveness, and public need for agricultural chemicals and in addition the consideration of availability of less hazardous substances or other means of control. This type of legislative action certainly is the most significant in reference to the use of agricultural chemicals. Generally speaking, other legislative actions deal with specific rules and regulations. For example, California is requiring that reports of poisoning due to agricultural chemicals be made to local health officers.[60] In several instances, action has been taken to classify certain compounds as restricted-use substances requiring that persons applying them be specifically certified and that they be required to report how much of the substance was used, what is was used upon, and how the excess material and used containers were handled or disposed. From the standpoint of the future, one can expect to see other states following with similar legislation in creating lists of restricted-use pesticides. Generally, these restricted-use substances are defined as described by the laws enacted by Florida and New York.[61,62] These laws define a restricted-use substance as one that persists in the environment, accumulates as a pesticide or metabolic degradation product in plant or animal tissues or their products, and is not excreted or eliminated in a reasonable length of time. In addition these laws go on to state that if persistence of such a compound creates or presents a future risk or harmful effect on any organism other than the target organism or if the compound is found to be hazardous to man or other forms of life, then restriction on its sale, purchase, use, or possession are in the public interest.

While practically all of the uniform state economic poison acts define an economic poison as a substance intended to prevent, destroy, repel, or mitigate pests, Mississippi has been the first state to enact legislation that includes attractants as economic poisons.[63]

Other legislation dealing with soil conservation in relation to problems of siltation, particularly prevalent in the Midwest, has been suggested. Attempts are being made to create and regulate better agricultural land-use practices. It is apparent that there must be better land-use management of agricultural lands to prevent siltation and the carrying of pesticides and fertilizer chemicals into ponds, streams, and rivers. It is almost a certainty that one can look forward to some sort of action that will curtail indiscriminate and excessive use of pesticides and fertilizers. In addition, society must be made aware that the expense of this effort will not only be borne by the vested interest groups, but also by the general public. In looking towards the future one can expect that agriculture will be called upon to define, explain, and defend its real needs for fertilizers, pesticides, and feed additives in the production of food for man. There is a growing concern that the hazards and deleterious side effects of convenience pesticides and fertilizers used in urban areas must be defined and divorced from the same compounds and their consideration in agriculture. In addition, there will be an increasing responsibility placed upon the manufacturers and dealers to insure the proper and safe use of these products. Society is calling upon industries associated with agriculture as well as publically supported agricultural research agencies to provide answers to questions that are being asked by an aroused and increasingly less sympathetic public.

The health effects from the mechanical efficiency created by technologic advancement is not unrecognized. Recent enactment of legislation concerning the employment of youth on farms as well as the enactment of new health and occupational safety law will have a decided impact on minimizing the numbers of agricultural workers injured by their mechanical helpmates.[64,65]

Problems of agriculture are perhaps best personified in an article by Hardin in which he develops a thesis that the population problem has no technical solution but requires a fundamental extension in morality. When we are confronted by the dilemma of providing an adequate food supply from a finite resource for perhaps an infinite population, one cannot help but express the same concern as revealed by Wiesner and York in which they concluded that a dilemma such as this has no technical solution and that if we continue to look for solutions only in the areas of science and technology, the result will be to worsen the situation.[66]

REFERENCES

1. Deevey, E. S., Jr., The human population. *In* "The Subversive Science" (P. Shepard and D. McKinley, eds.) pp. 42-54. Houghton and Mifflin, Boston, Massachusetts, 1969.
2. Malthus, T. R., "Essay on the Principle of Population," 1st ed., 1803, 9th ed. 1826. MacMillan, London.
3. von Hoerner, S., The general limits of space travel. *Science 137,* 18 (1962).
4. Hardin, G., The tragedy of the commons. *Science 162,* 1243 (1969).
5. Whitten, J. L., "That We May Live." Van Norstrand, New York, 1966.
6. Futrell, G., Agricultural outlook for 1971. *Iowa Farm Sci. 25* (4), 3 (1971).
7. Howell, H. B., Present and future changes in Iowa agriculture. *Iowa Farm Sci. 25 (4),* 7 (1971).
8. "Accident Facts," 1969 ed., National Safety Council, Chicago, Illinois.
9. Knapp, L. W., Jr., The farm tractor: Overturn and power take-off accident problem. Bulletin No. 11. Inst. Agr. Med., University of Iowa, Iowa City, Iowa, 1968.

10. Knapp, L. W., Jr., Man machine relationship in tractor accidents. *Trans. Amer. Soc. Agr. Eng. 9,* 178 (1966).
11. Simmons, S. W., The pesticides program activities of the Public Health Service. *Pub. Health Rep. 83,* 967 (1968).
12. Long, K. R., and Walden, J., Preliminary studies on economic poisons. Bulletin No. 4, Inst. Agr. Med., University of Iowa, Iowa City, Iowa, 1958.
13. Durham, W. F., Wolfe, H. R., and Quinby, G. E., Organophosphorus insecticides and mental alertness. *Arch. Environ. Health 10,* 55 (1965).
14. Grob, D., and Harvey, A. M., Effects and treatment of nerve gas poisoning. *Amer. J. Med. 14,* 52 (1953).
15. Metcalf, D. R., and Holmes, J. H., EEG, psychological and neurological alterations in humans with organophosphorus exposure. *Ann. N.Y. Acad. Sci. 160,* 357 (1969).
16. Spencer, M. C., Occupational dermatitis and eczema among farmers. *Ill. Med. J. 119,* 136 (1961).
17. Spencer, M. C., Herbicide dermatitis. *J. Amer. Med. Ass. 198,* 1307 (1966).
18. Bognisky, E., "Occupational Disease in California attributed to Pesticides and Other Agricultural Chemicals." State of California, Department of Public Health, Bureau of Occupational Health and Environmental Epidemiology, 1967.
19. Long, K. R., Unpublished data. Iowa Community Pesticides Study Prog. Rep. No. 16, 1969.
20. Long, K. R., Pesticide residues in the blood of migrant field workers in relation to occupational exposure. "Whither Rural Medicine," pp. 71–75, Proc. Cong. Rural Med. Published by Jap. Ass. Rural Med., Tokyo, 1970.
21. Tocci, P. M., Mann, J. B., Davies, J. E., and Edmundson, W. F., Biochemical differences found in persons chronically exposed to high levels of pesticides. *Ind. Med. 38,* 188 (1969).
22. Long, K. R., Beat, V. B., Gombart, A. K., Sheets, R. F., Hamilton, H. E., Falaballa, F., Bonderman, D. B., and Choi, U. Y., The epidemiology of pesticides in a rural area. *Amer. Ind. Hyg. Ass. J. 30,* 298 (1969).
23. Selby, L. A., Newell, K. W., Waggenspak, C., Hauser, G. A., and Junker, G., Estimating pesticide exposure in man as related to measurable intake; environmental versus chemical index. *Amer. J. Epidemiol. 89,* 241 (1969).
24. Sandifer, S. H., and Keil, J. E., Medical University of South Carolina, unpublished data.
25. Duggan, R. E., and Weatherwax, J. R., Dietary intake of pesticide chemicals. *Science 157,* 1006 (1967).
26. Report of Committee on Persistent Pesticides, Division of Biology and Agriculture, National Research Council to U. S. Department of Agriculture (Nat. Acad. Sci., Washington, D.C., 1969).
27. Dale, W. E., and Quinby, G. E., Chlorinated insecticides in the body fat of people in the United States. *Science 142,* 593 (1963).
28. Quinby, G. E., Hayes, W. J., Jr., Armstrong, J. F., and Durham, W. F., DDT storage in the U. S. population. *J. Amer. Med. Ass. 191,* 175 (1965).
29. Laug, E. P., Kunze, F. M., and Prickett, C. S., Occurrence of DDT in human fat and milk. *Arch. Ind. Hyg. 3,* 245 (1951).
30. Hoffman, W. S., Fishbein, W. I., and Andelman, M. B., Pesticide storage in human fat tissue. *J. Amer. Med. Ass. 188,* 819 (1964).
31. Morgan, D. P., and Roan, C. C., Chlorinated hydrocarbon pesticide residue in human tissues. *Arch. Environ. Health 20,* 452 (1970).
32. Quinby, G. E., Armstrong, J. F., and Durham, W. F., DDT in human milk. *Nature (London) 207,* 726 (1965).

33. Hayes, W. J., Jr., Dale, W. E., and Burse, V. W., Chlorinated hydrocarbon pesticides in the fat of people in New Orleans. *Life Sci. 4,* 1611 (1965).
34. Davies, J. E., Edmundson, W. F., Schneider, N. J., and Cassady, J. C., Pesticides in people. *Pestic. Monit. J. 2,* 80 (1968).
35. Fiserova-Bergerova, V. Radomski, J. L., Davies, J. E., and Davis, J. H., Levels of chlorinated hydrocarbon pesticides in human tissue. *Ind Med. Surg. 36,* 65 (1967).
36. Curley, A., and Kimbrough, R., Chlorinated hydrocarbon insecticides in plasma and milk of pregnant and lactating women. *Arch. Environ. Health 18,* 156 (1969).
37. Egan, H. Goulding, R., Roburn, J., and Tatton, J. O'G., Organo-chlorine pesticide residues in human fat and human milk. *Brit. Med. J. 2,* 66 (1965).
38. Edmundson, W. F., Davies, J. E., Cranmer, M., and Nachman, G. A., Levels of DDT and DDE in blood and DDA in urine of pesticide formulators following a single intensive exposure. *Ind. Med. Surg. 38,* 145 (1969).
39. Milby, T. H., Samuels, A. J., and Ottoboni, F., Human exposure to Lindane; blood Lindane levels as a function of exposure. *J. Occup. Med. 10,* 584 (1968).
40. Dale, W. E., Curley, A., and Cueto, C., Jr., Hexane extractable chlorinated insecticides in human blood. *Life Sci. 5,* 47 (1966).
41. Perron, R. C., and Barrentine, B. G., Human serum DDT concentration related to environmental DDT exposure. *Arch Environ. Health 20,* 368 (1970).
42. O'Leary, J. A., Davies, J. E., Edmundson, W. F., and Feldman, M., Correlation of prematurity and DDE levels in fetal whole blood. *Amer. J. Obstet. Gynecol. 106,* 939 (1970).
43. Durham, W. F., Body burden of pesticides in man. *Ann. N.Y. Acad. Sci 160,* 183 (1969).
44. Curley, A., Copeland, M. F., and Kimbrough, R. D., Chlorinated hydrocarbon insecticides in organs of stillborn and blood of newborn babies. *Arch, Environ. Health 19,* 628 (1969).
45. Zavon, M. R., Hine, C. H., and Parker, K. D., Chlorinated hydrocarbon insecticides in human body fat in the United States. *J. Amer. Med. Ass. 193,* 837 (1965).
46. Radomski, J. L., Deichmann, W. B., and Clizer, E. E., Pesticide concentrations in the liver, brain, and adipose tissue of terminal hospital patients. *Food Cosmet. Toxicol. 6,* 209 (1968).
47. Durham, W. F., Armstrong, J. F., Upholt, W. M., and Heller, C., Insecticide content of diet body fat of Alaskan natives. *Science 134,* 1880 (1961).
48. Dale, W. E., Curley, A., and Hayes, W. J., Jr., Determination of chlorinated insecticides in human blood. *Ind. Med. Surg. 36,* 275 (1967).
49. Davies, J. E., Edmundson, W. F., Carter, C. H., and Barquet, A., Effect of anticonvulsant drugs on dicophane (DDT) residues in man, *Lancet 297 (pt. 1),* 7 (1969).
50. Herrick, J. B., Animal health outlook – production without antibiotics. *Iowa Farm Sci. 25,* 31 (1971).
51. Herrick, J. B., Meat and chemical residues. *Iowa Farm Sci. 20,* 6 (1965).
52. Frawley, J. P., Fuyat, H. N., Hagan, E. C., Blake, J. R., and Fitzhugh, O. G., Marked potentiation in mammalian toxicity from simultaneous administration of two anticholinestease compounds. *J. Pharmacol. Exp. Ther. 121,* 96 (1957).
53. Murphy, S. D., and DuBois, K. P., Quantitative measurement of the inhibition of the enzymatic detoxification of malathion by EPN (ethyl *p*-nitrophenyl thionobenzenephosphonate). *Proc. Soc. Exp. Biol. Med. 96,* 813 (1957).
54. Eling, T. E., Harbison, R. D., Becker, B. A., and Fouts, J. R., Diphenylhydantoin effect on neonatal and adult rat hepatic drug metabolism. *J. Pharmacol. Exp. Ther. 171,* 127 (1970).

55. Bresnick, E., and Stevenson, J. G., Microsomal *N*-demethylase activity in developing rat liver after administration of 3-methylcholanthrene. *Biochem. Pharmacol. 17,* 1815 (1968).
56. Nakatsugawa, T., Tolman, N. M., and Dahm, P. A., Degradation and activation of Parathion analogs by microsomal enzymes. *Biochem. Pharmacol. 17,* 1517 (1968).
57. Conney, A. H., Schneidman, K., Jacobson, M., and Kuntzman, R., Drug-induced changes in steroid metabolism. *Ann. N.Y. Acad. Sci. 123,* 98 (1965).
58. Hart, L. G., and Fouts, J. R., Effects of acute and chronic DDT administration on hepatic microsomal drug metabolism in the rat. *Proc. Soc. Exp. Biol. Med. 114,* 388 (1963).
59. DuBois, K. P., Combined effects of pesticides. *Can. Med. Ass. J. 100,* 173 (1969).
60. Senate Bill No. 22, California State Legislature. Approved September 18, 1970.
61. House Bill 3188, Amendment to Florida Pesticide Law. Passed May 26, 1970.
62. Assembly Bill 5881-B, Amendment to New York Economic Poison Law. Passed May 12, 1970.
63. Senate Bill No. 1943, Amendment to Mississippi Economic Poisons Act of 1950. Approved April 6, 1970.
64. Revision of Part 15, Table 29, Code of Federal Regulations, Chapter 13, Bureau of Labor Standards, Dept. of Labor, Established January 7, 1970.
65. PL-91-596 Occupational Health and Safety Act U. S. Congress, 1970.
66. Wiesner, J. B., and York, H. F., National security and the nuclear-test ban. *Sci. Amer. 211,* 25 October (1969).

Author Index

Numbers in parentheses are reference numbers and indicate that an author's work is referred to, although his name is not cited in the text. Numbers in italics show the page on which the complete reference is listed.

A

D

G

L

M

N

O

P

Q

R

S

T

Y

Z

Subject Index

B

C

S

T